Krishna's

Discrete Mathematics

Krishna's

Discrete MATHEMATICS

(For Degree, Honours and Post-Graduate Students of All Indian Universities)

By

M.K. Gupta

M.Sc., M.Phil., Ph.D.

Professor & Head, Deptt. of Mathematics

C.C.S. University, Meerut

KRISHNA Prakashan Media (P) Ltd.

KRISHNA HOUSE, 11, Shivaji Road, Meerut-250 001 (U.P.), India

Krishna's

Discrete MATHEMATICS

First Edition : 2003

Twentieth Edition : 2022

Edition : 2025

Book Code No. : 250-20 (C)

Price : ₹ 595.00 Only

Published by : **Sugam Rastogi & Kanupriya Rastogi**
for **KRISHNA** Prakashan Media (P) Ltd.
11, Shivaji Road, Meerut – 250 001 (U.P.) India.
Phones : 91.121.2644766, 2642946, 4026111, 4026112, Fax : 91.121.2645855
Website : www.krishnaprakashan.com
E-mail : info@krishnaprakashan.com

Typesetting : V.S.S. Infotech, New Delhi. (S.A.J.)
Printed at : Citizen Offset, Meerut.

Jai Shri Radhey Shyam

Dedicated

to

Lord

Krishna

Author & Publishers

PREFACE

Discrete Mathematics is that part of Mathematics in which the concepts of limits play little role. The recent upsurge in the interest in Discrete Mathematics is mainly due to its applications in computer technology. However, Discrete Mathematics is also important in many other fields such as Operations Research, Electric Engineering and Economics.

Discrete Mathematics is a compulsory paper in most Computing Programmes (MCA, M.Sc., B E., B.Sc., DOEACC(B-level)) of universities in our country. This book has been written to meet the requirements of under-graduate and post-graduate students pursuing courses in Mathematics, Computer Science and Engineering. *It covers topics from Set Theory, Combinatorics, Graph Theory, Lattices, Boolean Algebra (including Switching Circuits) and Finite State Machines*.

Each topic in the book has been treated in an easy and lucid manner. The proofs of various theorems and examples have been given with enough details. Numerous solved examples have been included in the book to help the students in understanding various concepts easily. A set of exercises (with answers) has also been given at the end of each topic to test the student's comprehension.

The author and publishers acknowledge their indebtness to all those authors on the subject whose works have been consulted.

The author also feels thankful to the publishers and printers for their cooperation and patience during the preparation of the book.

Suggestions for the improvement of the book will be gratefully accepted and acknowledged.

Meerut — *M.K. Gupta*

January 1, 2003

BRIEF CONTENTS

Chapters

DETAILED CONTENTS

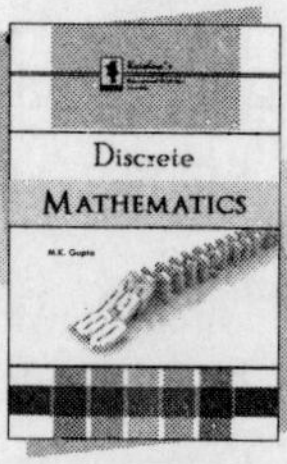

Chapter 3: Lattices ....(113–172)

Chapter 4: Discrete Numeric Functions and Generating Functions................................(173–222)

Chapter 5: Boolean Algebra..(223–276)

Chapter 8: Trees and Cut-Sets..................................(387–420)

CHAPTER ONE

Mathematical Logic

Logic is a field of study that deals with the method of reasoning. Logic provides rules by which we can determine whether a given argument or reasoning is valid (correct) or not. Logical reasoning is used in Mathematics to prove theorems. In computer science, logic is used to verify the correctness of programs.

§ 1.1. Statements

Definition: A statement (or proposition) is a sentence which is either true or false but not both.

Example 1.1. Which of the following are statements?

(a) Indira Gandhi was one of the Prime Ministers of India.

(b) 8 is greater than 10.

(c) $2 + 4 = 6$

(d) Blood is green.

(e) It is raining.

(f) The sun will come out tomorrow.

Solution:

(a) is a statement because it is true.

(b) is a statement because it is false.

(c) is a statement because it is true.

(d) is a statement because it is false.

(e) is a statement because the sentence "it is raining" is either true or false but not both at a given time.

(f) is a statement since it is either true or false but not both. Although, we would have to wait until tomorrow whether it is true or false.

If a sentence is a question (interrogative type) or a command or not free of ambiguity then the sentence cannot be answered as true or false and therefore such sentences are not statements.

Example 1.2. The following are not statements.

(a) Is the number 6 a prime? (b) $2 - x = 6$.

(c) What are you studying? (d) Open the door.

(e) This statement is false.

Explanation:

(a) is not a statement because it is a question.

(b) is not a statement because it is true or false depending on the value of x.

(c) is not a statement because it is a question.

(d) is a command and therefore it is not a statement.

(e) is not a statement because it is not possible to assign a definite true or false value to it. If we assume that sentence (e) is true then it says that statement (e) is false. On the other hand if we assume that sentence (e) is false then it implies that statement (e) is true. Hence it is not a statement.

Truth Value of a Statement

We know that a statement (or proposition) is either true or false but not both. Therefore a statement can have one and only one of two possible values namely true and false. The two values true and false, which we can assign to a given statement, are called truth values of the statement.

We shall use the letters p, q, r ... to represent arbitrary statements. We assign to a statement p the symbol T when the statement is true and the symbol F when it is false. Some authors use the symbols 1 and 0 instead of T and F.

§ 1.2. Negation of a Statement

Definition: Let p be a statement. We define the negation of p to be a statement, which is true when p is false and is false when p is true. The negation of a statement p is denoted by $\sim p$. In the literature, the symbols p′ and $\neg p$ are also used to denote the negation of a statement p. The negation of a statement is generally formed by using the word "not" at a proper place in the given statement or by the statement "it is false that p."

Example 1.3. Let p be the statement "Delhi is a city." Then $\sim p$ is the statement "Delhi is not a city." $\sim p$ can also be written as "it is false that Delhi is a city."

Example 1.4. Consider the statement:

p : Sleeping is pleasant.

Then $\sim p$ is any of the following statements:

1. It is false that sleeping is pleasant.
2. Sleeping is not pleasant.
3. Sleeping is unpleasant.

§ 1.3. Conjunction

Two statements can be combined by the connective "and" to form a new statement, which is called conjunction of the original statements.

Definition: The conjunction of two statements p and q is the statement $p \wedge q$ (read as 'p and q')which is true when both p and q are true and is false when either one or both of p and q are false.

The definition of the conjunction of two statements p and q is summarized in its truth table given below:

p	q	$p \wedge q$
T	T	T
T	F	F
F	T	F
F	F	F

Truth table for $p \wedge q$

Example 1.5. Let p be the statement "Sleeping is pleasant" and let q be the statement "I am strong." Then $p \wedge q$ is the statement "Sleeping is pleasant and I am strong."

Example 1.6. Form the conjunction of statements p and q where

p : Ram is an honest man.

q : Rahim is an honest man.

Solution: The conjunction of the given statement is

$p \wedge q$: Ram and Rahim are honest men.

§ 1.4. Disjunction

Definition: The disjunction of two statements p and q is the statement $p \vee q$ (read as p *or* q). The statement $p \vee q$ is true when either one or both of p and q are true, and $p \vee q$ is false when both p and q are false.

The definition of the disjunction of two statements p and q is described in the following truth-table.

p	q	$p \vee q$
T	T	T
T	F	T
F	T	T
F	F	F

Truth table for $p \vee q$

Example 1.7. Let

p : Sleeping is pleasant.

q : I am strong.

be the two statements. Then the disjunction of p and q is the statement

$p \vee q$: Sleeping is pleasant or I am strong.

Example 1.8. Let

p : Ice is cold.

q : Blood is green.

be the two statements. Then $p \vee q$ is the statement

$p \vee q$: Ice is cold or blood is green.

Note: In literature, pq and $p+q$ are also used to denote the conjunction and disjunction of two statements p and q, respectively.

§ 1.5. Truth Tables

A statement obtained from the combination of other statements is called compound statement. A statement which is not a combination of other statements is called an *atomic* or simple statement. In other words, a compound statement is a combination of atomic statements. For example, if p and q are statements then $p \vee q$ and $p \wedge q$ are compound statements.

Truth table for a compound statement is a table, which gives truth values of the compound statement for each possible combination of the truth values of the atomic statements of the compound statement. Truth tables for the compound statements $p \vee q$ and $p \wedge q$ have already been constructed in the definitions of $p \vee q$ & $p \wedge q$. These compound statements have two atomic statements namely p and q. Since each statement has only two possible truth values, there are $2^2 = 4$ possible combinations of truth values that must be considered. Thus each of the two tables for $p \vee q$ and $p \wedge q$ has 4 possible combinations. In general, if there are n distinct atomic statements in a compound statements then we need to consider 2^n possible combinations of truth values in order to obtain the truth table.

Example 1.9. Construct the truth table for $\sim p \vee q$.

Solution: We must consider all possible combinations of truth values of p and q. All possible combinations of the truth values of the statements p and q are listed in the first two columns of the table. The truth values of $\sim p$ are entered in the third column and the truth values of $\sim p \vee q$ are entered in the fourth column.

p	q	$\sim p$	$\sim p \vee q$
T	T	F	T
T	F	F	F
F	T	T	T
F	F	T	T

Truth table for $\sim p \vee q$

Example 1.10. Construct the truth table for $p \wedge \sim p$.

Solution: Since the statement $p \wedge \sim p$ has only one distinct atomic statement. We have to consider 2 possible combinations of truth values. The truth table for $p \wedge \sim p$ is given below.

p	$\sim p$	$p \wedge \sim p$
T	F	F
F	T	F

Truth table for $p \wedge \sim p$

Example 1.11. Construct the truth-table for $\sim(p \wedge \sim q)$.

Solution: In the first two columns, we list all the variables and the combinations of their truth values. In the third column, we write truth values for $\sim q$. The truth values of $p \wedge \sim q$ are listed in the next column. Finally we obtain the truth values of the proposition $\sim(p \wedge \sim q)$. Thus we have the following truth table:

p	q	$\sim q$	$p \wedge \sim q$	$\sim(p \wedge \sim q)$
T	T	F	F	T
T	F	T	T	F
F	T	F	F	T
F	F	T	F	T

Truth table for $\sim(p \wedge \sim q)$

Example 1.12. Construct the truth-table for $(p \vee q) \wedge (p \vee r)$.

Solution: Here, we have three atomic statements. Therefore we shall require eight rows to list all possible combinations of the truth values of statements p, q and r. Rest of the procedure will be the same as above. We shall proceed in steps and in the final column we will have the truth values of the given statement.

p	q	r	$p \vee q$	$p \vee r$	$(p \vee q) \wedge (p \vee r)$
T	T	T	T	T	T
T	T	F	T	T	T
T	F	T	T	T	T
T	F	F	T	T	T
F	T	T	T	T	T
F	T	F	T	F	F
F	F	T	F	T	F
F	F	F	F	F	F

Truth table for $(p \vee q) \wedge (p \vee r)$

Exercise

1. Which of the following are statements?
   (a) Is 3 a positive number?
   (b) $x^2 - 5x + 6 = 0$.
   (c) There will be snow in December.
   (d) Give me ten rupees.
   (e) Ramesh is poor but honest.
   (f) No triangles are squares.

2. Let p be the proposition "Mathematics is easy" and let q be the proposition "five is greater than four." Write in English the proposition, which corresponds to each of the following:
   (a) $p \wedge q$ (b) $p \vee q$
   (c) $\sim(p \wedge q)$ (d) $\sim p \vee \sim q$
   (e) $(p \wedge \sim q) \vee (\sim p \wedge q)$

3. Write the negation of each of the following statements:

(a) $2+7 \leq 13$

(b) 3 is an odd integer and 8 is an even integer.

(c) No nice people are dangerous.

4. Let p be the statement "Ravi is rich" and let q be the statement "Ravi is happy." Write the following statements in symbolic form:

(a) Ravi is poor but happy.

(b) Ravi is rich or unhappy.

(c) Ravi is neither rich nor unhappy.

(d) Ravi is poor or he is both rich and unhappy.

5. Construct the truth-table for the following functions:

(a) $(p'+q')'$ (b) $(p'q')'$

(c) $p(p+q)$ (d) $pqr+p'q'r'$

(e) $(p'+qr)'(pq+q'r)$

6. Given the truth values of p and q as true and those of r and s as false; find the truth values of the following:

(a) $p \vee (q \wedge r)$

(b) $(p \wedge (q \wedge r)) \vee \sim((p \vee q) \wedge (r \vee s))$

Answers

1. (c), (e) and (f) are statements.
2. (a) Mathematics is easy and five is greater than four.

   (b) Mathematics is easy or five is greater than four.

   (c) Either Mathematics is not easy or five is not greater than four.

   (d) Mathematics is not easy and five is not greater than four.

   (e) Either Mathematics is easy and five is not greater than four or Mathematics is not easy and five is greater than four.
3. (a) It is false that $2+7 \leq 13$

   (b) Either 3 is not an odd integer or 8 is not an even integer.

   (c) Some nice people are dangerous.

4. (a) $\sim p \wedge q$ (b) $p \vee \sim q$
(c) $\sim p \wedge q$ (d) $\sim p \vee (p \wedge \sim q)$
6. (a) True (b) True

§ 1.6. Conditional and Biconditional Statements

Definition 1: If p and q are given statements then the statements 'if p then q' denoted as $p \rightarrow q$, is called a conditional statement or *implication.* Here p is called **antecedent** or **hypothesis** and q is called consequent or **conclusion**. The conditional statement $p \rightarrow q$ is frequently read "p implies q" or "p only if q."

Example 1.13. If p is the statement "I am hungry" and q is the statement "I will eat", then $p \rightarrow q$ is the statement "if I am hungry then I will eat."

The following table defines the truth values of $p \rightarrow q$ in terms of truth values of p and q.

p	q	$p \rightarrow q$
T	T	T
T	F	F
F	T	T
F	F	T

Truth table for $p \rightarrow q$

It is clear that the statement $p \rightarrow q$ is false only when p is true and q is false.

Definition 2: If p and q be two statements then the statement "p if and only if q" written as '$p \leftrightarrow q$' is called a biconditional statement or equivalence. The truth values of $p \leftrightarrow q$ are defined in the following table.

p	q	$p \leftrightarrow q$
T	T	T
T	F	F
F	T	F
F	F	T

Truth table for $p \leftrightarrow q$

It is clear that $p \leftrightarrow q$ is true only when both p and q are true or when both p and q are false. $p \leftrightarrow q$ is also stated as p is necessary and sufficient condition for q.

Example 1.14. Is the following biconditional statement, a true statement? 5 is prime if and only if 2+2 = 4

Solution: Let p be the statement '5 is prime' and let q be the statement 2+2 = 4. Then the given statement can be expressed as $p \leftrightarrow q$. Since both p and q are true, we conclude that $p \leftrightarrow q$ is true.

Note: The symbols $\vee, \wedge, \sim, \rightarrow$ and $\leftrightarrow$ defined above are called connectives.

Converse, Inverse and Contrapositive of $p \rightarrow q$

Definition: Let $p \rightarrow q$ be any conditional statement. Then,

(a) the converse of $p \rightarrow q$ is the statement $q \rightarrow p$.

(b) the inverse of $p \rightarrow q$ is the statement $\sim p \rightarrow \sim q$.

(c) the contrapositive of $p \rightarrow q$ is the statement $\sim q \rightarrow \sim p$.

Example 1.15. Write the converse, inverse and contrapositive of the conditional statement "if 2+2=4 then I am not the Prime Minister of India."

Solution: Let p : 2+2 = 4 and

q : I am not the Prime Minister of India.

Then the given statement can be written as $p \rightarrow q$. Therefore, the converse is $q \rightarrow p$. That is, if I am not the Prime Minister of India then 2+2 = 4. The inverse of $p \rightarrow q$ is the statement $\sim p \rightarrow \sim q$. That is, if 2+2 $\neq$ 4 then I am Prime Minister of India. The contrapositive of $p \rightarrow q$ is the statement $\sim q \rightarrow \sim p$. That is, contrapositive of the given statement is "if I am Prime Minister of India then 2+2 $\neq$ 4."

§ 1.7. Propositional Functions and Propositional Variables

By a propositional variable, we mean a symbol which represents an arbitrary statement (proposition). Thus propositional variable is a variable that can be replaced by a statement. We shall use the symbols p, q, r, ... or $p_1, p_2, p_3, \ldots$ to denote propositional variables.

Propositional function is a function or statement which is formed by using propositional variables and connectives. For example, compound statements such as $p \vee q$, $p \wedge q$, $p \rightarrow q$ and $p \wedge (q \rightarrow r)$ are propositional functions. More formally,

Definition 1: A propositional function is an expression, which is a combination of propositional variables and connectives. Propositional function are denoted as $f(p, q, r, \ldots)$, where $p, q, r, \ldots$ are the variable used in forming the function f. A propositional function f in n variables $p_1, p_2, p_3, \ldots p_n$ will be denoted as $f(p_1, p_2, p_3, \ldots, p_n)$.

Illustrations:

(1) $f(p, q) = \sim(p \wedge q)$ is a propositional function in propositional variables p and q.

(2) $f(p, q, r) = p \rightarrow (q \rightarrow r)$ is a propositional function in propositional variables p, q and r.

(3) $f(p_1, p_2, p_3) = (p_1 \rightarrow p_2) \wedge (p_2 \rightarrow p_3)$ is a propositional function in propositional variables p_1, p_2, and p_3.

Thus we see that propositional functions are compound statements formed by using finite number of simple (atomic) statements and connectives. We shall often use the word statement for propositional function.

§ 1.8. Tautologies and Contradictions

Definition 1: A statement (or propositional function) which is true for all possible truth values of its propositional variables is called a tautology.

Definition 2: A statement which is always false is called a contradiction.

A simple method to determine whether a given statement is a tautology is to construct its truth table. If the statement is tautology then the column corresponding to the statement in the truth table contains only *T*. Similarly a statement is contradiction if the column corresponding to the statement contains only *F*.

Example 1.16. Show that each of the following is a tautology.

(a) $[p \wedge (p \rightarrow q)] \rightarrow q$

(b) $[(p \to q) \wedge (q \to r)] \to (p \to r)$

Solution:

(a) We shall construct truth-table for the function $p \wedge (p \to q) \to q$

p	q	$p \to q$	$p \wedge (p \to q)$	$[p \wedge (p \to q)] \to q$
T	T	T	T	T
T	F	F	F	T
F	T	T	F	T
F	F	T	F	T

Truth table for $p \wedge (p \to q)] \to q$

Since the column for $[p \wedge p \to q)] \to q$ contains only T, it is a tautology.

(b) Here we construct the truth-table for $[(p \to q) \wedge (q \to r)] \to (p \to r)]$.

p	q	r	$p \to q$	$q \to r$	$p \to r$	$(p \to r) \wedge (q \to r)$	$[(p \to r) \wedge (q \to r)] \to (p \to r)$
T	T	T	T	T	T	T	T
T	T	F	T	F	F	F	T
T	F	T	F	T	T	F	T
T	F	F	F	T	F	F	T
F	T	T	T	T	T	T	T
F	T	F	T	F	T	F	T
F	F	T	T	T	T	T	T
F	F	F	T	T	T	T	T

Truth table for $[(p \to r) \wedge (q \to r)] \to (p \to r)$

Since the last column corresponding to $[(p \to q) \wedge (q \to r)] \to (p \to r)$ contains only T, it is a tautology.

Example 1.17. Show that the statement $p \wedge \sim p$ is a contradiction.

Solution: Consider the truth table for $p \wedge \sim p$.

p	$\sim p$	$p \wedge \sim p$
T	F	F
F	T	F

Truth table for $p \wedge \sim q$

It follows from the table that $p \wedge \sim p$ is a contradicti .

§ 1.9. Equivalent Statements or Functions

Definition: Two statements p and q are said to be logically equivalent or equal if they have identical truth values. We shall denote logical equivalence or equality by the symbol '='.

One method to determine whether any two statements are equal is to construct a column for each statement in a truth table and compare these to see if they are identical.

Example 1.18. Prove that $p \to q = \sim p \vee q$

Solution: We shall construct truth table for statements $p \to q$ and $\sim p \vee q$.

p	q	$p \to q$	$\sim q$	$\sim p \vee q$
T	T	T	F	T
T	F	F	F	F
F	T	T	T	T
F	F	T	T	T

Truth table for $p \to q$ and $\sim p \vee q$

We observe that the truth values in the columns for $p \to q$ and $\sim p \vee q$ are identical. Hence $p \to q = \sim p \vee q$.

Example 1.19. Show that the statements $(p \wedge \sim p) \vee q$ and q are equal.

Solution: Consider the truth table for given statements.

p	q	$\sim p$	$p \wedge \sim p$	$(p \wedge \sim p) \vee q$
T	T	F	F	T
T	F	F	F	F
F	T	T	F	T
F	F	T	F	F

Truth table for $(p \wedge \sim p) \vee q$ and q

From the truth table we see that columns for $(p \wedge \sim p) \vee q$ and q are identical. Hence they are equal.

Example 1.20. Show that two statements p and q are equivalent if biconditional statement $p \leftrightarrow q$ is a tautology.

Solution: From the definition of biconditional statement we know that $p \leftrightarrow q$ is true whenever both p and q have the same truth values. Thus $p = q$ if $p \leftrightarrow q$ is a tautology.

Note: (1) Some authors have used the symbol '$\Leftrightarrow$' to denote equivalent or equal statements instead of the symbol '='. Students are advised to note the difference between symbols $\Leftrightarrow$ and $\leftrightarrow$. Symbol $\leftrightarrow$ is used for biconditional statement. From Example 1.20, we have $p \Leftrightarrow q$ if $p \leftrightarrow q$ is tautology.

(2) Two equivalent statements may contain different variables as is clear from Example 1.19 above.

Example 1.21. Show that $p \to (q \to r) = (p \wedge q) \to r$

Solution: Consider the following truth table.

p	q	r	$q \to r$	$p \wedge q$	$p \to (q \to r)$	$(p \wedge q) \to r$
T	T	T	T	T	T	T
T	T	F	F	T	F	F
T	F	T	T	F	T	T
T	F	F	T	F	T	T
F	T	T	T	F	T	T
F	T	F	F	F	T	T
F	F	T	T	F	T	T
F	F	F	T	F	T	T

Truth table for $p \to (q \to r)$ & $(p \wedge q) \to r$

We see that columns for $p \to (q \to r)$ and $(p \wedge q) \to r$ are identical, hence given statements are equal.

The following theorem contains various laws satisfied by propositions. We shall use these laws for simplification of propositions.

Theorem 1: The following laws are satisfied by statements:

1. **Commutative laws:**

   (a) $p \vee q = q \vee p$. (b) $p \wedge q = q \wedge p$.

2. **Associative laws:**

   (a) $p \vee (q \vee r) = (p \vee q) \vee r$ (b) $p \wedge (q \wedge r) = (p \wedge q) \wedge r$

3. **Distributive laws:**

   (a) $p \vee (q \wedge r) = (p \vee q) \wedge (p \vee r)$

   (b) $p \wedge (q \vee r) = (p \wedge q) \vee (p \wedge r)$

4. **Idempotent laws:**

   (a) $p \vee p = p$ (b) $p \wedge p = p$

5. **Laws of absorption:**

   (a) $p \vee (p \wedge q) = p$ (b) $p \wedge (p \vee q) = p$

6. **Involution laws:**

   (a) $\sim(\sim p) = p$

7. **Complement laws:**

   (a) $p \vee \sim p = T$ (b) $p \wedge \sim p = F$

8. **De Morgan's laws:**

   (a) $\sim(p \vee q) = \sim p \wedge \sim q$ (b) $\sim(p \wedge q) = \sim p \vee \sim q$

9. **Operations with T**

   (a) $p \vee T = T$ (b) $p \wedge T = p$

10. **Operations with F**

   (a) $F \vee p = p$ (b) $F \wedge p = F$

Here T and F denote statements, which are tautology and contradiction respectively.

Proof: We shall prove 3(a) and 8(a). The remaining laws can be proved exactly in the same way by constructing truth tables.

To prove 3(a), consider the following truth-table:

p	q	r	$p \vee q$	$p \vee r$	$q \wedge r$	$p \vee (q \wedge r)$	$(p \vee q) \wedge (p \vee r)$
T	T	T	T	T	T	T	T
T	T	F	T	T	F	T	T
T	F	T	T	T	F	T	T
T	F	F	T	T	F	T	T
F	T	T	T	T	T	T	T
F	T	F	T	F	F	F	F
F	F	T	F	T	F	F	F
F	F	F	F	F	F	F	F

Since columns for $p\vee(q\wedge r)$ and $(p\vee q)\wedge(p\vee r)$ are identical they are equal. This proves 3(a).

To prove 8(a), consider the following truth table:

p	q	$\sim p$	$\sim q$	$p\vee q$	$\sim(p\vee q)$	$\sim p\wedge\sim q$
T	T	F	F	T	F	F
T	F	F	T	T	F	F
F	T	T	F	T	F	F
F	F	T	T	F	T	T

It follows from the table that $\sim(p\vee q) = \sim p\wedge\sim q$.

Theorem 2: Show that

(a) $p\rightarrow q = \sim p\vee q$

(b) $p\leftrightarrow q = (p\rightarrow q)\wedge(q\rightarrow p)$

Proof: Using the definitions of $\rightarrow$ and $\leftrightarrow$ we have,

(1) p	(2) q	(3) $\sim p$	(4) $p\rightarrow q$	(5) $q\rightarrow p$	(6) $p\leftrightarrow q$	(7) $\sim p\vee q$	(8) $(p\rightarrow q)\wedge(q\rightarrow p)$
T	T	F	T	T	T	T	T
T	F	F	F	T	F	F	F
F	T	T	T	F	F	T	F
F	F	T	T	T	T	T	T

Since the truth values in columns (4) and (7) are identical, we have (a). Similarly, since the truth values in columns (6) and (8) are identical, we have (b).

Theorem 3: Show that

(a) $\sim(p\rightarrow q) = p\wedge\sim q$

(b) $\sim(p\leftrightarrow q) = p\leftrightarrow\sim q$

Proof: Using the definitions of $\rightarrow$ and $\leftrightarrow$, we construct the truth table (given on next page).

Since the truth values in columns (6) and (7) of the table are identical, we have,

$$\sim(p\rightarrow q) = p\wedge\sim q.$$

(1)	(2)	(3)	(4)	(5)	(6)	(7)	(8)	(9)	(10)
p	q	$\sim p$	$\sim q$	$p\to q$	$\sim(p\to q)$	$p\wedge\sim q$	$p\leftrightarrow q$	$\sim(p\leftrightarrow q)$	$p\leftrightarrow\sim q$
T	T	F	F	T	F	F	T	F	F
T	F	F	T	F	T	T	F	T	T
F	T	T	F	T	F	F	F	T	T
F	F	T	T	T	F	F	T	F	F

Similarly since columns (9) and (10) have identical truth values, we have,

$$\sim(p\leftrightarrow q) = p\leftrightarrow\sim q.$$

Example 1.22. Prove that $\sim(p\wedge q)\rightarrow(\sim p\vee(\sim p\vee q)) = \sim p\vee q$, without constructing truth table.

Solution: We shall use Theorems 1 and 2.

L.H.S.= $\sim(p\wedge q)\rightarrow(\sim p\vee(\sim p\vee q))$

$= (p\wedge q)\vee(\sim p\vee(\sim p\vee q))$ since $p\rightarrow q = \sim p\vee q$

$= (p\wedge q)\vee((\sim p\vee\sim p)\vee q)$ by associative law

$= (p\wedge q)\vee(\sim p\wedge q)$ by idempotent law

$= ((p\wedge q)\vee\sim p)\vee q$ by associative law

$= (\sim p\vee(p\wedge q))\vee q$ by commutative law

$= ((\sim p\vee p)\wedge(\sim p\vee q))\vee q$ by distributive law

$= (T\wedge(\sim p\vee q))\vee q$ by complement law

$= (\sim p\vee q)\vee q$ by 9(b) of Theorem 1

$= \sim p\vee(q\vee q)$ by associative law

$= \sim p\vee q$ by idempotent law

= R.H.S.

Exercise

1. Prove that each of the following is a tautology:

(a) $p\rightarrow p$ (b) $p\wedge q\rightarrow p$

(c) $p \to (p \vee q)$ (d) $(p \wedge (p \to q)) \to q$

(e) $(p \to q) \to [(p \vee (q \wedge r)) \leftrightarrow q \wedge (p \vee r)]$

2. Write in words the converse, inverse, contrapositive and negation of the implication "if she works then she will earn money."

3. Construct truth tables to determine whether each of the following is a tautology or a contradiction:

(a) $p \wedge \sim p$ (b) $p \to (q \to p)$

(c) $p \to (q \wedge p)$ (d) $q \vee (\sim q \wedge p)$

4. Prove the following:

(a) $p \vee q = q \vee p$ (b) $p \wedge (q \wedge \mathrm{r}) = (p \vee q) \wedge \mathrm{r}$

(c) $p \wedge (q \vee \mathrm{r}) = (p \wedge q) \vee (p \wedge \mathrm{r}$

(d) $p \vee p = p$ (e) $\sim (p \wedge q) = \sim p \vee \sim q$

(f) $\sim (p \leftrightarrow q) = \sim p \leftrightarrow q$

5. Write in English the negation of each of the following:

(a) The weather is bad and I will not go to work.

(b) I grow fat only if I eat too much.

6. Show the following equivalences:

(a) $p \to (q \to p) \Leftrightarrow \sim p \to (p \to q)$

(b) $\sim (p \leftrightarrow q) \Leftrightarrow (p \wedge \sim q) \vee (\sim p \wedge q)$

7. We define $p \Rightarrow q$ if and only if $p \to q$ is tautology. Prove the following:

(a) $p \to q \Rightarrow p \to (p \wedge q)$

(b) $(p \to q) \to q \Rightarrow p \vee q$

8. Prove that for any propositions p and q.

(a) $p \vee T = T$

(b) $p \wedge F = F$

(c) $\{(p \vee \sim q) \wedge (\sim p \vee \sim q)\} \vee q = T$

Answers

2. The converse of the statement is "if she earns money then she works." The inverse is "if she does not work then she will not earn money." The contrapositive is "if she does not earn money then she does not work" The negation of the statement is "she works and she will not earn money."

3. (a) Contradiction (b) Tautology

   (c) Neither tautology nor contradiction

   (d) Neither tautology nor contradiction

5. (a) The weather is bad but I will go to work.

   (b) I grow fat and (although) I don't eat too much.

§ 1.10. Law of Duality

In this section, we consider only those statements which contain the connectives $\wedge$, $\vee$ and ~ only.

Definition: Two statements p and p^* are said to be duals of each other if either one can be obtained from the other by replacing $\wedge$ by $\vee$, $\vee$ by $\wedge$, T by F and F by T.

Example 1.23. Find the dual of the following:

(a) $(p \vee q) \wedge r$ (b) $(p \wedge q) \vee T$

(c) $\sim(p \vee q) \wedge (p \vee \sim(p \wedge s))$

Solution:

(a) Interchanging $\vee$ and $\wedge$, we have. dual as $(p \wedge q) \vee r$.

(b) Dual is $(p \vee q) \wedge F$

(c) Dual is $\sim(p \wedge q) \vee (p \wedge \sim(p \vee s))$

It is obvious from the definition that dual of a statement is the statement itself. We now state (without proof) the principle of duality.

Principle of Duality

It states that if any two statements are equal then their duals are also equal.

Example 1.24. Prove the following:

(a) $\sim(p \wedge q) = \sim p \vee \sim q$ (b) $\sim(p \vee q) = \sim p \wedge \sim q$

Solution: We shall only prove (a). The result stated in (b) will follow by principle of duality.

To prove (a), consider the following table:

p	q	$\sim p$	$\sim q$	$p \wedge q$	$\sim(p \wedge q)$	$\sim p \vee \sim q$
T	T	F	F	T	F	F
T	F	F	T	F	T	T
F	T	T	F	F	T	T
F	F	T	T	F	T	T

From the table, it follows that $\sim(p \wedge q) = \sim p \vee \sim q$

§ 1.11. Functionally Complete Set of Operations

Definition: A set of operations (connectives) is said to be functionally complete if every statement can be expressed entirely in terms of the operations in the set.

It is assumed that such a functionally complete set does not contain any connective which can be expressed in terms of the other connectives in the set.

Recall that so far, we have defined the connectives $\wedge, \vee, \sim, \rightarrow$ and $\leftrightarrow$. We know that $p \rightarrow q$ is equal to $\sim p \vee q$. Thus it is possible to replace each occurrence of $\rightarrow$ in any statement with an equivalent expression involving $\sim$ and $\vee$. Similarly, we know that:

$$p \leftrightarrow q = (p \rightarrow q) \wedge (q \rightarrow p) = (\sim p \vee q) \wedge (\sim q \vee p)$$

Thus the symbol $\leftrightarrow$ can be replaced by connectives $\vee, \wedge$ and $\sim$.

By De Morgan's law, we know that

$$\sim(p \wedge q) = \sim p \vee \sim q$$

or $\quad p \wedge q = \sim(\sim p \vee \sim q)$

Similarly, we have, $p \vee q = \sim(\sim p \wedge \sim q)$.

Thus it is possible to replace $\wedge$ in any statement by connectives ~ and $\vee$. Hence any statement can be expressed in an equivalent statement containing ~ and $\wedge$ only. This shows that {$\wedge$, ~} is functionally complete set of operations. Similarly, any statement can be expressed in an equivalent statement containing $\vee$ and ~ only. Thus {$\vee$, ~} is also functionally complete set.

Example 1.25. Write the statement $(p \vee \sim q) \rightarrow (p \wedge r)$ in terms of $\vee$ and ~ only.

Solution: We know that

$$p \rightarrow q = \sim p \vee q$$

Therefore,

$$(p \vee \sim q) \rightarrow (p \wedge r) = \sim(p \vee \sim q) \vee (p \wedge r)$$
$$= \sim(p \vee \sim q) \vee \sim(\sim p \vee \sim r)$$

because $\quad p \wedge r = \sim(\sim p \vee \sim r)$

Example 1.26. Express $p \wedge (q \leftrightarrow r)$ in terms of $\wedge$ and ~.

Solution:

$$p \wedge (q \leftrightarrow r) = p \wedge [(q \rightarrow r) \wedge (r \rightarrow q)]$$
$$= p \wedge [(\sim q \vee r) \wedge (\sim r \vee q]$$
$$= p \wedge \sim(q \wedge \sim r) \wedge \sim(r \wedge \sim q) \qquad (\because p \vee q = \sim(\sim p \wedge \sim q))$$

Example 1.27. Show that {$\rightarrow$, ~} is functionally complete set.

Solution: We know that {$\vee$, ~} is functionally complete set. Thus any statement can be expressed in an equivalent statement containing $\vee$ and ~. Since

$$p \vee q = (\sim p) \rightarrow q$$

Therefore we can replace the connective $\vee$ in any statement by ~ and $\rightarrow$. Hence the statement can be expressed in terms of $\rightarrow$ and ~. Thus {$\rightarrow$, ~} is functionally complete.

Example 1.28. Show that {$\wedge$, $\vee$} is not functionally complete.

Solution: Since the connective'~' can not be expressed entirely in terms of $\vee$ & $\wedge$, therefore any statement containing ~ can not be expressed in an equivalent statement containing $\wedge$ & $\vee$ only. Hence $\{\wedge, \vee\}$ is not functionally complete.

Connectives ↓ and ↑

We now introduce two more connectives, which have useful applications in the design of computers. They are NAND and NOR. The connective NAND is denoted by the symbol ↑ and is defined by the following truth table:

p	q	$p \uparrow q$
T	T	F
T	F	T
F	T	T
F	F	T

Truth-table for $p \uparrow q$

The connective NOR is denoted by the symbol↓ and is defined by the following table:

p	q	$p \downarrow q$
T	T	F
T	F	F
F	T	F
F	F	T

Truth-table for $p \downarrow q$

For any two statements p and q, it can be shown by constructing truth tables that,

$$p \uparrow q = \sim(p \wedge q) \quad \text{and}$$
$$p \downarrow q = \sim(p \vee q).$$

Theorem 1: Show that each of the sets $\{\downarrow\}$ and $\{\uparrow\}$ is functionally complete.

Proof: We know that the sets of connectives $\{\vee, \sim\}$ and $\{\wedge, \sim\}$ are functionally complete. Therefore, in order to show that $\{\uparrow\}$ and $\{\downarrow\}$

are functionally complete, it is sufficient to show that the connective $\wedge$, $\vee$ and $\sim$ can be expressed either in terms of $\uparrow$ alone or in terms of $\downarrow$ alone. From the definition of $\uparrow$, we see that

$$p\uparrow p = \sim(p\wedge p)$$
$$= \sim p \quad \text{-------(1)}$$

Again,

$$(p\uparrow q)\uparrow(p\uparrow q) = \sim(p\uparrow q) \quad \text{since } p\uparrow p = \sim p \text{ by (1)}$$
$$= \sim(\sim(p\wedge q))$$
$$= p\vee q \quad \text{-------(2)}$$

Finally,

$$(p\uparrow p)\uparrow(q\uparrow q) = (\sim p)\uparrow(\sim q)$$
$$= \sim(\sim p\wedge\sim q)$$
$$= p\vee q \quad \text{-------(3)}$$

Thus $\sim$, $\vee$ and $\wedge$ have been expressed in term of $\uparrow$ alone. Hence $\{\uparrow\}$ is functionally complete. In a similar manner, we have

$$p\downarrow p = \sim(p\vee p)$$
$$= \sim p\,,$$
$$(p\downarrow q)\downarrow(p\downarrow q) = p\vee q$$

and $$(p\downarrow p)\downarrow(q\downarrow q) = p\wedge q$$

Thus $\{\downarrow\}$ is functionally complete.

§ 1.12. Quantifiers

In the discussion of logic, some very important statements contain quantifiers. The following are examples of statements which contain quantifiers:

(1) Some people are honest.

(2) No woman is a player.

(3) All Americans are crazy.

The words *some*, *no* and *all* are known as quantifiers. From quantifiers, we know "how many" of a certain set of things is being considered.

Universal Quantifier

Let p be a statement. We define the symbol $\forall_x p$ to mean that for every value of x in the given set, the statement p is true. The symbol $\forall$ is called the universal quantifier. $\forall$ can also be read as 'for all', 'for every' or 'for any.'

Illustration: The statement "for all natural numbers, $n+4>3$" can be expressed as $\forall_x p$, where x belongs to the set N of natural numbers and p is the statement '$n+4>3$.'

Existential Quantifier

Let p be a statement. We define the symbol $\exists_x p$ to mean that for one or more elements x of a certain set, the proposition p is true. The symbol $\exists$ is called existential quantifier and is usually read as "there exists" or "for at least one" or "for some".

Illustration:

(1) The statement "there exists a number x such that $x^2-4x = 16$" may be written as

$$\exists_x(x^2-4x=16)$$

(2) The statement $\exists_n(n+4<7)$, where n is in the set of natural numbers is true since there exists a natural number, namely 1, such that $n+4<7$ is true.

Negation of Statements Involving Quantifiers

It is important to know how the negations of statements having quantifiers are formed. Consider the statement

"All Americans are crazy"

The negation of this statement would be

"It is false that all Americans are crazy" or equivalently,

"There exists at least one American who is not crazy."

In general, we have

$$\sim(\exists_x p) = \forall_x(\sim p) \quad \text{and}$$

$$\sim(\forall_x p) = \exists_x(\sim p)$$

Example 1.29. Form the negation of each of the following:

(a) "For all positive integers x, we have $x+2>8$"

(b) "All men are honest or some man is a thief."

(c) "There is at least one person who is happy all the time."

(d) "The sum of any two integers is an even integer."

(e) At least one student does not live in the dormitories.

Solution: Negation of the statements are:

(a) There exists a positive integer x such that $x+2 \not> 8$.

(b) There exists a man who is not honest and all men are not thief.

(c) No person is happy all the time.

(d) There exists two integers such that their sum is not an even integer.

(e) All students live in the dormitories.

§ 1.13. Arguments

The main problem in logic is the investigation of the process of reasoning. In Mathematics, a certain set of statements (propositions) is assumed and from this set, other statements are derived by logical reasoning. In this section, we shall investigate those processes which can be accepted as valid in the derivation of a statement from the given set of statements. The given set of statements is called **premises** or **hypothesis** and the statement derived from the given statements is called **conclusion.**

Definition: An argument is a process by which a conclusion is formed from a given set of statements called premises.

An argument is said to be **valid argument** if and only if the conjunction of the premises implies the conclusion. That is, the agrument which yields a conclusion r from the premises $p_1, p_2, p_3, \text{-----}, p_n$ is valid if and only if the statement

$$(p_1 \wedge p_2 \wedge p_3 \text{ - - - } \wedge p_n) \rightarrow r$$

is a tautology.

An argument which is not valid is called a **fallacy**. An argument which is derived from the premises or hypothesis $p_1, p_2, \text{- - - -}, p_n$ is written as

$$\begin{array}{l} p_1 \\ p_2 \\ p_3 \\ - \\ - \\ p_n \\ \hline q \end{array}$$

That is, the premise or premises will be listed first and the conclusion will be written beneath a horizontal line.

Example 1.30. Prove that the following argument is valid:

$$\begin{array}{c} p \\ p \rightarrow q \\ \hline q \end{array}$$

Solution: Here p and $p \rightarrow q$ are two premises and q is the conclusion. To show that the argument is valid we show that conjunction of the given premises implies the conclusion is a tautology. That is, we show $[p \wedge (p \rightarrow q)] \rightarrow q$ is a tautology by constructing truth table.

p	q	$p \rightarrow q$	$p \wedge (p \rightarrow q)$	$[p \wedge (p \rightarrow q)] \rightarrow q$
T	T	T	T	T
T	F	F	F	T
F	T	T	F	T
F	F	T	F	T

It follows from the truth table that $[p \wedge (p \rightarrow q)] \rightarrow q$ is a tautology. Thus the argument is valid.

Rule of Detachment or Modus Ponens

The valid argument

$$\begin{array}{c} p \\ p \rightarrow q \\ \hline q \end{array}$$

is called rule of detachment. Rule of detachment is also known as **modus ponens.**

Law of Syllogism

The argument

$$\begin{array}{c} p \rightarrow q \\ q \rightarrow r \\ \hline p \rightarrow r \end{array}$$

is valid argument and is known as the **law of syllogism.**
The validity of the law of syllogism is proved by constructing truth table for $[(p \rightarrow q) \wedge (q \rightarrow r)] \rightarrow (p \rightarrow r)$.

p	q	r	$p \rightarrow q$	$q \rightarrow r$	$p \rightarrow r$	$(p \rightarrow q) \wedge (q \rightarrow r)$	$[(p \rightarrow q) \wedge (q \rightarrow r)] \rightarrow (p \rightarrow r)$
T	T	T	T	T	T	T	T
T	T	F	T	F	F	F	T
T	F	T	F	T	T	F	T
T	F	F	F	T	F	F	T
F	T	T	T	T	T	T	T
F	T	F	T	F	T	F	T
F	F	T	T	T	T	T	T
F	F	F	T	T	T	T	T

Truth table for $[(p \rightarrow q) \wedge (q \rightarrow r)] \rightarrow (p \rightarrow r)$

It follows from the truth table that the law of syllogism is a valid argument.

Given an argument, there are, in general, three methods to check the validity of the argument. These methods are given below.

1. Validity using Truth Table

In this method, we construct a truth table as follows: Let $p_1, p_2 ----, p_n$ be all the premises and let q be the conclusion in the given argument. We construct truth table for the statement

$$(p_1 \wedge p_2 \wedge p_3 - - - - - \wedge p_n) \rightarrow q$$

If we have all Ts in the column of this statement then the statement is tautology and so the argument used is valid otherwise the argument is not valid.

2. Validity using Simplification Methods

In this method, we convert all the implication statement $p \rightarrow q$ to the equivalent statement $\sim p \vee q$ in the argument involved and then we simplify the resulting statements using rules of statements (Theorem 1 of § 1.9). If the statement

$$(p_1 \wedge p_2 \wedge p_3 \wedge - - - - \wedge p_n) \rightarrow q$$

can be reduced to T, then we say that the argument is valid.

3. Validity using Rules of Inference

In this method, we reduce the given argument to a series of arguments each of which is known to be valid. Two of the most frequently used rules of inference (i.e. valid arguments) are the rule of detachment and the law of syllogism.

Example 1.31. Show that the following argument is valid.

$$\begin{array}{c} p \\ p \rightarrow q \\ q \rightarrow r \\ \hline r \end{array}$$

Solution: We shall show the validity of the argument by all three methods.

First solution: We construct the truth table for the statement

$$f = [p \wedge (p \rightarrow q) \wedge (q \rightarrow r)] \rightarrow r$$

p	q	r	$p\to q$	$q\to r$	$p\wedge(p\to q)$	$p\wedge(p\to q)\wedge(q\to r)$	f
T	T	T	T	T	T	T	T
T	T	F	T	F	T	F	T
T	F	T	F	T	F	F	T
T	F	F	F	T	F	F	T
F	T	T	T	T	F	F	T
F	T	F	T	F	F	F	T
F	F	T	T	T	F	F	T
F	F	F	T	T	F	F	T

Truth table for $f = [p\wedge(p\to q)\wedge(q\to r)]\to r$

Since the column for f contains only T's, the argument is valid.

Second solution: We shall simplify the statement

$f = [p\wedge(p\to q)\wedge(q\to r)]\to r$

$= \sim[(p\wedge(\sim p\vee q)\wedge(\sim q\vee r)]\vee r$ since $p\to q = \sim p\vee q$

$= \sim p\vee(p\wedge\sim q)\vee(q\wedge\sim r)\vee r$ using $\sim(p\wedge q) = \sim p\vee\sim q$

$= (\sim p\vee p)\wedge(\sim p\vee\sim q)\vee(q\wedge\sim r)\vee r$ using distributive law.

$= T\wedge(\sim p\vee\sim q)\vee\ r\ \vee(q\wedge\sim r)$ since $\sim p\vee p = T$

$= (\sim p\vee\sim q)\vee\{(r\vee q)\wedge(r\vee\sim r)\}$ since $p\wedge T = p$

$= (\sim p\vee\sim q)\vee\{(r\vee q)\wedge T\}$ since $r\vee\sim r = T$

$= (\sim p\vee\sim q)\vee(r\vee q)$ since $p\wedge T = p$

$= \{(\sim p\vee\sim q)\vee q\}\vee r$ since $p\vee q = q\vee p$

$= \{\sim p\vee(\sim q\vee q)\}\vee r$ by associative law

$= \{\sim p\vee T\}\vee r$ since $\sim p\vee p = T$

$= T$

Since f reduces to T, so the argument is valid.

Third solution: Consider the following sequence of arguments

p	a premise
$p\to q$	a premise
q	a conclusion by modus ponens
$q\to r$	a premise
r	a conclusion by modus ponens

This sequence of valid arguments shows that the given argument is valid.

Example 1.32. Check the validity of the argument

$$\begin{array}{c} p \to q \\ r \to \sim q \\ \hline p \to \sim r \end{array}$$

Solution: Since the statement $r \to \sim q$ is equal to $q \to \sim r$, we can replace the premise $r \to \sim q$ by $q \to \sim r$.

Now
$$\begin{array}{c} p \to q \\ q \to \sim r \\ \hline p \to \sim r \end{array}$$
is valid argument by the law of syllogism. Hence given argument is valid.

A Short Method for Invalidity of an Argument

In checking a validity of a given argument, if it is found or suspected that the argument is not valid, a proof of invalidity can be given more easily than by constructing the entire truth table related to the argument. For proving that the argument is invalid, it is sufficient to exhibit a particular set of truth values for the statements involved for which the premises are all true and the conclusion is false. This is equivalent to demonstrating that one row in the truth table would contain F and hence the argument is invalid.

Example 1.33. Show that the following argument is not valid

$$\begin{array}{c} p \\ \sim p \vee r \\ \sim p \to q \\ \hline r \end{array}$$

Solution: If p is true, q is false and r is false then each of the premises is true but the conclusion is false. Hence the argument is invalid.

Example 1.34. Given the following statements as premises, all referring to an arbitrary meal:

(a) If he takes coffee, he doesn't drink milk.

(b) He eats crackers only if he drinks milk.

(c) He does not take soup unless he eats crackers.

(d) At noon today, he had coffee.

Whether he took soup at noon today? If so, what is the correct conclusion?

Solution: Let

p : he takes coffee.

q : he drinks milk.

r : he eats crackers.

s : he takes soup.

Then we have,

by condition (a) that $p \to \sim q$

by condition (b), we have $r \to q$

by condition (c), we have $\sim r \to \sim s$ and

by condition (d), we have p

Since implication $r \to q$ is equivalent to its contrapositive $\sim q \to \sim r$, we have the following chain of arguments:

$p \to \sim q$	a premise
$\sim q \to \sim r$	contrapositive of premise (b)
$p \to \sim r$	a conclusion by law of syllogism
$\sim r \to \sim s$	a premise
$p \to \sim s$	a conclusion by law of syllogism
p	a premise
$\sim s$	a conclusion by modus ponens.

Hence $\sim s$ is the conclusion. That is, he did not take soup at noon today.

Indirect Method of Proof

An important poof technique called the indirect method follows from the fact that any implication $p \to q$ is equivalent to its contrapositive $\sim q \to \sim p$. Thus to prove $p \to q$ indirectly, we assume that q is false and show that p is then false.

More generally, to prove the validity of an argument with premises $p_1, p_2, p_3, ----, p_n$ and conclusion q by indirect method, we consider second argument with premises $\sim q, p_1, p_2 --- p_{i-1}, p_{i+1}, ----, p_n$ and conclusion $\sim p_i$ and prove the validity of this second argument.

Example 1.35. Show that the following argument is a valid argument.

$$\begin{array}{l} p \\ p \wedge q \to r \vee s \\ q \\ \sim s \\ \hline r \end{array}$$

Solution: We will take as premises for the indirect proof all given premises except $\sim s$ and the negation of the conclusion $\sim r$ *i.e.* we shall show that the following argument is valid:

$$\begin{array}{l} p \\ p \wedge q \to r \vee s \\ q \\ \sim r \\ \hline s \end{array}$$

Now,

p	a premise
q	a premise
$p \wedge q$	a conclusion because $p \wedge q \to p \wedge q$ is always a tautology.

Now,

$p \wedge q$	a valid conclusion
$p \wedge q \to r \vee s$	a premise

$r \vee s$	a valid conclusion by *modus ponens*
$\sim r$	a premise
s	a valid conclusion because $(r \vee s) \wedge \sim r \rightarrow s$ is a tautology.

Proof by Counter-Example

If a statement claims that a property holds for all objects of a certain type, then to prove it, we must use steps that are valid for all objects of that type. To disprove such a statement, we need only show one counter example. That is one particular object for which the statement is false. Such a proof is called a proof by counter-example.

Example 1.36. Prove that the statement "if n is an integer, then $n^2 - n + 41$ is a prime number" is false.

Solution: We need only find one example for which the statement is false. If $n = 41$, then $n^2 - n + 41 = 41^2$ which is not a prime. Hence the statement is false.

SOME MORE EXAMPLES ON STATEMENTS

Example 1.37. Rohan made the following two statements:

1. I love Vicky.
2. If I love Vicky then I also love Vivian.

Given that Rohan either told the truth or lied in both cases, determine whether Rohan really loves Vicky.

Solution: Let p : Rohan loves Vicky

and q : Rohan loves Vivian.

Consider the following truth table:

p	q	$p \rightarrow q$
T	T	T
T	F	F
F	T	T
F	F	T

We are given that both p and $p \to q$ are either true or both of them are false. From the table, it is possible for both p and $p \to q$ to be true (row 1) but not possible for both p and $p \to q$ to be false. Hence Rohan must have told the truth and we conclude that Rohan really loves Vicky.

Example 1.38. An island has two tribes of natives. Any native from first tribe always tells the truth while any native from the second tribe always lies. Suppose you arrive at the island and ask a native if there is gold on the island. He answers "There is a gold on the island if and only if I always tell the truth." Determine whether there is gold on the island.

Solution: Let p & q be the following propositions:

p : He (the native from whom question is being asked) always tells the truth.

q : There is gold on the island.

Then his answer is the statement '$p \leftrightarrow q$'. Suppose p is true then $p \leftrightarrow q$ is true. Consequently q must be true. If p is false then his statement $p \leftrightarrow q$ is false. Consequently q must be true. Thus in both cases we can conclude that there is gold on the island.

Example 1.39. A logician was captured by a certain gang. The leader of the gang blindfolded the logician and placed him in a locked room containing two boxes. He gave the following instruction "one box contains the key to the room and other a poisonous shake. You are to reach into either box you choose and if you find the key, you can use it to go free. To help you, you can ask my assistant a single question requiring a yes or no answer. However, he does not have to answer truthfully, he may lie if he chooses."After a moment of thought, the logician asked a question, reached into the box with the key and left. What question did the logician ask so that he was certain to go free?

Solution: Let p be the statement "the box on my left contains the key". Let q be the statement "you are telling the truth." Suppose we desire the answer "yes" if p is true and "no" if p is false. In the table given below, the first three columns represent the possible truth values of p and q and the desired answer. Then the required statement (i.e. a question requiring answer in 'yes' or 'no') must have column 4 as its truth values.

p	***q***	***Desired answer***	***Truth value of required statement***
T	*T*	Yes	*T*
T	*F*	Yes	*F*
F	*T*	No	*F*
F	*F*	No	*T*

We explain the reasoning used in forming the truth table by considering row 2. In this row truth value of statement p is T while that of q is F. Thus the key is in the left box and the man is lying. Consequently to obtain an affirmative answer the function must have the value F (because the value of the function in row 1 is T) The statement corresponding to this truth table is $p \leftrightarrow q$. Hence the proper question is "does the box on the left contain the key if and only if you are telling the truth?

Example 1.40. Given that the value of $p \rightarrow q$ is true, can you determine the value of $\sim p \vee (p \leftrightarrow q)$?

Solution: We shall construct the truth table having columns for $p \rightarrow q$ and $\sim p \vee (p \leftrightarrow q)$.

p	q	$p \rightarrow q$	$\sim p$	$p \leftrightarrow q$	$\sim p \vee (p \leftrightarrow q)$
T	*T*	*T*	*F*	*T*	*T*
T	*F*	*F*	*F*	*F*	*F*
F	*T*	*T*	*T*	*F*	*T*
F	*F*	*T*	*T*	*T*	*T*

From the table it follows that if $p \rightarrow q$ is true then the value of $\sim p \vee (p \leftrightarrow q)$ is true.

Note: In Example 1.40 above, we could determine the value of $\sim p \vee (p \leftrightarrow q)$ because corresponding to each possible choices of p and q for which the value of $p \rightarrow q$ is true, the value of $\sim p \vee (p \leftrightarrow q)$ is same as T.

Exercise

1. Express $p \to q$ and $p \wedge q$ in terms of $\downarrow$.

2. Express the statement $(p \vee \sim q) \to p \wedge r$ in terms of:
   (a) $\vee$ and $\sim$ only
   (b) $\to$ and $\sim$ only
   (c) $\downarrow$ only

3. Form the negation of the following statements:
   (a) The corresponding sides of two triangles are equal if and only if the triangles are congruent
   (b) If the number x is less than 10 then there is a number y such that $x^2 + y^2 - 100$ is positive.

4. Let p be the statement "x is wealthy" and let q be the statement "y is married." Further assume that both x and y represent members of the set of all men. Write the following statements in English (without using $\exists$ and $\forall$):
   (a) $\exists_x p$
   (b) $(\forall_x p) \vee (\exists_y q)$
   (c) $\forall_x(\sim p)$

5. Check the validity of the following arguments:
   (a)
$$\begin{array}{l} p \to q' \\ r' \to q' \\ \hline p \to r' \end{array}$$
   (b)
$$\begin{array}{l} p \\ q \\ \sim p \to r \\ q \to \sim r \\ \hline \sim r \end{array}$$
   (c)
$$\begin{array}{l} p \\ q \\ p \wedge q \to r \vee q \\ p \wedge q \to r \wedge q \\ \hline r \end{array}$$

6. Given that the value of $p \to q$ is false, determine the value of $(\sim p \vee \sim q) \to q$.

7. Is the following argument valid? "if the labour market is perfect then the wages of all persons in a particular employment will be equal. But it is always the case that wages for such persons are not equal. Therefore the labour market is not perfect."

8. Show that the following argument is a fallacy:

$$\begin{array}{c} p \to q \\ \sim p \\ \hline \sim q \end{array}$$

Answers

1. $p \to q = (\sim p \downarrow q) \downarrow (\sim p \downarrow q)$

   $p \wedge q = (p \downarrow p) \downarrow (q \downarrow q)$

2. (a) $\sim(p \vee \sim q) \vee \sim(\sim p \vee \sim r)$

   (b) $(q \to p) \to \sim(p \to \sim q)$

(c) $[\{(q \downarrow q) \downarrow ((p \downarrow p) \downarrow (p \downarrow p))\} \downarrow \{(p \downarrow p) \downarrow (r \downarrow r)\}] \downarrow$
$[\{(q \downarrow q) \downarrow ((p \downarrow p) \downarrow (p \downarrow p))\} \downarrow \{(p \downarrow p) \downarrow (r \downarrow r)\}]$

3. (a) The corresponding sides of two triangles are equal if and only if the triangles are not congruent.

   (b) The number x is less than 10 and there exists no y such that x^2+y^2-100 is positive.

4. (a) There exists a man who is wealthy.

   (b) All men are wealthy and some are married.

   (c) No man is wealthy.

5. (a) Valid (b) Valid (c) Valid

6. False

7. Valid

CHAPTER TWO

Sets, Relations and Functions

The concept of set is used in all branches of mathematics. A set is a well-defined collection of objects. By a well-defined collection we mean that there exists a rule with the help of which it is possible to determine whether a given object is a member of the given collection or not. Each object belonging to a set is called an element or a member of the set. We generally use capital letters A, B, C, X, Y, Z etc. to denote sets and lower case letters a, b, c, x, y etc. to denote elements of a set.

If an element x belongs to a set A. then we write.

$$x \in A$$

which is read as "x belongs to A" or "x is a member of A" or simply "x is in A". If an object y is not an element of a set A, then we write

$$x \notin A$$

which is read as "x does not belong to A" or "x is not a member of A".

§ 2.1. Methods of Describing Sets

There are essentially two methods to specify a given set.

1. **Roster Method:** A set may be described by listing all its elements. For example, the set of vowels in the English alphabet is

$$A = \{a, e, i, o, u\}.$$

Here the elements are separated by commas and are enclosed in a pair of braces {}. This method of describing a set is called the roster method or the tabular form of the set.

Sometimes, it is not convenient to list all the elements of a set. For example, the set of natural numbers may be written as

$$N = \{1, 2, 3, \ldots\}.$$

2. **Property Method (or Set Builder Form):** The roster method of specifying a set is not always convenient and sometimes it is not possible to use this method to describe a set. A set can also be defined by some property which characterizes all the elements of the set. For example,

$$A = \{x : x \text{ is a vowel in the English alphabet}\}$$

which reads "A is the set of x such that x is a vowel in the English alphabet. This method of describing a set is called property method (or set builders form).

Illustrations:

1. $N = \{n : n \text{ is a natural number}\}$
2. $E = \{x : x \text{ is an even integer}\}$

§ 2.2. Equality of Sets

Two sets A and B are said to be equal if they contain the same elements. This statement is also known as the *axiom of extension*. We write $A = B$ if the sets A and B are equal and $A \neq B$ if the sets A and B are not equal. For example,

1. $\{a, \text{b}, c\} = \{b, a, c\}$
2. $\{2, 3, 5\} \neq \{2, 3, 7\}$

Remark: If $A = \{a, a, b, c\}$ and $B = \{a, b, \text{c}\}$ then it is clear that $A = B$.

Thus $\{a, a, b, c\}$ is a redundant representation of the set $\{a, b, c\}$. For this reason, some authors defined a set to be a well-defined collection of distinct objects.

Subsets

Let A and B be two sets. If every element of A is also an element of B then A is called a subset of B. We also say that A is contained in B or that B contains A. In symbols, we write

'$A \subseteq B$' or '$B \supseteq A$'.

We say A is not a subset of B if at least one element of A does not belong to B and we write it as $A \not\subseteq B$. It is clear that two sets A and B are equal if and only if $A \subseteq B$ and $B \subseteq A$.

Illustration:

Consider the set $A = \{1, 3, 4, 5\}$, $B = \{1, 2, 3, 5\}$ and $C = \{2, 3\}$. Then $C \subseteq B$ but $C \not\subseteq A$.

Proper Subset of a Set

From the definition of a subset it is clear that every set is a subset of itself. A set B is called proper subset of A if B is a subset of A and B is not equal to A. Briefly, B is a proper subset of A if $B \subseteq A$ and $B \neq A$ and write $B \subsetneq A$.

Empty Set

The set which contains no element is called the empty set (or null set or the void set) and is denoted by $\{\}$. The empty set is also denoted by the symbol ϕ. Since ϕ has no element, therefore, empty set is subset of every set. A set which is not empty is called non-empty set.

Disjoint Sets

Two sets A and B are said to be disjoint if they have no elements in common. For example, the sets $A = \{1, 2, 3, 4\}$ and $B = \{0, 5, 6\}$ are disjoint while the sets $A = \{1, 2, 3, 4\}$ and $C = \{1, 2, 6\}$ are no disjoint.

Singleton

A set which contains exactly one element is called a singleton. For example, {2} is a singleton.

Universal Set

In any mathematical discussion, we usually consider all the sets to be subsets of a fixed set called the universal set. Universal set is sometimes referred to as the universe or the universe of discourse.

For example, in studying human population the universal set consists of all human in the world and while discussing plane geometry we may consider the universal set to be the set consisting of all the points in the plane.

Power Set

If X is any set then the set of all subsets of X is called the power set of X, denoted by $P(X)$. Thus

$$P(X) = \{A : A \subseteq X\}$$

If X contains n elements then $P(X)$ contains 2^n elements.

Example 2.1. Let $X = \{1, 2, 3\}$ then

$$P(X) = \{\phi, \{1\}, \{2\}, \{3\}, \{1, 2\}, \{1, 3\}, \{2, 3\}, \{1, 2, 3\}\}.$$

Since X has three elements, $P(X)$ has $2^3 = 8$ elements.

Finite and Infinite Set

A set is said to be finite if it contains a *finite* number of distinct elements. A set is said to be infinite if it is not finite.

Example 2.2. Let $A = \{1, 3, 5, 7, 9\}$. Then A is finite because it contains 5 distinct elements.

Example 2.3. Let $B = \{1, 2, 3, 4, \ldots\}$. Then B is an infinite set.

Venn Diagrams

A venn diagram is a pictorial representation of sets in which sets are represented by enclosed regions in the plane. The universal set U is represented by a region enclosed in a rectangle and the other sets are represented by closed curves drawn within the rectangle.

§ 2.3. Operations on Sets

We introduce and study some basic operations in the section. Using these operations, we can construct new sets by combining the elements of given sets.

Union of Sets

Let A and B be two sets. The *union* of A and B is the set of all elements which are in the set A or in the set B. The union of two sets A and B is denoted by the symbol $A \cup B$ which is read as 'A union B'. Symbolically,

$$A \cup B = \{x : x \in A \text{ or } x \in B\}$$

In the adjoining Venn-diagram, the union of A and B is shown by the shaded area.

Fig. 2.1: $A \cup B$ is shaded

Intersection of Sets

Let A and B be two sets. The *intersection* of A and B is the set of all elements which are both in A and B. We denote the intersection of A and B by $A \cap B$, which is read as 'A intersection B'. Symbolically,

$$A \cap B = \{x : x \in A \text{ and } x \in B\}$$

In the adjoining venn diagram the intersection of two sets A and B is shown by the shaded region.

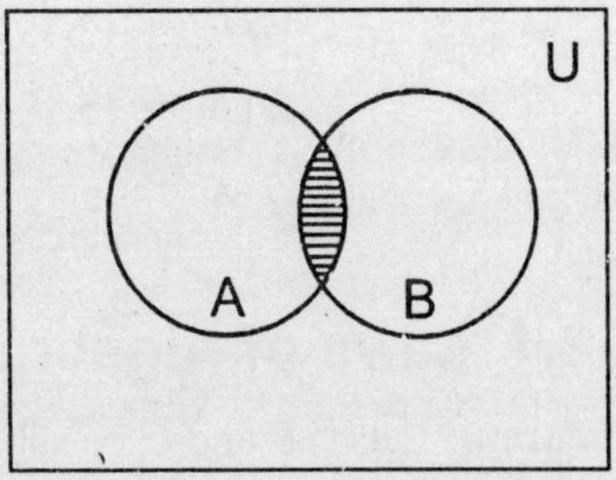

Fig. 2.2: $A \cap B$ is shaded

Complement of a Set

Let A be a subset of a universal set U. The set of all those elements of U which are not in A is called the complement of A and is denoted by U – A or simply A'. Symbolically,

$$A' = U - A = \{x : x \in U \text{ and } x \notin A\}$$

Difference of Sets

The difference $A-B$ of two sets A and B is the set of elements which belong to A but which do not belong to B. Thus,

$$A-B = \{x : x \in A \text{ and } x \notin B\}$$

The shaded region represents $A-B$ in the adjoining diagram.

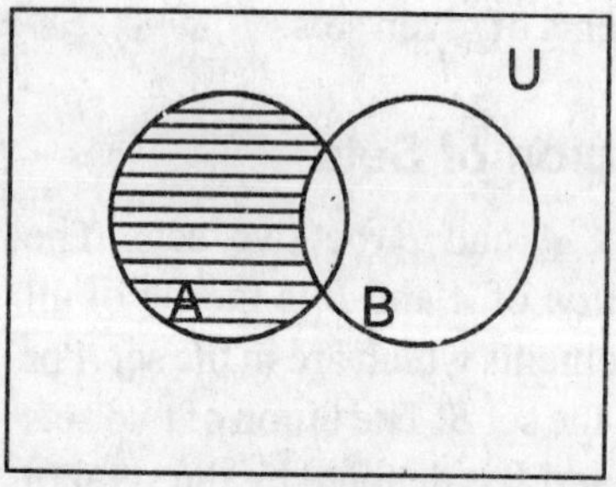

Fig. 2.3: A–B is shaded

Illustration:

Let $\quad U = \{a, b, c, d, e, f\}$

$\quad A = \{a, b, c, d\}$ and $B = \{b, d, f\}$.

Then $\quad A-B = \{a, c\}$

$\quad B-A = \{f\}$

$\quad A \cup B = \{a, b, c, d, f\}$

$\quad A \cap B = \{b, d\}$ and

$\quad U-A = \{e, f\}$

§ 2.4. Laws of Algebra of Sets

Theorem: Let A, B and C be subsets of an universal set U. Then the operations defined on sets satisfy the following properties:

1. *Commutative laws*:

(a) $A \cup B = B \cup A$ (b) $A \cap B = B \cap A$

2. *Associative laws*:

   (a) $A \cup (B \cup C) = (A \cup B) \cup C$

   (b) $A \cap (B \cap C) = (A \cap B) \cap C$

3. *Idempotent laws*:

   (a) $A \cup A = A$ (b) $A \cap A = A$

4. *Distributive laws*:

   (a) $A \cup (B \cap C) = (A \cup B) \cap (A \cup C)$

   (b) $A \cap (B \cup C) = (A \cap B) \cup (A \cap C)$

5. *Identity Laws*:

   (a) $A \cup \phi = A$ (b) $A \cap \phi = \phi$

6. *Involution laws*:

   $(A')' = A$

7. *De-Morgan's laws*:

   (a) $(A \cup B)' = A' \cap B'$ (b) $(A \cap B)' = A' \cup B'$

8. *Complement laws*:

   (a) $A \cup A' = U$ (b) $A \cap A' = \phi$

Proof: We will prove 4(a) and 7(a) only and leave proofs of remaining laws as exercises for the reader. We first prove 4(a), i.e.

$$A \cup (B \cap C) = (A \cup B) \cap (A \cup C)$$

Let $x \in A \cup (B \cap C)$

Then we have,

$$x \in A \cup (B \cap C) \Leftrightarrow x \in A \text{ or } x \in B \cap C$$

$$\Leftrightarrow x \in A \text{ or } (x \in B \text{ and } x \in C)$$

$$\Leftrightarrow (x \in A \text{ or } x \in B) \text{ and } (x \in A \text{ or } x \in C)$$

$$\Leftrightarrow x \in A \cup B \text{ and } x \in A \cup C$$

$$\Leftrightarrow x \in (A \cup B) \cap (A \cup C)$$

Hence

$$A \cup (B \cap C) = (A \cup B) \cap (A \cup C)$$

To prove 7(a), let

$$x \in (A \cup B)'$$

Then we have,

$$\begin{aligned} x \in (A \cup B)' &\Leftrightarrow x \notin A \cup B \\ &\Leftrightarrow x \notin A \text{ and } x \notin B \\ &\Leftrightarrow x \in A' \text{ and } x \in B' \\ &\Leftrightarrow x \in A' \cap B' \end{aligned}$$

Hence $\quad (A \cup B)' = A' \cap B'$

Exercise

1. Determine whether each of the following is true or false:

   (a) $\phi \subseteq \phi$

   (b) $\phi \in \phi$

   (c) $\phi \in \{\phi\}$

   (d) $\{\phi\} \in \{\phi\}$

   (e) $\{a,b\} \subseteq \{a,b,c, \{a,b,c\}\}$

   (f) $\{a,b\} \in \{a,b,c, \{a,b,c\}\}$

   (g) $\{a,b,c\} \in \{a,b,c,\{a,b,c\}\}$

   (h) $\{a,b,c\} \subseteq \{a,b,c,\{a,b,c\}\}$

2. Determine the following sets:

   (a) $\{\phi\} \cup \{a,\phi,\{\phi\}\}$

   (b) $\{\phi\} \cap \{a,\phi,\{\phi\}\}$

3. Give an example of sets A, B and C such that $A \in B$, $B \in C$ and $A \notin C$.

4. Write the power set of the set = { ϕ, { ϕ}}.

5. For $A = \{a, b, \{a, c\}, \phi\}$, determine the following sets:

   (a) $A-\{a\}$

   (b) $A-\{\phi\}$

   (c) $A-\{\{a, b\}\}$

   (d) $A-\{\{a, c\}\}$

   (e) $\{\phi\}-A$

6. Prove that $A-B = A \cap B'$.

7. Given that $A \cap B = A \cap C$ and $A' \cap B = A' \cap C$. Is it necessary that $B = C$? Justify your answer.

Answers

1. (a) True. (b) False. (c) True. (d) False.
   (e) True. (f) False. (g) True. (h) True.

2. (a) $\{a, \phi, \{\phi\}\}$ (b) $\{\phi\}$

3. $A = \{a, b\}, B = \{a, b, \{a, b\}\}, C = \{a, b, \{a, b, \{a, b\}\}\}$

4. $P(X) = \{ \phi, \{\phi\}, \{\{\phi\}\}, \{\phi, \{\phi\}\}\}$

5. (a) $\{b, \{a, c\}, \phi \}$ (b) $\{a, b, \{a, c\}\}$

   (c) A (d) $\{a, b, \phi\}$ (e) ϕ

7. Yes

§ 2.5. Cartesian Product of Sets

An ordered pair (a, b) consists of two objects a and b in a given fixed order. We call a and b as the first and second coordinates respectively of the ordered pair (a, b). a and b may have equal values.

Two ordered pairs (a, b) and (x, y) are said to be equal if $a = x$ and $b = y$ and we write $(a, b) = (x, y)$. Thus $(1, 2) \neq (2, 1)$. A distinction between the ordered pair $(1, 2)$ and the set $\{1, 2\}$ is that $(1, 2) \neq (2, 1)$ but $\{1, 2\} = \{2, 1\}$.

Definition: Let A and B be two sets. The set of all ordered pairs (a, b), where the first coordinate a of (a, b) is an element of A and the second coordinate b is an element of B is called the cartesian product or simply product of sets A and B and is written as $A \times B$.
Symbolically,

$$A \times B = \{(a, b) : a \in A \text{ and } b \in B\}$$

Example 2.4. Let $A = \{1, 2\}$ and $B = \{a, b, c\}$. Find $A \times B$, $B \times A$, and $A \times A$.

Solution:

$$A \times B = \{(1, a), (1, b), (1, c), (2, a), (2, b), (2, c)\}$$

$$B \times A = \{(a, 1), (a, 2), (b, 1), (b, 2), (c, 1), (c, 2)\}$$

and

$$A \times A = \{(1, 1,), (1, 2), (2, 1), (2, 2)\}$$

The cartesian product $A \times A$ is written as A^2 and we see from the above example that in general $A \times B \neq B \times A$

Generalizing the definition of the cartesian product of two sets, we define the cartesian product of n sets $A_1, A_2, \ldots, A_n$ as follows:

$$A_1 \times A_2 \times \ldots \times A_n = \{(a_1, a_2, \ldots, a_n) : a_1 \in A_1, a_2 \in A_2, \ldots, a_n \in A_n\}$$

Example 2.5. Find x, y if $(2x-3, 3y-1) = (5, 5)$

Solution: We know that

$$(2x-3, 3y-1) = (5, 5) \quad \text{if } 2x-3 = 5 \text{ and } 3y-1 = 5$$

$$\Rightarrow x = 4 \quad \text{and} \quad y = 2$$

Example 2.6. Let $A = \{1, 2, 4\}$, $B = \{2, 5, 7\}$ and $C = \{1, 3, 7\}$. Show that

$$A \times (B \cap C) = (A \times B) \cap (A \times C)$$

Solution: $B \cap C = \{7\}$

$$\therefore A \times (B \cap C) = \{(1,7), (2,7), (4,7)\}$$

Again,

$A\times B=\{(1,2), (1, 5), (1, 7), (2, 2), (2, 5), (2, 7), (4, 2), (4, 5), (4, 7)\}$

and

$A\times C=\{(1, 1), (1, 3), (1, 7), (2, 1), (2, 3), (2, 7), (4, 1),(4, 3), (4, 7)\}$

$\therefore\ (A\times B)\cap(A\times C) = \{(1, 7), (2, 7), (4, 7)\}$

Hence

$A\times(B\cap C) = (A\times B)\cap(A\cap C)$

§ 2.6. Functions

Definition: Let X and Y be two sets. A function f from X to Y is a rule which associates to each element x in X a unique element y of Y. The function f from X to Y is denoted by

$$f: X\rightarrow Y$$

The terms such as mapping, transformation or correspondence are also used for function.

If f is a function from X to Y, i.e., $f: X\rightarrow Y$, then the set X is called the domain of the function f and Y is called co-domain of f. The element $x\in X$ is called an argument of the function and the element $y\in Y$ which the function f associates to $x\in X$ is denoted by $f(x)$ and is called the image of x under f or the value of the function f at x. The set $\{f(x) : x\in X\}$ is called the range of f. If $A\subseteq X$ then $f(A) = \{f(x) : x\in A\}$.

Functions as Sets of Ordered Pairs

If X and Y be any two sets, then a function f from X to Y is a subset f of $X\times Y$ satisfying the following two conditions:

(1) For each $x\in X$, $(x, y)\in f$ for some $y\in Y$.

(2) If $(x, y)\in f$ and $(x, z)\in f$ then $y = z$

The first condition ensures that every x in X is associated with some y in Y and the second condition guarantees that the image of each x in X is unique.

Equal functions: Two functions f and g are said to be equal if their domains are same and $f(x) = g(x)$ for all x in the domain.

Illustrations

1. Let $X = \{a, b, c\}$ and $Y = \{1, 2, 3\}$. Then

$$f = \{(a, 1), (b, 2), (a, 3), (c, 3)\}$$

is not a function from X to Y because the second condition is not satisfied for $a \in X$ as both $(a, 1)$ and $(a, 3)$ are in f.

2. Let $X = \{1, 2, 3\}$. Then

$$f = \{(1, 2), (3, 1)\}$$

is not a function from X to X because $2 \in X$ is not the first coordinate of any pair in f and so f does not assign any image to 2.

3. Let $X = \{1, 2, 3, 4\}$, $Y = \{a, b, c, d\}$ and let $f = \{(1, a), (2, a), (3, d), (4, c)\}$.

Then f is a function from X to Y.

Here, we have

$f(1) = a$, $f(2) = a$,
$f(3) = d$, $f(4) = c$.

and range of $f = \{a, c, d\}$

We observe that the element $a \in Y$ appears as the second coordinate of two different ordered pairs in f. This does not violate the definition of a function.

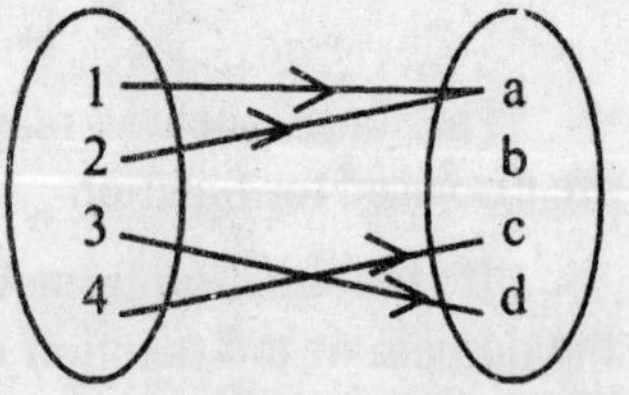

Fig. : 2.4

Types of Functions

Definition 1: A function $f: X \to Y$ is called *onto* (or surjective) if range of $f = Y$. That is, each element of y is the image of some element of X.

If $f: X \to Y$ is not onto then it is called *into*.

Definition 2: A function $f: x \to Y$ is called *one-to-one* (or injective or 1–1) if distinct elements of X have distinct images under f. In other words, f is one-to-one if

$$x_1 \neq x_2 \Rightarrow f(x_1) \neq f(x_2)$$

Or equivalently,

$$f(x_1) = f(x_2) \Rightarrow x_1 = x_2$$

A mapping $f: X \to Y$ is said to be *many-one* if at least two distinct elements in X have the same image in Y under f, i.e.,

$$x_1 \neq x_2 \text{ but } f(x_1) = f(x_2)$$

Definition 3. A mapping $f: X \to Y$ is said to be *bijective* if f is one-to-one and onto. A bijective mapping is also called one-to-one correspondence.

Example 2.7. Let $f: R \to R$ be defined by $f(x) = x^2$. Then f is neither one-to-one nor onto. f is not one-to-one because both 2 and -2 are mapped on 4. f is not onto because range of $f = \{x^2 : x \in R\} = R^+ \cup \{0\} \neq R$, where R^+ denotes the set of all positive real numbers.

Example 2.8. Let f: $\{1, 3, 5, 7, 9\} \to \{2, 4, 6, 8, 10\}$ be given by $f(x) = x + 1$. Then f is one-to-one and onto. Thus f is bijective.

***Identity function*:** A function $I : X \to X$ is called an identity function if

$$I(x) = x \ \forall \ x \in X.$$

Under identity function, each element is mapped on itself.

***Remainder function*:** Let m be any positive integer and k be any integer. We define a function f_m from the set of integers Z to $\{0, 1, 2, \ldots, m-1\}$ as follows:

$$f_m(k) = k(\text{mod } m)$$

where $k(\text{mod } m)$ denotes the remainder r, $0 \leqslant r < m$ when k is divided by m. This functions is known as remainder function (or the mod m- function)

Given any integer k and a fixed positive integer m, the value of $k(\text{mod } m)$ is obtained as follows:

If k is positive then divide k by m to obtain the remainder r. For example,

1. 25(mod 7) =4 because when we divide 25 by 7 then we get the remainder as 4.

2. 25(mod 5) =0 because when we divide 25 by 5, then the remainder is 0.
3. 3(mod 8) =3 because when we divide 3 by 8 the remainder is 3.

If k is negative then divide $|k|$ by m to obtain remainder r', then

$$k(\text{mod } m) = m-r'.$$

For example,

$$-26(\text{mod } 7) = 7-5=2 \text{ and } -371(\text{ mod } 8)=8-2=5.$$

Example 2.9. The remainder function is onto but not one-to-one.

Solution: Let m be any position integer. Define $f_m : Z \rightarrow \{0, 1, \ldots, m-1\}$ as follows:

$$f_m(k) = r = k(\text{mod } m).$$

Suppose $k \in Z$ is mapped on r. Then we see that $k+m$, $k+2m$, . . . are also mapped on r. Thus f_m is not one-to-one.

f_m is onto. Let $r \in \{0, 1, 2, \ldots, m\text{-}1\}$. Then $r \in Z$ and $f_m(r) = r$.

Thus f_m is onto.

Composite function: Let f be a function from X to Y and let g be a function from Y to Z. Then the composite of the functions f and g denoted by $g \circ f$ is a mapping from X to Z defined by

$$(g \circ f)(x) = g(f(x)) \quad \forall\, x \in X.$$

Example 2.10. Let $X = \{1, 2, 3\}$ and f and g be functions from X to X given by :

$$f = \{(1, 2), (2, 3), (3, 1\}) \text{ and}$$
$$g = \{(1, 1), (2, 2), (3, 1)\}$$

Find $f \circ g$ and $g \circ f$ and also show that $f \circ g \neq g \circ f$.

Solution: $f : X \rightarrow X$ and $g : X \rightarrow X$

$\therefore$ $g \circ f : X \rightarrow X$ defined by

$$(g \circ f)(x) = g(f(x))$$

$$\therefore \quad (g \circ f)(1) = g(f(1)) = g(2) = 2$$

$$(g \circ f)(2) = g(f(2)) = g(3) = 1$$
$$(g \circ f)(3) = g(f(3)) = g(1) = 1$$
$$\therefore \quad g \circ f = \{(1, 2), (2, 1), (3, 1)\}$$
Similarly, $f \circ g = \{(1, 2), (2, 3), (3, 2)\}$

We see that $g \circ f \neq f \circ g$

Invertible functions: Let $f: X \to Y$ be a one-to-one and onto mapping. Let y be any element of Y. Since the mapping f is onto, therefore there exists an element $x \in X$ such that

$$y = f(x)$$

Since the mapping f is also one-to-one, there will be only one element $x \in X$ such that $y = f(x)$. Let us denote x by $f^{-1}(y)$. Thus we see that if f: $X \to Y$ is one-to-one and onto then we can define a new mapping called the inverse of f and denoted by f^{-1} from Y to X which associates to each element y of Y a unique element in X. It can be seen that only one-to-one and onto (i.e. bijective) mappings possess inverse mappings. In other words, if $f: X \to Y$ is not bijective then f^{-1} does not exist.

Definition: Let $f: X \to Y$ be a bijective mapping. Then the mapping $f^{-1}: Y \to X$ defined by $f^{-1}(y) = x$, where $f(x) = y$, is called the inverse mapping of f.

It can be seen easily that if f is bijective then f^{-1} is also bijective. Moreover $f^{-1} o f$ and $f o f^{-1}$ both are identity mappings on X and Y respectively.

Example 2.11. Let Q be the set of rational numbers. Let $f: Q \to Q$ be defined by $f(x) = 2x+3$. Show that f is bijective. Also find a formula that defines the inverse function f^{-1}.

Solution: Let x_1 and x_2 be two distinct elements in Q.

Then $\quad x_1 \neq x_2 \Rightarrow 2x_1+3 \neq 2x_2+3$

$$\Rightarrow f(x_1) \neq f(x_2).$$

Hence f is one-to-one.

Let y be any element in Q. If $y = f(x)$ then $y = 2x+3$ and therefore $x = (y-3)/2$ is also a rational number. Thus

$$f\left(\frac{y-3}{2}\right)=2\left(\frac{y-3}{2}\right)+3=y$$

Hence f is onto.

Solving $y = f(x)$, we get $x = \frac{y-3}{2}$. Thus

$$f^{-1}(y)=x=\frac{y-3}{2}$$

defines the inverse function $f^{-1}: Q \to Q$.

Exercise

1. Find x and y such that $(2x, x+y) = (6, 2)$.
2. Let $A = \{a, b, c\}$ and $B = \{+, *\}$. Find $A \times B$.
3. For $A = \{1, 2, 4\}$, $B = \{2, 5, 7\}$ and $C = \{1, 3, 7\}$, show that $A\times(B\cup C)=(A\times B)\cup(A\times C)$.
4. Let $A = \{a, b, c, d\}$ and $B = \{1, 2, 3\}$. Determine whether the following are function from A to B. If f is a function, give its range.
   (a) $f = \{(a, 1), (b, 2), (c, 1), (d, 2)\}$
   (b) $f = \{(a, 1), (b, 1), (c, 1), (d, 1)\}$
   (c) $f = \{(a, 1), (b, 2), (a, 2), (c, 1), (d, 2)\}$
   (d) $f = \{(a, 3), (b, 2), (c, 1)\}$
5. Prove that if $f: X \to Y$ and $g: Y \to Z$ are bijective then $g \circ f$ is also bijective.
6. Let $f: R \to R$ and $g: R \to R$ be defined by $f(x) = x-1$ and $g(x) = x^2+1$. Find $(f\circ g)(2)$, $(g\circ f)(2)$, $(f\circ f)(2)$ and $(g\circ g)(2)$.

Answers

1. $x = 3, y = -1$
2. $A\times B = \{(a,+), (a,*), (b,+), (b,*), (c,+), (c, *)\}$
4. (a) f is a function and range of $f = \{1,2\}$

   (b) f is a function and range of $f = \{1\}$

   (c) f is not a function. (d) f is not a function.
5. $(f\circ g)(2) = 4$, $(g\circ f)(2) = 2$, $(f\circ f)(2) = 0$ and $(g\circ g)(2) = 26$.

§ 2.7. Binary Operations or Binary Compositions

We are familiar with operations like addition, subtraction and multiplication. Whenever we add two natural numbers we get a unique natural number. In other words, addition is a mapping from $N\times N \to N$, where N is the set of natural numbers.

Definition: Let X be a set. Then a mapping $f: X\times X \to X$ is called a binary operation (or binary composition) on X.

Thus binary operations on X is a mapping from $X\times X$ to X. In other words, binary operation on X is a mapping which associates to each pair a, b of elements of X, a unique element c of X. We use symbols like $*, +, \circ, \oplus, \wedge, \vee$ etc. for binary operations.

Illustrations

1. Addition is a binary operation on the set of integers because addition of two integers is again a unique integer.
2. Multiplication is a binary operation on the set N of natural numbers.
3. Subtraction is a binary operation on the set Z of integers because subtraction of two integers is again a unique integer but subtraction is not a binary operation on the set N of natural numbers because subtraction of two natural numbers is not necessarily a natural numbers. For example $3-4 \notin N$.

A set X is said to be closed under operation $*$ if $*$ is a binary operation on the set X. That is, $a*b$ is a unique element of X for all a, b in X. This property is often called the *closure property*.

Properties of Binary Operations

Now we shall discuss some general properties of binary operations.

Commutative operation: A binary operation $*$ on a set X is said to be commutative if

$$x*y = y*x \quad \forall\, x, y \in X.$$

Example 2.12. Addition and multiplication are commutative binary operations on the set R of real numbers. But subtraction is not a commutative binary operation on R because $a-b \neq b-a$ in general.

Associative operation: A binary operation $*$ on a set X is said to be associative if

$$x*(y*z) = (x*y)*z \quad \forall\, x, y, z \in X.$$

Example 2.13. Since $x+(y+z) = (x+y)+z$ $\forall x, y, z \in \mathbf{R}$. Therefore addition is associative binary operation on **R**. Similarly multiplication is also associative binary operation on **R**.

Distributive operations: Let $*$ and o be two binary operations on a set X. We say $*$ is left distributive over o if

$$a*(b \circ c) = (a*b)\circ(a*c) \quad \forall\, a, b, c \in X \ldots (1)$$

We say $*$ is right distributive over o if

$$(b \circ c)*a = (b*a)\circ(c*a) \quad \forall\, a, b, c \in X \ldots (2)$$

When both (1) and (2) hold, we say that $*$ is distributive over o.

Example 2.14. On the set R of real numbers, multiplication is distributive over +, since

$$a\cdot(b+c) = a\cdot b + a\cdot c$$

and

$$(b+c)\cdot a = b\cdot a + c\cdot a \quad \forall\, a, b, c \in R$$

But + is not distributive over $(\cdot)$ because

$$a+b\cdot c \neq (a+b)\cdot(a+c)$$

for every $a, b, c \in R$

Identity Element for a Binary Operation

Given a binary operation * on a set X, we now define identity element of X for the operation *. Identity element may or may not exist.

Definition: Let * be a binary operation on X. If there exists an element $e \in X$ such that

$$a * e = e * a = a \quad \forall\, a \in X$$

then e is called identity element for *.

Example 2.15. Addition is a binary operation on **R**, the set of real numbers. Since

$$a+0 = 0+a = a \quad \forall\, a \in R$$

therefore, the element 0 is the identity for +.

Example 2.16. We know that addition is a binary operation on N, the set of natural numbers. But N has no identity element for + because there exists no natural number e such that

$$a+e = e+a = a \quad \forall\, a \in N$$

Invertible elements: Let * be a binary operation on the set X with the identity element e. An element $a \in X$ is said to be invertible if there exists an element $b \in X$ such that:

$$a * b = b * a = e$$

Moreover, b is then said to be inverse of a and is denoted by a^{-1}. The identity element is always invertible because $e * e = e * e = e$.

Example 2.17. We know the operation of addition is a binary operation on R and 0 is the identity for +. Since for each $a \in R$, $-a$ also belongs to R such that:

$$a+(-a) = (-a)+a = 0$$

Therefore, every real number has inverse for the operation of addition.

Example 2.18. Consider the operation of multiplication on *R*. 1 is the identity for multiplication and for every non-zero real number $a \in R$, 1/a is also a real number such that:

$$a.\frac{1}{a} = \frac{1}{a}.a = 1$$

Hence every non-zero real number is invertible. But 0 is not invertible with respect to multiplication. Similarly the operation of multiplication on *N* has identity element , namely 1 but no element except 1 is invertible.

§ 2.8. Composition Table

A binary composition on a finite set can be defined by means of a table called *composition table* which may be described as follows:

Let $A = \{a_1, a_2, a_3, ..., a_n\}$ be a finite set. We write the elements of the set *A* in a horizontal row as well as in a vertical column. The element $a_i * a_j$ is entered at the intersection of the ith row and j^{th} column.

Example 2.19. Let $A = \{1, 5, 7, 11\}$. We define the operation of multiplication module 12 denoted $\times_{12}$ on *A* as follows:

$$a \times_{12} b = r, \qquad 0 \leqslant r < 12$$

where *r* is the least non-negative remainder when the product $a \cdot b$ is divided by 12.

The composition table for $\times_{12}$ on *A* is given below:

$\times_{12}$	1	5	7	11
1	1	5	7	11
5	5	1	11	7
7	7	11	1	5
11	11	7	5	1

In the body of the table, the second entry of the third row is 11 which is obtained as follows:

$$7 \times_{12} 5 = 11 \text{ since } 7 \times 5 = 35 = 2(12)+11$$

Similarly other entries of the table are obtained.

Solved Examples

Example 2.20. Let $*$: I×I→I, where I is the set of integers, be defined as

$$x*y = x+y-xy.$$

Show that the binary operation $*$ is commutative and associative. Find the identity element and indicate the inverse of each element.

Solution: Let x and y be two integers. Then $x+y$ and xy are also integers. Now since subtraction of two integers is also are integer, we have $x+y-xy$ is an integer. Thus $x*y \in \mathrm{I}$. Hence $*$ is binary operation on I.

$*$ is commutative:

Since
$$\begin{aligned} x*y &= x+y-xy \\ &= y+x-yx \\ &= y*x, \end{aligned}$$

the operation $*$ is commutative

$*$ is associative: For any x, y and z in I, we have,

$$\begin{aligned} x*(y*z) &= x*(y+z-yz) \\ &= x+(y+z-yz)-x(y+z-yz) \\ &= x+y+z-yz-xy-xz+xyz \end{aligned}$$

Similarly $(x*y)*z = x+y+z-yz-xy-xz+xyz$

Hence $x*(y*z) = (x*y)*z$ and $*$ is therefore associative.

Identity element: If e is the identity element for $*$ then

$$a*e = e*a = a \quad \forall\, a \in X$$

$$\Rightarrow \quad a+e-ae = a$$

$$\Rightarrow \quad e = 0$$

Inverse of an element: Let a be any element of X. Let b be the inverse of a. Then

$$a*b = 0$$

$$\Rightarrow \quad a+b-ab = 0$$

$$\Rightarrow \quad b = a/(a-1) \text{ if } a \neq 1$$

Thus if $a \neq 1$, then $a^{-1} = a/(a-1)$.

Example 2.21. Let I_+ be the set of positive integers. Let

$*: I_+ \times I_+ \rightarrow I_+$ be defined as follows:

$$x*y = \text{lcm of } x \text{ and } y.$$

Show that $*$ is commutative and associative. Find the identity element.

Solution: We know that the least common multiple of two positive integers is again a positive integer. Hence $*$ is binary operation on I_+.

$*$ is commutative: We know that lcm of x and y = lcm of y and x. Hence

$$x*y = y*x$$

Thus $*$ is commutative:

$*$ is associative: For any three positive integers x, y, and z,

$$\begin{aligned} x*(y*z) &= x*(\text{lcm of } y \text{ and } z) \\ &= \text{lcm of } x, y, \text{ and } z \\ &= (\text{lcm of } x \text{ and } y)*z \\ &= (x*y)*z \end{aligned}$$

Thus $*$ is associative.

Identity Element: We know that for any positive integer x, the least common multiple of x and $1 = x$. Hence $1 \in I_+$ such that $x*1 = 1*x \; \forall \; x \in I_+$. Thus 1 is the identity element.

Example 2.22. Let $X = \{0, 1, 2, 3, 4\}$ and let $\times_5$ be the operation "multiplication modulo 5". Give composition table for the operation $\times_5$. Show that $\times_5$ is commutative. Also indicate the identity element.

Solution: For $a, b \in X$,

$$a \times_5 b = ab(\text{mod}5) = r,$$

where r is the remainder when ab is divided by 5.

The composition table is given below:

$\times_5$	0	1	2	3	4
0	0	0	0	0	0
1	0	1	2	3	4
2	0	2	4	1	3
3	0	3	1	4	2
4	0	4	3	2	1

From the table, we see that $a \times_5 b$ for any $a, b \in X$ is again in the set X. Hence the set is closed under the operation $\times_5$. Since the entries in the table are symmetrical about main diagonal, the operation is commutative. Form the table it is evident that 1 is the identity element.

Exercise

1. Show that $x * y = x - y$ is not a binary operation over the set of natural numbers but it is binary operation on the set of integers. Is it commutative or associative?

2. How many distinct binary operations can be defined on the set $\{0, 1\}$.

[Hint: Every binary operation $*$ on $\{0, 1\}$ can be described by the table

*	0	1
0		
1		

].

3. Consider the binary operation $*$ defined on the set $A = \{a, b, c, d\}$ by the following table:

$*$	a	b	c	d
a	a	c	b	d
b	d	a	b	c
c	c	d	a	a
d	d	b	a	c

Compute $c*d$, $d*c$, $b*d$ and $d*b$. Is $*$ commutative?

4. Show that $x*y = xy$ is a binary operations on the set N of natural numbers. Determine whether $*$ is commutative or associative.

5. Fill in the following table so that the binary operation $*$ is commutative.

$*$	a	b	c
a	b		
b	c	b	a
c	a		c

6. Let $I_5 = \{0, 1, 2, 3, 4\}$. Give composition table for the operation $+_5$ defined by $x +_5 y = (x+y)(\text{mod } 5)$. Indicate the identity element.

7. Let $*$ denote a binary operation on N given by $x*y = x$. Show that $*$ is not commutative, but is associative.

Answers

1. It is neither commutative nor associative.
2. $2^4 = 16$
3. $c*d = a$, $d*c = a$, $b*d = c$, $d*b = b$, $*$ is not commutative.
4. $*$ is neither commutative nor associative.

§ 2.9. Relations

The word "relation" suggests some familiar examples of relations between two people such as the relation of father to son, mother to son, brother to sister, etc. If A is the set of real numbers then there are many commonly used relations between two real numbers such as "less than", "greater than" or that of "equality". These examples suggest relationship between two objects. A relation, which describes relationship between two objects is called a *binary relation*. If a relation describes relationship among three objects then it is called *ternary relation*. For example, the relation described as "three integers a, b, and c in the set $\{1, 2, 3, \ldots, 15\}$ are related if their sum $a+b+c$ is divisible by 5", is a ternary relation on the set $\{1, 2, 3, \ldots, 15\}$. In this relation, the integers 2, 3 and 5 are related because their sum is divisible by 5 but the integers 1, 2, and 4 are not related. The relation between parents and child on the set of human beings is also ternary relation. In general, an n-ary relation among the sets $A_1, A_2, A_3, \ldots A_n$ is a set of n-tuples in which first coordinate is an element of A_1, the second coordinate is an element of $A_2, \ldots$, and the nth element is an element of A_n.

Here, we shall only consider binary relations. Such a relation between two objects can be defined by listing the two objects as an ordered pair. A set of all such ordered pairs, in each of which first elements is related to the second elements describes a binary relation. We shall call a binary relation simply as relation.

Definition: Let A and B be two sets. A **binary relation** (or simply a relation) from A to B is a subset of $A \times B$. Symbolically, R is a relation from A to B if $R \subseteq A \times B$.

If R is a relation from A to B and if the ordered pair $(a, b) \in R$ then we say that the element a is related to the element b by R and we also write aRb which is read as "a is R-related to b". If x is not related to y, we write $(x, y) \notin R$.

Relation on a Set: Let R be a relation from A to B. If $A=B$ then we say that R is a relation on the set A instead of saying that R is a relation from A to A . Thus a relation on a set A is a subset of $A \times A$.

Example 2.23. Let $A = \{1, 2, 3\}$ and $B = \{p, q\}$

Then $R = \{(1, p), (2, q), (3, q)\}$ is a relation from A to B.

Example 2.24. Let I be the set of integers. Define the following relation (less than) on I.

$$xRy \text{ if and only if } x<y$$

Then $R = \{(x, y) : x, y \in \text{I and } x<y\}$.

Example 2.25. Let $A = \{1, 2, \ldots, 12\}$. Define the following relation R on A :

aRb if and only if a divides b. Then

$$R = \{(1, 1), (1, 2), (1, 3), (1, 4), (1, 5), (1, 6), (1, 7), (1, 8), (1, 9), (1, 10), (1, 11), (1, 12), (2, 2), (2, 4), (2, 6), (2, 8), (2, 10), (2, 12), (3, 3), (3, 6), (3, 9), (3, 12), (4, 4), (4, 8), (4, 12), (5, 5), (5, 10), (6, 6), (6, 12)\}.$$

Representation of a Relation by a Matrix and a Graph

A relation from a finite set A to a finite set B can also be represented by a matrix called *relation matrix* of R.

Let $A = \{a_1, a_2, \ldots, a_m\}$ and $B = \{b_1, b_2, \ldots, b_n\}$ and R be a relation from A to B. Define an $m \times n$ matrix whose m rows correspond to the m elements in A and the n columns correspond to the n elements in B as follows:

The matrix element

$$r_{ij} = \begin{cases} 1 & \text{if } a_i R b_j \\ 0 & \text{if } a_i R b_j \end{cases}$$

where r_{ij} is the element in the ith row and jth column.

Example 2.26. Let $A = \{a, b, c, d\}$ and $B = \{\alpha, \beta, \gamma\}$. Then $R = \{(a, \alpha), (b, \gamma), (c, \alpha), (c, \gamma), (d, \beta)\}$ can be represented in relation matrix as:

$$\begin{array}{c} \\ a \\ b \\ c \\ d \end{array}\begin{array}{c} \alpha \quad \beta \quad \gamma \\ \begin{bmatrix} 1 & 0 & 0 \\ 0 & 0 & 1 \\ 1 & 0 & 1 \\ 0 & 1 & 0 \end{bmatrix} \end{array}$$

A relation can also be represented pictorially by drawing its graph. Here we shall use the term graphs only as a tool to represent relations. Let R be a relation in a set $A = \{a_1, a_2, ..., a_m\}$. The elements of A are represented by small circles called vertices or nodes. The vertices corresponding to the element a_i is labeled as a_i. If $a_i R a_j$, that is, if $(a_i, a_j) \in R$, then we join vertices a_i and a_j by means of an arc called edge and put an arrow on the edge in the direction from a_i to a_j. When all the vertices corresponding to the ordered pairs in R are connected by edges with proper arrows, we get a graph of the relation R.

Example 2.27. Let $A = \{1, 2, 3, 4, 5\}$ and let $R = \{(1, 2), (1, 3), (1, 4), (1, 5), (2, 3), (2, 4), (2, 5), (3, 4), (3, 5), (4, 5)\}$. Then the relation R can be represented in graph as shown below.

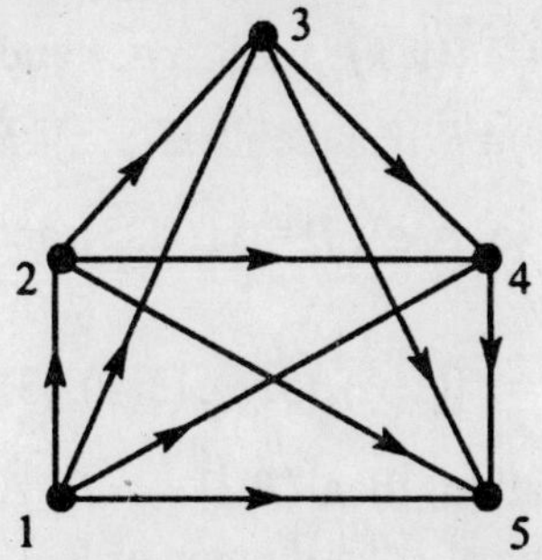

Figure 2.5 : Graph of relation R

Domain and Range of a Relation: Let R be a relation from A to B. That is, $R \subseteq A \times B$. The *domain* D of the relation R is the set of all first elements of the ordered pairs which belong to R. Symbolically,

$$D = \{x : x \in A \text{ and } (x, y) \in R \text{ for some } y \in B\}$$

The *range* E of the relation R is the set of all second elements of the ordered pairs which belong to R. Symbolically,

$$E = \{y : y \in B \text{ and } (x, y) \in R \text{ for some } x \in A\}.$$

Obviously, for any relation $R \subseteq A \times B$, domain $\subseteq A$ and range $\subseteq B$.

Example 2.28. Let $A = \{1, 2, 3, 4\}$ and $B = \{\alpha, \beta, \gamma\}$.

Let $R = \{(2, a), (4, a), (4, b)\}$ be a relation. Then domain of R is the set $\{2, 4\}$ and range of R is the set $\{\alpha, \beta\}$.

Total number of distinct relations from a set A to a set B

Suppose that the set A has m elements and the set B has n elements. Then the product $A \times B$ has mn elements. Therefore the power set of $A \times B$, that is, $P(A \times B)$ will have 2^{mn} elements. Thus $A \times B$ have 2^{mn} different subsets. Now since every subset of $A \times B$ is a relation from A to B, therefore we shall have 2^{mn} different relations from A to B.

Inverse relation: Let R be a relation from A to B. The inverse relation of R, denoted by R^{-1}, is a relation from B to A defined by:

$$R^{-1} = \{(y, x) : y \in B, x \in A \text{ and } (x, y) \in R\}$$

In other words, $(x, y) \in R$ if and only if $(y, x) \in R^{-1}$.

Example 2.29. Let $A = \{a, b, c\}$, $B = \{1, 2, 3\}$ and $R = \{(a, 1), (a, 3), (b, 3), (c, 3)\}$. Then

$$R^{-1} = \{(1, a), (3, a), (3, b), (3, c)\}.$$

Clearly if R is any relation, then $(R^{-1})^{-1} = R$. Also the domain of R^{-1} = range of R and range of R^{-1} = domain of R.

§ 2.10. Difference between relations and functions

Suppose A and B are two sets. Let f be a function from A to B. Then f is a subset of $A \times B$ satisfying the following two conditions:

1. For each $a \in A$, the ordered pair $(a, b) \in f$ for some $b \in B$.
2. If $(a, b) \in f$ and $(a, c) \in f$ then $b = c$.

On the other hand, every subset of $A \times B$ is a relation from A to B. Thus every function is a relation but every relation is not necessarily a function. In a relation from A to B an element of A may be related to more than one element in B. Also there may be some elements of A which may not be related to any element of B. But in a function from A to B each element of A must be associated to one and only one element of B.

Example 2.30. Let $A = \{1, 2, 3, 4\}$ and $B = \{a, b, c\}$.

Then $$R = \{(1,a), (2, a), (3, a), (4, a)\}$$

is both relation and function from A to B. But

$$S = \{(1, a), (2, b), (1, c), (3, a)\}$$

is a relation from A to B but not a function from A to B because the element 1 of A is associated to two different elements a and c of B.

Composition of Relations

Let A, B and C be sets and let R be a relation from A to B and let S be a relation from B to C, that is, $R \subseteq A \times B$ and $S \subseteq B \times C$. Then the composition of R and S, denoted by $R \circ S$, is the relation from A to C defined by setting $(a, c) \in R \circ S$ if and only if there exists an element $b \in B$ such that $(a, b) \in R$ and $(b, c) \in S$.

Suppose that R is a relation on a set A, that is $R \subseteq A \times A$. Then $R \circ R$, the composition of R with itself is always defined and $R \circ R$ is sometimes denoted by R^2. Similarly, $R^3 = R^2 \circ R = R \circ R \circ R$ and so on. Hence R^n is defined for all positive integers n.

Example 2.31. Let $A = \{1, 2, 3, 4\}$, $B = \{a, b, c, d\}$, $C = \{x, y, z\}$ and let $R = \{(1, a), (2, d), (3, a), (3, b), (3, d)\}$ and $S = \{(b, x), (b, z), (c, y), (d, z)\}$. Then $R \circ S = \{(2, z), (3, x), (3, z)\}$.

To find composition of Relations using Matrices

There is another way of finding RoS using matrix representations of R and S. Let M_R and M_S denote matrices of the relations R and S respectively. Consider the matrix M obtained by multiplying M_R and M_S. That is $M = M_R M_S$. Then the elements corresponding to the non-zero entries in M are related by $R \circ S$. In other words, $M = M_R M_S$ and $M_{R \circ S}$ have the same non-zero entries. That is, $(i, j)^{th}$ element in M is non-zero if and only if $(i, j)^{th}$ element in M_{RoS} is non-zero. Note that the non-zero entries in M and M_{RoS} may not have the same values.

Example 2.32. Let us consider the above example. Then

$$M_R = \begin{array}{c} \\ 1 \\ 2 \\ 3 \\ 4 \end{array}\!\!\begin{array}{c} \begin{array}{cccc} a & b & c & d \end{array} \\ \begin{bmatrix} 1 & 0 & 0 & 0 \\ 0 & 0 & 0 & 1 \\ 1 & 1 & 0 & 1 \\ 0 & 0 & 0 & 0 \end{bmatrix} \end{array} \text{ and } M_S = \begin{array}{c} \\ a \\ b \\ c \\ d \end{array}\!\!\begin{array}{c} \begin{array}{ccc} x & y & z \end{array} \\ \begin{bmatrix} 0 & 0 & 0 \\ 1 & 0 & 1 \\ 0 & 1 & 0 \\ 0 & 0 & 1 \end{bmatrix} \end{array}$$

Multiplying M_R and M_S, we obtain the matrix

$$M = M_R\, M_S = \begin{array}{c} \\ 1 \\ 2 \\ 3 \\ 4 \end{array}\!\!\begin{array}{c} \begin{array}{ccc} x & y & z \end{array} \\ \begin{bmatrix} 0 & 0 & 0 \\ 0 & 0 & 1 \\ 1 & 0 & 2 \\ 0 & 0 & 0 \end{bmatrix} \end{array}$$

Corresponding to non-zero entries in M, we obtain the relation

$$RoS = \{(2, z), (3, x), (3, z)\}.$$

§ 2.11. Properties of Relations on a Set

Consider the given set A. In this section we discuss a number of important types of relations, which are defined on A.

1. Reflexive relations: A relation R on a set is said to be reflexive relation if

$$(a, a) \in R \quad \forall\, a \in A.$$

Thus relation R is reflexive if we have aRa, $\forall\, a \in A$.

A relation is not reflexive if there exists an $a \in A$ such that $(a, a) \notin R$.

Example 2.33. Let A be the set of positive integers. Define a relation R on A as follows:

aRb if and only if a divides b.

Since every integer always divides itself, R is a reflexive relation.

Example 2.34. Let $A = \{1, 2, 3, 4\}$. Then $R = \{(1, 1), (2, 3), (2, 4), (3, 3), (4,1), (4,4)\}$ on A is not reflexive since $2 \in A$ but $(2,2) \notin R$.

Clearly a relation on a set A represented by a matrix will be reflexive if and only if all the entries on the main diagonal of the matrix are 1.

2. Symmetric relations: A relation R on a set A is said to be symmetric relation if $(a,b) \in R \Rightarrow (b,a) \in R$.

Thus R is symmetric if whenever aRb then bRa.

A relation R is not symmetric if there exists $a, b \in A$ such that $(a,b) \in R$ but $(b, a) \notin R$.

Example 2.35. Let

$$\begin{aligned} A &= \{a, b, c\} \text{ and let} \\ R_1 &= \{\ \} \\ R_2 &= \{(a, a), (b, b)\} \\ R_3 &= \{(a, b), (b, a)\} \text{ and} \\ R_4 &= A \times A \end{aligned}$$

Then all four R_1, R_2, R_3 and R_4 are symmetric relations. The relations R_1 is called empty relation or the void relation on A, while the relation R_4 which is equal to $A \times A$ is called universal relation on A.

Example 2.36. Let A be the set of positive integers. Define a relation R on A as follows: $(a,b) \in R$ if and only if $a \geqslant b$.

Then the relation R is not symmetric because $(10, 9) \in R$ but $(9, 10)$ is not in R.

Clearly, a relation on a set A represented by a matrix will be symmetric if the entries in the matrix are symmetrical with respect to the main diagonal.

Anti-symmetric relations: A relation R on a set A is said to be anti-symmetric relation if $(a, b) \in R$ and $(b, a) \in R$ then $a = b$.

Thus R is anti-symmetric if (a, b) in R implies that (b, a) is not in R unless $a = b$.

A relation R on a set A is not anti-symmetric if there exists elements a, b in A such that aRb, bRa but $a \neq b$.

Example 2.37. Let N be the set of natural numbers. Let R be the relation on N defined by "x is a divisor of y". Then R is anti-symmetric since a divides b and b divides a implies $a = b$.

Example 2.38. Let A be the set of positive integers and R be a relation on A such that (a,b) is in R if and only if $a \geqslant b$. Then R is an anti-symmetric relation.

Example 2.39. Let $A = \{a, b, c\}$ and let relation R on A is given by $R = \{(a, b), (a, c), (c, a)\}$. Then R is not anti-symmetric.

Transitive relations:

A relation R on a set A is said to be transitive relation if

$$(a, b) \in R \text{ and } (b, c) \in R \Rightarrow (a, c) \in R.$$

Thus R is transitive if aRb, bRc then aRc.

A relation R on a set A is not transitive if there exists elements a, b and $c \in A$, not necessarily distinct, such that:

$$(a, b) \in R, (b, c) \in R \text{ but } (a, c) \notin R.$$

If R is transitive relation on A then it is clear that $RoR \subseteq R$.

Example 2.40. The relation $\leqslant$ is a transitive relation on the set of real numbers.

Example 2.41. Let A be set of positive integers. Define a relation R on A as follows: $a R b$ if and only if a divides b.

Then R is a transitive relation because if a divides b and b divides c then a divides c.

Example 2.42. Let A = the set of all lines in the plane. Define the following relation R on A:

$l_1 R l_2$ if and only if l_1 is perpendicular to l_2, where l_1 and l_2 are lines in the plane. Then R is not a transitive relation.

Transitive Extension and Transitive Closure of a Relation

Let R be a relation on A. Then the *transitive extension* of R is a relation R_1 on A defined by $\quad R_1 = R \cup (R \circ R)$, where $R \circ R$ is the composition of relation R with itself. Thus R_1 contains R and moreover if (a, b) and (b, c) are in R then (a, c) is also in R_1.

It is obvious that a relation R on A is transitive if and only if the transitive extension R_1 of R is equal to R. That is

$$R_1 = R.$$

Definition: Let R be a relation on A. The transitive closure R^* of R is the relation on A such that

1. R^* is transitive.
2. $R \subseteq R^*$
3. R^* has the fewest possible elements.

Thus the transitive closure of a relation R is the smallest transitive relation on A containing R.

Given a relation R on a set A, the transitive closure R^* can be determined as follows:

Step1. Calculate $R_1 = R \cup (R o R)$ and $R_{i+1} = R_i \cup (R_i o R_i)$, where the relation R_{i+1} is the transition extension of the relation R_i and so on.

Step2. If $R_1 = R$ then R is transitive and relation R itself is the transitive closure of R. If $R_1 \neq R$, calculate R_2.

Step3. If $R_{i+1} \neq R_i$, calculate next R_i.

Step4. If $R_{i+1} = R_i$, stop. The transitive closure R^* of R is the relation R_i. That is, $R^* = R_i$.

Example 2.43. Let $A = \{1, 2, 3, 4\}$ and let R be a relation on A defined by $R = \{(1, 1), (1, 2), (2, 4), (3, 2), (4, 3)\}$. Find the transitive closure of R.

Solution: The matrix of the relation R is given by

$$M_R = \begin{array}{c} \\ 1 \\ 2 \\ 3 \\ 4 \end{array} \begin{array}{c} \begin{array}{cccc} 1 & 2 & 3 & 4 \end{array} \\ \begin{bmatrix} 1 & 1 & 0 & 0 \\ 0 & 0 & 0 & 1 \\ 0 & 1 & 0 & 0 \\ 0 & 0 & 1 & 0 \end{bmatrix} \end{array}$$

Multiplying M_R with itself, we obtain,

$$M = M_R M_R = \begin{bmatrix} 1 & 1 & 0 & 0 \\ 0 & 0 & 0 & 1 \\ 0 & 1 & 0 & 0 \\ 0 & 0 & 1 & 0 \end{bmatrix} \begin{bmatrix} 1 & 1 & 0 & 0 \\ 0 & 0 & 0 & 1 \\ 0 & 1 & 0 & 0 \\ 0 & 0 & 1 & 0 \end{bmatrix} = \begin{array}{c} \\ 1 \\ 2 \\ 3 \\ 4 \end{array} \begin{array}{c} \begin{array}{cccc} 1 & 2 & 3 & 4 \end{array} \\ \begin{bmatrix} 1 & 1 & 0 & 1 \\ 0 & 0 & 1 & 0 \\ 0 & 0 & 0 & 1 \\ 0 & 1 & 0 & 0 \end{bmatrix} \end{array}$$

Corresponding to non-zero entries in M, we obtain

$$RoR = \{(1, 1), (1, 2), (1, 4), (2, 3), (3, 4), (4, 2)\}$$

$\therefore R_1 = R \cup (RoR) = \{(1, 1), (1, 2), (1, 4), (2, 3), (2, 4), (3, 2), (3, 4), (4, 2), (4, 3)\}$.

Since $R_1 \neq R$, we will calculate $R_2 = R_1 \cup (R_1 o R_1)$.

The matrix of relation R_1 is given by

$$M_{R_1} = \begin{bmatrix} 1 & 1 & 0 & 1 \\ 0 & 0 & 1 & 1 \\ 0 & 1 & 0 & 1 \\ 0 & 1 & 1 & 0 \end{bmatrix}$$

Multiplying M_{R_1} with itself, we obtain

$$M' = M_{R_1} M_{R_1} = \begin{bmatrix} 1 & 1 & 0 & 1 \\ 0 & 0 & 1 & 1 \\ 0 & 1 & 0 & 1 \\ 0 & 1 & 1 & 0 \end{bmatrix} \begin{bmatrix} 1 & 1 & 0 & 1 \\ 0 & 0 & 1 & 1 \\ 0 & 1 & 0 & 1 \\ 0 & 1 & 1 & 0 \end{bmatrix} = \begin{bmatrix} 1 & 2 & 2 & 2 \\ 0 & 1 & 1 & 1 \\ 0 & 1 & 2 & 1 \\ 0 & 1 & 1 & 1 \end{bmatrix}$$

Corresponding to non-zero entries in M', we obtain

$$R_1 o R_1 = \{(1, 1), (1, 2), (1, 3), (1, 4), (2, 2), (2, 3), (2, 4), (3, 2), (3, 3), (3, 4), (4, 2), (4, 3), (4, 4)\}$$

$$\therefore R_2 = R_1 \cup (R_1 o R_1) = \{(1, 1), (1, 2), (1, 3), (1, 4), (2, 2), (2, 3), (2, 4), (3, 2), (3, 3), (3, 4), (4, 2), (4, 3), (4, 4)\}$$

Again since $R_2 \neq R_1$, we will calculate $R_3 = R_2 \cup (R_2 o R_2)$.

The matrix of relation R_2 is given by

$$M_{R_2} = \begin{bmatrix} 1 & 1 & 1 & 1 \\ 0 & 1 & 1 & 1 \\ 0 & 1 & 1 & 1 \\ 0 & 1 & 1 & 1 \end{bmatrix}$$

Multiplying M_{R_2} with itself, we obtain,

$$M'' = M_{R_2} \cdot M_{R_2} = \begin{bmatrix} 1 & 4 & 4 & 4 \\ 0 & 3 & 3 & 3 \\ 0 & 3 & 3 & 3 \\ 0 & 3 & 3 & 3 \end{bmatrix}$$

Since non-zero entries in M_{R_2} and M'' are same, therefore $R_2 o R_2 = R_2$.

Thus $R_3 = R_2$.

Since $R_3 = R_2$, the transitive closure R^* of R is the relation R_2.

Hence $R^* = \{(1, 1), (1, 2), (1, 3), (1, 4), (2, 2), (2, 3), (2, 4), (3, 2), (3, 3), (3, 4), (4, 2), (4, 3), (4, 4)\}$.

§ 2.12. Equivalence Relations

Let A be a non-empty set and let R be a relation on A. Then R is said to be equivalent relation if it is

1. *Reflexive*. That is, for every $a \in A$, aRa.
2. *Symmetric*. That is, aRb, then bRa.
3. *Transitive*. That is, if aRb and bRc then aRc.

The equivalence relation is usually denoted by the symbol $\sim$.

Illustrations:

The following are some examples of equivalence relations:

1. Equality of numbers on the set of real numbers.
2. Equality of subsets of a universal set.
3. Congruency of triangles on the set of triangles.
4. Relation of lines being parallel on the set of lines in a plane.

Example 2.44. Let I be the set of integers. Define a relation R on I as follows: xRy if and only if $x-y$ is divisible by 5, $\forall\, x, y \in$ I.
Show that R is an equivalence relation on I.

Solution:

1. R is *reflexive*:

   Since for any $a \in$ I, $a-a = 0$ which is divisible by 5, therefore aRa. Thus R is reflexive.

2. R is *symmetric*:

   For any a, $b \in$ I, if $a-b$ is divisible by 5, then $b-a$ is also divisible by 5. Thus, $aRb \Rightarrow bRa$. Hence R is symmetric.

3. R is *transitive*:

   Suppose aRb and bRc for any $a,b,c \in$ I

   $\Rightarrow$ both $a-b$ and $b-c$ are divisible by 5

   $\Rightarrow$ $(a-b)+(b-c)$ are divisible by 5

   $\Rightarrow$ $a-c$ is divisible by 5

   $\Rightarrow$ aRc

   Thus R is transitive.

Since R is reflexive, symmetric and transitive, therefore R is an equivalence relation.

Definition: Let m be a fixed integer. Two integers a and b are said to be congruent module m, written as

$$a \equiv b (\text{mod } m)$$

if and only if m divides $a-b$. That is, $a \equiv b(\text{mod } m)$ if and only if $a-b = km$ for some integer k.

The relation $a \equiv b\,(\text{mod } m)$ defined on the set of integers is called the relation of 'congruence modulo m'.

Example 2.45. Show that the relation of congruence modulo m' on the set I of integers is an equivalence relation.

Solution: Let a, $b \in$ I. Then the relation R of congruence modulo m is defined as follows: aRb if and only if m divides $a-b$

1. R is *reflexive*:

   Since $a-a = 0 = 0 \cdot m$, $\forall\ a \in I$, aRa.

   Hence R is reflexive.

2. R is *symmetric*:

   Suppose aRb

   $\Rightarrow a-b$ is divisible by m

   $\Rightarrow a-b = k \cdot m$ for some integer k,

   $\Rightarrow b-a = (-k) \cdot m$ for some integer $-k$.

   $\Rightarrow bRa$

   Hence R is symmetric.

3. R is *transitive*:

   Suppose aRb and bRc

   $\Rightarrow a-b = k_1 m$ and $b-c = k_2 m$ for some integers k_1 and k_2.

   $\Rightarrow (a-b)+(b-c) = (k_1+k_2)m$

   $\Rightarrow a-c = (k_1+k_2)m$ for the integer k_1+k_2.

   $\Rightarrow aRc$

   Hence R is transitive.

   This proves that R is an equivalence relation on I.

§ 2.13. Equivalence Classes

Definition: Let R be an equivalence relation on A. For any $a \in A$, the set of all $x \in A$ which are related to a is called equivalence class of a. We denote equivalence class of a by the symbol $[a]$. Thus symbolically,

$$[a] = \{x : x \in A \text{ and } aRx\}$$

Example 2.46. Let $A = \{1, 2, 3, 4\}$ and let relation R on A be given by $R = \{(1, 1), (1, 2), (1, 3), (2, 1), (2, 2), (2, 3), (3, 1), (3, 2), (3, 3), (4, 4)\}$. Find the equivalence classes of relation R.

Solution: It is easy to verify that R is an equivalence relation. The equivalences class of $1 \in A$ is

$$[1] = \{x : x \in A \text{ and } 1Rx\}$$
$$= \{1,2,3\}$$

Similarly, $[2] = \{1, 2, 3\}$

$$[3] = \{1,2,3\}$$

and $[4] = \{4\}$

Example 2.47. Let R be the relation of 'congruence module 3' on the set I of integers. Find equivalence classes of R.

Solution: We start with the element $0 \in \mathrm{I}$. The equivalence class of 0 is

$$[0] = \{x : x \in \mathrm{I} \text{ and } 0Rx\}$$
$$= \{x : x \in \mathrm{I} \text{ and } x \equiv 0 \pmod 3\}$$
$$= \{x : x \in \mathrm{I} \text{ and } x \text{ is multiple of } 3\}$$
$$= \{..., -6, -3, 0, 3, 6, ...\}$$

Since $[0] \neq \mathrm{I}$, we take an element in I which is not in $[0]$. Let us choose $1 \in \mathrm{I}$. Now we compute, equivalence class of 1,

$$[1] = \{x : x \in \mathrm{I} \text{ and } 1Rx\}$$
$$= \{x : x \in \mathrm{I} \text{ and } x \equiv 1 \pmod 3\}$$
$$= \{x : x \in \mathrm{I} \text{ and } x = 3k+1, \text{ when } k \in \mathrm{I}\}$$
$$= \{..., -5, -2, 1, 4, 7, ...\}$$

Since $[0] \cup [1] \neq \mathrm{I}$, we choose an element in I which is not in $[0] \cup [1]$. Let us choose $2 \in \mathrm{I}$. The equivalence class of 2 is

$$[2] = \{... -4, -1, 2, 5, 8, . . .\}.$$

Since $[0] \cup [1] \cup [2] = \mathrm{I}$, we see that these are the only distinct equivalence classes with respect to the relation 'congruence modul 3'.

Properties of Equivalence Classes

The following theorem contains some important properties of an equivalence relation.

Theorem: Let A be a non-empty set and let R be an equivalence relation on A. For a, $b \in A$,

1. $a \in [a]$
2. if $b \in [a]$ then $[b] = [a]$,
3. $[a] = [b]$ if and only if $(a, b) \in R$
4. either $[a] = [b]$ or $[a] \cap [b] = \phi$. That is, two equivalence classes are either disjoint or identical.

Proof:

1. Since R is reflexive, we have aRa. Thus $a \in [a]$.
2. Suppose $b \in [a]$. Then bRa.

Let x be any element of $[a]$. Then xRb, but R is transitive, therefore xRb and $bRa \Rightarrow xRa \Rightarrow x \in [a]$. Thus $[b] \subseteq [a]$.

Again let y be any element of $[a]$. Then yRa. Since R is symmetric, therefore

$$bRa \Rightarrow aRb.$$

Now yRa and $aRb \Rightarrow yRb \Rightarrow y \in [b]$

Thus $[a] \subseteq [b]$

Hence $[a] \subseteq [b]$ and $[b] \subseteq [a] \Rightarrow [a] = [b]$

3. Suppose $[a] = [b]$. We show that aRb.

Since R is reflexive, we have aRa

Now

$$aRa \Rightarrow a \in [a] \Rightarrow a \in [b] \qquad \text{(since } [a] = [b]\text{)}$$
$$\Rightarrow aRb$$

Conversely, suppose that aRb. Then we show that $[a] = [b]$.

Let x be any element of $[a]$. Then xRa. But it is given that aRb. Therefore,

$$xRa \text{ and } aRb \Rightarrow xRb \qquad \text{[since } R \text{ is transitive]}$$
$$\Rightarrow x \in [b]$$

Therefore, $[a] \subseteq [b]$

Again let y be any element of $[b]$. Then

$$y \in [b] \Rightarrow yRb.$$

Now we are given that aRb.

$\Rightarrow bRa$ [since R is symmetric]

Now, yRb and $bRa \Rightarrow yRa$ [since R is transitive]

$\Rightarrow y \in [a]$

Therefore $[b] \subseteq [a]$.

Finally, since $[a] \subseteq [b]$ and $[b] \subseteq [a]$, we have $[a] = [b]$.

4. If $[a] \cap [b] = \phi$, then we are nothing to prove. So let us suppose that $[a] \cap [b] \neq \phi$.

Since $[a] \cap [b] \neq \phi$, suppose $x \in [a] \cap [b]$

Now $x \in [a] \cap [b] \Rightarrow x \in [a]$ and $x \in [b]$

$\Rightarrow xRa$ and xRb

$\Rightarrow aRx$ and xRb

$\Rightarrow aRb$

$\Rightarrow [a] = [b]$ [by part (iii)]

Thus $[a] \cap [b] \neq \phi \Rightarrow [a] = [b]$

Thus two equivalence classes $[a]$ and $[b]$ are either identical or disjoint.

§ 2.14. Partitions

Definition: Let S be a non-empty set. A collection $P = \{A_1, A_2, ...\}$ of non-empty subsets of S is called a partition of S if

1. $A_1 \cup A_2 \cup A_3 ... = S$.
2. If $A_i \neq A_j$ then $A_i \cap A_j = \phi$.

Example 2.48. Let I = {..., –5, –4, –2, –1, 0, 1, 2, 3, 4, ...}. Then the collection $\{A_1, A_2, A_3\}$, where

$A_1 = \{..., -6, -3, 0, 3, 6, 9,...\}$

$A_2 = \{..., -5, -2, 1, 4, 7, 10, ...\}$ and

$A_3 = \{..., -4, -1, 2, 5, 8, 11,...\}$

is a partition of I because $A_1 \cup A_2 \cup A_3 = \text{I}$ and $A_i \cap A_j = \phi$ when $A_i \neq A_j$.

It may be seen that every equivalence relation on a set determines a unique partition of the set and every partition of a set defines an equivalence relation on the set.

Example 2.49. Let $A = \{1, 2, 3, 4\}$ and consider the partition $P = \{\{1, 2\}, \{3, 4\}\}$ of A. Find the equivalence relation R on A determined by P.

Solution: The disjoint sets in P are $\{1, 2\}$ and $\{3, 4\}$. Each elements in a disjoint set is related to every other element in the same disjoint set and only to those elements. Thus

$$R = \{(1, 1), (1, 2), (2, 1), (2, 2), (3, 3), (3, 4), (4, 3), (4, 4)\}.$$

Quotient Set: Let A be a non-empty set and R be an equivalence relation on A. The set of all equivalence classes is called the quotient set of A modulo R and is denoted by A/R.

Exercise

1. Give an example of a relation on the set $\{a,b,c\}$ which is

   (a) Reflexive but is neither symmetric nor transitive.

   (b) Reflexive, symmetric but not transitive.

   (c) Symmetric and transitive but not reflexive.

2. (a) Let $A = \{1, 2, 3\}$ and $B = \{r, s\}$. Then write the matrix of the relation R from A to B given by

   $$R = \{(1, r), (2, s), (3, r)\}$$

   (b) Let $A = \{1, 2, 3, 4\}$. Give the relation R on A that has relation matrix

   $$M_R = \begin{bmatrix} 1 & 1 & 0 & 1 \\ 0 & 1 & 1 & 0 \\ 0 & 0 & 1 & 1 \\ 1 & 0 & 0 & 0 \end{bmatrix}$$

   Also draw the graph of R.

3. Let $A = \{1, 2, 3, 4\}, B = \{1, 4, 6, 8, 9\}$. Let R be a relation from A to B defined by aRb if and only if $b = a^2$. Find the domain and range of R.

4. Let A be the set of real numbers. Consider the following relation R on A; $(a,b) \in R$ if and only if $a^2+b^2 = 25$. Find the domain and range of R.

5. (a) Let A be a set with 10 distinct elements. How many relations are there on A? How many of them are reflexive?

   (b) Let $A = \{1, 2, 3\} \times \{a, b\}$. How many relations are there on A?

6. Let R be the following relation on $A = \{1, 2, 3, 4\}$:

   $$R = \{(1, 3), (1, 4), (3, 2), (3, 3), (3, 4)\}$$

   (a) Find the matrix M_R of R.

   (b) Find the domain and range of R.

   (c) Find the R^{-1}.

   (d) Draw the graph of R.

   (e) Find the composition relation $R \circ R$.

7. If relations R and S are reflexive, symmetric and transitive then show that $R \cap S$ is also reflexive, symmetric and transitive.

8. Given $A = \{a, b, c\}$ and a relation R on A is defined by

   $R = \{(a, a), (a, b), (b, c), (c, c)\}$. Find transitive closure of R.

9. Let $A = \{1, 2, ..., 9\}$ and let $\sim$ be the relation on $A \times A$ defined by

   $$(a, b) \sim (c, d) \text{ if } a+d = b+c.$$

   Prove that $\sim$ is an equivalence relation. Also find the equivalence class of (2, 5).

10. Let R be the relation on N defined by the equation $x+3y = 12$ That is, $R = \{(x, y) : x+3y = 12\}$. Write R as a set of ordered pairs. Also find $R \circ R$.

11. Let $A = \{1, 2, 3, 4\}$ and $R = \{(x, y) : x > y\}$. Draw the graph of R and also give its matrix.

12. If $\{\{1, 2, 3\}, \{4\}\}$ is a partition of the set $A = \{1, 2, 3, 4\}$ then determine the corresponding equivalence relation R on A.

13. For a given partition of a set A, define a relation R on A such that R is an equivalence relation corresponding to the partition.

Answers

1. (a) $R = \{(a, a), (b, b), (c, c), (a, b), (b, c)\}$

   (b) $R = \{(a, a), (b, b), (c, c), (a, b), (b, a), (b, c), (c, b)\}$

   (c) $R = \{(a, b), (b, a), (a, a), (b, b)\}$

2. (a)

$$M_R = \begin{array}{c} \\ 1 \\ 2 \\ 3 \end{array}\begin{array}{c} \begin{array}{cc} r & s \end{array} \\ \begin{bmatrix} 1 & 0 \\ 0 & 1 \\ 1 & 0 \end{bmatrix} \end{array}$$

(b) $R = \{(1, 1), (1, 2), (1, 4), (2, 2), (2, 3), (3, 3), (3, 4), (4, 1)\}$. Graph of R is

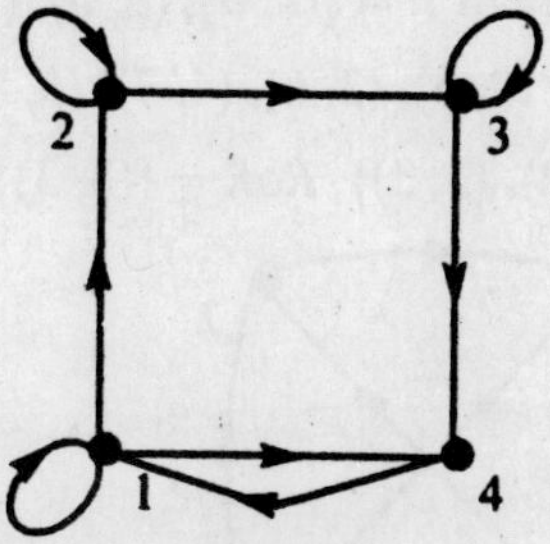

Fig. : 2.6

3. Domain of $R = \{1, 2, 3\}$ and range of $R = \{1, 4, 9\}$

4. Domain of $R = \{x : -5 \leqslant x \leqslant 5\}$, Range of $R = \{x : -5 \leqslant x \leqslant 5\}$

5. (a) Number of relations on A=2^{100}, number of reflexive relation on $A = 2^{90}$.

(b) Since A contains 6 elements, number of relations on $A = 2^{36}$.

6. (a)

$$M_R = \begin{bmatrix} 0 & 0 & 1 & 1 \\ 0 & 0 & 0 & 0 \\ 0 & 1 & 1 & 1 \\ 0 & 0 & 0 & 0 \end{bmatrix}$$

(b) Domain = {1, 3}, *Range* ={2 ,3, 4};

(c) $R^{-1} = \{(3, 1), (4, 1), (2, 3), (3, 3), (4, 3)\}$;

(d) Graph of R is

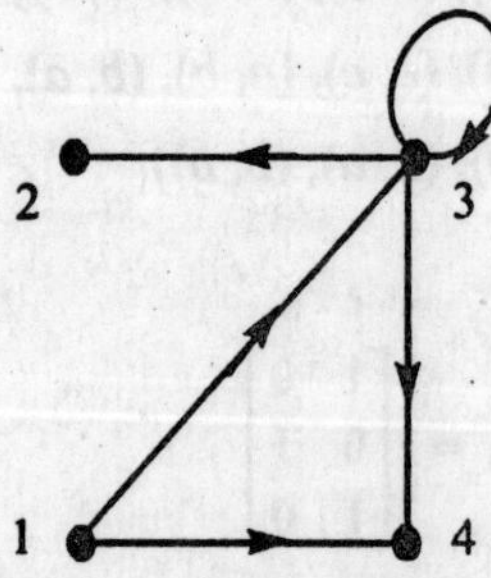

Fig. : 2.7

(e) $RoR = \{(1, 2), (1, 3), (1, 4), (3, 2), (3, 3), (3, 4)\}$

8. Transitive closure of $R = \{(a, a), (a, b), (b, c), (c, c), (a, c)\}$.

9. $[(2, 5)] = \{(1, 4), (2, 5), (3, 6), (4, 7), (5, 8), (6, 9)\}$

10. $R = \{(9, 1), (6, 2), (3, 3)\}$, $RoR = \{(3, 3)\}$

11.

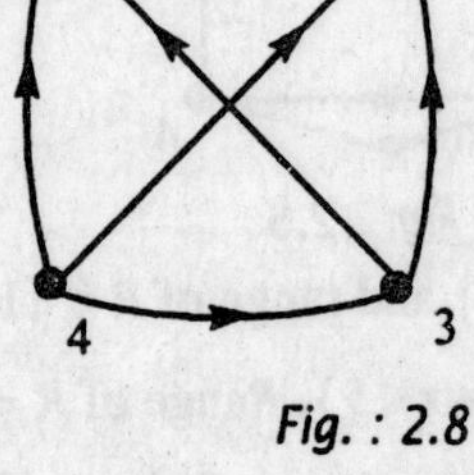

Graph of R

$$\begin{bmatrix} 0 & 0 & 0 & 0 \\ 1 & 0 & 0 & 0 \\ 1 & 1 & 0 & 0 \\ 1 & 1 & 1 & 1 \end{bmatrix}$$

Matrix of R

Fig. : 2.8

12. $R = \{(1, 1), (1, 2), (1, 3), (2, 1), (2, 2), (2, 3), (3, 1), (3, 2), (3, 3), (4, 4)\}$

§ 2.15. Partially Ordered Sets

Definition 1: A relation R on a set A is said to be *partial order* if R is reflexive, anti-symmetric and transitive. Thus a relation R on a set A is partial order relation if the following conditions hold:

1. Reflexivity : $aRa \ \forall \ a \in A$
2. Anti-symmetry : If aRb, bRa then $a = b$
3. Transitivity : If aRb, bRc then aRc

The set A together with the partial order R is called partially ordered set.

For convenience, we generally denote a partial order by the symbol $\leqslant$ in place of R. We read $\leqslant$ as less than or equal to. The symbol $\leqslant$ does not necessarily mean usual "less than or equal to" as is used for real numbers. If $\leqslant$ is partial order on A, then the ordered pair $(A, \leqslant)$ is called a partially ordered set or simply a *poset*.

Example 2.50. Let A be a collection of subsets of a set S. The relation $\subseteq$ of set inclusion is a partial order on A. Thus $(A, \subseteq)$ is a poset.

Example 2.51. The relation $\leqslant$ of divisibility is a partial order on the set N of natural numbers. Here $a \leqslant b$ means $a \mid b$ (a divides b).

Example 2.52. The relation $<$ on N is not a partial order, since it is not reflexive.

Definition 2: Let $(A, \leqslant)$ be a poset. The elements a and b of A are said to be comparable if $a \leqslant b$ or $b \leqslant a$.

Thus a and b are non-comparable if neither $a \leqslant b$ nor $b \leqslant a$.

Example 2.53. Consider the poset in Example 2.51. The elements 2 and 5 are not comparable since neither 2 divides 5 nor 5 divides 2. Thus in a poset every pair of elements need not be comparable.

Let $(A, \leqslant)$ be a poset. We say $a < b$ if $a \leqslant b$ and $a \neq b$. We also say b in larger than or equal to a and write $b \geqslant a$ if $a \leqslant b$.

Totally ordered set or chain: Let $(A, \leqslant)$ be a poset. $(A, \leqslant)$ is called *chain* or *totally ordered set* if every two elements in A are comparable. That is, if $a, b \in A$ then $a \leqslant b$ or $b \leqslant a$. Totally ordered sets are also called linearly orderly sets.

Illustrations:

Let R the set of real numbers and let $\leqslant$ be the usual less than or equal to relation on R. Then $(R, \leqslant)$ is poset which is a chain whereas the posets in Example 2.50 and Example 2.51 are not chains.

Representation of Posets: Hasse Diagrams

A partial ordering $\leqslant$ on a poset P can be represented by a diagram known as a Hasse diagram of $(P, \leqslant)$. In such diagram we represent each element by a small circle or by a dot and any two comparable elements are joined by lines in such a way that if $a \leqslant b$ then a lies below b in the diagram. Non-comparable elements are not joined. Thus there will not be any horizontal lines in the diagram of a poset. If Hasse diagram of a poset is given then we can obtain the set of ordered pairs in $\leqslant$ from the diagram.

Example 2.54. Let $X = \{2, 3, 6, 12, 24, 36\}$ and the relation $\leqslant$ be such that $x \leqslant y$ if x divides y (written as $x|y$). Draw the Hasse diagram of $(X, \leqslant)$.

Solution: The Hasse diagram is given below:

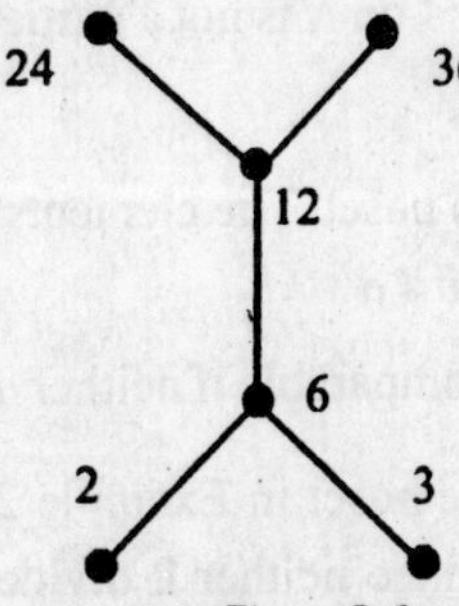

Fig. : 2.9

Example 2.55. Let = $\{1, 2, 3\}$. Then power set of X, $P(X) = \{\phi, \{1\},$

{2}, {3}, {1, 2}, {1, 3}, {2, 3}, {1, 2, 3}}. Let $\subseteq$ be the inclusion relation on P(X). Draw Hasse diagram of (P(X), $\subseteq$).

Solution: The Hasse diagram of (P(X), $\subseteq$) is given below.

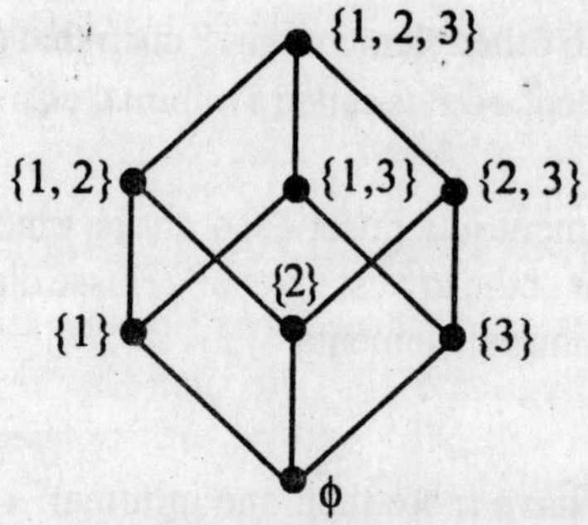

Fig. : 2.10

Example 2.56. Let X be the set of factors of 12 and let $\leqslant$ be the relation divides, i.e., $x \leqslant y$ if and only if $x \mid y$. Draw the Hasse diagram of (X, $\leqslant$).

Solution: Here X = {1, 2, 3, 4, 6, 12}. The Hasse diagram is given below:

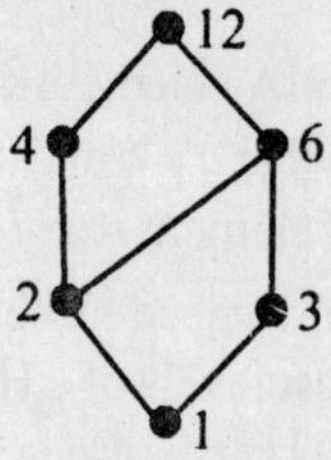

Fig. : 2.11

Example 2.57. Let P = {1, 2, 3, 4, 5} and $\leqslant$ be the relation "less than or equal to". Draw the Hasse diagram of (P, $\leqslant$).

Solution: The Hasse diagram is as shown in fig 2.12.

Fig. : 2.12

Since any two elements of P are comparable, (P,$\leqslant$) is a totally ordered set. A totally ordered set would always have the Hasse diagram consisting of circles (or dots) one below the other as shown in the Hasse-diagram of this poset. This justifies the name 'chain' for a totally ordered set.

Minimal and Maximal Elements

Definition 1: Let (P, $\leqslant$) be a poset. An element a in P is called a *minimal element* if there is no other element b in P such that $b \neq a$ and $b \leqslant a$. In other words, an element $a \in$ P is called a minimal element in P if $b < a$ for no b in P.

A minimal element in a poset need not be unique. All those elements, which appear at the lowest level of a Hasse diagram of a partially ordered set are minimal elements.

Illustrations:

1. A poset may have more than one minimal element. For example, in the poset {2, 3, 6, 12, 24, 36} with divisibility relation, both 2 and 3 are minimal elements.
2. A poset may not have a minimal element. For example, the set of integers with usual $\leqslant$ relation is a poset which has no minimal element.
3. Every finite poset has at least one minimal element.

Definition 2: Let (P, $\leqslant$) be a poset. An element a in P is called a *maximal element* if there is no b in P such that $a \neq b$ and $a \leqslant b$. In other words, an element $a \in$ P is called maximal element if $a<b$ for no b in P.

A maximal element in a poset need not be unique. All those elements which appear at the highest levels of a Hasse diagram of a poset are maximal elements.

Illustrations:

1. Let X = {2, 3, 6, 12, 24, 36} and the relation $\leqslant$ be such that $a \leqslant b$ if $a | b$. Then (X, $\leqslant$) is a poset. In this poset, both 24 and 36 are maximal elements.
2. A poset may not have a maximal element. For example, the set I of integers with usual $\leqslant$ relation is a poset in which no element is a maximal element.

Least and Greatest Elements

Definition 1: Let (P, $\leqslant$) be a poset. If there exists an element $a \in$P such that $a \leqslant x$ for all $x \in$ P, then a is called the *least element* in P.

The least element is also called first element or zero element of P.

From the definition it is clear that the least element, if it exists, is unique. It may happen that the least element does not exist. The least element is generally denoted by 0.

Illustrations:

1. Let $X = \{1, 2, 3, 4, 6, 12\}$ and let $\leqslant$ be defined as follows:

   $x \leqslant y$ if and only if $x \mid y$, i.e., x divides y.

   Then (X, $\leqslant$) is a poset. In this poset, 1 is the least elements.
2. Consider the poset {2, 3, 6, 12, 24, 36} with divisibility relation. This poset has two minimal elements 2 and 3 but it has no least element.
3. Let A be any non-empty set and let P(A) be its powerset. Let $\subseteq$ be the inclusion relation on P(A). Then (P(A), $\subseteq$) is a poset. In this poset, the least element is ϕ which is also a minimal element.

Definition: Let (P, $\leqslant$) be a poset. If there exists an elements $a \in$ P such that $x \leqslant a$ for all $x \in$ P, then a is called the *greatest element* in P.

The greatest element is also called last element or *unit element* of P and is usually denoted by 1. From the definition, it is clear that the greatest element, if it exists, is unique. It may happen that the greatest element does not exist.

Illustrations:

1. Let $X = \{1, 2, 3, 4, 6, 12\}$ and let $\leqslant$ be defined as follows : $x \leqslant y$ if and only if $x \mid y$. Then (X, $\leqslant$) is a poset. In this poset, 12 is the greatest element. This can also be seen from the Hasse diagram of this poset.
2. Let X = {2, 3, 6, 12, 24, 36} and let $\leqslant$ be the divisibility

relation on X. Then $(X, \leqslant)$ is a poset. This poset has two maximal elements, 24 and 36 but it has no greatest element.

3. The power set of any non-empty set A with set inclusion relation $\subseteq$ is a poset. In this poset, A is the greatest element.

Upper Bounds and Least Upper Bound

Definition: Let $(P, \leqslant)$ be a poset and let A be a subset of P. An element $x \in P$ is called an upper bound of A if $a \leqslant x \ \forall \ a \in A$.

Illustrations:

1. Let $X = \{1, 2, 3\}$. Then $(P(X), \subseteq)$ is a poset.

   Let $A = \{\phi, \{1, 2\}, \{2\}, \{3\}\}$.

   Then $\{1, 2, 3\}$ is an upper bound of A because every element of A is contained in $\{1, 2, 3\}$.

   For the subset $B = \{\{3\}, \{1,3\}\}$, the upper bound of B are $\{1,3\}$ and $\{1,2,3\}$.

2. Let $X = \{2, 3, 6, 12, 24, 36\}$ and let $\leqslant$ be the divisibility relation on X. For the subset $A = \{2, 3, 6\}$ of X, the upper bounds are 6, 12, 24, 36.

3. Consider the set $N = \{1, 2, 3, ...\}$ with usual $\leqslant$ relation. Then $(N, \leqslant)$ is poset. For $A = \{2, 3, 4\}$, all natural numbers greater than or equal to 4 are upper bounds of A.

Definition: Let $(P, \leqslant)$ be a poset and let $A \subseteq P$. An element $x \in P$ is said to be a *least upper bound* or *supremum* of A if x is an upper bound of A and $x \leqslant y$ for all upper bounds y of A.

A least upper bound, if it exists, is unique and will be denoted by "lub" or "sup". That is, supremum of A is denoted by $\sup(A)$.

Illustrations:

1. Let $P = \{a, b, c, d, e, f\}$. Let $\leqslant$ on P be represented by Hasse diagram as shown in fig. 2.12.

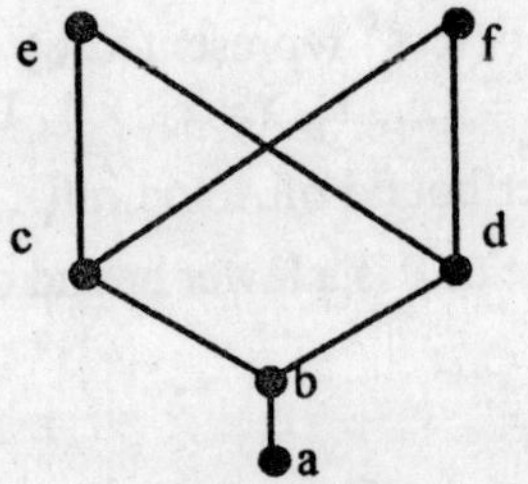

Fig. : 2.12

Let $A = \{b, c, d\}$. The upper bounds of A are e and f but sup (A) does not exist because upper bounds of A are non-comparable.

2. Let P be the set of natural numbers and $\leqslant$ be a relation on P such that $a \leqslant b$ if $a \mid b$. Then (P, $\leqslant$) is a poset. Let $A = \{a, b\}$ where a, b are any two natural numbers. Then the least common multiple (LCM) of a and b is the supremum of A.

3. Consider the following Hasse diagram of a poset (P, $\leqslant$), where P $=\{1, 2, 3, ..., 8\}$. Let $A = \{4, 5, 7\}$. The upper bounds of A are 1, 2 and 3 and sup (A) = 3.

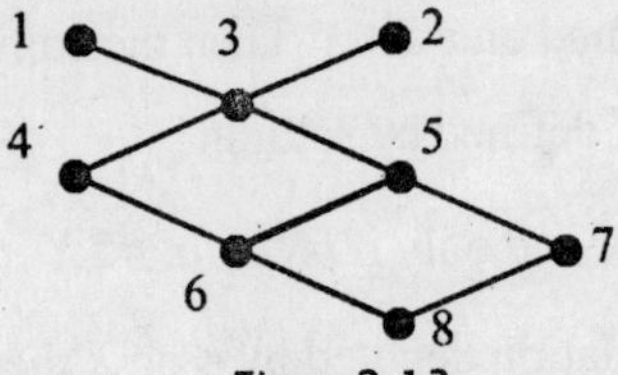

Fig. : 2.13

Lower Bounds and Greatest Lower Bound

Definition 1: Let (P, $\leqslant$) be a poset and let A be a subset of P. An element $x \in$ P is said to be a *lower bound* of A if $x \leq a \ \forall \ a \in A$.

Definition 2: Let (P, $\leqslant$) be a poset and $A \subseteq$ P. An element $x \in$ P is said to be a *greatest lower bond* or infimum of A, written as glb(A) or inf(A), if x is a lower bound and $y \leqslant x$ for all lower bounds y of A.

A greatest lower bound, if it exists, is unique.

Illustrations:

1. Let (N, $\leqslant$) be the poset of natural numbers. Let $A = \{5, 7, 9\}$. Then 1, 2, 3, 4 and 5 are lower bounds of A and inf(A) = 5.

2. Consider the poset (P, $\leqslant$) represented by the following Hasse diagram, where P = $\{x_1, x_2, x_3, x_4, x_5\}$. Let $A = \{x_1, x_2, x_3\}$. x_4 is the only lower bound of A and $\inf(A) = x_4$. If $B = \{x_3, x_4, x_5\}$ then no element of P is a lower bound of B and hence $\inf(B)$ does not exist.

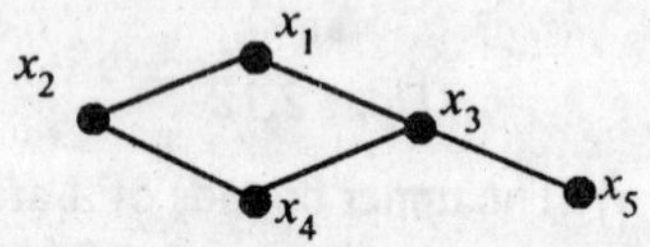

Fig. : 2.14

3. Let P be the set of natural number and $\leqslant$ be defined by $x \leqslant y$ if and only if $x|y$. Then (P, $\leqslant$) is a poset. Let $A = \{a, b\}$ be any subset of P containing two natural numbers. Then the greatest common division of a and b is the infimum of A. If $A = \{6, 9\}$ then $\inf\{6, 9\} = 3$.

Dual of a Poset

Let R be a relation defined on a set X. Then the converse of R, denoted by $\overline{R}$, is a relation on X defined by setting.

$$a \overline{R} b \text{ if and only if } bRa, \; a, b \in X$$

If R is a partial order relation denoted by $\leqslant$ on X then $\overline{R}$, the converse of R, is denoted by $\geqslant$.

Theorem: If (X, R) is a poset then $(X, \overline{R})$ is also poset.

Proof: We show $\overline{R}$ is reflexive, anti-symmetric and transitive.

$\overline{R}$ **is reflexive:** Let $a \in X$

Since R is reflexive, we have,

$$aRa \; \forall \; a \in X \Rightarrow a \overline{R} a \; \forall \; a \in X$$

$$\Rightarrow \overline{R} \text{ is reflexive.}$$

$\overline{R}$ **is anti-symmetric:** Suppose $a \overline{R} b$ and $b \overline{R} a$.

$$\text{Now } a\overline{R}\,b \Rightarrow bRa \qquad \text{and}$$

$$b\,\overline{R}\,a \Rightarrow aRb$$

$$\text{But } aRb \text{ and } bRa \Rightarrow a=b \qquad [R \text{ is anit-symmetric}]$$

$$\text{Thus } a\,\overline{R}\,b \text{ and } b\,\overline{R}\,a \Rightarrow a=b$$

Hence $\overline{R}$ is anti-symmetric.

$\overline{R}$ is transitive

$$\text{Let } a\overline{R}\,b \text{ and } b\overline{R}\,c \Rightarrow bRa \text{ and } cRb$$

$$\text{Now } cRb \text{ and } bRa \Rightarrow cRa \qquad [R \text{ is transitive}]$$

$$\Rightarrow a\,\overline{R}\,c$$

$$\text{Thus } a\,\overline{R}\,b \text{ and } b\,\overline{R}\,c \Rightarrow a\,\overline{R}\,c$$

Hence $\overline{R}$ is transitive.

Therefore, $\overline{R}$ is a partial order relation.

In view of this theorem, we define dual of a poset as follows:

Definition: Let (P, $\leqslant$) be a poset, then (P, $\geqslant$) is also a poset. (P, $\geqslant$) is called the *dual* of (P,$\leqslant$).

It may be seen that the least element of poset (P, $\leqslant$) is equal to the greatest element of (P, $\geqslant$) and vice versa. Moreover the maximal and minimal elements are interchanged and for any subset A of P, the glb(A) in (P ,$\leqslant$) is equal to the lub(A) in (P, $\geqslant$).

Product of Two Posets

Let A and B be two posets. Then we can show that the set $A \times B = \{(a,b) : a \in A\,,\, b \in B\}$ forms a poset under the relation defined by

$$(a_1, b_1) \leqslant (a_2, b_2) \text{ if } a_1 \leqslant a_2 \text{ on } A \text{ and } b_1 \leqslant b_2 \text{ on } B.$$

Note that we have denoted three relations on $A\times B$, A and B by the symbol $\leqslant$ but they are different.

Theorem: If A and B are two posets then $A\times B$ is also a poset under the relation $\leqslant$ defined above.

Proof:

Reflexivity: Let (a, b) be any element of $A \times B$. Then $a \in A$ and $b \in B$

$\Rightarrow$ $a \leqslant a$ and $b \leqslant b$ [A and B are posets]

$\Rightarrow$ $(a, b) \leqslant (a, b)$

$\Rightarrow$ Relation $\leqslant$ on $A \times B$ is reflexive.

Anti-symmetry: Let $(a_1, b_1) \leqslant (a_2, b_2)$ and $(a_2, b_2) \leqslant (a_1, b_1)$

$(a_1, b_1) \leqslant (a_2, b_2) \Rightarrow a_1 \leqslant a_2$ and $b_1 \leqslant b_2$

$(a_2, b_2) \leqslant (a_1, b_1) \Rightarrow a_2 \leqslant a_1$ and $b_2 \leqslant b_1$

Now $a_1 \leqslant a_2$ and $a_2 \leqslant a_1 \Rightarrow a_1 = a_2$ and

$b_1 \leqslant b_2$ and $b_2 \leqslant b_1 \Rightarrow b_1 = b_2$ [A and B are posets]

Thus $(a_1, b_1) = (a_2, b_2)$.

Hence $\leqslant$ on $A \times B$ is anti-symmetric.

Transitivity: Let $(a_1, b_1) \leqslant (a_2, b_2)$ and $(a_2, b_2) \leqslant (a_3, b_3)$

Then $a_1 \leqslant a_2$, $b_1 \leqslant b_2$ and $a_2 \leqslant a_3$, $b_2 \leqslant b_3$

$\Rightarrow$ $a_1 \leqslant a_2$ and $a_2 \leqslant a_3$ and $b_1 \leqslant b_2$ and $b_2 \leqslant b_3$

$\Rightarrow$ $a_1 \leqslant a_3$ and $b_1 \leqslant b_3$

$\Rightarrow$ $(a_1, b_1) \leqslant (a_3, b_3)$.

Hence $\leqslant$ on $A \times B$ is transitive.

Thus $A \times B$ is a poset.

Isomorphic Posets

Definition: Let (A, R) and (B, R') be two posets. A one-to-one mapping $f: A \to B$ is called isomorphism if

$$xRy \Leftrightarrow f(x)R'f(y), \forall x, y \in A.$$

That is, if f preserves the partial order relation.

If there is no scope of confusion, we can use the same symbol $\leqslant$ for both the relations R and R'. Thus, we have

A mapping $f: A \to B$ is called isomorphism if f is one-to-one and

$$x \leqslant y \Leftrightarrow f(x) \leqslant f(y)$$

Definition: Two posets A and B are said to be isomorphic if there exists a one-to-one mapping from A onto B which is isomorphism. That is, f is bijective and

$$x \leqslant y \Leftrightarrow f(x) \leqslant f(y).$$

It is easy to see that the relation of being isomorphic is an equivalence relation on the set of all posets.

Example 2.58. Let $A = \{1, 2, 3, 4, 6, 12\}$. Let R be the partial order relation on A given by xRy if and only if $x|y$. Then (A, R) is a poset. Consider another poset $(A, \leqslant)$ in which $\leqslant$ denotes the usual "less than equal to" relation on A. Then (A, R) and $(A, \leqslant)$ are not isomorphic. The Hasse diagrams of (A, R) and $(A, \leqslant)$ are given below:

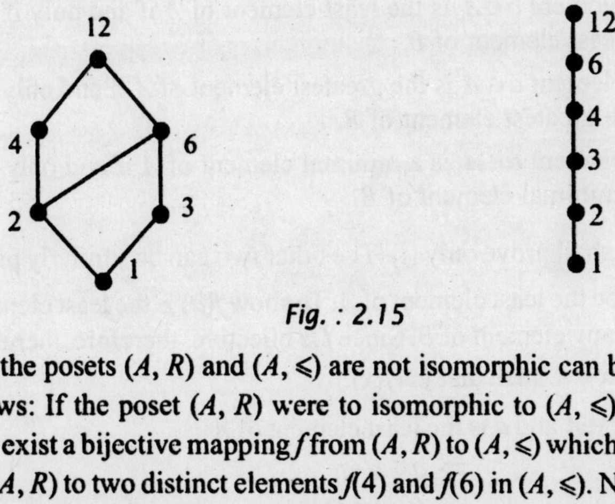

Fig. : 2.15

That the posets (A, R) and $(A, \leqslant)$ are not isomorphic can be shown as follows: If the poset (A, R) were to isomorphic to $(A, \leqslant)$, then there must exist a bijective mapping f from (A, R) to $(A, \leqslant)$ which takes 4 and 6 in (A, R) to two distinct elements $f(4)$ and $f(6)$ in $(A, \leqslant)$. Now $f(4)$ and $f(6)$ are comparable in $(A, \leqslant)$, while 4 and 6 are non-comparable.

Example 2.59. Let $A = \{1, 2, 3, 6\}$ and let $\leqslant$ the divisibility relation on A. Let $B = \{\phi, \{a\}, \{b\}, \{a, b\}\}$ and let $\leqslant'$ be the relation $\subseteq$.

If $f: A \to B$ is defined by

$$f(1) = \phi, f(2) = \{a\}, f(3) = \{b\}, f(4) = \{a, b\}.$$

Then f is bijective isomorphism from $(A, \leqslant)$ to $(B, \leqslant')$. Hence $(A, \leqslant)$ and $(B, \leqslant')$ are isomorphic.

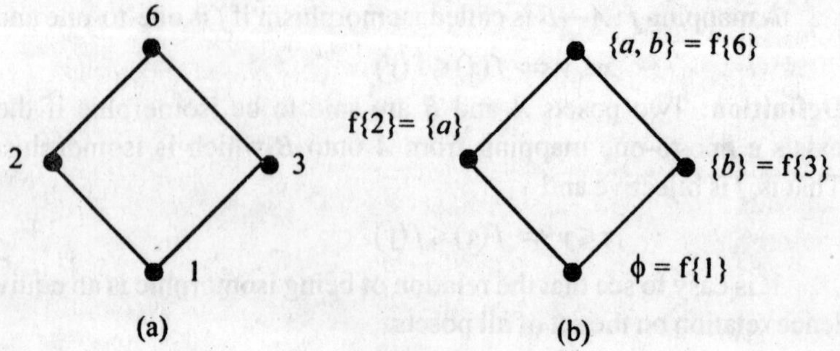

Fig. : 2.16

Figure (a) and (b) are the Hasse diagrams of $(A, \leqslant)$ and $(B, \leqslant')$ respectively.

Theorem: Let a poset A be isomorphic to a poset B and $f: A \to B$ be isomorphism from A onto B. Then

1. An element $a \in A$ is the least element of A if and only if $f(a)$ is the least element of B.
2. An element $a \in A$ is the greatest element of A if and only if $f(a)$ is the greatest element of B.
3. An element $a \in A$ is a minimal element of A if and only if $f(a)$ is a minimal element of B

Proof: We shall prove only (i). The other two can be similarly proved.

Let a be the least element of A. To show $f(a)$ is the least element of B. Let y be any element of B. Since f is bijective, therefore, there exists an element $x \in A$ such that $y = f(x)$.

Now $x \in A$ and a is the least element of A

$$\Rightarrow a \leqslant x$$
$$\Rightarrow f(a) \leqslant f(x)$$
$$\Rightarrow f(a) \leqslant y$$

Since y is any element of B, $\therefore f(a) \leqslant y \ \forall \ y \in B$

Thus $f(a)$ is the least element of B.

Conversely, suppose $f(a)$ is the least element of B.

$\Rightarrow \quad f(a) \leqslant y \ \forall \ y \in B$

We show $a \leqslant x \ \forall \ x \in A$

Let x be any element of A. Then $f(x) \in B$.

$\therefore \quad f(a) \leqslant f(x)$

$\Rightarrow a \leqslant x$ since f is isomorphism.

$\Rightarrow a$ is the least element of A.

Example 2.60. Show that the posets $(A, \leqslant)$ and $(B, \leqslant')$, whose Hasse diagrams are given below in figure (a) and (b), respectively, are not isomorphic.

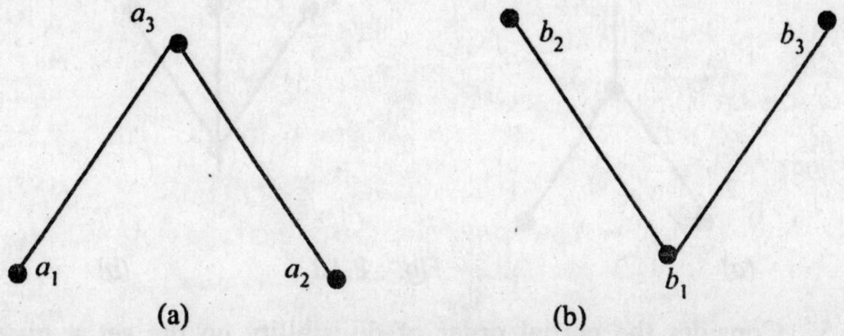

Fig. : 2.17

Solution: The two posets are not isomorphic because $(A, \leqslant)$ does not have a least element, while $(B, \leqslant')$ does have a least element.

Exercise

1. Give a relation on $A = \{a, b, c\}$ which is both partial order relation and equivalence relation.
2. Determine whether the relation R is a partial order on the set where
   (a) $A = Z$, the set of integers and aRb if and only if $b^2 | a$.
   (b) $A = Z$, the set of integers and aRb if and only if $a = 2b$.
   (c) $A = N \times N$, the cartesian product of N with itself and $(a, b) R (c, d)$ if and only if $\max \{a, b\} \leqslant \max \{c, d\}$.
3. Draw the Hasse diagram of the relation R, where
   (a) $A = \{1, 2, 3, 4\}$, $R = \{(1, 1), (1, 2), (2, 2), (2, 4), (1, 3), (3, 3), (3, 4), (1, 4), (4, 4)\}$

(b) $A = \{1, 2, 3, 4, 5\}, R = \{(1,1), (2,2), (3,3), (1,3), (3,4), (3,5), (1,4), (4,4), (1,5), (2,3), (2,4), (2,5), (5,5)\}$

4. Describe the ordered pairs in the relation determined by the Hasse diagram on the set A in figures 2.18(a) and (b).

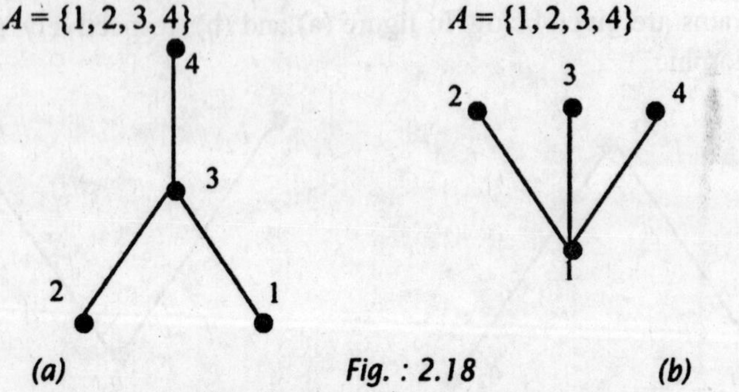

Fig. : 2.18

5. Consider the partial order of divisibility on the set A given below. Draw the Hasse diagram of the poset.

(a) $A = \{1, 2, 3, 5, 6, 10, 15, 30\}$

(b) $A = \{2, 4, 8, 16, 32\}$

6. Let $A = \{1, 2, 3, 5, 6, 10, 15, 30\}$ and consider the partial order $\leqslant$ of divisibility on A. Let $B = P(S)$, the power set of S, where $S = \{a, b, c\}$ be the poset with partial order $\subseteq$. Show that $(A, \leqslant)$ and $(B, \subseteq)$ are isomorphic.

7. Find all the maximal and minimal elements of posets whose Hasse diagrams are given in the figure below.

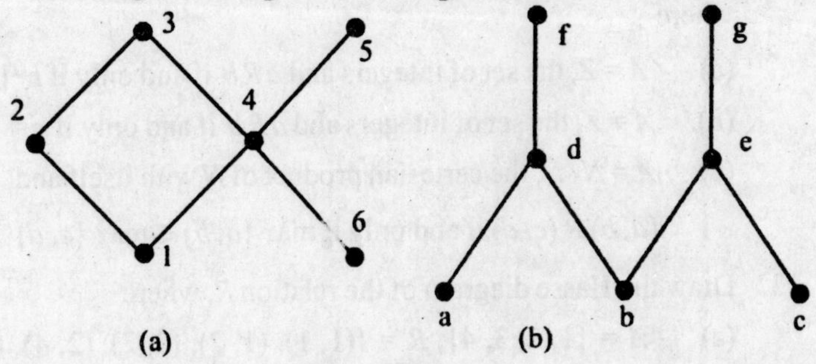

Fig. : 2.19

8. Find the greatest and the least elements, if they exist, of the following posets.
   (a) The set $A = \{2, 4, 6, 8, 12, 18, 24, 36, 72\}$ with the partial order of divisibility.
   (b) The set $A = \{x : x \text{ is a real number and } a<x<1\}$ with the usual $\leqslant$ relation.

9. Let $A = \{2, 3, 4, 6, 8, 12, 24, 48\}$ and $\leqslant$ be the partial order of divisibility. Let $B = \{4, 6, 12\}$ be a subset of A. Find
   (a) all upper bounds of B.
   (b) all lower bounds of B
   (c) the least upper bound of B
   (d) the greatest lower bound of B.

10. Consider the poset whose Hasse diagram is given below.

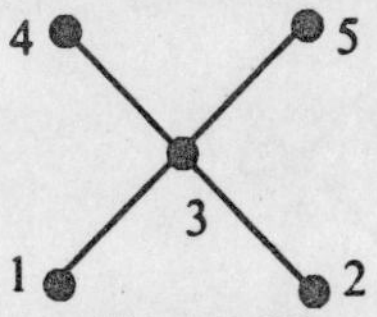

Fig. : 2.20

Find, if they exist, all upper bounds, all lower bounds, the supremum and the infimum of the poset.

11. Find, if they exist, all upper bounds, all lower bounds, the supremum and the infimum of the set $B = \{c, d, e\}$ in the poset whose Hasse diagram is given in fig. 2.21.

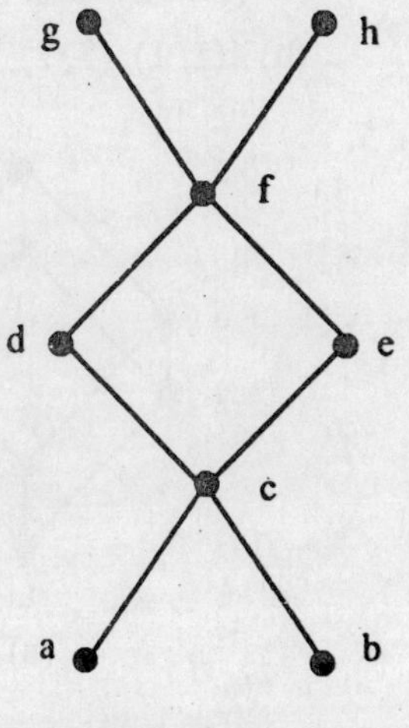

Fig. : 2.21

12. Show that there are only five distinct Hasse diagrams for posets that contain three elements.

(Hint: Let $P = \{a, b, c\}$. Then

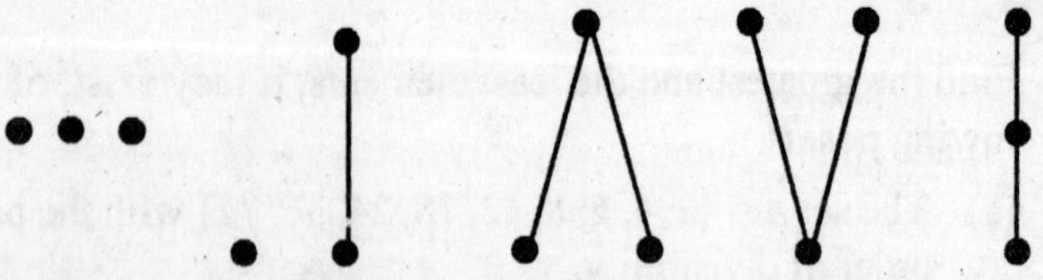

(1) (2) (3) (4) (5)

are distinct Hasse diagrams).

Answers

1. $\{(a, a), (b, b), (c, c)\}$
2. (a) No; (b) No; (c) No.
3.

(a)

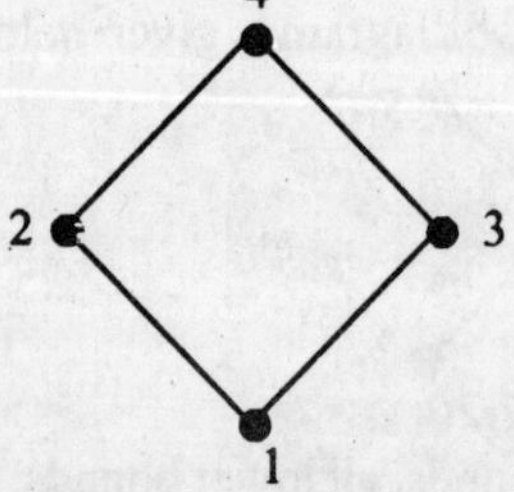

(b)

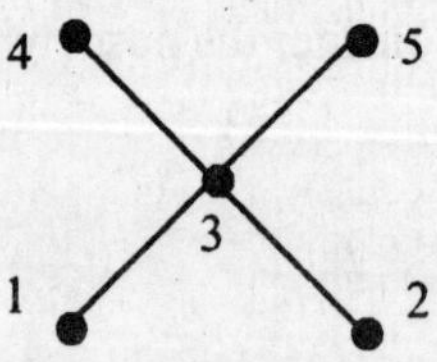

4.(a) $\{(1, 1), (2, 2), (3, 3), (4, 4), (1, 3), (1, 4), (2, 3), (2, 4), (3, 4)\}$

(b) $\{(1, 1), (2, 2), (3, 3), (4, 4), (1, 2), (1, 3), (1, 4)\}$

5.

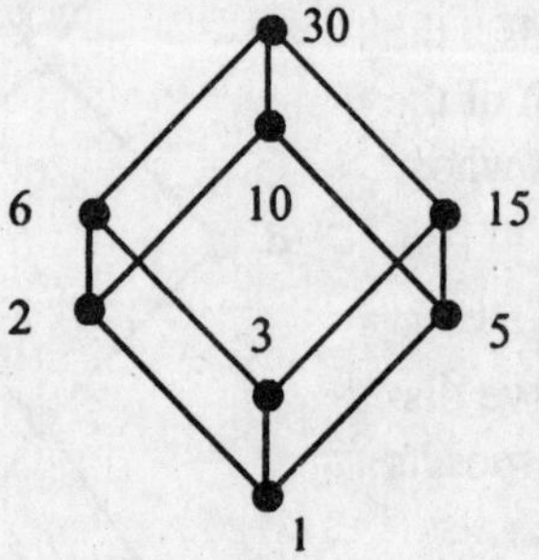

(a)

(b)

7. (a) Maximal elements = {3, 5}, minimal elements = {1, 6}

   (b) Maximal elements = {f, g}, minimal elements = {a, b, c}

8. (a) The greatest element = 72, the least element = 2

   (b) The greatest element, none, the least element, none.

9. (a) {12, 24, 48}; (b){2}; (c) 12; (d) 2.

10. Upper bound none, lower bound none, supremum none an infimum none.

11. {f, g, h}, {a, b, c}, sup (B) = f, inf(B) = c

§ 2.16. Countable and Uncountable Sets

Definition: A set A is said to be *finite* if either it is empty or there exists a one-to-one mapping from the set {1, 2, ..., n} onto the set A for some natural number n. In other words, a set A is finite if either it is empty or there is one-to-one correspondence between the elements of the set{1, 2, ..., n} and A for some natural number n. A set which is not finite is called infinite.

Illustrations:

1. The set ϕ is a finite set.
2. The set {a, b, c} is a finite set because there exists one-to-one mapping from the set {1, 2, 3} onto the set {a, b, c}. Consider, for example a mapping which takes $1 \to a$, $2 \to b$, $3 \to c$. This map is one-to-one and onto.
3. The set N of all natural numbers is infinite.
4. The set Q of all rational numbers in infinite.

It is obvious that every subset of a finite set is finite and union of two finite sets is again finite.

Equivalent sets: Two sets A and B are said to be *equivalent* or *equipotent* if there exists a bijective map $f: A \to B$. If A and B are equivalent then we write $A \sim B$.

Examples 2.61.

1. Let $A = \{1, 2, 3, ..., n\}$ and $B = \{x_1, x_2,..., x_n\}$. Then A and B are equipotent because the map $f: A \rightarrow B$ defined by $f(i) = x_i$, $i = 1,...,n$ is bijective.
2. Sets $A = \{1, 2, 3, 4, ...\}$ and $B = \{2, 4, 6, 8, ...\}$ are equipotent because the map $f: A \rightarrow B$ defined by $f(n) = 2n$ is bijective map.
3. Sets $A = [1, 2]$ and $B = [1, 4]$ are equipotents because the map defined by $f(x) = 3x-2$ is a bijective mapping from A onto B.
4. Any two closed intervals $[a, b]$ and $[c, d]$ are equipotent. Consider a mapping $f: [a, b] \rightarrow [c, d]$ defined as

$$f(x) = c + \left(\frac{d-c}{b-a}\right)(x-a)$$

Then f is a bijective mapping.

Theorem1. Prove that the relation of equipotence on the family of sets is an equivalence relation.

Proof: We say two sets A and B are equipotent, written as, $A \sim B$ if there is a bijective map $f: A \rightarrow B$.

We first show

1. **~ is reflexive**

   Since the identity map $I: A \rightarrow A$ defined by $I(x) = x$ is one-to-one and onto (i.e. bijective), therefore $A \sim A$

2. **~ is symmetric**

   Let $A \sim B$.

   $A \sim B \Rightarrow$ there is a bijective map $f: A \rightarrow B$

   $\Rightarrow f^{-1}: B \rightarrow A$ is also bijective

   $\Rightarrow B \sim A$

   Hence ~ is symmetric.

3. **~ is transitive**

   Let $A \sim B$ and $B \sim C$. To show that $A \sim C$.

$A \sim B \Rightarrow \exists$ a bijective map $f: A \to B$

$B \sim C \Rightarrow \exists$ a bijective map $g: B \to C$

Now $A \sim B$ and $B \sim C \Rightarrow gof: A \to C$ is bijective

$\Rightarrow A \sim C$

Hence the relation ~ of equipotence is equivalence relation.

Denumerable set: A set A is said to be *denumerable* if A is equipotent to the set N of all natural numbers. In other words, A is denumerable if there exists a bijective map from the set N of natural numbers onto the set A.

Countable set: A set A is said to be *countable* if it is either finite set or denumerable. A set which is not countable is said to be uncountable.

Illustrations:

1. The set N of all natural numbers is denumerable, the identity mapping is the desired one-to-one and onto mapping.
2. The set $A = \{2, 4, 6, 8, ...\}$ is also denumerable because $f(n) = 2n$ is a bijective map from the set N onto the set A.
3. The empty set is countable because it is finite.
4. The set I of all integers in a countable set. For, we can write I as $\{0, -1, 1, -2, 2, -3, 3, -4, 4, ...\}$. Now define a map $f: N \to I$ by setting:

$$f(n) = \begin{cases} \frac{1}{2}(n-1) & \text{if } n \text{ is odd,} \\ \frac{n}{2} & \text{if } n \text{ is even.} \end{cases}$$

Then f is bijective map. Thus I is countable.

5. The set $A = \{1/2, 2/3, 3/4, 4/5, ...\}$ is denumerable and therefore countable. A is denumerable because the mapping $f: N \to A$ given by $f(n) = n/(n+1)\ \forall\ n \in N$, is bijective.

Theorem 2. Every subset of a countable set is countable.

Proof: Let A be a countable set and let B be a subset of A. If B is finite then we have nothing to prove. We may, therefore, assume that B is infinite subset of A.

Since A is countable, let $A = \{a_1, a_2, a_3, a_4, ...\}$. Since B is a subset of A, each element of B is an a_i for some i. Let n_1 be the smallest positive integer such that $a_{n_1} \in$B. Consider now the set $A- \{a_{n_1}\}$. Let n_2 be the smallest positive integer such that a_{n_2} belongs to B as well as to A$- \{a_{n_1}\}$. Consider now the set A $- \{a_{n_1}, a_{n_2}\}$. Let n_3 be the smallest number for which a_{n_3} belongs to B as well as to $A-\{a_{n_1}, a_{n_2}\}$. Continuing in this manner, we find that $B = \{a_{n_1}, a_{n_2}, a_{n_3}, ...\}$. Then $k \to a_{n_k}$ an is a bijective function from N onto B. Thus B is countable.

Theorem 3. Any denumerable set can be put into one-to-one correspondence with a proper subset of itself.

Proof: Let A be a denumerable set. Then A can be written as

$$\{a_1, a_2, a_3, a_4, ...\}.$$

Let $B = A - \{a_1\}$. Then B is a proper subset of A. Now the mapping $f: A \to B$ defined by $f(a_i) = a_{i+1}$, $i = 1, 2, ...$ is a bijective mapping. Hence $A \sim B$.

Theorem 4. The union of countable family of countable sets in countable.

Proof: Let $\{A_1, A_2, ..., A_n, ...\}$ be a countable family of sets such that each A_n is countable. We can write.

$$A_1 = \{a_{11}, a_{12}, a_{13}, a_{14}, ...\}$$
$$A_2 = \{a_{21}, a_{22}, a_{23}, a_{24}, ...\}$$
$$\cdots\cdots\cdots\cdots\cdots\cdots$$
$$\cdots\cdots\cdots\cdots\cdots\cdots$$
$$A_n = \{a_{n1}, a_{n2}, a_{n3}, a_{n4}, ...\}$$
$$\cdots\cdots\cdots\cdots\cdots\cdots$$

It is evident that a_{ij} is the jth element of the ith set A_i. Let us define the height of the element a_{ij} to be $i+j$. With this definition, the height of each element is unique. Now we arrange the elements of sets A_1, A_2, A_3, ... according to their heights as $a_{11}, a_{21}, a_{12}, a_{31}, a_{22}, a_{13}, ...$ leaving out any element that has already occurred. In this way, all elements will be counted out and consequently $\bigcup_{n=1}^{\infty} A_n$ is countable. The following diagram demonstrates the above counting process:

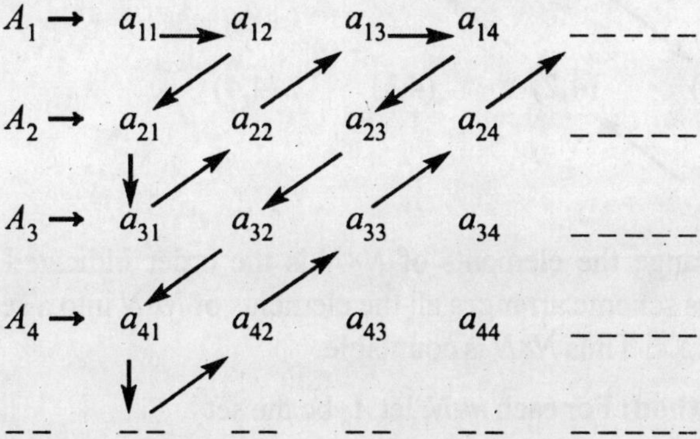

We now list the elements of $\cup A_n$ as follows:

a_{11}

a_{21}, a_{12}

a_{31}, a_{22}, a_{13}

$a_{41}, a_{32}, a_{23}, a_{14}$

....

....

$a_{n1}, a_{(n-1)2}, ...,a_{1n}$

....

....

In this listing, it is clear that a_{ij} is the jth element of $(i+j-1)^{th}$ row. Thus all the elements have been counted out. Thus $\cup A_n$ is countable.

Example 2.62. Show that the set $N \times N$ is countable.

Solution: We shall arrange the set $N \times N$ as shown below:

First method:

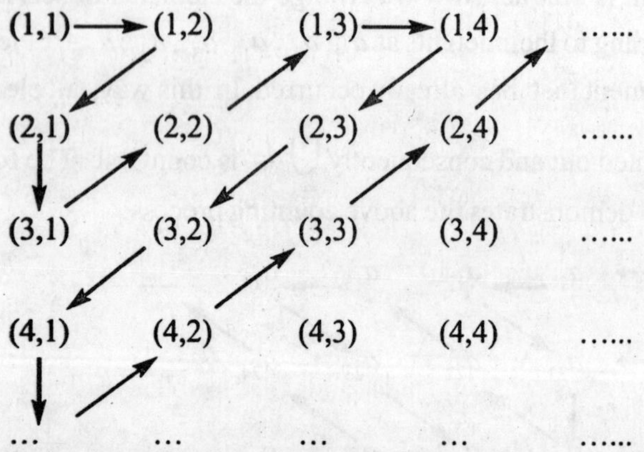

We can arrange the elements of $N \times N$ is the order indicated by the arrows. This scheme arranges all the elements of $N \times N$ into a sequence $< a_1, a_2, \ldots\ldots >$. Thus $N \times N$ is countable.

Second method: For each $n \in N$, let A_n be the set

$$A_n = \{(n, 1), (n, 2), (n, 3), (n, 4), \ldots\ldots\}$$

That is,

$$A_1 = \{(1, 1), (1, 2), (1, 3), (1, 4), \ldots\ldots\}$$

$$A_2 = \{(2, 1), (2, 2), (2, 3), (2, 4) \ldots\ldots\} \text{ and so on.}$$

The set A_n is countable of each $n \in N$. Since

$$N \times N = \bigcup_{n=1}^{\infty} A_n$$

therefore, $N \times N$ is countable by Theorem 4 above.

Third method: We list the elements of $N \times N$ as follows:

(1, 1)

(2, 1), (1, 2)

(3, 1), (2, 2), (1, 3)

(4, 1), (3, 2), (2, 3), (1, 4)

....

....

$(n, 1), (n-1, 2), \ldots\ldots\ldots\ldots (1, n)$

....

....

Define the mapping $f: N \times N \to N$ as follows

$$f(i, j) = \frac{1}{2}(i+j-1)(i+j-2)+j$$

That is, under f

$$(i, j) \to \frac{1}{2}(i+j-1)(i+j-2)+j$$

To show $N \times N$ is countable, it is enough to show that f is one-to-one.

We observe that in the above listing, the element (i, j) is the jth element of $(i+j-1)^{th}$ row.

Let (i, j) and (p, q) be two distinct elements of $N \times N$. Since $(i, j) \neq (p, q)$, the following two cases arise.

(i) $i+j \neq p+q$ or

(ii) $i+j = p+q$ but $j \neq q$

Let us first consider the case when $i+j \neq p+q$. Without loss of generality, we assume that $p+q > i+j$. We shall use the inequality

$$\frac{(n+1)(n+2)}{2} > \frac{n(n+1)}{2} + n$$

which is true for all positive integers.

Since $p+q > i+j$, we have

$$\frac{(p+q-1)(p+q-2)}{2} > \frac{(i+j-1)(i+j-2)}{2} + (i+j-2)$$

$$\Rightarrow \frac{(p+q-1)(p+q-2)}{2} + q > \frac{(i+j-1)(i+j-2)}{2} + j + (i+q-2)$$

But $i+q-2 \geq 0 \quad \because i, q \geq 1$

$$\Rightarrow \frac{(p+q-1)(p+q-2)}{2}+q > \frac{(i+j-1)(i+j-2)}{2}$$

Thus $f(p, q) \neq f(i, j)$.

Now consider the case (ii).

Since $j \neq q$, we may assume that $q > j$

Now, since $i+j = p+q$, we have

$$\frac{(p+q-1)(p+q-2)}{2} = \frac{(i+j-1)(i+j-2)}{2}$$

$$\Rightarrow \frac{(p+q-1)(p+q-2)}{2}+q > \frac{(i+j-1)(i+j-2)}{2}+j \qquad \because q \geq j$$

$$\Rightarrow f(p,q) \neq f(i,j)$$

Thus f is one-to-one. Hence $N\times N$ is countable.

Note: Although the mapping f defined above is onto also, but since we are considering f from $N\times N$ to N, we are not required to show that it is onto for showing $N\times N$ to be countable.

Example 2.63. If A and B are countable then show that $A\times B$ is countable.

Solution: Since A and B are countable, we can write

$$A = \{a_1, a_2, a_3, a_4, \ldots\} \text{ and}$$

$$B = \{b_1, b_2, b_3, b_4, \ldots\}$$

Let $$A_n = \{a_n\}\times B = \{(a_n, b) : b \in B\}$$

Then A_n is countable for each n because B is so. Now

$$A\times B = \bigcup_{n=1}^{\infty} A_n$$

Hence $A\times B$ is countable by theorem 4.

Example 2.64. Show that the set Q of all rational numbers is denumerable.

Solution: Let us define

$$A_n = \{0/n, -1/n, 1/n, -2/n, 2/n, -3/n, 3/n, ...\} \text{ for each } n \in N$$

We show each A_n is countable. Define the function $f: N \to A_n$ by setting

$$f(r) = \begin{cases} \dfrac{r-1}{n} & \text{if } r \text{ is odd} \\ -\dfrac{r}{2n} & \text{if } r \text{ is even} \end{cases}$$

Then f is one-to-one and onto. Thus each A_n is denumerable. Now $Q = \bigcup_{n=1}^{\infty} A_n$ is the countable union of denumerable sets. Therefore Q is denumerable by Theorem 4.

Example 2.65. Show that the set of all rational numbers in [0, 1] is countable.

Solution: The set of all rational numbers in [0, 1] is a subset of Q, the set of all rational numbers. Therefore by Theorem 2 and Example 2.64, the set of all rational numbers in [0, 1] is countable (denumerable).

Theorem 5. The set [0, 1] is uncountable.

Proof: We shall prove the result by contradiction. Suppose if possible, the interval [0, 1] is countable. Then there exists a bijective mapping f from N onto [0, 1]. This means that we can write the set [0, 1] as $\{f(1), f(2), f(3), ..., f(n), ...\}$. Thus we can write

$$\begin{aligned} f(1) &= 0 \cdot a_{11}\, a_{12}\, a_{13}\, a_{14} \ldots \\ f(2) &= 0 \cdot a_{21}\, a_{22}\, a_{23}\, a_{24} \ldots \\ f(3) &= 0 \cdot a_{31}\, a_{32}\, a_{33}\, a_{34} \ldots \\ & \ldots\ldots\ldots\ldots\ldots\ldots \\ & \ldots\ldots\ldots\ldots\ldots\ldots \\ f(n) &= 0 \cdot a_{n1}\, a_{n2}\, a_{n3}\, a_{n4} \ldots \\ & \ldots\ldots\ldots\ldots\ldots\ldots \end{aligned}$$

where $a_{ij} \in \{0, 1, 2, ..., 9\}$

We choose a positive integer b_n, for each $n \in N$, as follows:

$$b_n = 1 \text{ if } a_{nn} \neq 1 \text{ and}$$

$$b_n = 2 \text{ if } a_{nn} = 1$$

For example, if $a_{11} = 1$ then we take $b_1 = 2$ and if $a_{11} \neq 1$ we take $b_1 = 1$.

Let $y = 0 \cdot b_1 b_2 b_3 b_4 ...$

The decimal expansion of y is unique since no b_n is equal to 0 or 9. Moreover $y \in [0, 1]$ and y is not equal to $f(n)$ for any n. In fact, y differs from $f(1)$ in the first decimal place, it differs from $f(2)$ in the second decimal place and so on. But this contradicts the assumptions that every real number in [0, 1] is in the set $\{f(1), f(2), ...\}$. Hence [0, 1] is uncountable.

Corollary1. The set R of real numbers is uncountable.

Proof: If R were countable then [0, 1] would also be countable by Theorem 4. But [0, 1] is not countable shows R is not countable. Hence R is uncountable.

Corollary2. The set of irrational numbers is uncountable.

Proof: Let S be the set of irrational numbers. If S were countable then $S \cup Q$ would also be countable. But since $S \cup Q = R$ is not countable, we have, S is uncountable.

§ 2.17. Cardinality of Sets

Recall that two sets A and B are said to be equipotent (or equivalent) if there exists bijective mapping $f : A \to B$ and we write $A \sim B$ if A and B are equipotent. We have shown in §2.16 that the relation of equipotence $\sim$ is an equivalence relation on the family of sets. Hence relation $\sim$ decomposes the family of sets into equivalence classes. *Cardinal number* of a set A, denoted by $|A|$ or card (A), is defined in such a way that for any two sets A and B

$|A| = |B|$ if and only if there exists a bijective mapping $f : A \to B$,

and

$|A| \leqslant |B|$ if and only if there exists a one-to-one function f from A into B.

Definitions :

1. Two sets A and B are said to have the same cardinality if and only if A and B are equipotent. In other words, $|A| = |B|$ if and only if there exists a bijective mapping $f : A \to B$.
2. let $\mathfrak{I}$ be a family of sets and let ~ be the relation of equipotence on $\mathfrak{I}$. Then the equivalence classes of $\mathfrak{I}$ under the relation ~ are called *cardinal numbers*.

Thus, two sets A and B have same cardinal numbers if they belong to same equivalence class. In other words, $|A| = |B|$ if and only if there is a bijective mapping $f : A \to B$. We denote the cardinality of empty set by 0 and cardinality of any finite set A is denoted by a natural number n if the set A is equipotent to the set $\{1, 2, 3, ..., n\}$ having n elements. Thus cardinality of any finite set is the number of elements it contains. In other words, the cardinal numbers of a finite set is the number of element in the set.

The cardinals of a denumerable set is denoted by the symbol $\aleph_0$ called aleph-naught. Thus if N be the set of natural number then $|N| = \aleph_0$

The cardinality of the set of real numbers is denoted by c and is called power of continuum. Thus if R denotes the set of real number then $|R| = c$.

Since any two sets have the same cardinality if and only if they are equipotent, therefore any set which has one-to-one correspondence with the set N of natural numbers will have cardinality $\aleph_0$. Similarly any set, which has one-to-one correspondence with the set R of real numbers will have cardinality c.

Example 2.66. Find the cardinal number of each of the following sets:

(a) $A = \{a, b, c, d\}$

(b) $B = \{2, 4, 6, 8, 10, ...\}$

(c) $C = \{3, 4, 5, 6, ...\}$

(d) $D =]0, 1[$, the set of all real numbers between 0 and 1.

Solution:

(a) $|A| = 4$ because A has 4 elements.

(b) The cardinality of B is $\aleph_0$ because the function $f : N \rightarrow B$ defined by $f(x) = 2x$ is a bijective mapping.

(c) The cardinality of the set C is $\aleph_0$ because the function $f : N \rightarrow C$ defined by $f(x) = x+2$ is a bijective mapping.

(d) The cardinality of the set $D =]0, 1[$ is c, the power of continuum, because the function $f:]0,1[\rightarrow R$ defined by

$$f(x) = \begin{cases} \dfrac{1}{2x} - 1 & \text{if } 0 < x \leq \dfrac{1}{2} \\ \dfrac{1}{2(x-1)} + 1 & \text{if } \dfrac{1}{2} \leq x < 1 \end{cases}$$

is a bijective mapping.

We now introduce a partial order relation on the family of subsets of the universal set and a corresponding partial order relation on the set of cardinal numbers.

Definition: Let A and B be two sets. We say A precedes B, written as, $A \leqslant B$ if A is equipotent to a subset of B. That is,

$A \leqslant B$ if and only if there exists a bijective map from A onto a subset of B. We also say $|A|$ is less than or equal to $|B|$ if and only if A precedes B. Symbolically,

$$|A| \leqslant |B| \text{ if and only if } A \leqslant B$$

From the definition, it is clear that the relations defined above for two sets A and B and for two cardinal numbers are reflexive and transitive. We now state a theorem known as the *Schroder-Bernstein theorem* without proof which shows that the relations defined above are anti-symmetric and we can thus say that these relations are partial order relations.

Theorem1 (Schroder-Bernstein): Let A and B be two sets such that A precedes B and B precedes A. Then A is equipotent to B.

It can be seen easily that the relation $\leqslant$ defined on the set of cardinal numbers is a total order relation. Thus for any two sets A and B, we have either $|A|\leqslant|B|$ or $|B|\leqslant|A|$. We also define a relation < given by

$|A|<|B|$ if and only if $|A|\leqslant|B|$ and $|A|\neq|B|$

That is, $|A|<|B|$ if and only if $|A|\leqslant|B|$ and A is not equipotent to B.

Example 2.67. Show that $\aleph_0 < c$.

Solution: We know $|N| = \aleph_0$ and $|R| = c$.

To show that $|N|<|R|$, we show $|N|\leqslant|R|$ and N is not equipotent to R.

Since N is a subset of R and identity mapping $f: N\rightarrow N$ is a bijective mapping from N to a subset N of R, we have, $|N|\leqslant|R|$.

Since N is countable and R is uncountable, we can not find any bijective mapping from N to R. Hence N and R are not equipotent sets. Thus $|N|\neq|R|$.

Therefore $|N|<|R|$ or $\aleph_0 < c$.

Theorem2: No set A can be equipotent to its power set $P(A)$. In particular, $|A|<|P(A)|$.

Proof : Define a mapping $f: A\rightarrow P(A)$ by setting

$$f(a) = \{a\}\ \forall\, a\in A.$$

Clearly, f is one-to-one. Thus $A\leqslant P(A)$.

We shall show A is not equipotent to $P(A)$. If possible, assume that there exists a mapping $g : A\rightarrow P(A)$ which is one-to-one and onto. Let us define

$$B = \{x : x\in A \text{ and } x\notin g(x)\}$$

Clearly, B is a subset of A and hence $B\in P(A)$. Now since g is onto, there must be an element $b\in A$ such that $g(b) = B$. Now two cases arise:

(i) $b\in B$

(ii) $b\notin B$

Case (i) If $b\in B$, then by definition of B, $b\notin g(b) = B$, a contradiction.

Case (ii) If $b \notin B$, then again by definition of *B*, $b \in g(b) = B$, which is again a contradiction.

Hence $g : A \rightarrow P(A)$ is not one-to-one and onto. Therefore, *A* and *P*(*A*) are not equipotent and so $|A| < |P(A)|$.

Exercise

1. If $A = \{x : x \in R \text{ and } x > 1\}$, prove that $|A| = c$.
2. If *A* and *B* are disjoint finite sets there show that $|A \cup B| = |A| \cup |B|$. What can you say if *A* and *B* are not disjoint?
3. Find the cardinality of the set $A = \{n^7 : n \in N\}$.
4. If *A* is a finite set then show that $|P(A)| = 2^{|A|}$.
5. If *A* is uncountable and *B* is countable sets then show that *A*–*B* is uncountable.
6. Find the cardinal number of the set *A* of all relations on the set $\{1, 2, 3, 4\}$.
7. Find the cardinal number of the set *A* of all function from {a, *b*, *c*, *d*} into $\{1, 2, 3, 4, 5\}$.
8. Show that cardinality of the set [0, 1] is *c*.

Answers

3. $\aleph_0$
6. 2^{16}
7. $5^4 = 625$

CHAPTER THREE

Lattices

In this chapter we shall study lattices. A lattice will be first introduced as a partially ordered set satisfying certain properties. Partially ordered sets, their properties and associated terminology given in section 15 of Chapter 2 will be used throughout in our discussion here. In particular, the concept of the supremum (lub) and the infimum (glb) of a subset of a poset will be used repeatedly. We shall also define a lattice as an algebraic system and the prove the equivalence of the two definitions.

§ 3.1. Lattices as Partially Ordered Sets

Definition: A **lattice** is a poset $(L, \leqslant)$ in which every subset $\{a, b\}$ of two elements of L has a greatest lower bound (inf) and a least upper bound (sup). In other words, poset $(L, \leqslant)$ is a lattice if for every $a,b \in L$, sup $\{a,b\}$ and inf$\{a,b\}$ exist in L.

We denote sup$\{a, b\}$ by $a \vee b$ and call it the **join** of a and b and inf$\{a, b\}$ by $a \wedge b$ and call it the **meet** of a and b. Other notations like $a + b$ and $a \cdot b$ or $a \cup b$ and $a \cap b$ are also used for sup$\{a, b\}$ and inf$\{a, b\}$ respectively.

Example 3.1. Let S be any non-empty set and $P(S)$ be its power set. The partially ordered set $(P(S), \subseteq)$ is a lattice.

Solution: We know from chapter 2 that $(P(S), \subseteq)$ is a poset. Also for any two sets $A, B \in P(S)$, we have,

$$A \vee B = \sup\{A, B\} = A \cup B$$

and $$A \wedge B = \inf\{A, B\} = A \cap B$$

Since both $A \cup B$ and $A \cap B$ also belong to $P(S)$, therefore $(P(S), \subseteq)$ is a lattice.

We illustrate this example by taking $S = \{1, 2, 3\}$.

Then

$$P(S) = \{\phi, \{1\}, \{2\}, \{3\}, \{1, 2\}, \{1, 3\}, \{2, 3\}, \{1, 2, 3\}\}$$

The lattice is represented by the diagram given below:

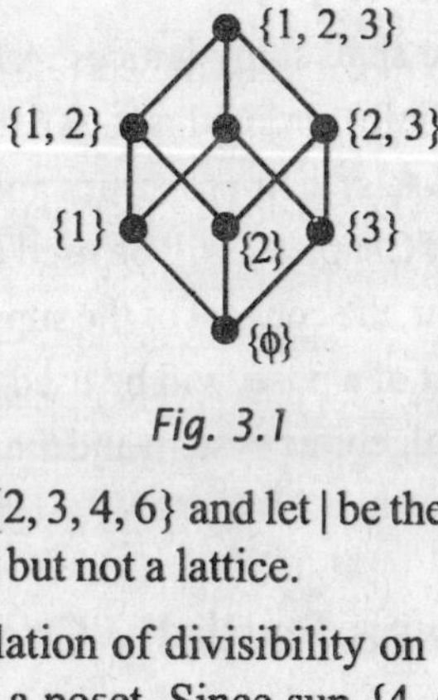

Fig. 3.1

Example 3.2. Let $A = \{2, 3, 4, 6\}$ and let | be the divisibility relation on A. Then $(A, |)$ is a poset but not a lattice.

Solution: Since the relation of divisibility on $\{2, 3, 4, 6\}$ is a partial order relation, $(A, |)$ is a poset. Since sup $\{4, 6\}$ does not exist in A, $(A, |)$ is not a lattice.

Example 3.3. The set L of all factors of 12 under divisibility forms a lattice.

Solution: Here $L = \{1, 2, 3, 4, 6, 12\}$. Let a,b be any two elements in L, then sup $\{a, b\}$ under divisibility relation is least positive integer such that $a \mid c$ and $b \mid c$. Thus $\sup\{a, b\}$ = lcm of and b. Similarly, inf $\{a, b\}$ is the greatest positive integer d such that $d \mid a$ and $d \mid b$. Hence inf $\{a, b\}$ = gcd of a and b. This lattice is represented by the diagram as shown in figure 3.2.

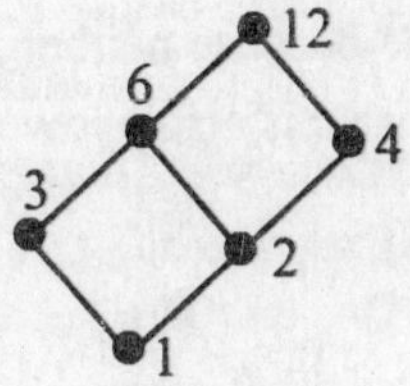

Fig. 3.2

Example 3.4. Let $(N, \leqslant)$ be a partial ordered set, where N is the set of natural numbers and $\leqslant$ is the "less than or equal to" relationship among natural numbers. Then $(N, \leqslant)$ is a lattice.

Solution: Let a and b be any two natural numbers. Then $\sup\{a, b\}$ is the least natural numbers c such that $a \leq c$ and $b \leq c$. Clearly $c = \max\{a, b\}$. Thus $\sup\{a, b\} = \max\{a, b\}$. Similarly $\inf\{a, b\} = \min\{a, b\}$. Since $\max\{a, b\}$ and $\min\{a, b\}$ belong to N, $(N, \leqslant)$ is a lattice.

Example 3.5. The set N of natural numbers under divisibility forms a lattice in which $a \vee b$ = lcm of a and b and $a \wedge b$ = gcd of a and b.

Example 3.6. Let L be the set of all subgroups of a group G. Then L is a lattice under set inclusion relation $\subseteq$, where for any $H, K \in L$.

$$H \wedge K = H \cap K, \quad \text{and}$$

$$H \vee K = \langle H \cup K \rangle$$

where $\langle H \cup K \rangle$ means the smallest subgroup of G containing both H and G.

Example 3.7. Every chain is a lattice.

Solution: Let $(L, \leqslant)$ be a chain. Let $a, b \in L$. Then since L is a chain, we have, $a \leqslant b$ or $b \leqslant a$. Without loss of generality, we assume $a \leqslant b$. Then $a \vee b = b$ and $a \wedge b = a$. Thus both $a \vee b$ and $a \wedge b$ exist in L. Hence every chain is a lattice.

Example 3.8. Explain why the partial ordered sets given in the figure 3.3 are not lattices.

Solution: The poset given in figure (a) is not a lattice because $d \vee e$ does not exist. Moreover, $b \vee c$ does not exist. The poset given in figure (b) is not a lattice because neither $e \wedge d$ nor $b \vee c$ exist. Note that $e \wedge d$ is not b because c is also a lower bound of $\{e, d\}$ and $c \leqslant b$ is not true.

Fig. 3.3

Let $(L, \leqslant)$ be a poset and let $(L, \geq)$ be the dual poset. If $(L, \leqslant)$ is a lattice then $(L, \geq)$ is also a lattice. For if $a, b \in L$ then least upper bound of a and b in $(L, \leqslant)$ exist. Then the greatest lower bound of a and b in $(L, \geq)$ is equal to the least upper bound of a and b in $(L, \leqslant)$. Hence the greatest lower bound of a and b exists in $(L, \geq)$. Similarly the greatest lower bound of a and b in $(L, \leqslant)$ is equal to the least upper bound of a and b in $(L, \geq)$.

Principle of duality: Any statement about lattices involving the operations $\vee$ and $\wedge$ and relations $\leqslant$ and $\geq$ remains true if $\wedge$ is replaced by $\vee$ and $\vee$ by $\wedge$, $\leqslant$ by $\geq$ and $\geq$ by $\leqslant$.

The lattices $(L, \leqslant)$ and $(L, \geq)$ are called duals of each other.

§ 3.2. Some Properties of Lattices

Theorem 1: If $(L, \leqslant)$ is a lattice then for any $a, b, c \in L$, the following results hold:

1. $a \wedge a = a$ and $a \vee a = a$ (Indempotent)
2. $a \wedge b = b \wedge a$ and $a \vee b = b \vee a$ (Commutative)
3. $(a \wedge b) \wedge c = a \wedge (b \wedge c)$ and $(a \vee b) \vee c = a \vee (b \vee c)$ Associative)
4. $a \wedge (a \vee b) = a$ and $a \vee (a \wedge b) = a$ (Absorption)

Proof: We prove the results for the meet operation. The results for join operations are left for the readers to prove similarly.

1. $a \wedge a = \inf\{a, a\} = a$. Thus $a \wedge a = a$

2. $a \wedge b = \inf\{a, b\} = \inf\{b, a\} = b \wedge a$

3. Let $b \wedge c = d$ then $d = \inf\{b, c\}$

$\Rightarrow d \leqslant b, d \leqslant c$

Now, let $a \wedge d = e$ then $e = \inf\{a, d\}$

$\Rightarrow e \leqslant a$ and $e \leqslant d$

$\Rightarrow e \leqslant a$ and $e \leqslant b$ and $d \leqslant c$

[$\because \leqslant$ is transitive and $d \leqslant b, d \leqslant c$]

$\Rightarrow e$ is a lower bound of $\{a, b, c\}$

If f is any lower bound of $\{a, b, c\}$ then

$f \leqslant a, f \leqslant b, f \leqslant c$

Now,

$f \leqslant b, f \leqslant c$ and $d = \inf\{b, c\}$ give $f \leqslant d$

$f \leqslant a$ and $e = \inf\{a, d\}$ give $f \leqslant e$

Hence $e = \inf\{a, b, c\}$

Similarly, we can show

$(a \wedge b) \wedge c = \inf\{a, b, c\}$

Thus

$(a \wedge b) \wedge c = a \wedge (b \wedge c)$

4. To Prove $a \wedge (a \vee b) = a$

We know, $a \leqslant a$ and $a \leqslant a \vee b$

$\Rightarrow a \leqslant a \wedge (a \vee b)$ ---- (i)

Again

$a \wedge (a \vee b) = \inf\{a, a \vee b) \leqslant a$ ---- (ii)

From (i) and (ii), we have,

$a \wedge (a \vee b) = a$

Theorem 2: Let $(L, \leqslant)$ be a lattice in which $\wedge$ and $\vee$ denote the operations of meet and joint. Then for any $a, b \in L$, show that

(a) $a \leqslant b \Leftrightarrow a \wedge b = a$

(b) $a \leqslant b \Leftrightarrow a \vee b = b$

Proof: (a) Suppose that $a \wedge b = a$.

Since $a \wedge b = \inf\{a, b\}$, therefore $a \wedge b \leqslant b$

$\Rightarrow a \leqslant b$ $[\because a \wedge b = a]$

Conversely, suppose that $a \leqslant b$. Since $\leqslant$ is reflexive, we have,

$a \leqslant a$.

Now,

$a \leqslant b$ and $a \leqslant a \Rightarrow a$ is a lower bound of $\{a, b\}$

$\Rightarrow a \leqslant \inf\{a, b\} = a \wedge b$

Since $a \wedge b$ is infimum of $\{a,b\}$, $a \wedge b \leqslant a$

Hence $a = a \wedge b$

(b) Suppose that $a \vee b = b$

Since

$a \leqslant a \vee b$ and $a \vee b = b$

We have

$a \leqslant b$

Conversely, suppose that $a \leqslant b$. We know by reflexivity of $\leqslant$ that $b \leqslant b$. Therefore, $\sup\{a, b\} \leqslant b$. Thus $a \vee b \leqslant b$. But from the definition of $a \vee b = \sup\{a, b\}$, we have $b \leqslant a \vee b$. Now $a \vee b \leqslant b$ and $b \leqslant a \vee b$ give $a \vee b = b$.

Corollary: Let L be a lattice and $a, b \in L$. Then

$a \vee b = a$ if and only if $a \vee b = b$

Proof: The proof follows from parts (a) and (b) of the above result.

Theorem3: Let (L, ≤) be a lattice and $a, b, c \in$ L. Then the following implications hold:

(a) $a \leqslant b$ and $a \leqslant c \Rightarrow a \leqslant b \vee c$

(b) $a \leqslant b$ and $a \leqslant c \Rightarrow a \leqslant b \wedge c$

Proof: (a) We know from the definition of join operation that $b \vee c$ is the supremum of b and c. Hence $b \leqslant b \vee c$.

Thus $a \leqslant b$ and $b \leqslant b \vee c \Rightarrow a \leqslant b \vee c$ by transitivity.

(b) Suppose $a \leqslant b$ and $a \leqslant c$.

$\Rightarrow a$ is a lower bound of $\{b, c\}$

$\Rightarrow a \leqslant b \wedge c$, the greatest lower bound of $\{b, c\}$.

Hence proved.

Corollary: Let $(L, \leqslant)$ be a lattice and $(L, \geq)$ be its dual. Then for $a, b, c \in L$

(a) $a \geq b$ and $a \geq c \Rightarrow a \geq b \wedge c$

(b) $a \geq b$ and $a \geq c \Rightarrow a \geq b \vee c$

Proof: Applying principle of duality on Theorem 3, we get the results.

Theorem 4: For any a, b, c, d in a lattice $(L, \leqslant)$, the following properties hold:

(a) $a \leqslant b$ and $c \leqslant d \Rightarrow a \vee c \leqslant b \vee \mathrm{d}$

(b) $a \leqslant b$ and $c \leqslant d \Rightarrow a \wedge c \leqslant b \wedge \mathrm{d}$

Proof: (a) Suppose that $a \leqslant b$ and $c \leqslant \mathrm{d}$.

By definition of join operation $\vee$, we have

$b \leqslant b \vee \mathrm{d}$ and $\mathrm{d} \leqslant b \vee \mathrm{d}$

Now $a \leqslant b$ and $b \leqslant b \vee \mathrm{d} \Rightarrow a \leqslant b \vee \mathrm{d}$ by transitivity.

Similarly $c \leqslant d$ and $d \leqslant b \vee d \Rightarrow c \leqslant b \vee d$. Thus $b \leqslant d$ is an upper bound of a and c. Since $a \vee c$ is the least upper bound of a and c, we have, $a \vee c \leqslant b \vee \mathrm{d}$.

(b) Suppose that $a \leqslant b$ and $c \leqslant \mathrm{d}$.

By definition of meet operation $\wedge$, we have

$a \wedge c \leqslant a$ and $a \wedge c \leqslant c$

By transitivity,

$a \wedge c \leqslant a$ and $a \leqslant b \Rightarrow a \wedge c \leqslant b$ and

$a \wedge c \leqslant c$ and $c \leqslant d \Rightarrow a \wedge c \leqslant d$

Thus $a \wedge c$ is a lower bound of b and d. Hence $a \wedge c \leqslant b \wedge \mathrm{d}$, since $b \wedge \mathrm{d}$ is the greatest lower bound of b and d.

Corollary: For any a, b, c in a lattice $(L, \leqslant)$, the following properties hold.

(a) $a \leqslant b \Rightarrow a \vee c \leqslant b \vee c$

(b) $a \leqslant b \Rightarrow a \wedge c \leqslant b \wedge c$

Proof: The proofs of (a) and (b) follow by taking $d = c$ in the above results. For example, to prove (a), we have

$a \leqslant b$ and $c \leqslant c \Rightarrow a \vee c \leqslant b \vee c$ from above theorem.

Similarly, to prove (b)

$a \leqslant b$ and $c \leqslant c \Rightarrow a \wedge c \leqslant b \wedge c$.

Theorem 5: Let $(L, \leqslant)$ be a lattice. For any $a, b, c \in L$, the following inequalities hold:

(a) $a \wedge (b \vee c) \geq (a \wedge b) \vee (a \wedge c)$

(b) $a \vee (b \wedge c) \leqslant (a \vee b) \wedge (a \vee c)$

Proof: (a) we know that

$$a \wedge b \leqslant a$$

and $$a \wedge b \leqslant b \leqslant b \vee c$$

$\Rightarrow a \wedge b$ is a lower bound of $\{a, b \vee c\}$

$$\Rightarrow a \wedge b \leqslant a \wedge (b \vee c) \qquad \text{---- (1)}$$

because $a \wedge (b \vee c)$ is the greatest lower bound of $\{a, b \vee c\}$.

Again,

$$a \wedge c \leqslant a$$

and $$a \wedge c \leqslant c \leqslant b \vee c$$

$$\Rightarrow a \wedge c \leqslant a \wedge (b \vee c) \qquad \text{---- (2)}$$

From (1) and (2), $a \wedge (b \vee c)$ is an upper bound of $\{a \wedge b, a \wedge c\}$.

$$\Rightarrow (a \wedge b) \vee (a \wedge c) \leqslant a \wedge (b \vee c).$$

(b) This inequality can be proved in a similar manner or using the principle of duality. If we apply the principle of duality, we immediately get

$$a \vee (b \wedge c) \leqslant (a \vee b) \wedge (a \vee c)$$

by interchanging $\vee$ and $\wedge$ and replacing $\leqslant$ by $\geq$ in (a)

Note: The inequalities in Theorem 5 above are known as distributive inequalities. They are also called semi distributive laws.

Theorem 6: Let $(L, \leqslant)$ be a lattice. For any $a, b, c \in L$, the following holds.

$$a \leqslant c \Leftrightarrow a \vee (b \wedge c) \leqslant (a \vee b) \wedge c.$$

This is known as the *modular* inequality.

Proof: We know from Theorem 2(b) that

$$a \leqslant c \Leftrightarrow a \vee c = c$$

Substituting c for $a \vee c$ in the *R.H.S* of the inequality given in part (b) of Theorem 5, we get

$$a \leqslant c \Leftrightarrow a \vee (b \wedge c) \leqslant (a \vee b) \wedge c$$

Hence proved.

Solved Examples

Example 3.9. Show that the idempotent laws follow from the absorption laws.

Solution: By absorption laws (given in Theorem 1), we have,

$$a \wedge (a \vee b) = a \quad \text{---- (1)}$$

and $a \vee (a \wedge b) = a$ ---- (2)

Substituting $a \wedge b$ for b in (1), we get

$$a \wedge (a \vee (a \wedge b)) = a$$

Now using (2), we get

$$a \wedge a = a \qquad [\because a \vee (a \wedge b) = a]$$

Similarly, we can show $a \vee a = a$

Thus idempotent laws follow from the absorption laws.

Example 3.10. Show that the inequality

$$a \leqslant c \Leftrightarrow a \vee (b \wedge c) \leqslant (a \vee b) \wedge c$$

in a lattice is self dual.

Solution: The dual of the inequality

$$a \leqslant c \Leftrightarrow a \vee (b \wedge c) \leqslant (a \vee b) \wedge c \qquad \text{---- (1)}$$

is obtained by interchanging $\wedge$ and $\vee$ and replacing $\leqslant$ by $\geq$. Hence the dual of given in equality is

$$a \geq c \Leftrightarrow a \wedge (b \vee c) \geq (a \wedge b) \vee c$$

which can be written as

$$c \leqslant a \Leftrightarrow (a \wedge b) \vee c \leqslant a \wedge (b \vee c)$$

Replacing a by c and c by a, we get

$$a \leqslant c \Leftrightarrow (c \wedge b) \vee a \leqslant c \wedge (b \vee a)$$

$$\Rightarrow \qquad a \leqslant c \Leftrightarrow a \vee (b \wedge c) \leqslant (a \vee b) \wedge c$$

which is same as (1). Hence the inequality is self dual.

Example 3.11. Let $(L, \leqslant)$ be a lattice and $a, b, c, d \in L$. Then show that

(a) $(a \wedge b) \vee (c \wedge d) \leqslant (a \vee c) \wedge (b \vee d)$

(b) $(a \wedge b) \vee (b \wedge c) \vee (c \wedge a) \leqslant (a \vee b) \wedge (b \vee c) \wedge (c \vee a)$.

Solution: (*a*) We know that

$$a \wedge b \leqslant a \text{ and } c \wedge d \leqslant c$$

$$\Rightarrow (a \wedge b) \vee (c \wedge d) \leqslant a \vee c \qquad \text{. . . (1) [by Theorem 4]}$$

Again,

$$a \wedge b \leqslant b \text{ and } c \wedge d \leqslant d$$

$$\Rightarrow (a \wedge b) \vee (c \wedge d) \leqslant b \vee d \qquad \text{. . . (2) [by Theorem 4]}$$

From (1)and (2), we get

$$(a \wedge b) \vee (c \wedge d) \leqslant (a \vee c) \wedge (b \vee d)$$

by applying Theorem 3(b). Hence proved.

(b) We know that

$$a \wedge b \leqslant a \vee b$$

$$a \wedge b \leqslant b \leqslant b \vee c \text{ and}$$

$$a \wedge b \leqslant a \leqslant c \vee a$$

$$\Rightarrow \qquad a \wedge b \leqslant (a \vee b) \wedge (b \vee c) \wedge (c \vee a) \qquad \text{---- (1)}$$

Similarly,

$$b \wedge c \leqslant (a \vee b) \wedge (b \vee c) \wedge (c \vee a) \qquad \text{---- (2)}$$

and $$(c \wedge a) \leqslant (a \vee b) \wedge (b \vee c) \wedge (c \vee a) \qquad \text{---- (3)}$$

From (1), (2), and (3), we have

$$(a \wedge b) \vee (b \wedge c) \vee (c \wedge a) \leqslant (a \vee b) \wedge (b \vee c) \wedge (c \vee a).$$

Hence proved.

Example 3.12. Dual of a lattice is a lattice.

Solution: Let (L, R) be a lattice and let $(L, \bar{R})$ be its dual, where $\bar{R}$ is defined as $x \bar{R} y$ if and only if $y R x$. We have already proved that $(L, \bar{R})$ is a poset.

Let $x, y \in L$. Then since (L, R) is a lattice, $\sup\{x,y\}$ exists in (L,R).

Let $x \vee y = \sup\{x, y\}$ in (L, R).

Then $xR(x \vee y)$ and $yR(x \vee y)$

$\Rightarrow (x \vee y) \bar{R} x$ and $(x \vee y) \bar{R} y$

$\Rightarrow x \vee y$ is a lower bound of $\{x, y\}$ in $(L, \bar{R})$

We shall show $x \vee y$ is greatest lower bound of $\{x,y\}$ in $\{L, \bar{R}\}$

Let z be any lower bound of $\{x, y\}$ in $(L, \bar{R})$.

Then $z \bar{R} x$ and $z \bar{R} y$

$\Rightarrow x R z$ and $z R y$

$\Rightarrow z$ is an upper bound of $\{x, y\}$ in (L, R)

$\Rightarrow (x \vee y) R z$ because $x \vee y = \sup\{x, y\}$ in (L,R)

$\Rightarrow z \bar{R} (x \vee y)$

$\Rightarrow x \vee y$ is greatest lower bound of $\{x,y\}$ in $(L, \bar{R})$.

Similarly, we can show $x \wedge y$ is least upper bound of $\{x, y\}$ in $(L, \bar{R})$.

Hence $(L, \bar{R})$ is a lattice.

Example 3.13. For any positive integers a and b, show that

$$\max\{a, \min\{a, b\}\} = a$$

$$\min\{a, \max\{a, b\}\} = a$$

Solution: Consider the lattice $(N, \leqslant)$ where N is the set of natural numbers and $\leqslant$ is the "less than or equal to" relation on N. In this lattice, join $\vee$ and meet $\wedge$ are given by

$$a \vee b = \max\{a, b\}$$

and $$a \wedge b = \min\{a, b\}$$

Now, applying absorption properties

$$a \wedge (a \vee b) = a$$

$$a \vee (a \wedge b) = a$$

in this lattice, we get

$$\min\{a, \max\{a, b\}\} = a \quad \text{and}$$

$$\max\{a, \min\{a, b\}\} = a$$

for any positive integers a and b.

Example 3.14. For any positive integers a and b, show that

$$\text{lcm}(a, \text{gcd}(a,b)) = a \quad \text{and}$$

$$\text{gcd}(a, \text{lcm}(a,b)) = a$$

Solution: We know that $(N, \leqslant)$ is a lattice where N is the set of natural numbers and $\leqslant$ is relation defined as $a \leqslant b$ if and only if $a|b$.

In this lattice, *join* $\vee$ and *meet* $\wedge$ are given by

$$a \vee b = \text{lcm}(a, b)$$

and $$a \wedge b = \text{gcd}(a, b)$$

Applying absorption laws

$$a \wedge (a \vee b) = a$$

and $$a \vee (a \wedge b) = a$$

we get

$$\text{gcd}(a, \text{lcm}(a,b)) = a \quad \text{and}$$

$$\text{lcm}(a, \text{gcd}(a,b)) = a$$

Exercise

1. Show that in a poset greatest elements, if exist, is unique.

2. Show that in a lattice if $a \leqslant b$ and $c \leqslant d$ then $a \vee c \leqslant b \vee d$.

3. Show that in a lattice if $a \leqslant b \leqslant c$ then,

$$a \vee b = b \wedge c \text{ and}$$
$$(a \wedge b) \vee (b \wedge c) = b = (a \vee b) \wedge (a \vee c)$$

4. In a lattice, show that

$$(a \wedge b) \vee (a \wedge c) \leqslant a \wedge [b \vee (a \wedge c)] \text{ and}$$
$$(a \vee b) \wedge (a \vee c) \geq a \vee [b \wedge (a \vee c)]$$

5. Prove that a poset $(L, \leqslant)$ is a lattice if and only if every non-empty finite subset of L has supremum and infimum.

6. Every finite lattice has least and greatest elements.

7. Show that gcd(lcm(a, b), lcm(c, d)) is divisible by lcm(gcd(a, c), gcd(b, c)),where a, b, c, d are natural numbers.

   (Hint: Apply the result of solved example 3.11 on the lattice $(N, |)$)

8. Consider the poset $\{a,b,c,d,e,f\}$, where $d<f$, $e<f$, $b<d$, $b<e$, $c<e$, $a<b$, $a<c$. Determine whether this is a lattice.

§ 3.3. Lattice as Algebraic Systems (Another Definition of Lattice)

Here, we shall define a lattice as an algebra with two binary compositions *meet* $\wedge$ and *join* $\vee$.

Definition: A non-empty set L together with two binary compositions (operations) $\wedge$ and $\vee$ is said to be a lattice if the following conditions are satisfied.

1. **Commutative properties:** For any a, $b \in L$,

$$a \wedge b = b \wedge a \text{ and } a \vee b = b \vee a$$

2. **Associative properties:** For any $a, b, c \in L$,

$$a \wedge (b \wedge c) = (a \wedge b) \wedge c \text{ and } a \vee (b \vee c) = (a \vee b) \vee c$$

3. **Absorption properties:** For any $a, b \in L$

$$a \wedge (a \vee b) = a \text{ and } a \vee (a \wedge b) = a$$

The lattice L together with operations $\vee$, $\wedge$ is denoted as $(L, \wedge, \vee)$.

We have already seen in solved example 3.9 of the last section that idempotent laws follow from the absorption laws. Hence the lattice L as defined above satisfies idempotent properties. That is,

$$a \wedge a = a \text{ and } a \vee a = a \; \forall a \in L.$$

Theorem1: Establish the equivalence of the two definitions of a lattice.

Proof: Definition I (as posets) $\Rightarrow$ definition II(as algebras).
According to definition I, if L is a lattice then for any $b, b \in L$ both inf $\{a, b\} = a \wedge b$ and $\sup\{a, b\} = a \vee b$ exist and are unique in L. Thus $\wedge$ and $\vee$ are binary compositions on L. That the conditions (1), (2) and (3) hold for $(L, \wedge, \vee)$ follow from Theorem 1 of §2. Hence definition of a lattice as poset implies the definition of a lattice as algebra.

Definition II (as algebras) $\Rightarrow$ Definition I(as posets)
Suppose that L is a lattice according to definition II. We first show that L forms a poset. For this, we define a relation $\leqslant$ on L such that for $a, b \in L$.

$$a \leqslant b \Leftrightarrow a \wedge b = a$$

We shall show that the relation $\leqslant$ as defined above is a partial order relation.

1. $\leqslant$ is **reflexive:** For any $a \in L$ we have,

$$a \wedge a = a$$
$$\Rightarrow a \leqslant a$$

Hence $\leqslant$ is reflexive.

2. $\leqslant$ is **antisymmetric:** For any $a, b \in L$, suppose that

$$a \leqslant b \text{ and } b \leqslant a$$

$\Rightarrow \quad a \wedge b = a$ and $b \wedge a = b$

But, $\quad a \wedge b = b \wedge a \Rightarrow a = b$

Hence $\leqslant$ is antisymmetric.

3. $\leqslant$ is **transitive:** Suppose that $a \leqslant b$ and $b \leqslant c$

$$\Rightarrow \quad a \wedge b = a \text{ and } b \wedge c = b$$

$$\begin{aligned} \Rightarrow \quad a \wedge c &= (a \wedge b) \wedge c \\ &= a \wedge (b \wedge c) && \text{[by associativity]} \\ &= a \wedge b && [\text{since } b \wedge c = b] \\ &= a && [\text{since } a \wedge b = a] \end{aligned}$$

$$\Rightarrow \quad a \leqslant c$$

Hence $\leqslant$ is transitive. Thus $\leqslant$ is a partial order relation on L.

We now show $\inf\{a,b\}$ exists in L by showing

$$\inf\{a, b\} = a \wedge b.$$

We have

$$\begin{aligned} (a \wedge b) \wedge a &= a \wedge (b \wedge a) && \text{(by associative property)} \\ &= a \wedge (a \wedge b) && \text{(by commutative property)} \\ &= (a \wedge a) \wedge b && \text{(by associative property)} \\ &= a \wedge b && \text{(by idempotent property)} \end{aligned}$$

$$\Rightarrow \quad a \wedge b \leqslant a \qquad \text{(by definition of } \leqslant)$$

Similarly,

$$a \wedge b \leqslant b$$

$\Rightarrow a \wedge b$ is a lower bound of $\{a, b\}$

We shall show $a \wedge b$ is greatest lower bound of $\{a, b\}$. Suppose c be any lower bound of $\{a, b\}$. Then $c \leqslant a$ and $c \leqslant b$.

$$\Rightarrow \quad c \wedge a = c \text{ and } c \wedge b = c$$

Thus

$$\begin{aligned} c \wedge (a \wedge b) &= (c \wedge a) \wedge b \\ &= c \wedge b \\ &= c \end{aligned}$$

$\Rightarrow \quad c \leqslant a \wedge b$ by definition of $\leqslant$.

Thus $a \wedge b$ in glb$\{a, b\}$. This prove that inf$\{a, b\} = a \wedge b$.

We now show that

$$a \leqslant b \Leftrightarrow a \wedge b = a \Leftrightarrow a \vee b = b$$

Now

$$a \wedge b = a \;\Rightarrow\; (a \wedge b) \vee b = a \vee b$$

$$\Rightarrow\; b = a \vee b \qquad \text{by absorption property}$$

Again

$$a \vee b = b \;\Rightarrow\; a \wedge (a \vee b) = a \wedge b$$

$$\Rightarrow\; a = a \wedge b \qquad \text{by absorption property}$$

Hence

$$a \wedge b = a \Leftrightarrow a \vee b = b$$

Thus

$$a \leqslant b \Leftrightarrow a \wedge b = a \Leftrightarrow a \vee b = b$$

Therefore, by duality we can say that sup$\{a, b\}$ will be $a \vee b$. Hence definition I holds.

This establishes the equivalence of the two definitions.

Remarks: After proving $a \wedge b =$ inf$\{a, b\}$ in the above theorem we could not say that similarly $a \vee b =$ sup$\{a, b\}$ as our definition of $\leqslant$ depended only on $\wedge$ operation. Only after proving the equivalence of $a \wedge b = a$ and $a \vee b = b$ we could claim that sup$\{a, b\} = a \vee b$.

§ 3.4. Sublattices

Definition: let $(L, \wedge, \vee)$ be a lattice. A non-empty subset $S \subseteq L$ of L is called **sublattice** of L if S is closed under both operations $\wedge$ and $\vee$. In other words, S is called sublattice of L if $a, b \in S \Rightarrow a \wedge b$ and $a \vee b$ both are in S.

From the definition, it is clear that a sublattice itself is a lattice. Let L be a lattice and let $a, b \in L$ such that $a \leqslant b$. The **interval** $[a, b]$ in lattice L is defined as the set of all $x \in L$ such that $a \leqslant x \leqslant b$.

Example 3.15. Let $S = \{a, b, c\}$ and $P(S)$ be its power set. Then $(P(S), \subseteq)$ is a lattice in which the meet and join operations are the usual operation of intersection and union respectively. The subset $R = \{\phi, \{a\}, \{b\}, \{a, b\}\}$ of $P(S)$ is a sublattice of $(P(S), \subseteq)$, while the subset $U = \{\phi, \{a\}, \{b\}\}$ of $P(S)$ is not a sublattice because $\{a\} \cup \{b\}$ does not belong to U.

Example 3.16. Consider the lattice $(N, |)$ under the relation of divisibility, where N is the set of natural numbers. Let n be any positive integer and let S_n be the set of all divisors of n. Then S_n is a sublattice of $(N, |)$. For example, $n = 6$, $S_6 = \{1, 2, 3, 6\}$ is a sublattice of $(N, |)$.

Example 3.17. Consider the lattice L shown in the figure 3.4(a) given below. The subset S_1 shown in figure (b) is not a sublattice of L since $a \wedge b \notin S_1$, and $a \vee b \notin S_1$. However, the subset S_2 of L shown in figure (c) is a sublattice of L.

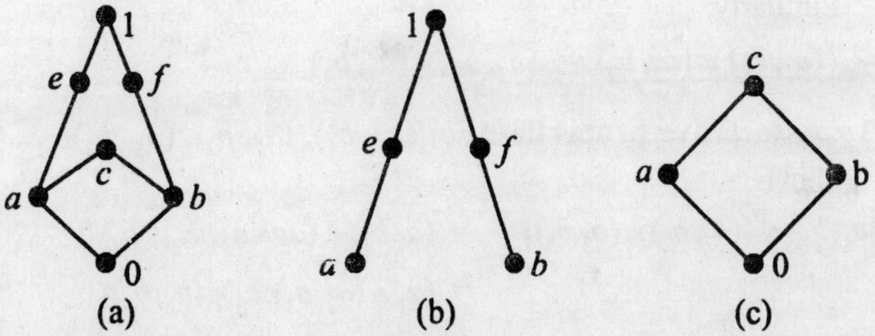

Fig. 3.4

Example 3.18. Show that the interval $[a, b]$, where $a \leqslant b$ and a, b are element of a lattice L is a sublattice.

Proof: Let $x, y \in [a, b]$. Then $a \leqslant x \leqslant b$ and $a \leqslant y \leqslant b$. Therefore $a \leqslant x \vee y \leqslant b$ and $a \leqslant x \wedge y \leqslant b$. This implies $x \vee y, x \wedge y$ both are in $[a, b]$. Hence $[a, b]$ is a sublattice of L.

§ 3.5. Direct Product of Lattices

Theorem: Let $(L_1, \wedge, \vee)$ and $(L_2, \wedge, \vee)$ be two lattices and let $L = L_1 \times L_2$. Show that L is lattice under the binary operations $\vee$ and $\wedge$ defined on L as follows:

$$(a_1, b_1) \wedge (a_2, b_2) = (a_1 \wedge a_2, b_1 \wedge b_2) \text{ and}$$

$$(a_1, b_1) \vee (a_2, b_2) = (a_1 \vee a_2, b_1 \vee b_2) \ \forall (a_1, b_1), (a_2, b_2) \in L$$

The lattice $(L, \wedge, \vee)$ is called the direct product of lattices $(L_1, \wedge, \vee)$ and $(L_2, \wedge, \vee)$.

Proof: Let $(a_1, b_1), (a_2, b_2)$ be any two elements of L. Then $a_1, a_2 \in L_1$ and $b_1, b_2 \in L_2$. Now L_1 is a lattice, therefore $a_1 \wedge a_2$ and $a_1 \vee a_2$ both are in L_1. Similarly, $b_1 \wedge b_2$ and $b_2 \vee b_2$ are in L_2. Thus $(a_1 \wedge a_2, b_1 \wedge b_2) \in L$ and $(a_2 \vee a_2, b_1 \vee b_2) \in L$. Hence both $\wedge$ and $\vee$ defined on L are binary operations.

1. **Commutative properties:** Let $(a_1, b_1), (a_2, b_2) \in L$. Then

$$\begin{aligned}(a_1, b_1) \wedge (a_2, b_2) &= (a_1 \wedge a_2, b_1 \wedge b_2)\\ &= (a_2 \wedge a_1, b_2 \wedge b_1) \quad [\because L_1 \text{ and } L_2 \text{ are lattice}]\\ &= (a_2, b_2) \wedge (a_1, b_1).\end{aligned}$$

Similarly,

$$(a_1, b_1) \vee (a_2, b_2) = (a_2, b_2) \vee (a_1, b_1)$$

2. **Associative properties:** For $(a_1, b_1), (a_2, b_2), (a_3, b_3) \in L$, we have

$$\begin{aligned}(a_1, b_1) \wedge \{(a_2, b_2) \wedge (a_3, b_3)\} &= (a_1, b_1) \wedge (a_2 \wedge a_3, b_2 \wedge b_3)\\ &= (a_1 \wedge (a_2, a_3), b_1 \wedge (b_2, b_3))\\ &= ((a_1 \wedge a_2) \wedge a_3, (b_1 \wedge b_2) \wedge b_3)\\ &\qquad [\text{since both } L_1 \text{ and } L_2 \text{ are lattices}]\\ &= (a_1 \wedge a_2, b_1 \wedge b_2) \wedge (a_3, b_3)\\ &= \{(a_1, b_1) \wedge (a_2, b_2)\} \wedge (a_3, b_3)\end{aligned}$$

$\Rightarrow$ the operation $\wedge$ is associative.

Similarly, the operation $\vee$ is associative.

3. **Absorption properties:** For $(a_1, b_1), (a_2, b_2) \in L$, we have

$$\begin{aligned}(a_1, b_1) \wedge \{(a_1, b_1) \vee (a_2, b_2)\} &= (a_1, b_1) \wedge (a_1 \vee a_2, b_1 \vee b_2)\\ &= (a_1 \wedge (a_1 \vee a_2), b_1 \wedge (b_1 \vee b_2))\\ &= (a_1, b_1)\\ &\qquad [\text{since both } L_1 \text{ and } L_2 \text{ are lattice}]\end{aligned}$$

Similarly,

$$(a_1, b_1) \vee \{((a_1, b_1) \wedge (a_2, b_2))\} = (a_1, b_1)$$

Thus L is a lattice.

Example 3.19. Let L_1 and L_2 be the lattices shown in figure (a) and (b) below respectively. Then $L = L_1 \times L_2$ is the lattice shown in figure (c):

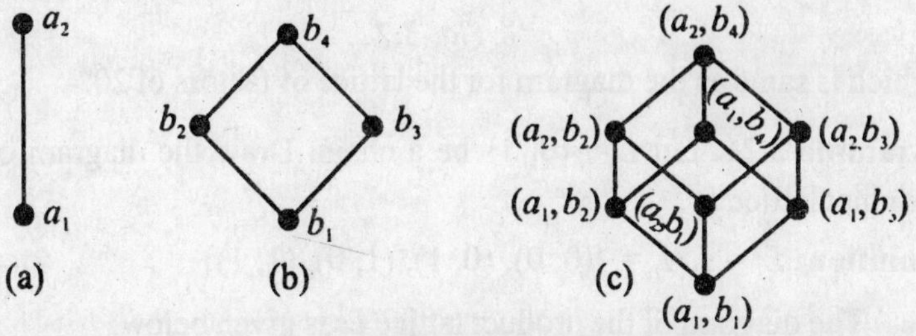

Fig. 3.5

Example 3.20. Draw the diagram of the lattice of factors of 20 under divisibility and show that it is same as that of the product of two chains with three and two elements.

Solution: The set of factors of 20 is $\{1, 2, 4, 5, 10, 20\}$. The diagram representing the lattice of factors of 20 is shown below.

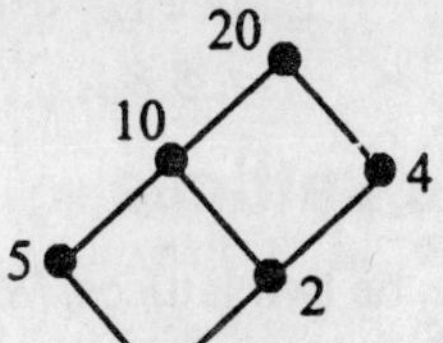

Fig. 3.6

Now we consider the two chains $C_1 = \{0, 1, 2\}$ and $C_2 = \{0, 1\}$ containing three and two elements. Then

$$C_1 \times C_2 = \{(0, 0), (0, 1), (1, 0), (1, 1), (2, 0), (2, 1)\}.$$

Now considering the operations on $C_1 \times C_2$ as defined in the above theorem, the diagram for $C_1 \times C_2$ is represented by figure 3.7.

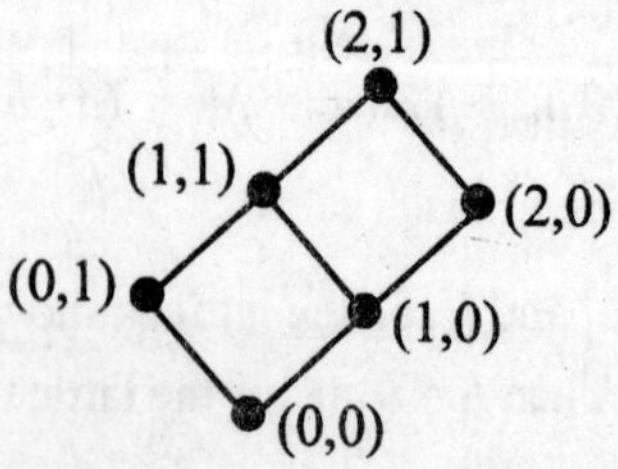

Fig. 3.7

which is same as the diagram for the lattice of factors of 20.

Example 3.21. Let $L = \{0, 1\}$ be a chain. Draw the diagram of the product lattice L^2.

Solution: $L^2 = L \times L = \{(0, 0), (0, 1), (1, 0), (1, 1)\}$.

The diagram of the product lattice L^2 is given below:

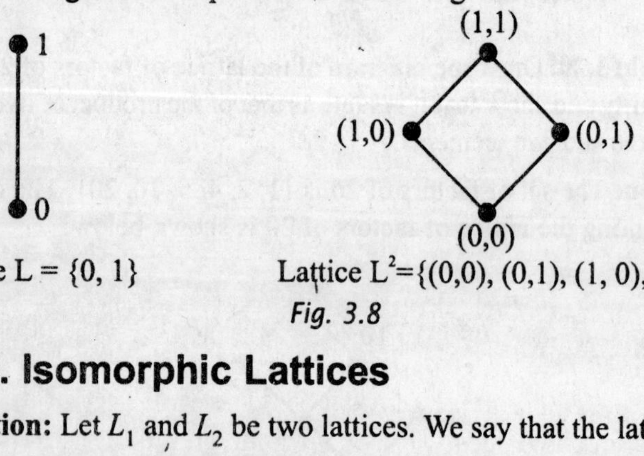

Lattice L = {0, 1} Lattice L^2={(0,0), (0,1), (1, 0),(1,1)}

Fig. 3.8

§ 3.6. Isomorphic Lattices

Definition: Let L_1 and L_2 be two lattices. We say that the lattice L_1 is isomorphic to the lattice L_2 if there exists a one-to-one mapping f from L_1 onto L_2 such that

$$f(a \wedge b) = f(a) \wedge f(b) \quad \text{and}$$
$$f(a \vee b) = f(a) \vee f(b) \;\; \forall\, a, b \in L_1$$

Remarks:

1. In the above definition, we have denoted the compositions in L_1 and L_2 with the same symbols. The reader can use different symbols to denote the operations on L_1 and L_2.

2. The map f in the above definition is called isomorphism. If $L_1 = L_2$ then f is called automorphism.

Example 3.22. Consider the lattice $L = \{1, 2, 3, 6\}$ under divisibility relation and the lattice $(P(S), \subseteq)$, where $S = \{a, b\}$. Then the lattices L and $P(S)$ are isomorphic.

Solution: Define a mapping $f: L \rightarrow P(S)$ as follows:

$$f(1) = \phi$$
$$f(2) = \{a\}$$
$$f(3) = \{b\}$$
$$f(6) = \{a, b\}$$

Then f is one-to-one and onto. Moreover,

$$f(a \wedge b) = f(a) \wedge f(b)$$

and $$f(a \vee b) = f(a) \vee f(b) \ \forall \ a, b \in L.$$

Thus f is an isomorphism and lattice $(L, |)$ is isomorphic to the lattice $(P(S), \subseteq)$.

Example 3.23. Let $L = \{1, 2, 3, 4, 6, 12\}$. Now consider the lattices $(L, |)$ and $(L, \leqslant)$ where $|$ is the 'divisibility' relation on L and $\leqslant$ is the usual 'less than or equal to' relation on L. The lattice $(L, |)$ and $(L, \leqslant)$ are not isomorphic.

Solution: Consider any mapping $f: (L, |) \rightarrow (L, \leqslant)$. Then, in $(L, |)$,

$$3 \wedge 4 = 1 \quad \Rightarrow \quad f(3 \wedge 4) = f(1).$$

But $f(3) \wedge f(4) = f(3)$ or $f(4)$ depending upon whether $f(3)$ is less than or equal to $f(4)$ or $f(4)$ is less than or equal to $f(3)$. In any case, $f(3 \wedge 4) \neq f(3) \wedge f(4)$.

Thus $(L, |)$ and $(L, \leqslant)$ are not isomorphic.

Theorem 1: If the lattices L and M are isomorphic and f denotes an isomorphism then f preserves the ordering relation. That is, for any $a, b \in L$

$$a \leqslant b \quad \Rightarrow \quad f(a) \leqslant f(b)$$

Proof: Let $f: L \rightarrow M$ be an isomorphism.

Let $a, b \in L$ such that $a \leqslant b$.

Then

$$a = a \wedge b$$
$$\Rightarrow f(a) = f(a \wedge b) = f(a) \wedge f(b)$$
$$\Rightarrow f(a) \leqslant f(b).$$

Thus f preserves ordering relation.

Theorem 2; Let L and M be two lattices. If $f: L \rightarrow M$ is an isomorphism of L onto M and a is the least element of L then $f(a)$ is the least element of M.

Proof: Let a be least element of L. Then

$$a \leqslant x \; \forall \; x \in L \Rightarrow f(a) \leqslant f(x) \quad \forall \; f(a) \in M$$

Since f is onto, any element $y \in M$ is of the form $f(x)$ for some $x \in L$. Thus $f(a)$ is least element of M.

Theorem 3: If $f: L \rightarrow M$ is an isomorphism and $x, y \in M$ such that $x < y$ then there exist $a, b \in L$ such that $f(a) = x, f(b) = y$ and $a < b$.

Proof: Since f is onto and $x, y \in M$, therefore there exist a and c in L such that

$$f(a) = x \text{ and } f(c) = y$$

Now

$$f(a \vee c) = f(a) \vee f(c) \qquad [\because f \text{ is isomorphism}]$$
$$= x \vee y$$
$$= y \qquad [\text{since } x < y]$$

We know that

$$a \leqslant a \vee c$$

If

$$a = a \vee c \text{ then } f(a) = f(a) \vee f(c) = y$$
$$\Rightarrow \quad x = y, \qquad \text{a contradiction}$$

Hence

$$a \neq a \vee c$$

Thus $a < a \vee c$.

Now we take $b = a \vee c$. Then

$$a<b \text{ and } f(a)=x, \ f(b)=y.$$

Hence proved.

Note: In view of above theorems, it is clear that two lattices can be represented by the same diagram if and only if they are isomorphic.

Exercise

1. Prove that the non-empty intersection of two sublattices is again a sublattice of a given lattice.
2. Show that the union of two sublattices is not necessarily a sublattice.
3. If L is a lattice and $a, b \in L$ then show that the subset $\{a, b\}$ of L is sublattice if and only if a and b are comparable.
4. Consider the chains of divisors of 4 and 9. Draw the diagram of the product lattice of these lattices.
5. Let $S = \{a, b, c\}$ and $L = P(S)$, the power set of S. Prove that the lattice $(L, \subseteq)$ is isomorphic to the lattice of factors of 42 under divisibility relation.
6. Let $(L, \vee, \wedge)$ be an algebraic system, where $\vee$ and $\wedge$ are binary operations satisfying the absorption law. Show that $\vee$ and $\wedge$ also satisfy the idempotent law.
7. Let f : $L \rightarrow M$ be an isomorphic from a lattice L onto a lattice M. If a is the greatest element of L then $f(a)$ is the greatest element of M.
8. Let f : $L \rightarrow M$ be an isomorphism from a lattice L onto a lattice M. If $f(a) \leqslant f(b)$ then show that $a \leqslant b$.

§ 3.7. Bounded Lattices

We recall from Chapter 2 that an element a of a poset $(P, \leqslant)$ is called a greatest element of the poset P if $x \leqslant a$ for all $x \in P$. An element $a \in P$ is called a least element of P if $a \leqslant x$ for all $x \in P$. Since a lattice is also a partially ordered set, these definitions can be extended to a lattice.

Definition: An element a of a lattice L is called a **greatest element** of the lattice L if $x \leqslant a$ for all $x \in L$. Similarly, an element a of a lattice L is called a least element of L if $a \leqslant x \ \forall \ x \in L$.

The least and the greatest elements of a lattice, if they exist, are also called universal lower bound and universal upper bound respectively.

The least element is denoted by 0 and the greatest element is denoted by 1.

Bounded lattice: A lattice L is said to be **bounded lattice** if L contains both greatest element and least element. Thus a lattice which has both least element 0 and greater element 1 is called bounded lattice.

Example 3.24. Let $L = \{1, 2, 3, 6\}$ and let $\leqslant$ be the divisibility relation on L. Then $(L, \leqslant)$ is a bounded lattice. In this lattice, least element is 1 and greatest element is 6.

Example 3.25. The lattice $(N, \leqslant)$, where N is the set of natural numbers and $\leqslant$ is the 'usual less than or equal to relation on N, is not bounded because this lattice has no greatest element. Although $(N, \leqslant)$ has least element namely 1.

Theorem1: Every finite lattice is a bounded lattice.

Proof: let $L = \{a_1, a_2, a_3, ----, a_n\}$ be any finite lattice. We will show L contains both least element and greatest element. Put

$$b_1 = a_1$$

$$b_2 = a_2 \wedge b_1$$

$$b_3 = a_3 \wedge b_2$$

....

....

$$b_n = a_n \wedge b_{n-1}$$

Then $b_1, b_2, ----, b_n$ all belong to L and

$$b_n \leqslant a_i \ \forall \ i = 1, 2, ----, n$$

$\Rightarrow b_n$ is the least element of L.

By associativity of the operation $\wedge$, we write

$$b_n = a_1 \wedge a_2 \wedge a_3 ---- \wedge a_n.$$

Similarly, we can show $a_1 \vee a_2 \vee ---- \vee a_n$ is the greatest element of L. Thus L is bounded.

Theorem 2: Let $(L, \leqslant)$ be a lattice with least element 0 and greatest element 1. For any element $a \in L$, show that

(i) $a \vee 1 = 1$ and $a \wedge 1 = a$

(ii) $a \vee 0 = a$ and $a \wedge 0 = 0$

Proof: (i) Let a be any element of a lattice L. Since 1 is the greatest element of L, we have

$$a \vee 1 \leqslant 1 \qquad ---- (1)$$

Since $a \vee 1$ is the supremum of a and 1, we have

$$1 \leqslant a \vee 1 \qquad ---- (2)$$

From (1) and (2) we get,

$$a \vee 1 = 1$$

Again, $a \wedge 1$ is the infimum of a and 1, therefore

$$a \wedge 1 \leqslant a \qquad ---- (3)$$

Also, since $a \leqslant a$ and $a \leqslant 1$, Theorem 3(b) of section 3.2 gives

$$a \leqslant a \wedge 1 \qquad ---- (4)$$

From (3) and (4), we get

$$a \wedge 1 = a$$

(ii) Let a be any element of L and 0 be the least element of L.

$$\therefore 0 \leqslant a \text{ and } a \leqslant a$$

$$\Rightarrow \qquad a \vee 0 \leqslant a$$

Because a is upper bound of 0 and a while $a \vee 0$ is the least upper bound of $\{0, a\}$.

Also, from the definition of $\vee$, we have

$$a \leqslant a \vee 0$$

Now,

$$a \vee 0 \leqslant a \quad \text{and} \quad a \leqslant a \vee 0$$

$$\Rightarrow \qquad a \vee 0 = a$$

Again, $a \wedge 0$ is the infimum of a and 0, therefore

$$a \wedge 0 \leqslant 0$$

Since $0 \leqslant a$ and $0 \leqslant 0$, we have

$$0 \leqslant a \wedge 0$$

Now

$$a \wedge 0 \leqslant 0 \quad \text{and} \quad 0 \leqslant a \wedge 0$$

$$\Rightarrow \qquad a \wedge 0 = 0$$

Corollary: 0 is the identity of the operation $\vee$ and 1 is the identity of the operation $\wedge$ in a bounded lattice $(L, \wedge, \vee)$.

Complete Lattices

Definition: A lattice L is called complete lattice if every non-empty subset of L has its supremum and infimum in L.

Theorem1: If L is a complete lattice then it is bounded lattice.

Proof: Let L be a complete lattice. Then every non-empty subset of L has least upper bound and greatest lower bound. In particular, L itself has least upper bound and greatest lower bound. Thus L is bounded lattice.

Example 3.26. Consider the lattice $(L, \leqslant)$, where L is the set of all rational numbers x such that $0 \leqslant x \leqslant 2$ and $\leqslant$ is the usual 'less than or equal to' relation on L. Then $(L, \leqslant)$ is a bounded lattice. The least element of L is 0 and the greatest element is 2. But $(L, \leqslant)$ is not a complete lattice because the set S of all those rational numbers in L whose squares are less than 2 does not have least upper bound. That is, the set

$S = \{x : x$ is a rational number such that $0 \leqslant x^2 < 2\}$
does not have any rational number as its least upper bound.

Let x be any rational number in L. We shall show x cannot be least upper bound of S. The following cases arise:

(i) $x = 0$. Since $1 \in S$, therefore x is not an upper bound of S and consequently it cannot be the least upper bound of S.

(ii) $x > 0$ and $x^2 < 2$

Consider $$y = \frac{4+3x}{3+2x} \quad \text{......(1)}$$

Then $$y^2 - 2 = \frac{x^2-2}{(3+2x)^2} \quad \text{......(2)}$$

and $$y - x = \frac{2(2-x^2)}{3+2x} \quad \text{.......(3)}$$

Since x is a positive rational number, therefore from (1) we find that y is also a positive rational number. Since $x^2 < 2$, therefore from (2), we find that $y^2 < 2$ and from (3), we find that $y > x$.
Since y is a positive rational number such that $0 < y^2 < 2$, therefore it follows that $y \in S$. Also, since $x < y$, therefore it follows that x cannot be an upper bound of S.

(iii) $x > 0$ and $x^2 = 2$

This is not possible because there exists no rational number x whose square is 2.

(iv) $x > 0$ and $x^2 > 2$.

Let $$y = \frac{4+3x}{3+2x}$$

Then, as in case (ii), we find that y is a positive rational number and $y^2 > 2$. From (3), we find $y < x$. If p be any number of S, then we have $0 \leqslant p^2 < 2 < y^2$ and $y < x$. That is, $0 < p < y < x$. This shows that x and y are upper bounds of S and that x is not the least upper bound because $y < x$.

Since x is any rational number in L, we conclude that no rational number can be the least upper bound of S. Thus L is not complete.

Example 3.27 Consider the lattice $L = \{0, 1, 2, 3, 6\}$ under the divisibility relation. Then $(L, |)$ is complete lattice.

Theorem 2: Every finite lattice is complete.

Proof: Let $(L, \wedge, \vee)$ be any finite lattice. Let S be any non-empty subset of L. Then S is also a finite set. Let

$S = \{x_1, x_2, \text{----}, x_m\}$. Then

$x_1 \wedge x_2 \wedge x_3 \wedge x_4 \text{----} \wedge x_m$ and $x_1 \vee x_2 \vee x_3 \vee x_4 \text{----} \vee x_m$

are the supremum and infimum of S in L. Hence L is complete.

Theorem 3: Dual of a complete lattice is complete.

Proof: Let (L, R) be a complete lattice with relation R and let $(L, \bar{R})$ be its dual. Then $(L, \bar{R})$ is a lattice.

Let S be any non-empty subset of $(L, \bar{R})$. Since (L, R) is complete, sup S and inf S exist in (L, R).

Let $a = \inf S$ in (L, R)

Then $aRx \ \forall \ x \in L$

$\Rightarrow$ $x\bar{R}a \ \forall \ x \in L$

$\Rightarrow a$ is an upper bound of S in $(L, \bar{R})$

Let b be any other upper bound of S in $(L, \bar{R})$

Then $x \bar{R} b \ \forall \ x \in S$

$\Rightarrow$ $b \bar{R} a \ \forall \ x \in S$

$\Rightarrow$ bRa as $a = \inf S$ in (L, R)

$\Rightarrow$ $a \bar{R} b$

$\Rightarrow a$ is lub of S in $(L, \bar{R})$

Similarly, we can show that sup S in (L, R) will be inf S in $(L, \bar{R})$. Hence $(L, \bar{R})$ is complete.

Theorem 4: If $(P, \leqslant)$ is a poset with greatest element 1 such that every non-empty subset S of P has infimum, then $(P, \leqslant)$ is a complete lattice.

Proof: Let S be any non-empty subset of P. We need to prove that sup S exists.

Since 1 is greatest element of P,

$\therefore x \leqslant 1 \ \forall \ x \in P$ and thus $s \leqslant 1 \ \forall \ s \in S$.

$\Rightarrow$ 1 is an upper bound of S.

Let U be the set of all upper bounds of S. then U is non-empty because $1 \in \text{U}$. Therefore by given condition, inf U exists.

Let $\qquad k = \inf \text{U}$

Now $\qquad s \in S \Rightarrow s \leqslant x \quad \forall \ x \in \text{U}$

$\Rightarrow$ each element of S is a lower bound of U

$\Rightarrow \qquad s \leqslant k \ \ \forall \ \ s \leqslant S$

$\Rightarrow k$ is an upper bound of S.

But k is being a lower bound of U means $k \leqslant x \ \forall \ x \in \text{U}$.

That is, $k \leqslant x$ for all upper bounds of S.

$\Rightarrow \qquad k = \sup S$

Thus P is a poset in which every non-empty subset has supremum and infimum and therefore P is complete.

Theorem 5: If $(P, \leqslant)$ is a poset with least element 0 such that every non-empty subsets of P has supremum then P is a complete lattice.

Proof: The proof follows by principle of duality from the above theorem.

§ 3.8. Complements, Complemented Lattices

For bounded lattices, it is possible to introduce the notion of a complement of an element in the following manner.

Definition: Let L be a bounded lattice with universal lower and upper bounds 0 and 1. For an element a in L, an element b in L is said to be a *complement* of a if

$$a \vee b = 1 \quad \text{and} \quad a \wedge b = 0$$

From the definition, it is clear that if a is a complement of b then b

is also a complement of a. Also, an element might have more than one complement. That is complements need not be unique.

Example 3.28. Consider the lattice whose diagram is given below.

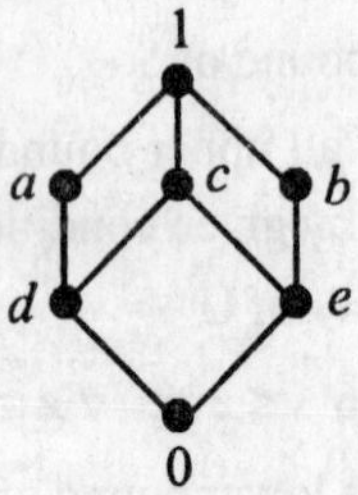

Fig. 3.9

In this lattice, b and e are complements of a. Moreover element c has no complement. Also, 1 is the complement of 0 and 0 is the complement of 1.

Example 3.29. Consider the lattice $L = P(S)$, the power set of S. In this lattice, every element has a complement. Let $A \in P(S)$, then $S-A$ is the complement of A.

Thus Examples 3.28 and 3.29 show that an element a in a lattice need not have a complement and it may have more than one complement. Also 1 is the unique complement of 0 and 0 is the unique complement of 1.

Complemented Lattices

Definition: A bounded lattice L is said to be **complemented lattice** if every element in L has a complement.

Example 3.30. The lattice $(P(S), \subseteq)$ of the power set of any set S is complemented. In this lattice each element has unique complement.

Example 3.31. Any chain having 3 or more elements is not a complemented lattice.

Example 3.32. The lattice with the diagram as shown in figure 3.10 is complemented. In this lattice, complements are not unique. Both a and b are complements of c.

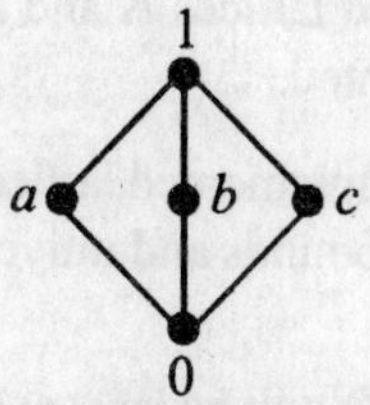

Fig. 3.10

Theorem 1: Dual of a complemented lattice is complemented.

Proof: Let (L, R) be a complemented lattice with 0 and 1 as least and greatest elements. Let $(L, \bar{R})$ be the dual of (L, R). Then 1 and 0 are least and greatest elements of $(L, \bar{R})$

Let $a \in L$ be any element.

Since (L, R) is complemented, therefore there exists $a' \in L$ such that

$a \wedge a' = 0$ and $a \vee a' = 1$ in the lattice (L, R)

That is, $0 = \inf\{a, a'\}$ and $1 = \text{Sup}\{a, a'\}$ in (L, R)

$\Rightarrow$ $0Ra, 0Ra'$ and $aR1, a'R1$

$\Rightarrow$ $a\,\bar{R}0, a'\,\bar{R}0$ and $1\,\bar{R}a, 1\,\bar{R}a'$

$\Rightarrow$ 0 is an upper bound of $\{a, a'\}$ in $(L, \bar{R})$ and 1 is a lower bound of $\{a, a'\}$ in $(L, \bar{R})$.

Let k be any upper bound of $\{a, a'\}$ in $(L, \bar{R})$. Then

$a\,\bar{R}k$ and $a'\,\bar{R}k$

$\Rightarrow$ kRa and kRa'

$\Rightarrow$ $kR0$ because 0 is infimum of $\{a, a'\}$ in (L, R)

$\Rightarrow$ $0\,\bar{R}k$

Thus 0 is lub of $\{a, a'\}$ in $(L, \bar{R})$. Hence $a \vee a' = 0$ in $(L, \bar{R})$.

Similarly, $a \wedge a' = 1$ in $(L, \bar{R})$.

Thus a' is complement of a in $(L, \bar{R})$.

Hence $(L, \bar{R})$ is complemented.

Theorem 2: Two bounded lattices A and B are complemented if and only if A×B is complemented.

Proof: Let A and B be complemented lattices and suppose that 0, 1 and 0′, 1′ are universal lower bounds and universal upper bounds of A and B respectively.

Then (0, 0′) and (1, 1′) will be least and greatest elements of $A \times B$. Let (a, b) be any element of $A \times B$.

$$\Rightarrow \qquad a \in A \text{ and } b \in B$$

Since A and B are complemented therefore $\exists a' \in A$ and $b' \in B$ such that

$$a \wedge a' = 0, a \vee a' = 1 \text{ and } \quad b \wedge b' = 0', b \vee b' = 1'.$$

We shall show (a', b') is complement of (a, b) in $A \times B$.

$$\begin{aligned}(a, b) \wedge (a', b') &= (a \wedge a', b \wedge b') \\ &= (0, 0')\end{aligned}$$

Similarly,

$$\begin{aligned}(a, b) \vee (a', b') &= (a \vee a', b \vee b') \\ &= (1, 1')\end{aligned}$$

Thus (a', b') is complement of (a, b) in $A \times B$.

Hence $A \times B$ is complemented.

Conversely, let $A \times B$ be complemented. We shall show both A and B are complemented.

Let $a \in A$ and $b \in B$ be any elements. Then $(a, b) \in A \times B$.

Since $A \times B$ is complemented, $\exists (a', b') \in A \times B$ such that

$$\begin{aligned}&(a, b) \wedge (a', b') = (0,0') && \text{and} \quad (a, b) \vee (a' b') = (1, 1') \\ &\Rightarrow (a \wedge b', b \wedge b') = (0, 0') && \text{and} \quad (a \vee a', b \vee b') = (1, 1') \\ &\Rightarrow a \wedge a' = 0, b \wedge b' = 0' && \text{and} \quad a \vee a' = 1, b \vee b' = 1' \\ &\Rightarrow a \wedge a' = 0, a \vee a' = 1 && \text{and} \quad b \wedge b' = 0', b \vee b' = 1'\end{aligned}$$

$\Rightarrow a'$ and b' are complements of a and b respectively. Hence A and B are complemented.

Theorem 3. In a lattice L with universal lower bound 0 and universal upper bound 1, show that 0 is the unique complement of 1 and 1 is the unique complement of 0.

Proof: It is clear that

$$0 \wedge 1 = 0 \text{ and } 0 \vee 1 = 1$$

Hence 0 and 1 are complements of each other. We shall first show that 1 is the only complement of 0. Suppose a be any other complement of 0. Then

$$0 \wedge a = 0 \quad \text{and} \quad 0 \vee a = 1$$

But we know that

$$0 \vee a = a$$

Thus

$$a = 1$$

Thus the complemented of 0 is unique. In a similar manner we can show that 0 is the only complement of 1.

§ 3.9. Cover of an Element, Atoms and Irreducible Elements

Cover of an Element

Let a and b be two elements in a lattice. a is said to **cover** b if $b < a$ and there is no element c such that $b < c < a$.

Atoms

Let L be a lattice with a least element 0. An element a in L is called an **atom** if a covers 0. That is, $0 < a$ and there is no element c such that $0 < c < a$.

Irreducible Elements

Let L be a lattice. An element a in L is said to be **join irreducible** if

$$a = x \vee y \Rightarrow a = x \quad \text{or} \quad a = y$$

An element a in L is said to be **meet irreducible** if

$$a = x \wedge y \Rightarrow a = x \quad \text{or} \quad a = y$$

Example 3.33. Consider the lattice whose diagram is given in figure 3.11.

Fig. 3.11

In this lattice, a and b are atoms. The element c is not an atom because $0< a< c$. Moreover c is cover of a, while 1 is cover of both b and c. Further, a, b and c are join irreducible.

Example 3.34. Consider the lattice as shown in the figure given below.

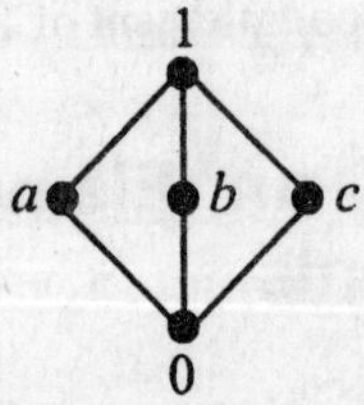

Fig. 3.12

In this lattice, a, b and c are atoms. The elements a, b and c are join-irreducible as well as meet irreducible elements. 0 is join irreducible but not meet-irreducible and 1 is meet-irreducible but not join-irreducible.

Theorem 1: Let L be a complemented lattice with unique complements. Then the join-irreducible elements of L other than 0 are its atoms.

Proof: Let L be a complemented lattice with unique complements. Without loss of generality, we may assume that L has more than two elements. For if complemented lattice L contains two elements only then these two elements are 0 and 1 and 1 is clearly an atom in this lattice. Let a be any join-irreducible element of L which is not 0. We shall show a is an atom. If possible, suppose a is not an atom. Then $\exists$ an element x in L such that:

$$0 < x < a$$

Let a' be the complement of a.

Now $\quad 0 < x < a$

$\Rightarrow \quad x \wedge a' \leqslant a \wedge a'$

$\Rightarrow \quad 0 \leqslant x \wedge a' \leqslant 0 \qquad [\because a \wedge a' = 0]$

$\Rightarrow \quad x \wedge a' = 0 \qquad$ ------(1)

Now if $x \leqslant a'$ then $x \wedge a' = x$

$\Rightarrow \quad 0 = x \qquad$ by (1)

But this is not true. Hence $x \leqslant a'$ is not possible.

So, we suppose that $x \nleqslant a'$

If $a' \leqslant x$ then by (1), we have

$$a' = x \wedge a' = 0$$

$\Rightarrow a = 1$ which is not join-irreducible in a complemented lattice having more than 2 elements.

If a' and x are not comparable then

$$x \vee a' = 1$$

Thus $x \wedge a' = 0$ and $x \vee a' = 1$

$\Rightarrow x = a$, again a contradiction.

Hence a is an atom.

Example 3.35. Consider the lattice L in figure given below.

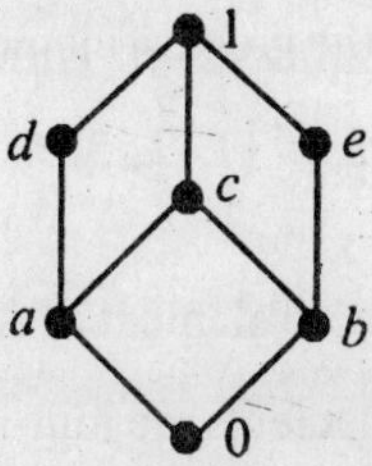

Fig. 3.13

(a) Which non-zero elements are join-irreducible?

(b) Which elements are meet-irreducible?

(c) Which elements are atoms?

(d) Find complements, if they exist, for the elements a, b, c.

(e) Is L a complemented lattice?

(f) Is L a complete lattice?

Solution:

(a) The non-zero elements a,b,d and e are join-irreducible. Note that $c = a \vee b$. Hence c is not a join-irreducible element.

(b) The elements c, d, e and 1 are meet-irreducible element.

(c) Only a and b are the atoms.

(d) Since $a \wedge e = 0$ and $a \vee e = 1$. $\therefore$ a and e are complements of each other. Also, b and d are complements of each other. But the element c has no complement.

(e) L is not complemented since the element c has no complement.

(f) Since L is finite, it is complete.

Exercise

1. Consider the lattice $L = \{1, 2, 3, 5, 6, 10, 15, 30\}$ ordered by divisibility.

   (a) Which elements are join-irreducible?

   (b) Which elements are atoms?

   (c) Which element is the greatest element?

2. Consider the lattice $L = \{1, 2, 3, 6, 9, 18\}$ under divisibility relation.

   (a) Which element is zero element of L and which element is 1?

   (b) Which non-zero elements are join-irreducible?

   (c) Which element are atoms?

   (d) Which elements are cover of 3?

   (e) Find complements, if they exist, for the element 2, 3, 6.

   (f) Is L a complemented lattice?

3. If A and B are two finite chains then show that an element $(a, b) \in A \times B$ is join irreducible if and only if a is least element of A or b is least element of B.

4. Find the complements of every element of the lattice of factors of 75 under divisibility relation.

5. Show that a chain having three or more elements is not complemented.

6. Show that the greatest element 1 in a complemented lattice having more than 2 elements is not join-irreducible.

7. Show that in a lattice with two or more elements, no element is its own complement.

8. Show that in any chain, if a covers $a \wedge b$ then $a \vee b$ will cover b.

9. Consider the lattice $L = \{0, 1, 2, 3, 6, 9, 18\}$ under divisibility relation. What is the greatest element of this lattice? Is L complete? Is L complemented?

10. Give an example to show that a sublattice of a complemented lattice need not be a complemented lattice.

Answers

1. (a) 2, 3 and 5 (b) 2, 3 and 5 (c) 30

2. (a) The least element zero is 1 and the greatest element 1 is 18.

   (b) The non-zero elements 2, 3 and 9 are join-irreducible.

   (c) The elements 2 and 3 are atoms.

   (d) The elements 6 and 9 are covers of 3.

   (e) Complement of 2 is 9 while 3 and 6 have no complement.

   (f) No.

4. The elements 1 and 75 are complements of each other. The

elements 3 and 25 are also complements of each other. The elements 5 and 15 have no complement.

9. The greatest element is 0. L is complete but not complemented.

§ 3.10. Modular Lattices

Definition: A lattice L is said to be *modular* if for all a, b, c in L

$$a \leqslant c \Rightarrow a \vee (b \wedge c) = (a \vee b) \wedge c$$

Remarks:

(i) If we put $a = c$, then we find

$$a \vee (b \wedge c) = a \vee (b \wedge a) = a$$

and $(a \vee b) \wedge c = (a \vee b) \wedge a = a$

Hence above condition is satisfied.

(ii) If $b \leqslant c$. Then

$$a \leqslant c \text{ and } b \leqslant c \Rightarrow a \wedge b \leqslant c \text{ and } a \vee b \leqslant c$$

$\therefore$ $a \vee (b \wedge c) = a \vee c = c$

and $(a \vee b) \wedge c = c \because a \vee b \leqslant c$

Hence above condition is satisfied.

(iii) If $c \leqslant b$. Then

$$a \leqslant c \text{ and } c \leqslant b \Rightarrow b \wedge c = c \text{ and } a \vee b = b$$

$\therefore$ $a \vee (b \wedge c) = a \vee c = c \quad \because a \leqslant c$

and $(a \vee b) \wedge c = b \wedge c = c \quad \because c \leqslant b$

Hence above condition is satisfied.

(iv) If $b \leqslant a$. Then

$$a \leqslant c \text{ and } b \leqslant a \Rightarrow b \leqslant c$$

$\therefore$ $a \vee (b \wedge c) = a \vee b = a \because b \leqslant a$

and $(a \vee b) \wedge c = a \wedge c = a$

Hence above condition is satisfied.

(v) If $a \leqslant b$. Then

$$a \leqslant c \text{ and } a \leqslant b \Rightarrow a \leqslant b \wedge c$$

$$\therefore \quad a \vee (b \wedge c) = b \wedge c \quad \because \quad a \leqslant b \wedge c$$

and $$(a \vee b) \wedge c = b \wedge c \quad \because \quad a \leqslant b$$

Hence above condition is satisfied.

Thus from (i) to (v), we conclude that whenever $a = c$ or b is comparable with a or c, the condition of modularity is satisfied. Hence to check modularity of a lattice, we need to check it only for triplets a,b,c where $a > c$ and b is not comparable with a or c.

Example 3.36. The lattices given by the following diagrams are modular. The lattice in figure (b) will be denoted as lattice M_5.

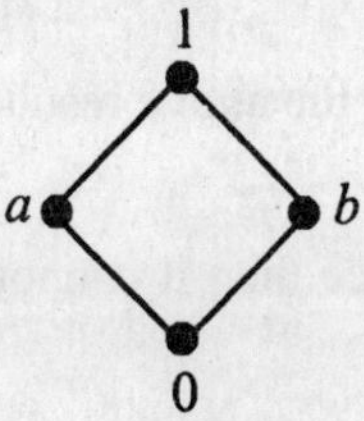

(a)

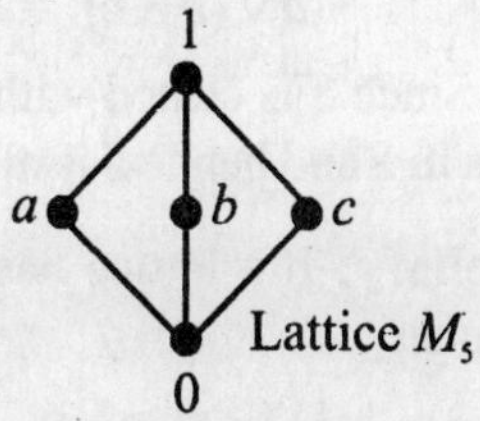

(b)

Fig. 3.14

Solution: In view of the remark given above, we need to check the lattice only for triplets a, b, c where $a>c$ and b is not comparable with a or c. In the first case, we cannot find any triplet a,b,c such that $a>c$ and b is not comparable with a or c. Hence it is modular.

A similar argument shows that the second lattice is also modular.

Example 3.37. Every chain is a modular lattice.

Solution: Since we cannot find any triplet a, b, c in a lattice such that $a>c$ and b is not comparable with a or c, therefore, every chain is modular.

Example 3.38. The lattice given by the diagram as shown in figure 3.15 is not modular. This lattice called **pentagonal** lattice.

Solution: The pentagonal lattice is not modular because here $a \leqslant c$ but

$$a \vee (b \wedge c) = a \neq (a \vee b) \wedge c = c$$

Fig. 3.15

Theorem 1: A sublattice of a modular lattice is modular.

Proof: Let S be a sublattice of a modular lattice L. If $a, b, c \in S$ with $a \leqslant c$ then as $S \subseteq L$, we have $a, b, c \in L$ and therefore

$$a \vee (b \wedge c) = (a \vee b) \wedge c \qquad \text{---- (1)}$$

Since S is closed with respect to $\wedge$ and $\vee$, the above result (1) holds in S and hence S is modular.

Corollary: If a lattice has a pentagonal sublattice then it cannot be modular.

Proof: Let L be a lattice and let L has a sublattice, which is pentagonal. Then the sublattice is not modular as shown in Example. Now L cannot be modular in view of Theorem 1.

Theorem 2: Two lattices L and M are modular if and only if $L \times M$ is modular.

Proof: Let L and M be modular and let $(a_1, b_1), (a_2, b_2), (a_3, b_3) \in L \times M$ be three element with $(a_1, b_1) \geq (a_3, b_3)$.

Then $a_1, a_2, a_3 \in L$ and $a_1 \geq a_3$, $b_1, b_2, b_3 \in L$ and $b_1 \geq b_3$

and since L and M are modular, we get

$$a_1 \vee (a_2 \wedge a_3) = (a_1 \vee a_2) \wedge a_3,$$

$$b_1 \vee (b_2 \wedge b_3) = (b_1 \vee b_2) \wedge b_3$$

$$\Rightarrow \quad (a_1, b_1) \vee [(a_2, b_2) \wedge (a_3, b_3)]$$

$$= (a_1, b_1) \vee (a_2 \wedge a_3,\ b_2 \wedge b_3)$$

$$= (a_1 \vee (a_2 \wedge a_3),\ b_1 \vee (b_2 \wedge b_3))$$

$$= ((a_1 \vee a_2) \wedge a_3, (b_1 \vee b_2) \wedge b_3)$$
$$= (a_1 \vee a_2, b_1 \vee b_2) \wedge (a_3, b_3)$$
$$= ((a_1, b_1) \vee (a_2, b_2)) \wedge (a_3, b_3)$$

Hence $L \times M$ is modular.

Conversely, let $L \times M$ be modular. We shall show both L & M are modular. Let

$a_1, a_2, a_3 \in L$ with $a_1 \geq a_3$ and $b_1, b_2, b_3 \in L$ with $b_1 \geq b_3$

Then $(a_1, b_1), (a_2, b_2), (a_3, b_3) \in L \times M$ and $(a_1, b_1) \geq (a_3, b_3)$

Since $L \times M$ is modular, we have

$$(a_1, b_1) \vee [(a_2, b_2) \wedge (a_3, b_3)] = [(a_1, b_1) \vee (a_2, b_2)] \wedge (a_3, b_3)$$

or $$(a_1, b_1) \vee (a_2 \wedge a_3, b_2 \wedge b_3) = (a_1 \vee a_2, b_1 \vee b_2) \wedge (a_3, b_3)$$

or $$(a_1 \vee (a_2 \wedge a_3), b_1 \vee (b_2 \wedge b_3)) = ((a_1 \vee a_2) \wedge a_3, (b_1 \vee b_2) \wedge b_3)$$

$$\Rightarrow \quad a_1 \vee (a_2 \wedge a_3) = (a_1 \vee a_2) \wedge a_3$$

and

$$b_1 \vee (b_2 \wedge b_3) = (b_1 \vee b_2) \wedge b_3$$

$\Rightarrow L$ and M are modular.

Theorem 3: The dual of a modular lattice is modular.

Proof: Let $(L, \leqslant)$ be a modular lattice. We show its dual $(L, \geq)$ is also a modular lattice.

Let $a, b, c \in L$ and let $a \geq c$. We shall show that

$$a \wedge (b \vee c) = (a \wedge b) \vee c \qquad \text{---- (1)}$$

Now, since $(L, \leqslant)$ is modular and $c \leqslant a$, we have

$$c \vee (b \wedge a) = (c \vee b) \wedge a$$
$$\Rightarrow \quad c \vee (a \wedge b) = (b \vee c) \wedge a$$
$$\Rightarrow \quad a \wedge (b \wedge c) = (a \wedge b) \vee c$$

which is same as (1). Hence $(L, \geq)$ is modular.

Theorem 4: A lattice L is modular if and only if it contains no sublattice isomorphic to the pentagonal lattice (defined in Ex. 3.38)

Proof: The pentagonal lattice is non-modular since

$a \leqslant c$ but $a \vee (b \wedge c) = a < c = (a \vee b) \wedge c$.

Therefore if a lattice L contains a sublattice isomorphic to the pentagonal lattice then it cannot be modular. Thus if lattice L is modular then it contains no sublattice isomorphic to the pentagonal lattice. Conversely suppose L is not a modular lattice. Then there exist at least three elements a, b, c in L with $a<c$ such that

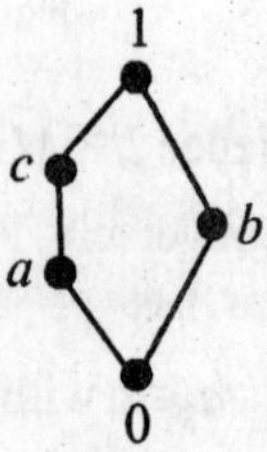

Fig. 3.16

$$a \vee (b \wedge c) \neq (a \vee b) \wedge c \quad \text{---- (1)}$$

But by Theorem 6 of section 3.2, we have

$$a \leqslant c \Leftrightarrow a \vee (b \wedge c) \leqslant (a \vee b) \wedge c \quad \text{---- (2)}$$

From (1) and (2), we have

$$a \vee (b \wedge c) < (a \vee b) \wedge c \quad \text{---- (3)}$$

Consider the chain,

$$b \wedge c \leqslant a \vee (b \wedge c) < (a \vee b) \wedge c \leqslant a \vee b \quad \text{---- (4)}$$

We shall now show that in this chain strict inequality holds at all places. That is, we shall show $b \wedge c \neq a \vee (b \wedge c)$ and $(a \vee b) \wedge c \neq a \vee b$.

Suppose if possible,

$$b \wedge c = a \vee (b \wedge c)$$

Then $a \leqslant b \wedge c$

$\Rightarrow$ $a \vee b \leqslant (b \wedge c) \vee b = b$

$\Rightarrow$ $a \vee b \leqslant b \leqslant a \vee b$

$\Rightarrow$ $a \vee b = b$

$\Rightarrow$ $(a \vee b) \wedge c = b \wedge c$

which is impossible by (4).

Hence $b \wedge c < a \vee (b \wedge c)$

Again, if possible, suppose $a \vee b = (a \vee b) \wedge c$. Then

$$a \vee b \leqslant c$$

$\Rightarrow$ $(a \vee b) \wedge b \leqslant c \wedge b$

$$\Rightarrow \qquad b \leqslant b \wedge c$$

$$\Rightarrow \qquad b = b \wedge c \qquad [\because b \wedge c \leqslant b]$$

$$\Rightarrow \qquad a \vee b = a \vee (b \wedge c)$$

which is impossible by (4) $[\because a \vee (b \wedge c) < a \vee b]$

Hence the chain (4) becomes

$$b \wedge c < a \vee (b \wedge c) < (a \vee b) \wedge c < a \vee b \qquad \text{---- (5)}$$

Similarly, we can show that $b \wedge c \neq b$ and $b \neq a \vee b$. Thus

$$b \wedge c < b < a \vee b \qquad \text{---- (6)}$$

Thus we have two chains, namely (5) and (6), with same end points in L. We now show that b does not lie in chain (5). For this it is sufficient to prove that b is not comparable with $(a \vee b) \wedge c$.

Suppose, if possible, $(a \vee b) \wedge c \leqslant b$

Then

$$((a \vee b) \wedge c) \wedge c \leqslant b \wedge c$$

$$\Rightarrow \qquad (a \vee b) \wedge c \leqslant b \wedge c$$

which is not possible by (5)

Again, if possible, suppose $b \leqslant (a \vee b) \wedge c$

Then

$$b \leqslant c$$

$$\Rightarrow \qquad b \wedge c = b$$

which is not possible by (6)

Hence the chain (5) and (6) form a pentagonal subset

$S = \{b \wedge c, a \vee (b \wedge c), (a \vee b) \wedge c, a \vee b, b\}$ of L.

We shall show that S is a sublattice . For this we need to show that meet and join of non-comparable elements in S is again in S.

Now

$$[a \vee (b \wedge c)] \wedge b \leqslant [(a \vee b) \wedge c] \wedge b \qquad \text{by (5)}$$

$$= (a \vee b) \wedge b \wedge c = b \wedge c$$

Again, we have,

$$b \wedge c \leqslant a \vee (b \wedge c) \text{ and } b \wedge c \leqslant b$$

$$\Rightarrow \quad b \wedge c \leqslant [a \vee (b \wedge c)] \wedge b$$

Thus

$$[a \vee (b \wedge c)] \wedge b = b \wedge c \in S$$

Also

$$[a \vee (b \wedge c)] \vee b = a \vee [(b \wedge c) \vee b] = a \vee b \in S$$

Similarly, we can show that

$$[(a \vee b) \wedge c] \wedge b = b \wedge c \in S$$

$$[(a \vee b) \wedge c] \vee b = a \vee b \in S$$

Hence S forms a sublattice of L. Thus if L contains no sublattice isomorphic to the pentagonal lattice then L is modular. This completes the proof.

Corollary: Every lattice having four or less elements is always modular.

Remark: It is important to note that in the above theorem, the assertion is that a modular lattice does not contain a pentagonal sublattice. It may contain a pentagonal subset. For instance, the lattice L of factors of 120 contains a subset $S = \{2, 6, 10, 12, 60\}$ having five elements but S is not a sublattice.

Theorem 5: If for all a, b, c in a lattice L,

$a \wedge b = a \wedge c, a \vee b = a \vee c \Rightarrow b = c$. Then lattice L is modular.

Proof: Suppose the lattice L is not modular. Then by Theorem 4, there exists a pentagonal sublattice as shown in the figure below.

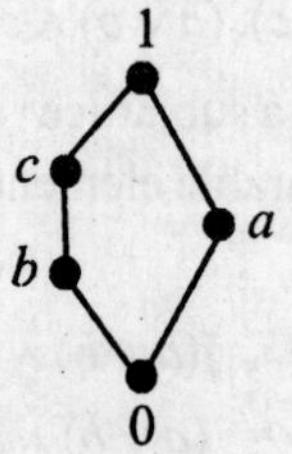

Fig. 3.17

In this sublattice of lattice L, we have

$a \wedge b = a \wedge c$ and $a \vee b = a \vee c$ but $b \neq c$.

Hence given condition does not hold. Thus L must be modular.

Theorem 6: In a modular lattice L, if two elements b and c are comparable and $a \wedge b = a \wedge c, a \vee b = a \vee c \; \forall \; a \in L$ then $b = c$.

Proof: Without loss of generality, we assume that $b \geq c$. Then

$$b = b \wedge (a \vee b) = b \wedge (a \vee c) = (b \wedge a) \vee c \quad [\because L \text{ is modular}]$$

$$= (a \wedge c) \vee c \quad \text{by given condition}$$

$$= c$$

$$\Rightarrow \quad b = c$$

Solved Examples

Example 3.39. Show that $N \times N$ is modular lattice where N is the lattice under usual $\leqslant$ relation.

Solution: Let $(a_1, a_2), (b_1, b_2), (c_1, c_2) \in N \times N$ be any three elements. Suppose that

$$(a_1, a_2) \wedge (b_1, b_2) = (a_1, a_2) \wedge (c_1, c_2) \quad \text{---- (1)}$$

and
$$(a_1, a_2) \vee (b_1, b_2) = (a_1, a_2) \vee (c_1, c_2) \quad \text{---- (2)}$$

We shall show $(b_1, b_2) = (c_1, c_2)$

By (1), $a_1 \wedge b_1 = a_1 \wedge c_1$ and $a_2 \wedge b_2 = a_2 \wedge c_2$ ---- (3)

By (2), $a_1 \vee b_1 = a_1 \vee c_1$ and $a_2 \vee b_2 = a_2 \vee c_2$ ---- (4)

From (3) and (4), we have

$a_1 \wedge b_1 = a_1 \wedge c_1$ and $a_1 \vee b_1 = a_1 \vee c_1$ ---- (5)

and $a_2 \wedge b_2 = a_2 \wedge c_2$ and $a_2 \vee b_2 = a_2 \vee c_2$ ---- (6)

Since $(N, \leqslant)$ is a chain, it is modular. Therefore by Theorem 6 and the fact that any two natural numbers are either equal or they are comparable, we have

$$b_1 = c_1 \quad \text{and} \quad b_2 = c_2$$

$$\Rightarrow \quad (b_1, b_2) = (c_1, c_2).$$

Thus $N \times N$ is modular by Theorem 5.

Example 3.40. Show that a lattice L is modular if and only if for any $a, b, c \in L$,

$$a \leqslant c \leqslant a \vee b \quad \Rightarrow \quad a \vee (b \wedge c) = (b \wedge c) = c$$

Solution: Suppose L is a modular lattice and $a, b, c \in L$.

Then by definition of modular lattice, we have

$$a \leqslant c \Rightarrow a \vee (b \wedge c) = (a \vee b) \wedge c \quad \text{---- (1)}$$

Now we show

$$a \leqslant c \leqslant a \vee b \Rightarrow a \vee (b \wedge c) = c$$

Since $\quad c \leqslant a \vee b$

$\therefore \quad (a \vee b) \wedge c = c$

Thus from (1), we have

$$a \leqslant c \leqslant a \vee b \Rightarrow a \vee (b \wedge c) = c$$

Conversely, suppose that for any $a, b, c \in L$

$$a \leqslant c \leqslant a \vee b \Rightarrow a \vee (b \wedge c) = c$$

If L is not modular then L contain a pentagonal sublattice as shown in the figure below:

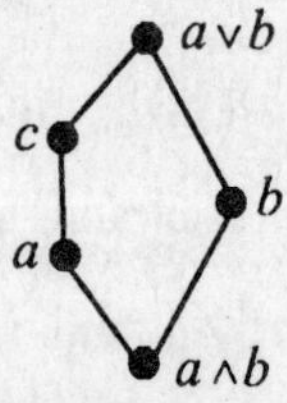

Fig. 3.18

In this sublattice, $a \leqslant c \leqslant a \vee b$ but

$$a \vee (b \wedge c) = a \neq c.$$

Hence given condition does not hold.

Thus L must be modular.

Example 3.41. Show that a lattice L is modular if and only if for any a, $b, c \in L$, the following condition holds:

$$a \vee (b \wedge (a \vee c)) = (a \vee b) \wedge (a \vee c).$$

Solution: First suppose that L is a modular lattice. Then

$$a \leqslant c \Rightarrow a \vee (b \wedge c) = (a \vee b) \wedge c \quad \text{---- (1)}$$

Since $a \leqslant a \vee c$, we have from (1),

$$a \vee (b \wedge (a \vee c)) = (a \vee b) \wedge (a \vee c)$$

Thus the given condition holds.

Conversely suppose that for all a, b, $c \in L$,

$$a \vee (b \wedge (a \vee c)) = (a \vee b) \wedge (a \vee c) \quad \text{---- (2)}$$

We show L is modular. Let a, b, $c \in L$, $a \leqslant c$.

Then $\quad a \vee c = c$

Hence from (2), we have

$$a \leqslant c \Rightarrow a \vee (b \wedge c) = (a \vee b) \wedge c$$

Thus L is modular.

§ 3.11. Distributive lattices

Before defining distributive lattices, we show that in a lattice L the following two equalities are equivalent.

$$a \wedge (b \vee c) = (a \wedge b) \vee (a \wedge c) \quad \text{---- (1)}$$

$$a \vee (b \wedge c) = (a \vee b) \wedge (a \vee c) \quad \text{---- (2)}$$

First we will show (1) $\Rightarrow$ (2)

$$\begin{aligned}
(a \vee b) \wedge (a \vee c) &= [(a \vee b) \wedge a] \vee [(a \vee b) \wedge c] \quad \text{by (1)} \\
&= a \vee [(a \vee b) \wedge c] \; \text{[by absorption property]} \\
&= a \vee [c \wedge (a \vee b)] \\
&= a \vee [(c \wedge a) \vee (c \wedge b)] \qquad \text{by (1)} \\
&= [a \vee (c \wedge a)] \vee (b \wedge c) \\
&= a \vee (b \wedge c) \quad \text{[by absorption property]}
\end{aligned}$$

Hence $a \vee (b \wedge c) = (a \vee b) \wedge (a \vee c)$ which is (2)

Now we will show (2) $\Rightarrow$ (1)

$$(a \wedge b) \vee (a \wedge c) = [(a \wedge b) \vee a] \wedge [(a \wedge b) \vee c] \quad \text{[by (2)]}$$
$$= a \wedge [(a \wedge b) \vee c] \quad \text{[by absorption property]}$$
$$= a \wedge [(c \vee a) \wedge (c \vee b)] \quad \text{[by (2)]}$$
$$= [a \wedge (c \vee a)] \wedge (c \vee b)]$$
$$= a \wedge (c \vee b) \quad \text{[by absorption property]}$$
$$= a \wedge (b \vee c)$$

Hence $a \wedge (b \vee c) = (a \wedge b) \vee (a \wedge c)$ which is (1)

Thus (1) and (2) are equivalent.

Definition: A lattice L is said to be **distributive lattice** if for any a, b, $c \in L$

$$a \wedge (b \vee c) = (a \wedge b) \vee (a \wedge c)$$

Since (1) and (2) are equivalent, therefore a lattice L is distributive if for any a, b, $c \in L$ any one of the above equalities holds.

Example 3.42. The lattice $(P(S), \subseteq)$ is a distributive lattice because

$$A \cap (B \cup C) = (A \cap B) \cup (A \cap C)$$

Example 3.43. Every chain is distributive lattice.

Solution: Let L be a chain and let a, b, $c \in L$. Since L is a chain, any two of a, b and c are comparable. Consider the following possible cases:

1. $a \leqslant b$ or $a \leqslant c$ and
2. $a \geq b$ and $a \geq c$.

For case (1), that is, $a \leqslant b$ or $a \leqslant c$

$$a \wedge (b \vee c) = a \quad \text{and} \quad (a \wedge b) \vee (a \wedge c) = a \vee a = a$$

Hence $a \wedge (b \vee c) = (a \wedge b) \vee (a \wedge c)$

For case (2), that is, $a \geq b$ and $a \geq c$

$$a \wedge (b \vee c) = b \vee c \text{ and}$$
$$(a \wedge b) \vee (a \wedge c) = b \vee c$$

Hence $a \wedge (b \vee c) = (a \wedge b) \vee (a \wedge c)$

Example 3.44. Consider the lattice shown in the figure given below.

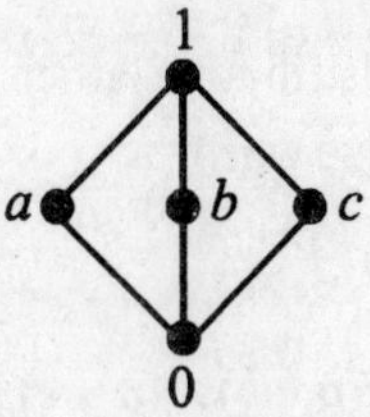

Fig. 3.19

This lattice is also not distributive because

$$a \wedge (b \vee c) = a, \text{ whereas}$$
$$(a \wedge b) \vee (a \wedge c) = 0$$

This lattice will be referred as M_5.

Example 3.45. Consider the lattice shown in the adjoining figure.

,This lattice is not also distributive because

$$c \wedge (a \vee b) = c, \text{ whereas}$$
$$(c \wedge a) \vee (c \wedge b) = a \vee 0 = a$$

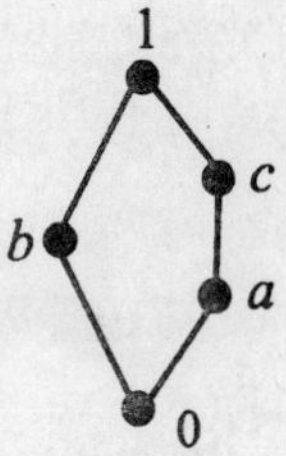

Fig. 3.20

Theorem 1: Every distributive lattice is modular.

Proof: Let L be a distributive lattice and $a, b, c \in L$. We will show

$$a \leqslant c \Rightarrow a \vee (b \wedge c) = (a \vee b) \wedge c$$

Since L is distributive, we have

$$a \vee (b \wedge c) = (a \vee b) \wedge (a \vee c) \quad \text{---- (1)}$$

Also, $$a \leqslant c \Rightarrow a \vee c = c \quad \text{---- (2)}$$

Hence from (1) and (2), we have

$$a \leqslant c \Rightarrow a \vee (b \wedge c) = (a \vee b) \wedge c$$

Thus L is modular lattice.

Example 3.46. The converse of the above theorem is not true. The lattice in example 3.44 above is modular but not distributive.

Theorem 2: The dual of distributive lattice is a distributive lattice.

Proof: Let $(L, \wedge, \vee)$ be a distributive lattice. We will show that its dual $(L, \wedge^*, \vee^*)$ is also distributive where $\wedge^* = \vee$ and $\vee^* = \wedge$. Let $a, b, c \in L$. Then we show

$$a \wedge^* (b \vee^* c) = (a \wedge^* b) \vee (a \wedge^* c)$$

or

$$a \vee (b \wedge c) = (a \vee b) \wedge (a \vee c)$$

But the last equality is true because $(L, \wedge, \vee)$ is distributive lattice. Hence $(L, \wedge^*, \vee^*)$ is also distributive.

Theorem 3: A lattice L is distributive if and only if

$$(a \vee b) \wedge (b \vee c) \wedge (c \vee a) = (a \wedge b) \vee (b \wedge c) \vee (c \wedge a) \quad \forall\, a, b, c \in L.$$

Proof: We suppose that L is distributive lattice.

$$\begin{aligned}
(a\vee b)\wedge(b\vee c)\wedge(c\vee a) &= a\wedge[(b\vee c)\wedge(c\wedge a)]\}\vee\{b\wedge[(b\vee c)\wedge(c\vee a)]\} \\
&\qquad [\because L \text{ is distributive}] \\
&= [(a\wedge(c\wedge a)\wedge(b\vee c)] \vee [(b\wedge(b\vee c))\wedge(c\vee a)] \\
&= [(a\wedge(b\vee c)] \vee [b\wedge(c\vee a)] \\
&\qquad \text{by absorption property} \\
&= [(a\wedge b)\vee(a\wedge c)] \vee [(b\wedge c)\vee(b\wedge a)] \\
&\qquad [\because L \text{ is distributive}] \\
&= (a\wedge b)\vee(b\wedge c)\vee(c\wedge a)
\end{aligned}$$

Conversely, we first show that L is modular.

Let x, y, z be any three elements of L with $x \leqslant z$

$$\begin{aligned}
x\vee(y\wedge z) &= [x\vee(x\wedge y)]\vee(y\wedge z) \quad [\text{by absorption}] \\
&= (x\wedge z)\vee(x\wedge y)\vee(y\wedge z) \\
&\qquad [\because x\leqslant z \Rightarrow x\wedge z = x] \\
&= (x\vee y)\wedge(y\vee z)\wedge(z\vee x) \\
&\qquad [\text{by given condition}] \\
&= (x\vee y)\wedge[(y\vee z)\wedge z] \qquad [\because x\leqslant z] \\
&= (x\vee y)\wedge z \quad [\text{by absorption property}]
\end{aligned}$$

Thus, L is modular.

Now for any $a, b, c \in L$

$$\begin{aligned} a \wedge (b \vee c) &= [a \wedge (a \vee c)] \wedge (b \vee c) \\ &\qquad \text{[by absorption property]} \\ &= a \wedge (a \vee b) \wedge (a \vee c) \wedge (b \vee c) \\ &\qquad \text{[by absorption property]} \\ &= [(a \vee b) \wedge (b \vee c) \wedge (c \vee a)] \wedge a \\ &= [(a \wedge b) \vee (b \wedge c) \vee (c \wedge a)] \wedge a \\ &\qquad \text{[by given condition]} \\ &= [((a \wedge b) \vee (c \wedge a)) \vee (b \wedge c)] \wedge a \qquad \text{---- (1)} \end{aligned}$$

Since

$$a \wedge b \leqslant a, a \wedge c \leqslant a \Rightarrow (a \wedge b) \vee (a \wedge c) \leqslant a,$$

we can apply modular identity on R.H.S of (1) to get

$$\begin{aligned} a \wedge (b \vee c) &= [(a \wedge b) \vee (c \wedge a)] \vee [(b \wedge c) \wedge a] \\ &= (a \wedge b) \vee [(c \wedge a) \vee (b \wedge (c \wedge a))] \\ &= (a \wedge b) \vee (c \wedge a) \\ &\qquad \text{[by absorption property]} \end{aligned}$$

$\Rightarrow$ L is distributive.

Theorem 4: Every sublattice of a distributive lattice is distributive.

Proof: Let S be a sublattice of a distributive lattice L. Let $a, b, c \in S$. Then $a, b, c \in L$.
Thus

$$a \wedge (b \vee c) = (a \wedge b) \vee (a \wedge c) \qquad \text{in } L.$$

Since S is closed under $\wedge$ and $\vee$, this equation holds in S. Hence S is distributive.

Theorem 5. A distributive lattice cannot contain a sublattice isomorphic with pentagonal lattice or with lattice M_5 (given in Ex.3.44).

Proof: In examples 3.44 and 3.45, we have seen that pentagonal lattice and lattice M_5 are not distributive and by Theorem 4, sublattice of a distributive lattice is distributive. Hence the result.

We know that every distributive lattice is modular but converse is not true in general. Since a modular lattice cannot have a sublattice isomorphic to the pentagonal lattice, therefore a distributive lattice also cannot have a sublattice isomorphic to the pentagonal lattice. The next theorem enables us to distinguish between distributive and non-distributive modular lattices.

Theorem 6: A modular lattice L is distributive if and only if it has no sublattice isomorphic to the lattice M_5.

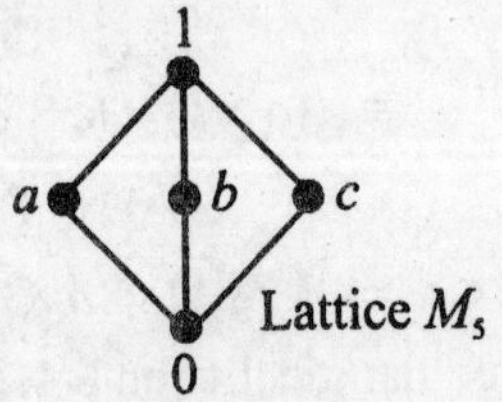

Fig. 3.21

Proof: The lattice M_5 is modular but not distributive because

$$a \wedge (b \vee c) = a$$

whereas

$$(a \wedge b) \vee (a \wedge c) = 0$$

Thus M_5 is not distributive. Since every sublattice of a distributive lattice is distributive, therefore the lattice L cannot be distributive. Hence a distributive lattice cannot contain sublattice isomorphic to lattice M_5.

Conversely, suppose the lattice L is modular. We shall prove the result by contradiction. Suppose, if possible, the lattice is not distributive. Then for some $a, b, c \in L$, we must have

$$(a \wedge b) \vee (b \wedge c) \vee (c \wedge a) \neq (a \vee b) \wedge (b \vee c) \wedge (c \vee a) \quad \text{---- (1)}$$

in view of Theorem 3.

We also know that in any lattice L, the inequality

$$(a \wedge b) \vee (b \wedge c) \vee (c \wedge a) \leqslant (a \vee b) \wedge (b \vee c) \wedge (c \vee a) \quad \text{---- (2)}$$

holds. From (1) and (2), we have

$$(a \wedge b) \vee (b \wedge c) \vee (c \wedge a) < (a \vee b) \wedge (b \vee c) \wedge (c \vee a) \quad \text{---- (3)}$$

Let

$$p = (a \vee b) \wedge (b \vee c) \wedge (c \vee a)$$
$$q = (a \wedge b) \vee (b \wedge c) \vee (c \wedge a)$$

Then $q < p$ in view of (3)

Put

$$x = p \wedge (q \vee a) = (p \wedge a) \vee q \quad \text{[by modular law as } q < p]$$
$$y = p \wedge (q \vee b) = (p \wedge b) \vee q \quad \text{[by modular law as } q < p]$$
$$z = p \wedge (q \wedge c) = (p \wedge c) \vee q \quad \text{[by modular law as } q < p]$$

From the definition of x, y, z, we have

$$q \leqslant x \leqslant p, \quad q \leqslant y \leqslant p, \quad q \leqslant z \leqslant p \quad \text{---- (4)}$$

We shall show $\{x, y, z, p, q\}$ forms a sublattice of L which is isomorphic to M_5.

Now

$$x \wedge y = [p \wedge (q \vee a)] \wedge [p \wedge (q \vee b)]$$
$$= p \wedge (q \vee a) \wedge (q \vee b)$$

Since

$$q = (a \wedge b) \vee (b \wedge c) \vee (c \wedge a),$$

$$\therefore \quad q \vee a = a \vee (b \wedge c) \quad \text{and} \quad q \vee b = b \vee (c \wedge a)$$

Substituting $q \vee a$ and $q \vee b$, we have

$$x \wedge y = p \wedge [\{a \vee (b \wedge c)\} \wedge \{b \vee (c \wedge a)\}]$$

Since

$$a \wedge c \leqslant a \leqslant a \vee (b \wedge c)$$

$\therefore$ using modular law, we get

$$x \wedge y = p \wedge [(c \wedge a) \vee \{(a \vee (b \wedge c)) \wedge b\}]$$

$$\Rightarrow \quad x \wedge y = p \wedge [(c \wedge a) \vee \{((b \wedge c) \vee a) \wedge b\}]$$

Since

$$b \wedge c \leqslant b \quad \therefore \quad \text{using modular law, we get}$$

$$x \wedge y = p \wedge [(c \wedge a) \vee \{(b \wedge c) \vee (a \wedge b)\}]$$

$$= p \wedge q = q \qquad [\because q \leqslant p]$$

By duality, we have

$$x \vee y = p \vee q = p$$

Thus

$$x \wedge y = q < p = x \vee y$$

Also, we claim

$$x \neq y$$

For

$$x = y \Rightarrow x \wedge y = x = q \quad \text{and}$$

$$x \vee y = x = p$$

$$\Rightarrow \quad p = q, \text{ a contradiction.}$$

Hence $\quad x \neq y$

Thus $\quad x \wedge y = q < p = x \vee y$ and $x \neq y$

By symmetry, we can show that

$$y \wedge z = q < p = y \vee z \text{ and } y \neq z$$

and

$$z \wedge x = q < p = z \vee x \text{ and } x \neq z$$

We now show that in (4), strict inequality holds.

Suppose $q = x$. Then since $q \leqslant y \leqslant p$, we get

$$x \leqslant y$$

$$\Rightarrow \quad x \wedge y = x \text{ and } x \vee y = y$$

$$\Rightarrow \quad p = y$$

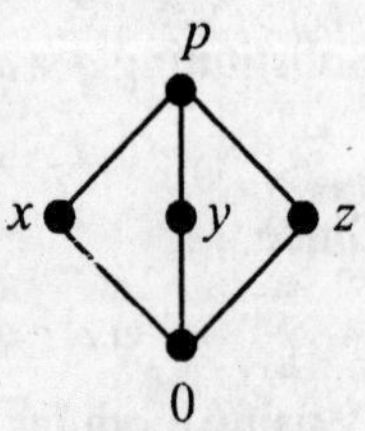

Fig. 3.22

Similarly, $q \leqslant z \leqslant p$ gives $x \leqslant z$

$\Rightarrow$ $p = z$

Now $p = y$ and $p = z$ gives $y = z$, which is not true.

Similarly other equalities do not hold in (4).

Hence $q < x < p,\ q < y < p,\ q < z < p.$

Combining all the results proved above it is obvious that $\{x, y, z, p, q\}$ forms a sublattice which is isomorphic to M_5. This completes the proof.

Using Theorem 4 of §3.10. and the above theorem, we obtain an important result as a corollary.

Corollary: A lattice is distributive if and only if it has no sublattice isomorphic to pentagonal lattice or M_5 lattice.

Theorem 7: In a distributive lattice, if an element has a complement then this complement is unique.

Proof: Let L be a distributive lattice. Suppose that an element a in L has two complements b and c. We shall show $b = c$.

Since b is a complement of a, we have

$$a \vee b = 1 \qquad \text{---- (1)}$$

and

$$a \wedge b = 0 \qquad \text{---- (2)}$$

Again c is a complement of a, we have

$$a \vee c = 1 \qquad \text{---- (3)}$$

and

$$a \wedge c = 0 \qquad \text{---- (4)}$$

Now

$$\begin{aligned} b &= b \wedge 1 \\ &= b \wedge (a \vee c) && \text{by (3)} \\ &= (b \wedge a) \vee (b \wedge c) && [\because L \text{ is distributive}] \\ &= 0 \vee (b \wedge c) && \text{by (2)} \\ &= (a \wedge c) \vee (b \wedge c) && \text{by (4)} \end{aligned}$$

$$= (a \vee b) \wedge c \quad \text{[by distributive property]}$$

$$= 1 \wedge c \quad \text{by (1)}$$

$$= c$$

Thus the complement of a is unique.

Theorem 8: Let L be a distributive lattice and let c' be the complement of an element c in L. If $b \wedge c' = 0$ then show that $b \leqslant c$.

Proof: Since $b \wedge c' = 0$, we have

$$(b \wedge c') \vee c = 0 \vee c = c$$

According to distributive law, we have

$$(b \vee c) \wedge (c' \vee c) = c$$

$$\Rightarrow \quad b \vee c = c$$

$$\Rightarrow \quad b \leqslant c$$

This completes the proof.

Solved Examples

Example 3.47. A lattice L is distributive if and only if the equations $a \wedge c = b \wedge c$ and $a \vee c = b \vee c$ imply $a = b$ for all $a, b, c \in L$.

Solution: Let L be a distributive lattice and let

$$a \wedge c = b \wedge c \text{ and } \quad a \vee c = b \vee c$$

Then

$$\begin{aligned} a &= a \wedge (a \vee c) \\ &= a \wedge (b \vee c) \\ &= (a \wedge b) \vee (a \wedge c) \\ &= (a \wedge b) \vee (b \wedge c) \\ &= b \wedge (a \vee c) \\ &= b \wedge (b \vee c) \\ &= b \end{aligned}$$

Conversely, let L be a lattice in which

$a \wedge c = b \wedge c$ and $a \vee c = b \vee c \Rightarrow a = b \; \forall \; a, b, c \in L$

By Theorem 5 of section 3.11, L is a modular lattice. Now suppose L is not distributive. Then L contains a sublattice $\{x, y, z, u, v\}$ isomorphic with the lattice given below.

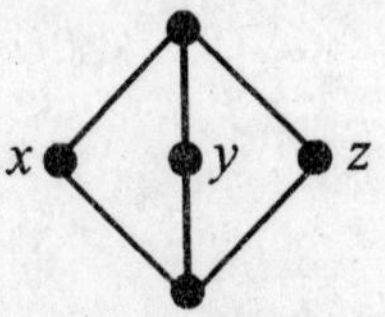

Fig. 3.23

Here $x \vee y = y \vee z$ and

$$x \wedge y = y \wedge z$$

Now by given condition, we find

$$x = z, \text{ a contradiction.}$$

Hence L is distributive.

Example 3.48. Let L be a complemented and distributive lattice. Then prove that De Morgan's laws given by

$$(a \vee b)' = a' \wedge b' \qquad \text{---- (1)}$$

and

$$(a \wedge b)' = a' \vee b' \qquad \text{---- (2)}$$

hold in L where a' denotes the complement of a.

Solution: To prove (1), we show that

$$(a' \wedge b') \vee (a \vee b) = 1$$

and

$$(a' \wedge b') \wedge (a \vee b) = 0$$

Now

$$\begin{aligned}(a' \wedge b') \vee (a \vee b) &= (a' \vee a \vee b) \wedge (b' \vee a \vee b)\\ &= (1 \vee b) \wedge (a \vee b \vee b')\\ &= (1 \vee b) \wedge (a \vee 1)\\ &= 1 \wedge 1\\ &= 1\end{aligned}$$

and

$$\begin{aligned}(a' \wedge b') \wedge (a \vee b) &= (a' \wedge b' \wedge a) \vee (a' \wedge b' \wedge b)\\ &= (a' \wedge a \wedge b') \vee (a' \wedge b' \wedge b)\\ &= (0 \wedge b') \vee (a' \wedge 0)\\ &= 0 \vee 0\\ &= 0\end{aligned}$$

Now by uniqueness of complement, we have

$$(a \vee b)' = a' \wedge b'$$

To prove (2), we will show

$$(a' \vee b') \vee (a \wedge b) = 1$$

and

$$(a' \vee b') \wedge (a \wedge b) = 0$$

Now

$$\begin{aligned}(a' \vee b') \vee (a \wedge b) &= (a' \vee b' \vee a) \wedge (a' \vee b' \vee b) \\ &= (a' \vee a \vee b') \wedge (a' \vee b' \vee b) \\ &= (1 \vee b') \wedge (a' \vee 1) \\ &= 1 \wedge 1 \\ &= 1\end{aligned}$$

Similarly

$$(a' \vee b') \wedge (a \wedge b) = 0$$

By uniqueness of complement, we have

$$(a \wedge b)' = a' \vee b'$$

Example 3.49. Consider the lattices shown in the following figures.

(a) (b)

Fig. 3.24

Show that the lattice in figure (a) is not distributive while the lattice in figures (b) is distributive.

Solution: The lattice in figure (a) is not distributive since $S = \{0, a, d, e, 1\}$ is a sublattice which is isomorphic to the non-distributive lattice.

The lattice in figure (b) is distributive since this lattice does not have a sublattice isomorphic to pentagonal lattice or the lattice of figure. 3.25

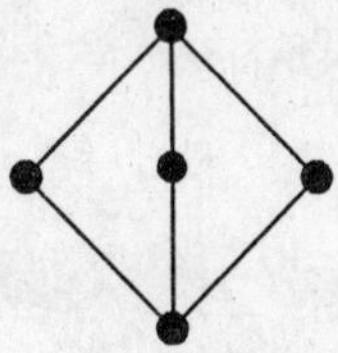

Fig. 3.25

Exercise

1. Show that a lattice L is modular if and only if for any elements a, $b, c \in L$,

   $$(a \wedge b) \vee (a \wedge c) = a \wedge ((a \wedge b) \vee c)$$

2. Show that in a bounded distributive lattice, the elements which have complements form a sublattice.

3. Prove that product $L \times M$ of two distributive lattices L and M is distributive.

4. Prove that the set of all normal subgroups of a group ordered by set inclusion $\subseteq$ forms a modular lattice.

5. Show that the lattice of factors of 45 under divisibility is a distributive lattice.

6. Show that a modular lattice is non-distributive if and only if it contains a sublattice isomorphic with M_5.

7. Define a distributive lattice. Consider the lattice $\{a, b, c, d, e\}$ where $a<b$, $a<c$, $b<d$, $d<e$, $c<e$. Show that this lattice is not distributive.

[illegible]

Exercises

1. Show that [illegible]

$f(a \wedge b) = f(a) \wedge f(b)$ [illegible]

2. Show that [illegible] have complements [illegible]

3. Prove that product [illegible] of two distributive lattices [illegible] distributive.

4. Prove that the set of all normal subgroups of a group ordered by set inclusion $\subseteq$ forms a modular lattice.

5. Show that [illegible] distributive lattice.

6. Show that a modular lattice is non-distributive if and only if it contains a sublattice isomorphic to [illegible].

7. Define a distributive lattice. Consider the lattice [illegible] where [illegible] Show that [illegible] distributive.

CHAPTER FOUR

Discrete Numeric Functions And Generating Functions

We recall that a real sequence is a function whose domain is the set N of natural numbers and whose range is a subset of the set R of real numbers. In the present chapter, we shall study the class of functions whose domain is the set {0, 1, 2, 3, ...} of non-negative integers and whose range is a subset of the set R of real numbers. We shall call such functions as discrete numeric functions or briefly, numeric functions. Numeric functions are useful because most of the counting problems involve non-negative integers.

§ 4.1. Discrete Numeric Functions

A function whose domain is the set of non-negative integers and whose range is a subset of the set of real numbers, is called **discrete numeric** functions or briefly, **numeric functions**. In other words, a discrete numeric function is a function from the set {0, 1, 2, 3, ...} into the set R of real numbers.

From the definition, it is obvious that a numeric function assigns to each non-negative integer *r*, a unique real number. It is usual to denote by a_r, the real number assigned by a numeric function **a**. This notation differs from the usual notations for functions in general. Thus, for a numeric function **a** we use $a_0, a_1, a_2, ..., a_r, ...$ to denote the values of the function at 0, 1, 2, ..., r, ...instead of **a**(0), **a**(1), **a**(2), ..., **a**(r), ...

From the definitions of a real sequence and a numeric function, it is clear that a numeric function can be viewed simply as a sequence of real numbers and conversely.

A numeric function can be described in several different ways.

We can describe a numeric function by listing in order its values ($a_0, a_1, a_2, ..., a_r, ...$). For example (1, 5, 10, 17, ..., r^2+1, ...) is the numeric function **a** whose r^{th} term a_r is r^2+1.

Another method for describing a numeric function is to use a representation (formula) for its rth term. For example, the numeric function (1, 5, 10, 17, ...) can also be specified as

$$a_r = r^2+1, \qquad r \geqslant 0$$

As another example, the numeric function (0, 2, 4, 26, 80, 342, ...) can be written as

$$a_r = \begin{cases} 2r, & 0 \le r \le 2 \\ 3^r - 1, & r \ge 3 \end{cases}$$

Sometimes, it is convenient to describe a numeric function ($a_0, a_1, a_2, ..., a_r, ...$) by specifying explicitly some initial terms a_0, a_1, a_2, etc and a relation for obtaining a_r in terms of $a_0, a_1, ..., a_{r-1}$. Such a relation is called *recurrence relation*. For example,

$$a_0 = 1 \text{ and } a_r = 3a_{r-1}, \qquad r \geqslant 1$$

describe the numeric function ($1, 3, 3^2, 3^3, ...$).

As another example, the recurrence relation

$$a^r = a_{r-1} + a_{r-2}, \qquad r \geqslant 2$$

together with

$$a_0 = 1, a_1 = 1$$

describe the numeric function (1, 1, 2, 3, 5, 8, 13,...). This numeric function is known as **Fibonacci numbers**.

It is clear that a recurrence relation determines the rth term of a numeric function indirectly. To compute a_r, we must first compute the terms $a_0, a_1, a_2, \ldots, a_{r-1}$. The explicitly given values of a numeric function are called initial conditions.

Sum of Numeric Functions

If **a** and **b** are two numeric functions, then their sum **a+b** is a numeric function whose value at r is equal to the sum of the values of the two numeric function **a** and **b** at r. For example, consider the two numeric functions **a** and **b** where

$$a_r = \begin{cases} 0, & 0 \le r \le 2 \\ 2^{-r} + 5, & r \ge 3 \end{cases}$$

and

$$b_r = \begin{cases} 3 - 2^r, & 0 \le r \le 1 \\ r + 2, & r \ge 2 \end{cases}$$

Let **c** be the sum of numeric functions **a** and **b** which is denoted by **a+b**. Then

$$c_r = a_r + b_r = \begin{cases} 3 - 2^r, & 0 \le r \le 1 \\ 4, & r = 2 \\ 2^{-r} + r + 7, & r \ge 3 \end{cases}$$

Example 4.1. Let **a** be a numeric function such that a_r is equal to the remainder when the integer **r** is divided by 17. Let **b** be a numeric function such that b_r is equal to 0 if the integer **r** is divisible by 3 and is equal to 1 otherwise. Let $c_r = a_r + b_r$. For what values of r is $c_r = 0$?

Solution: We have

$$a_r = r \ (\text{mod } 17)$$

and

$$b_r = \begin{cases} 0 & \text{if } r = 3n, n = 0, 1, 2, \ldots. \\ 1 & \text{if } r \ne 3n, n = 0, 1, 2, \ldots. \end{cases}$$

Since both a_r and b_r are non-negative, therefore their sum $c_r = a_r + b_r$ will be zero when both a_r and b_r are zero. From the definition of a_r, it is obvious that

$a_r = 0$ if $r = 0, 17, 34, \ldots$

That is, $a_r = 0$ when r is multiple of 17.

Similarly, $b_r = 0$ when r is multiple of 3.

Thus, $c_r = a_r + b_r$ is zero when r is multiple of 51, the lcm of 17 and 3. Hence $c_r = 0$ if $r = 51k$, $k = 0, 1, 2, \ldots$

Multiplication of Numeric Function by a Real Number

Let **a** be a numeric function and α be any real number. The multiplication of numeric function **a** with real number α is a numeric function whose value at r is equal to α times a_r and is denoted by $\alpha\mathbf{a}$. The numeric function $\alpha\mathbf{a}$ is called a scaled version of **a** with scaling factor α. For example, consider the numeric function **a**, where

$$a_r = \begin{cases} 0, & 0 \le r \le 2 \\ 2^{-r} + 5, & r \ge 2 \end{cases}$$

Then 2**a** is a numeric function whose value at **r** is equal to

$$2a_r = \begin{cases} 0, & \text{if } 0 \le r \le 2 \\ 2(2^{-r} + 5), & \text{if } r \ge 2 \end{cases}$$

Product of Numeric Functions

Let **a** and **b** be two numeric functions. The product of **a** and **b**, to be denoted by **ab**, is a numeric function whose value at r is equal to the product of the values of the two numeric functions **a** and **b** at r. For example, consider the two numeric functions **a** and **b**, where

$$a_r = \begin{cases} 0, & 0 \le r \le 2 \\ 2^{-r} + 5, & r \ge 3 \end{cases}$$

and

$$b_r = \begin{cases} 3 - 2^r, & 0 \le r \le 1 \\ r + 2, & r \ge 2 \end{cases}$$

If **c** be the product of **a** and **b**, then

$$c_r = a_r b_r = \begin{cases} 0, & 0 \le r \le 2 \\ (2^{-r} + 5)(r + 2), & r \ge 3 \end{cases}$$

Numeric Function $S^i a$ and $S^{-i} a$

Let **a** be a numeric function and *i* a positive integer. Then $S^i\mathbf{a}$ is a numeric function whose value at *r* is 0 for $r = 0, 1, 2, ..., i-1$ and is a_{r-i} for $r \geqslant i$. If we denote the numeric function $S^i\mathbf{a}$ by **b**, then

$$b_r = \begin{cases} 0, & 0 \le r \le i-1 \\ a_{r-i}, & r \ge i \end{cases}$$

denotes the value of $S^i\mathbf{a}$ at **r**. For example, consider the numeric function **a** where

$$a_r = \begin{cases} 1, & 0 \le r \le 10 \\ 2, & r \ge 11 \end{cases}$$

Then $S^5\mathbf{a}$ is a numeric function whose value at *r* is given by

$$b_r = \begin{cases} 0, & 0 \le r \le 4 \\ a_{r-5}, & r \ge 5 \end{cases} = \begin{cases} 0, & 0 \le r \le 4 \\ 1, & 5 \le r \le 15 \\ 2, & r \ge 16 \end{cases}$$

Let **a** be a numeric function and *i* a positive integer. Then $S^{-i}\mathbf{a}$ is a numeric function whose value at **r** is a_{r+i} for $r \geqslant 0$. In other words, $S^{-i}\mathbf{a}$ is a numeric function whose value at **r** is given by b_r, where

$$b_r = a_{r+i}, \quad r \geqslant 0$$

For example, let **a** be a numeric function , where

$$a_r = \begin{cases} 1, & 0 \le r \le 10 \\ 2, & r \ge 11 \end{cases}$$

Then $S^{-7}\mathbf{a}$ is a numeric function defined as follows:

$$b_r = \begin{cases} 1, & 0 \le r \le 3 \\ 2, & r \ge 4 \end{cases}$$

where b_r denoted the value of $S^{-7}\mathbf{a}$ at *r*.

Accumulated Sum of a Numeric Function

Let **a** be a numeric function. Then the **accumulated sum** of numeric function **a** is a numeric function **b** whose value at **r** is equal to

$$\sum_{i=0}^{r} a_i$$

In other words, accumulated sum of **a** is given by

$$b_r = \sum_{i=0}^{r} a_i, \qquad r \geq 0$$

Example 4.2. Suppose that Ravi deposits Rs. 100 each year in a saving account at an annual compound interest rate of 8 percent. What is the total amount in Ravi's saving account after **r** years?

Solution: The numeric function which describes the amount in the saving account at the end of the rth year if Rs. 100 was deposited into the account at the beginning is given by **a**, where

$$a_r = 100(1.08)^r, r \geqslant 0$$

It is clear that total amount in Ravi's saving account after r years is given by the accumulated sum of the numeric function **a**. Let **b** denote the accumulated sum of **a**. Then,

$$b_r = \sum_{i=0}^{r} a_i = \sum_{i=0}^{r} 100(1.08)^i = \frac{10{,}000}{8}[(1.08)^{r+1} - 1], \ r \geq 0$$

gives the total amount in Ravi's saving account after **r** years.

Convolution of Numeric Functions

Let **a** and **b** be two numeric functions. The convolution of **a** and **b**, denoted by **a*b**, is a numeric function **c** such that

$$c_r = a_0 b_r + a_1 b_{r-1} + \ldots.. + a_{r-1} b_1 + a_r b_0 = \sum_{i=0}^{r} a_i \, b_{r-i}$$

Example 4.3. Let **a** and **b** be two numeric functions such that

$$a_r = 3^r, r \geqslant 0$$

and

$$b_r = 2^r, r \geqslant 0$$

Then convolution **a*b** of **a** and **b** is the numeric function **c** such that

$$c_r = \sum_{i=0}^{r} a_i b_{r-i} = \sum_{i=0}^{r} 3^i 2^{r-i}, \ r \geq 0$$

Example 4.4. Determine **a∗b** where

$$a_r = \begin{cases} 1, & 0 \le r \le 2 \\ 0, & r \ge 3 \end{cases}$$

and

$$b_r = \begin{cases} 1, & 0 \le r \le 2 \\ 0, & r \ge 3 \end{cases}$$

Solution: Let **c** = **a∗b**. Then

$$\begin{aligned} c_0 &= a_0 b_0 = 1 \\ c_1 &= a_0 b_1 + a_1 b_0 = 2 \\ c_2 &= a_2 b_0 + a_1 b_1 + a_0 b_2 = 3 \\ c_3 &= a_3 b_0 + a_2 b_1 + a_1 b_2 + a_0 b_3 = 2 \\ c_4 &= a_4 b_0 + a_3 b_1 + a_2 b_2 + a_1 b_3 + a_0 b_4 = 1 \end{aligned}$$

and $c_r = 0$ if $r \geqslant 5$

Thus **c** = **a∗b** is given by

$$c_r = \begin{cases} 1, & r = 0, 4 \\ 2, & r = 1, 3 \\ 3, & r = 2 \\ 0, & r \ge 5 \end{cases}$$

§ 4.2. Asymptotic Behavior of Numeric Functions

A major concern of theoretical computer science is the design of algorithms for performing certain tasks, which minimize the length of computer time required. Thus, in order to assess the efficiency of an algorithm, we are interested in knowing the amount of time it takes for a computer to perform the task with large size (of inputs). For example, let **a** be a numeric function in which a_r is the time it takes for a certain sorting algorithm to sort **r** numbers. We are interested in how a_r increases when we want to use the algorithm to sort tens of thousands of names in the preparation of a telephone directory.

Let **a** be a numeric function. By the asymptotic behavior of the numeric function **a**, we mean how the value of the function varies for large *r*. For example, the value of the numeric function **a**, where

$$a_r = 7, \qquad r \geqslant 0$$

remains constant for increasing values of *r*. The value of the numeric function **b**, where

$$b_r = 5r, \qquad r \geqslant 0$$

increases for increasing values of **r** and is proportional to **r**, while for the numeric function **c**, where

$$c_r = 5/r, \qquad r \geqslant 0$$

the value of the function decreases for increasing **r** and is proportional to 1/**r** which approaches 0 as limit.

In many applications, we are interested in comparing the asymptotic behavior of two numeric functions (out of which behavior of one function is generally known). For this purpose, we introduce and study the notion of asymptotic dominance.

Asymptotic Dominance

Let **a** and **b** be two numeric functions. We say that **a** symptotically dominates **b** or **b** is asymptotically dominated by **a**, if there exist positive constants *k* and *n* such that

$$|b_r| \leqslant k\, a_r \qquad \text{whenever } r \geqslant n$$

In other words, $|b_r|/a_r$ is bounded by *k* throughout $[n, \infty[$.

It may be observed that the condition $|b_r| \leqslant ka_r$ for $r \geqslant n$ insists that a_r must be positive.

Example 4.5. Let **a** and **b** be two numeric functions such that

$$a_r = r + 2, \qquad r \geq 0$$

and

$$b_r = \frac{2}{r} + 7, \qquad r \geq 0$$

Then ***a*** asymptotically dominates ***b*** because for $n = 6$ and $k = 1$,

$$|b_r| \leqslant a_r \qquad \text{whenever } r \geqslant 6$$

Example 4.6. Let **a** and **b** be two numeric functions such that

$$a_r = r^2+r, \quad r \geqslant 0$$

and

$$b_r = r^2, \quad r \geqslant 0$$

Show that **a** asymptotically dominates **b** and **b** is also asymptotically dominates **a**.

Solution: We first show that **a** asymptotically dominates **b**. For $n = 1$ and $k = 1$, we have

$$|b_r| = r^2 \leqslant r^2+r = a_r \qquad \text{for } r \geqslant 1$$

Hence **a** asymptotically dominates **b**.

We now show **b** asymptotically dominates **a**. for $n = 1$ and $k = 2$, we have

$$|r^2+r| \leqslant r^2+r^2 = 2r^2 \text{ for } r \geqslant 1$$

Thus $|a_r| \leqslant 2b_r$ whenever $r \geqslant 1$. Hence, **b** asymptotically dominates **a**.

Example 4.7. In Example 4.5 above, show that **b** does not dominate **a**.

Solution: If possible, suppose that **b** dominates **a**. Then there exist positive constants k and n such that

$$|a_r| \leqslant kb_r \qquad \text{whenever } r \geqslant n.$$

Then, we have

$$r+2 \leqslant 2k/r + 7k \qquad \text{for all } r \geqslant n.$$

Let r_0 be the positive integer greater than or equal to the maximum of n and $7k$. Then we see that

$$r_0 + 2 > \frac{2k}{7k} + 7k \geq \frac{2k}{r_0} + 7k$$

But it gives $|a_r| > kb_r$ whenever $r \geqslant r_0$, a contradiction.

Hence **b** does not dominate **a**.

Example 4.8. Let **a** and **b** be two numeric functions such that

$$a_r = \frac{r^2}{10} - 10$$

and

$$a_r = \frac{r^2}{10} - 10$$

Then show that **a** asymptotically dominates **b**.

Solution: For $k = 1$ and $n = 10^3+1 = 1001$, we have

$$|b_r| = 1001 \times 100 \leq \frac{(1001)^2}{10} - 10 = a_r, \qquad \text{whenever } r \geq 1001$$

Hence **a** asymptotically dominates **b**.

Theorem 1: The following statements hold for numeric functions **a**, **b** and **c**:

(i) For any numeric function *a*, |*a*| asymptotically dominates *a*.

(ii) If **b** is asymptotically dominated by **a**, then for any real constant α, α**b** is also asymptotically dominated by **a**.

(iii) If **b** is asymptotically dominated by **a**, then for any integer *i*, S^i**b** is asymptotically dominated by $S^i a$.

(iv) If both **b** and **c** are asymptotically dominated by **a** then **b**+**c** is also asymptotically dominated by **a**.

(v) If **c** is asymptotically dominated by **b** and **b** is asymptotically dominated by **a**, then **c** is asymptotically dominated by **a**.

Proof:

(i) Since $|a_r| \leqslant |a_r|$ for every $r \geqslant 1$, it follows from the definition that |**a**| asymptotically dominates **a**.

(ii) Suppose that numeric function **b** is asymptotically dominated by **a**. It means there exist positive constants *k* and *n* such that

$$|b_r| \leqslant ka_r \qquad \text{whenever } r \geqslant n$$

$$\Rightarrow \quad |\alpha b_r| \leqslant |\alpha| ka_r \qquad \text{whenever } r \geqslant r$$

Let,

$$k_1 = |\alpha| k.$$

Then, we have

$$|\alpha b_r| \leqslant k_1 a_r \qquad \text{whenever } r \geqslant n$$

Hence $\alpha\mathbf{b}$ is asymptotically dominated by **a**.

(iii) Suppose that **b** is asymptotically dominated by **a**. Then there exist positive constants k and n such that

$$|b_r| \leqslant ka_r \qquad \text{whenever } r \geqslant n$$

$$\Rightarrow \qquad |b_{r-i}| \leqslant ka_{r-i} \text{ whenever } r \geqslant n+i \text{ for any } i.$$

Hence, we have found two positive constants k and $n_0 = n+i$ such that

$$|b_{r-i}| \leqslant a_{r-i} \qquad \text{whenever } r \geqslant n_0$$

Hence, $S^i\mathbf{b}$ is asymptotically dominated by $S^i\mathbf{a}$.

(iv) Suppose that numeric functions **b** and **c** both are asymptotically dominated by **a**. It means that there exist two pairs of constants k_1, n_1 and k_2, n_2 such that

$$|b_r| \leqslant k_1a_r \qquad \text{for all } r \geqslant n_1$$

and

$$|c_r| \leqslant k_2a_r \qquad \text{for all } r \geqslant n_2$$

Let

$$n = \max\{n_1, n_2\}$$

and

$$k = k_1 + k_2$$

Then

$$|b_r + c_r| \leqslant ka_r \qquad \text{whenever } r \geqslant n$$

Hence **b**+**c** is asymptotically dominated by **a**.

(v) Let **a**, **b** and **c** be numeric functions such that **c** is asymptotically dominated by **b** and **b** is asymptotically dominated by **a**. We will show that **c** is asymptotically dominated by **a**. Since **c** is asymptotically dominated by **b**, there exist positive constants k_1 and n_1, such that

$$|c_r| \leqslant k_1b_r \qquad \text{whenever } r \geqslant n_1 \qquad \dots (1)$$

Again, **b** is asymptotically dominated by **a** implies, there exist positive constants k_2 and n_2 such that,

$$|b_r| \leqslant k_2a_r \qquad \text{whenever } r \geqslant n_2 \qquad \dots (2)$$

Let $k = k_1k_2$ and $n = \max\{n_1, n_2\}$. Then from (1) and (2) we have

$$|c_r| \leqslant ka_r \qquad \text{whenever } r \geqslant n$$

Hence **c** is asymptotically dominated by **a**.

Example 4.9. Give two numeric functions **a** and **b** which are not equal such that **a** asymptotically dominates **b** and **b** also asymptotically dominates **a**.

Solution: Consider the following numeric functions **a** and **b** where

$$a_r = r^2+r+1, \qquad r \geqslant 0$$

$$b_r = \frac{r^2}{2} - r - 2 \qquad r \geq 0$$

For $k = 1$ and $n = 1$, we have

$$|b_r| \leqslant ka_r \qquad \text{whenever } r \geqslant 1$$

Thus $\boldsymbol{a}$ asymptotically dominates $\boldsymbol{b}$.

Again, for $k = 4$ and $n = 7$, we have

$$|a_r| \leqslant kb_r \qquad \text{whenever } r \geqslant 7$$

because $\quad r^2+r+1 \leqslant 4(r^2/2-r-2)$ whenever $\quad r \geq \sqrt{5r+9}$

and we know $\quad r \geq \sqrt{5r+9} \qquad$ when $r \geqslant 7$.

Thus **b** also asymptotically dominates **a**.

Note: The above example justifies the constant k in the definition of asymptotic dominance.

Example 4.10. Give two numeric functions **a** and **b** such that neither **a** asymptotically dominates **b**, nor **b** asymptotically dominates **a**.

Solution: Consider the numeric functions **a** and **b**, where

$$a_r = \begin{cases} 1, & r \text{ is even} \\ 0, & r \text{ is odd} \end{cases}$$

and

$$b_r = \begin{cases} 0, & r \text{ is even} \\ 1, & r \text{ is odd} \end{cases}$$

Then **a** does not asymptotically dominate **b** nor does **b** asymptotically dominate **a**.

§ 4.3. The Big O-Notation

For a given discrete numeric function **a**, we denote the set of all numeric functions that are asymptotically dominated by **a** by O(*a*). O(*a*) is read "order *a*" or "big-oh of *a*". Thus O(***a***) is a set consisting of all numeric functions which are asymptotically dominated by ***a***. Symbolically, $b \in O(a)$ if we can find two positive constants k and n such that

$$|b_r| \leqslant k\, a_r \qquad \text{whenever } r \geqslant n.$$

If **b** is in the set O(***a***) then we often simply say **b** is O(***a***) instead of saying that **b** is in the set O(***a***). Observe that **b** is O(***a***) only means that **b** does not grow faster than **a**. However it may be that **b** could grow much slower than **a**. For example, the numeric function **b** defined by $b_r = r$ is O(***a***), where $a_r = r^2$, which grows much slower than **a**.

When numeric function **a** can be expressed in a single closed form such as $a = r^2$ we shall also write O(**a**) as O(r^2). Thus when we say that a numeric function **b** is O(r^2), we mean that **b** is asymptotically dominated by the numeric function ***a*** = r^2. In particular, a numeric function **b** is O(1) means that **b** is asymptotically dominated by the numeric function ***a*** = 1 which implies ***b*** is bounded numeric function.

Example 4.11. Let

$$\mathbf{a} = r^2$$

$$\mathbf{b} = 10r^2 + 25$$

$$\mathbf{c} = 5 - r$$

and

$$d = \frac{1}{2}r^2 \log r - r^2$$

Show that **b** is O(**a**), **c** is O(**a**) but **d** is not O(**a**).

Solution: We shall show that both **b** and **c** are asymptotically dominated by **a** and **d** is not asymptotically dominated by **a**. **b** is asymptotically dominated by **a** because for $k = 11$ and $n = 5$ we have

$$|b_r| \leqslant ka_r \qquad \text{for all integers } r \geqslant n.$$

c is asymptotically dominated by **a** because for $k_1 = 1$ and $n_1 = 2$, we have

$$|c_r| \leqslant k_1 a_r \qquad \text{for all integers } r \geqslant n_1$$

Thus both **b** and **c** are O(**a**)

We now show that **d** is not dominated by **a** asymptotically. If possible suppose there exist positive constants k and n such that

$$|d_r| \leqslant ka_r \qquad \text{whenever } r \geqslant n.$$

$$\Rightarrow \quad \frac{1}{2}r^2 \log r - r^2 \leq kr^2 \qquad \text{for all integers } r \geqslant n.$$

$$\Rightarrow \quad \log r \leqslant 2(k+1) \qquad \text{for all integers } r \geqslant n.$$

But this is not true because there exists $r_0 \geqslant n$ such that $\log r_0 > 2(k+1)$.

Hence **d** is not O(**a**).

Example 4.12. Show that the numeric function **a** where

$$a_r = \alpha_0 + \alpha_1 r + \alpha_2 \alpha^2 + \ldots + \alpha_n r^n,$$

where α_i's are constants is $O(r^n)$.

Solution: We have

$$|a_r| = |\alpha_0 + \alpha_1 r + \alpha_2 r^2 + \ldots + \alpha_n r^n| \leqslant |\alpha_0| + |\alpha_1|\, r + |\alpha_2| r^2 + \ldots + |\alpha_n| r^n$$
$$\leqslant (|\alpha_0| + |\alpha_1| + \ldots + |\alpha_n|) r^n$$

Now for

$$k = |\alpha_0| + |\alpha_1| + \ldots + |\alpha_n| \text{ and } n = 1,$$

we have

$$|a_r| \leqslant kr^n \qquad \text{whenever } r \geqslant 1$$

Thus **a** is asymptotically dominated by r^n. Consequently, **a** is $O(r^n)$

Example 4.13. Show that for any i and j with $i \leqslant j$, the numeric function $a = \alpha r^i$, where **a** is a real constant is $O(r^j)$.

Solution: We know that for $i \leqslant j$

$$|r^i| \leqslant r^j \qquad \text{for every } r \geqslant 1$$

$$\Rightarrow \qquad r^i \text{ is } O(r^j)$$

$$\Rightarrow \qquad \alpha r^i \text{ is } O(r^j) \qquad \text{by Theorem 1(ii)}$$

Example 4.14. Show that for any $\alpha > 1$, $O(r^i) \subset O(\alpha^r)$

Solution: Since $\alpha > 1$, we can write

$$\alpha = 1 + \delta \text{ with } \delta > 0$$

Since r is a positive integer, therefore by binomial theorem,

$$\alpha^r = (1+\delta)^r$$

$$= 1 + r\delta + \frac{r(r-1)}{\lfloor 2}\delta^2 + \ldots + \frac{r(r-1)(r-2)\ldots(r-n+1)}{\lfloor n}\delta^n + \ldots$$

$$\geq \frac{r(r-1)\ldots\ldots(r-n+1)}{\lfloor n}\delta^n$$

Let n be a positive integer greater than i. Then if $r \geqslant n$

$$\frac{r^i}{\alpha^r} \leq \frac{\lfloor n}{\delta^n}\frac{r^i}{r(r-1)\ldots(r-n+1)} \leq \frac{\lfloor n}{\delta^n}\frac{r^n}{r(r-1)\ldots(r-n+1)}$$

because $r^i \leqslant r^n$ as $n > i$

This gives

$$\left|r^i\right| \leq k\alpha^r \qquad \text{where } k = \frac{\lfloor n}{\delta^n\left(1-\frac{1}{n}\right)\left(1-\frac{2}{n}\right)\ldots\left(\frac{1}{n}\right)} \text{ and } r \geq n$$

Since n is fixed, k is a constant.

Hence $O(r^i) \subset O(\alpha^r)$

Example 4.15. Show that $O(\log r) \subseteq O(r)$.

Solution: We know from elementary calculus that

$$\lim_{r\to\infty} \frac{\log r}{r} = 0$$

This implies that for any positive constant k there exists a positive integer n such that

$$\left|\frac{\log r}{r}\right| \leq k \text{ whenever } r \geq n$$

$$\Rightarrow \quad |\log r| \leqslant kr \text{ whenever } r \geqslant n$$

$$\Rightarrow \quad \log r \text{ is } O(\mathbf{r})$$

$$\Rightarrow \quad O(\log r) \subseteq O(\mathbf{r})$$

We now restate Theorem 1 in terms of the "big-oh" notations.

Theorem 2: For numeric functions **a**, **b** and **c**, the following results hold:

(i) For any numeric function **a**, **a** is O(|**a**|).

(ii) If **b** is O(**a**) then α**b** is also O(**a**) for any real constant α.

(iii) If **b** is O(**a**) then for any integer i, $S^i\mathbf{b}$ is $O(S^i\mathbf{a})$.

(iv) If both **b** and **c** are O(**a**) then **b**+**c** is also O(**a**).

(v) If **c** is O(**b**) and **b** is O(**a**) then **c** is O(**a**).

Let A and B be two sets of numeric functions. We use the following symbols to denote sets of numeric functions:

$$A+B = \{\mathbf{a}+\mathbf{b} : \mathbf{a}\in A, \mathbf{b}\in B\}$$

$$\alpha A = \{\alpha\mathbf{a} : \mathbf{a}\in A\}, \text{ where } \alpha \text{ is real constant}$$

$$A\cdot B = \{\mathbf{ab} : \mathbf{a}\in A, \mathbf{b}\in B\}$$

Theorem 3: If **a** and **b** are numeric functions, then the following results hold:

(i) If **b** is O(**a**) then O(**b**)⊆O(**a**). In particular, if **b** is O(**a**) and **a** is O(**b**), then O(**a**) = O(**b**)

(ii) For any **a**, O(**a**)+O(**a**) = O(**a**)

(iii) If **b**∈O(**a**), then O(**a**)+O(**b**) = O(**a**)

(iv) For any real constant α, αO(**a**)= O(α**a**) = O(**a**)

(v) For any **a** and **b**, O(**a**)O(**b**) = O(**ab**)

Proof: (i) Let **c** be in the set O(**b**). Then **c** is asymptotically dominated by **b**. Again **b** is O(**a**) implies **b** is a asymptotically dominated by **a**. Now by Theorem 1 (v), **c** is asymptotically dominated by **a**. Hence we have, **c** is in the set O(**a**). Thus O(**b**) $\subseteq$ O(**a**). In particular if **b** is O(**a**) and **a** is O(**b**) then O(**b**) $\subseteq$ O(**a**) and O(**a**) $\subseteq$ O(**b**) which implies O(**a**) = O(**b**).

(ii) To show O(**a**) + O(**a**) = O(**a**). Let **b**$\in$O(**a**)

$\Rightarrow$ **b** $\in$ O(**a**), **b** $\in$ O(**a**)

$\Rightarrow$ **b**+**b** $\in$ O(**a**)+O(**a**) by definition of A+B

$\Rightarrow$ 2**b** $\in$ O(**a**)+O(**a**)

$\Rightarrow$ **b** $\in$ O(**a**)+ O(**a**) by Theorem 2(ii)

(iii) We first show that O(**a**)+O(**b**) $\subseteq$ O(**a**). Let **c**$\in$O(**a**)+O(**b**).

$\Rightarrow$ **c** = **c**′+**c**″ where **c**′ is in O(**a**) and **c**″ is in O(**b**)

$\Rightarrow$ **c**′and **c**″ are O(**a**) because **b**$\in$O(**a**) $\Rightarrow$ O(**b**) $\subseteq$ O(**a**)

$\Rightarrow$ $c' + c''$ is O(**a**) by Theorem 2(iv)

$\Rightarrow$ **c** $\in$ O(**a**)

Hence

O(**a**) + O(**b**) $\subseteq$ O(**a**)

Again, let **d** $\in$ O(**a**)

$\Rightarrow$ **d**+0 $\in$ O(**a**) + O(**b**)

($\because$ 0$\in$O(**b**) where 0 is the numeric function whose each term is zero.)

$\Rightarrow$ **d** $\in$ O(**a**)+O(**b**)

$\Rightarrow$ O(**a**) $\subseteq$ O(**a**)+O(**b**)

Hence

O(**a**) + O(**b**)=O(**a**)

(iv) Let α be any real constant and let **b** $\in$ O(**a**). Then α**b** $\in$ O(**a**) by theorem 2 (ii).

$$\Rightarrow \quad \alpha\, O(\mathbf{a}) \subseteq O(\mathbf{a})$$

Again, let

$$\mathbf{c} \in \alpha\, O(\mathbf{a})$$

$$\Rightarrow \quad \mathbf{c} = \alpha\mathbf{b} \quad \text{where } \mathbf{b} \in O(\mathbf{a})$$

$$\Rightarrow \quad \alpha\mathbf{b} \in O(\mathbf{a}) \quad (\because \mathbf{b} \in O(\mathbf{a}) \text{ and in view of theorem 2 (ii))}$$

$$\Rightarrow \quad \mathbf{c} \in O(\mathbf{a})$$

$$\Rightarrow \quad \alpha O(\mathbf{a}) \subseteq O(\mathbf{a})$$

Hence

$$\alpha\, O(\mathbf{a}) = O(\mathbf{a})$$

Similarly, we can show

$$O(\alpha\mathbf{a}) = O(\mathbf{a})$$

Thus,

$$\alpha\, O(\mathbf{a}) = O(\alpha\mathbf{a}) = O(\mathbf{a})$$

(v) We have to show $O(\mathbf{a})O(\mathbf{b}) = O(\mathbf{ab})$

Let

$$\mathbf{c} \in O(\mathbf{a})O(\mathbf{b})$$

$$\Rightarrow \quad \mathbf{c} = \mathbf{d}\cdot\mathbf{e}, \quad \text{where } \mathbf{d} \in O(\mathbf{a}) \text{ and } \mathbf{e} \in O(\mathbf{b})$$

$\Rightarrow \exists$ positive constants k_1, n_1 and k_2, n_2 such that

$$|d_r| \leqslant k_1 a_r \quad \text{for every } r \geqslant n_1$$

and

$$|e_r| \leqslant k_2 b_r \quad \text{for every } r \geqslant n_2$$

Let $n = \max\{n_1, n_2\}$ and $k = k_1 k_2$. Then

$$|d_r e_r| \leqslant k a_r b_r \quad \text{for every } r \geqslant n$$

Hence

$$\mathbf{d}\cdot\mathbf{e} \in O(\mathbf{ab})$$

Thus

$$O(\mathbf{a})\, O(\mathbf{b}) \subseteq O(\mathbf{ab}) \qquad \dots (1)$$

Again, let

$$\mathbf{f} \in O(\mathbf{ab})$$

$\Rightarrow \exists$ positive constants k and n such that

$$|f_r| \leqslant k a_r b_r \quad \text{for every } r \geqslant n$$

Now since $\mathbf{a} \in O(\mathbf{a})$ and $\mathbf{b} \in O(\mathbf{b})$, therefore $\mathbf{ab} \in O(\mathbf{a})\, O(\mathbf{b})$.

$$|f_r| \leqslant k a_r b_r \quad \text{for all } r \geqslant n \text{ and } \mathbf{ab} \in O(\mathbf{a})\, O(\mathbf{b})$$

$$\Rightarrow \qquad f \in O(\mathbf{a})\, O(\mathbf{b})$$

Thus

$$O(\mathbf{ab}) \subseteq O(\mathbf{a})\, O(\mathbf{b}) \qquad \ldots (2)$$

From (1) and (2) we have

$$O(\mathbf{ab}) = O(\mathbf{a})\, O(\mathbf{b})$$

Example 4.16. Let

$$a = (r^3/2) + r^2 + r$$

Clearly, **a** is $O(r^3)$. Also, **a** is in the set of numeric functions $\left\{\frac{r^3}{2}\right\} + O(r^2)$ as well as in the set of numeric functions $\left\{\frac{r^3}{2} + r^2\right\} + O(r)$.

Observe that $\left\{\frac{r^3}{2}\right\} + O(r^2)$ denotes the set of numeric functions $A+B$, where A is the set $\{r^3/2\}$ and B is the set $O(r^2)$.

Notation: In the literature, we frequently write

$$\mathbf{a} = O(r^3)$$

to mean that **a** is $O(r^3)$. For example we write $\left\{\frac{r^2}{2}\right\} + r = O(r^2)$

While using this notation, certain important differences must be kept in mind. The most important consideration is the idea of one-way equalities. We write $(r^2/2)+r = O(r^2)$ but we never write $O(r^2) = (r^2/2)+r$ (otherwise, since $r^2/4 = O(r^2)$, we might come up with the absurd relation $r^2/4 = (r^2/2)+r$).

Example 4.17:

$$\text{Let} \quad \mathbf{a} = \{r\} + O\left(\frac{1}{r}\right)$$

and

$$\text{Let}\quad \mathbf{b} = \{\sqrt{r}\} + O\left(\frac{1}{\sqrt{r}}\right)$$

Then show that $\quad \mathbf{ab} = \{r^{3/2}\} + O(\sqrt{r})$

Solution: We have

$$\begin{aligned}
\mathbf{ab} &= \left(\{r\}+O\left(\frac{1}{r}\right)\right)\left(\{\sqrt{r}\}+O\left(\frac{1}{\sqrt{r}}\right)\right)\\
&= \{r\}\{\sqrt{r}\}+\{r\}O\left(\frac{1}{\sqrt{r}}\right)+\{\sqrt{r}\}O\left(\frac{1}{r}\right)+O\left(\frac{1}{r}\right)\cdot O\left(\frac{1}{\sqrt{r}}\right)\\
&= \{r\cdot\sqrt{r}\}+O\left(r\cdot\frac{1}{\sqrt{r}}\right)+O\left(\sqrt{r}\cdot\frac{1}{r}\right)+O\left(\frac{1}{r}\cdot\frac{1}{\sqrt{r}}\right)
\end{aligned}$$

$$(\because O(\mathbf{a})O(\mathbf{b}) = O(\mathbf{ab}))$$

$$\begin{aligned}
&= \{r^{3/2}\}+O(\sqrt{r})+O\left(\frac{1}{\sqrt{r}}\right)+O\left(\frac{1}{r\sqrt{r}}\right)\\
&= \{r^{3/2}\}+O(\sqrt{r})
\end{aligned}$$

$$\left(\because O\left(\frac{1}{r\sqrt{r}}\right)\subseteq O\left(\frac{1}{\sqrt{r}}\right)\subseteq O(\sqrt{r})\right)$$

Example 4.18: Simplify the expression

$$\left[\{\log n\}+O\left(\frac{1}{n}\right)\right]\left[\{n\}+O(\sqrt{n})\right]$$

Solution:

$$\left[\{\log n\}+O\left(\frac{1}{n}\right)\right]\left[\{n\}+O(\sqrt{n})\right]$$

$$=\{n\}\{\log n\}+\{n\}O\left(\frac{1}{n}\right)+\{\log n\}O\{\sqrt{n}\}+O\left(\frac{1}{n}\right)O(\sqrt{n})$$

$$= \{n \log n\} + O\{1\} + O\left(\sqrt{n}\ \log n\right) + O\left(\frac{1}{\sqrt{n}}\right)$$

$$= \{n \log n\} + O\left(\sqrt{n}\ \log n\right) + O(1) \qquad \left(\because O\left(\frac{1}{\sqrt{n}}\right) \subseteq O(1)\right)$$

$$= \{n \log n\} + O\left(\sqrt{n}\ \log n\right) \qquad \left(\because O(1) \subseteq O\left(\sqrt{n}\ \log n\right)\right)$$

Note: Some authors write {r}+O(**a**) as r+O(**a**). That is, {r} +O(**a**) and r + O(**a**) have the same meaning.

Exercise

1. A ball is dropped to the floor from a height 40 m. Suppose that the ball always rebounds to reach half of the height from it falls. Find the numeric function **a**, where a_r denotes the height it reaches in the rth rebound.

2. Let **a** be a numeric function such that

$$a_r = \begin{cases} 2, & 0 \le r \le 3 \\ 2^{-r} + 5, & r \ge 4 \end{cases}$$

Find S^2**a** and S^{-2}**a**.

3. Determine the convolution **a**∗**b** where

$a_r = 1$ for all $r \geqslant 0$ and $b_r = 2^r$ for all $r \geqslant 0$

4. Let $a = 3^r$ and $b = 2^r$. Show that **a** dominates **b** asymptotically. Does **b** dominate **a** asymptotically. Show further that **a**∗**b** dominates both **a** and **b** asymptotically.

5. Let $$a_r = \sum_{i=0}^{r} i^2$$

Show that **a** is O(r^3).

6. Show that every bounded numeric function is O(1)

7. Show that $(r+1)^2 = O(r^2)$. Also show that $(r+1)^2 = r^2 + O(r)$

8. Let **a**, **b** and **c** be numeric functions such that **a**∗**b** = **c**. Given that

$$\mathbf{a}_r = \begin{cases} 1, & r = 0 \\ 2, & r = 1 \\ 0, & r \geq 2 \end{cases} \quad \text{and} \quad \mathbf{c}_r = \begin{cases} 1, & r = 0 \\ 0, & r \geq 1 \end{cases}$$

determine **b**.

9. Let **a** and **b** be numeric functions such that

$$\mathbf{a}_r = \begin{cases} r^2, & \text{if } r \text{ is even} \\ 2r, & \text{if } r \text{ is odd} \end{cases} \quad \text{and} \quad \mathbf{b}_r = r^2$$

Show that **b** asymptotically dominates **a** but **a** does not asymptotically dominate **b**.

10. What is the mistake in the result O(**a**)–O(**a**) = 0.

Answers

1. $a_r = 40(1/2)^r, r \geqslant 0$

2.
$$S^2\mathbf{a} = \begin{cases} 0, & 0 \leq r \leq 1 \\ 2, & 2 \leq r \leq 5 \\ 2^{-r} + 5, & r \geq 6 \end{cases} \quad \text{and} \quad S^{-2}\mathbf{a} = \begin{cases} 2, & 0 \leq r \leq 1 \\ 2^{-r} + 5, & r \geq 2 \end{cases}$$

3. $\mathbf{a}*\mathbf{b} = 2^{r+1} - 1, r \geqslant 0$

4. **b** does not dominate **a**.

8. $b_r = (-2)^r, r \geqslant 0$.

10. Correct formula is O(**a**)–O(**a**) = O(**a**)

§ 4.4. Generating Functions

Generating functions provide an alternative way for representing and studying numeric functions and for solving recurrence relations as well as combinatorial problems.

Let $(a_0, a_1, a_2, ..., a_r, ...)$ be a numeric function **a**. Then the infinite series in terms of a parameter z,

$$A(z) = a_0 + a_1 z + a_2 z^2 + ... + a_r z^r + \quad ... (1)$$

is called *generating function* of the numeric function **a**. For a numeric function **a**, we shall use the corresponding upper case letter and write A(z) to denote the generating function of **a**. Clearly given a numeric function, we can obtain its generating function and conversely. Thus we can say generating function is an alternative way for representing a numeric function. For example, the generating function of the numeric function $(1, 2^1, 2^2, 2^3, ...)$ is

$$1+2z+2^2z^2+... +2^rz^r+ \quad ... (2)$$

Since the series in (2) can be written as

$$\frac{1}{1-2z}$$

therefore the numeric functions $(1, 2^1, 2^2, 2^3, ...)$ can be represented as

$$\frac{1}{1-2z}$$

The following table gives generating functions for some simple numeric functions:

Sl. No	General term of numeric function (a_r)	Generating function A(z)
1.	$a_r = 1$	$A(z) = 1/(1-z)$
2.	$a_r = r$	$A(z) = z/(1-z)^2$
3.	$a_r = r(r+1)$	$A(z) = 2z/(1-z)^3$
4.	$a_r = \alpha^r$	$A(z) = 1/(1-\alpha z)$
5.	$a_r = {}^nC_r$	$A(z) = (1+z)^n$
6.	$a_r = r^2$	$A(z) = z(z+1)/(1-z)^3$

Theorem 1: Let **a**, **b** and **c** be numeric functions and let A(z), B(z) and C(z) be respeetively the generating functions of **a**, **b** and **c**. Show that

(i) If $b_r = \alpha\, a_r$, where α is a real constant, then $B(z) = \alpha \cdot A(z)$

(ii) If $c_r = a_r + b_r$, then $C(z) = A(z) + B(z)$

(iii) If **c** is the convolution of **a** and **b**, i.e. **c** = **a**∗**b**, then C(z) = A(z)B(z)

(iv) If $b_r = \alpha^r a_r$, where α is a constant, then $B(z) = A(\alpha z)$

Proof:

(i) Let

$$a = (a_0, a_1, a_2, ..., a_r, ...)$$

and

$$b = (b_0, b_1, b_2, ..., b_r, ...)$$

Then

$$A(z) = a_0 + a_1 z + a_2 z^2 + ... + a_r z^r + ...$$

and

$$B(z) = b_0 + b_1 z + b_2 z^2 + ... + b_r z^r + ...$$

$$= \alpha a_0 + \alpha a_1 z + \alpha b_1 z^2 + ... + \alpha a_r z^r + ...$$

where $b_r = \alpha a_r$

$$= \alpha(a_0 + a_1 z + a_2 z^2 + ...- + a_r z^r + ...)$$

$$= \alpha A(z)$$

(ii) Let $c_r = a_r + b_r$ and let

$$C(z) = c_0 + c_1 z + c_2 z^2 + ... + c_r z^r + ...$$

be the generating function of **c**.

$$\Rightarrow \quad C(z) = (a_0+b_0)+(a_1+b_1)z+(a_2+b_2)z^2+... +(a_r+b_r)z^r+...$$

$$= (a_0+a_1z+a_2z^2+......) + (b_0+b_1z+b_2z^2+......)$$
$$= A(z)+B(z)$$

where A(z) and B(z) are generating functions for numeric function **a** and **b** respectively.

(iii) Let **c** = **a*b.** Then

Now

$$A(z)B(z) = (a_0+a_1z+a_2z^2+... +a_rz^r+...)$$
$$(b_0+b_1z+b_2z^2+... +b_rz^r+...)$$

Multiplying, we get,

$$A(z)B(z) = a_0b_0+(a_1b_0+a_0b_1)z+...$$
$$+(a_rb_0+a_{r-1}b_1+a_{r-i}b_i+ -- a_0b_r)z^r+...$$
$$= c_0+c_1z+.... +c_rz^r+....$$

because $c_r = \sum_{i=0}^{r} a_ib_{r-i}$

$$= C(z)$$

Hence generating function C(z) is equal to A(z)B(z).

(iv) Let $b_r = \alpha^r a_r$. Then generating function of **b** is

$$B(z) = b_0+b_1z+b_2z^2+... +b_rz^r+...$$
$$= \alpha^0a_0+\alpha a_1z+\alpha^2a_2z^2+ ... +\alpha^ra_rz^r+...$$

because $b_r = \alpha a_r$

$$= a_0+a_1(\alpha z)+a_2(\alpha z)^2+ ... +a_r(\alpha z)^r+ ...$$
$$= A(\alpha z)$$

This completes the proof.

Example 4.19. Find the generating functions of the following numeric functions.

(i) $a_r = 7\cdot 2^r, r \geqslant 0$

(ii) $a_r = 2^r+3^r, r \geqslant 0$

(iii) $a_r = 2^r \cdot r, r \geqslant 0$

Solution:

(i) We have

$$a_r = 7 \cdot 2^r$$

We know that the generating function for the numeric function 2^r is given by

$$1+2z+2^2z^2+\ldots +2^rz^r+\ldots$$

which can be written as

$$\frac{1}{1-2z}$$

Now applying Theorem 1(i), we find that

$$A(z) = 7 \cdot \frac{1}{1-2z} = \frac{7}{1-2z}$$

is the generating function for $a_r = 7 \cdot 2^r$

(ii) Let $a_r = 2^r + 3^r, r \geqslant 0$

Applying theorem 1(ii), we have the generating function A(z) for a_r

A(z) = generating function of 2^r + generating function of 3^r

$$= \frac{1}{1-2z} + \frac{1}{1-3z}$$

$$= \frac{2-5z}{(1-2z)(1-3z)}$$

(iii) Let $a_r = 2^r \cdot r$

We know that

$$\frac{1}{1-z} = 1 + z + z^2 + z^3 + \ldots + z^r + \ldots$$

$$\Rightarrow \frac{1}{(1-z)^2} = (1 + z + z^2 + \ldots)(1 + z + z^2 + \ldots)$$

$$= 1 + 2z + 3z^2 + 4z^3 + \ldots + r\,z^{r-1} + \ldots$$

$$\Rightarrow \qquad \frac{z}{(1-z)^2} = z + 2z^2 + 3z^3 + 4z^4 + \ldots + r z^r + \ldots$$

Thus $z/(1-z)^2$ is the generating function of the numeric function whose general term is r. Now applying Theorem 1(iv), the generating function for $a_r = 2^r r$ is given by

$$A(z) = \frac{2z}{(1-2z)^2}$$

Example 4.20. Find a simple expression for the generating function of each of the following discrete numeric functions:

(a) $1, -2, 3, -4, 5, -6, \ldots, (-1)^r(r+1), \ldots$

(b) $1, 2/3, 3/9, 4/27, \ldots, (r+1)/3^r, \ldots$

(c) $0\times1, 1\times2, 2\times3, 3\times4, \ldots$

(d) $0\times5^0, 1\times5^1, 2\times5^2, \ldots, r\times5^r, \ldots$

Solution:

(a) The generating function corresponding to the numeric function $(1, -2, 3, -4, \ldots)$ is given by

$$A(z) = 1-2z+3z^2+\ldots +(-1)^r(r+1)z^r+\ldots$$
$$= 1/(1+z)^2$$

(b) The generating function corresponding to the given numeric function

$$\left(1, \frac{2}{3}, \frac{3}{9}, \frac{4}{27}, \ldots, \frac{r+1}{3^r}, \ldots\right) \text{ is given by}$$

$$A(z) = 1 + \frac{2}{3}z + \frac{3}{3^2}z^2 + \frac{4}{3^3}z^3 + \ldots + \frac{r+1}{3^r}z^r + \ldots$$

$$= 1 + 2\left(\frac{z}{3}\right) + 3\left(\frac{z}{3}\right)^2 + \left(\frac{z}{3}\right)^3 + \ldots + (r+1)\left(\frac{z}{3}\right)^r + \ldots$$

$$= \frac{1}{\left(1-\frac{z}{3}\right)^2}$$

$$= \frac{9}{(3-z)^2}$$

(c) We know (by long division) that

$$\frac{1}{(1-z)^3} = 1+3z+6z^2+\ldots+\frac{1}{2}(r+1)(r+2)z^r+\ldots$$

Multiplying both sides by 2z, we get

$$\frac{2z}{(1-z)^3} = 2z+2\cdot 3z^2+3\cdot 4z^3+\ldots+r(r+1)z^r+\ldots$$

$$\Rightarrow \quad \frac{2z}{(1-z)^3} = 0\times 1+1\times 2z+1\times 2z^2+\ldots+r(r+1)z^r+\ldots$$

$\Rightarrow$ Generating function corresponding to the numeric function $(0\times 1, 1\times 2, 2\times 3, \ldots, r\times(r+1), \ldots)$ is given by $2z/(1-z)^3$

(d) By long division, we have

$$\frac{1}{(1-z)^2} = 1+2z+3z^2+\ldots(r+1)z^r+\ldots$$

$$\Rightarrow \quad \frac{z}{(1-z)^2} = 0+z+2z^2+3z^3+\ldots+rz^r+\ldots$$

$\Rightarrow$ $z/(1-z)^2$ is the generating function for the numeric function $(0, 1, 2, 3, \ldots, r, \ldots)$.

Now applying Theorem 1(iv), we get

$5z/(1-5z)^2$ is the generating function for the given numeric function $(0\times 5^0, 1\times 5^1, 2\times 5^2, \ldots, r\times 5^r, \ldots)$

Example 4.21. Determine the discrete numeric function corresponding to each of the following generating functions:

(a) $A(z) = (1+z)^n + (1-z)^n$, where n is positive integer

(b) $A(z) = (1+z)^2/(1-z)^4$

(c) $A(z) = 1/(5-6z+z^2)$

Solution:

(a) $A(z) = (1+z)^n + (1-z)^n$

$= (1+{}^nC_1 z + {}^nC_2 z^2 + \ldots + {}^nC_r z^r + \ldots)$

$$+(1-{}^nC_1z+{}^nC_2z^2+\ldots+(-1)^r\,{}^nC_rz^r+\ldots)$$

$$=2+2\,{}^nC_2\,z^2+2\,{}^nC_4\,z^4+\ldots+2\,{}^nC_{2r}\,z^{2r}+\ldots$$

$\therefore$ Numeric function a_r corresponding to A(z) is

$$a^r=\begin{cases}0 & \text{if } r \text{ is odd or } r>n\\ 2\,{}^nC_r & \text{if } r \text{ is even}\end{cases}$$

because ${}^nC_r=0$ if $r>n$

(b) $$A(z)=\frac{(1+z)^2}{(1-z)^4}$$

$$=\left(1+2z+z^2\right)\left(1+4z+\frac{4\cdot5}{\underline{|2}}z^2+\ldots+\frac{4\cdot5\cdot6\cdot\ldots(r+3)}{\underline{|r}}z^2+\ldots\right)$$

$\Rightarrow$ Coefficient of z^r in $A(z)$

$$=\frac{(r+1)(r+2)(r+3)}{1\cdot2\cdot3}+\frac{2r(r+1)(r+2)}{1\cdot2\cdot3}+\frac{r(r-1)(r+1)}{1\cdot2\cdot3}$$

$$=\frac{(r+1)\left(2r^2+4r+3\right)}{3}$$

Thus numeric function corresponding to A(z) is

$$a_r=\frac{(r+1)\left(2r^2+4r+3\right)}{3},\qquad r\geq0$$

(c) $$A(z)=\frac{1}{5-6z+z^2}=\frac{1}{(5-z)(1-z)}=\frac{1}{4(1-z)}-\frac{1}{4(5-z)}$$

$$=\frac{1}{4}\left(\frac{1}{(1-z)}\right)-\frac{1}{20}\left(\frac{1}{1-\frac{z}{5}}\right)$$

$\therefore$ Numeric function a_r corresponding to A(z) is

$$a_r = \frac{1}{4} - \frac{1}{20}\left(\frac{1}{5}\right)^r, \qquad r \geq 0$$

$$= \frac{1}{4}\left[1 - \left(\frac{1}{5}\right)^{r+1}\right], \qquad r \geq 0$$

Theorem 2: If $A(z) = a_0 + a_1 z + a_2 z^2 + ... + a_r z^r + ...$ be the generating function of the numeric function $a = (a_0, a_1, ..., a_r, ...)$ then show that

$$\frac{1}{1-z} A(z)$$

is the generating function of the numeric function **b** which is accumulated sum of **a**.

Solution: We know that

$$\frac{1}{1-z} = 1 + z + z^2 + . . . + z^r + . . . \qquad ...(1)$$

Hence $1/(1-z)$ is the generating function for the numeric function

$$(1, 1, 1, 1, ...).$$

Also we are given that

$$A(z) = a_0 + a_1 + a_2 z^2 + ... + a^2 z^r + ... \qquad ...(2)$$

is the generating function for **a**

Multiplying (1) and (2), we get

$$\frac{1}{1-z} A(z) = a_0 + (a_0 + a_1) z + (a_0 + a_1 + a_2) z^2 + . . . + (a_0 + a_1 + . . . + a_r) z^r + ...$$

$\Rightarrow \frac{1}{1-z} A(z)$ is the generating function for the numeric function **b**

where

$$b_r = \sum_{i=0}^{r} a_i$$

Thus **b** is accumulated sum of **a** and $\frac{1}{1-z} A(z)$ is the generating function of **b**.

Example 4.22. Evaluate the sum

$$1^2+2^2+3^2+ \ldots. +r^2$$

using generating functions.

Solution: We know that

$$\frac{1}{1-z} = 1+z+z^2+z^3+\ldots+z^r+\ldots$$

Differentiating both sides of the above identity w.r.to z, we get

$$\frac{1}{(1-z)^2} = 1+2z+3z^2+\ldots+rz^{r-1}+\ldots$$

Multiplying both sides of z, we have

$$\frac{z}{(1-z)^2} = z+2z^2+3z^3+4z^4\ldots+rz^r+\ldots$$

Differentiating both sides w.r. to z, we get

$$\frac{1+z}{(1-z)^3} = 1^2+2^2z+3^2z^2+\ldots+r^2z^{r-1}+\ldots$$

Multiplying both sides by z, we get

$$\frac{z(1+z)}{(1-z)^3} = 0^2+1^2z+2^2z^2+3^2z^3+\ldots+r^2z^r+\ldots$$

Now applying Theorem 2, we have

$$\frac{1}{(1-z)}\left[\frac{z(1+z)}{(1-z)^3}\right] = 0^2+\left(0^2+1^2\right)z+\left(0^2+1^2+2^2\right)z^2+$$

$$\ldots+\left(0^2+1^2+2^2\ldots+r^2\right)z^2+\ldots\ldots$$

Thus $0^2+1^2+2^2+\ldots+r^2 =$ Coefficient of z^r in $\frac{z(1+z)}{(1-z)^4}$.

From the binomial theorem, we know that the coefficient of z_r

in $\frac{1}{(1-z)^4}$ is

$$\frac{(r+1)(r+2)(r+3)}{1\times 2\times 3}$$

Therefore, the coefficient of z^r in $\dfrac{z(1+z)}{(1-z)^4}$ is

$$= \frac{r(r+1)(r+2)}{1\cdot 2\cdot 3} + \frac{r(r-1)(r+1)}{1\cdot 2\cdot 3}$$

$$= \frac{r(r+1)(2r+1)}{6}$$

$$\text{Hence } 1^2 + 2^2 + \ldots + r^2 = \frac{r(r+1)(2r+1)}{6}$$

Example 4.23. Using generating functions, evaluate the sum

$$3\cdot 2\cdot 1 + 4\cdot 3\cdot 2 + 5\cdot 4\cdot 3 + \ldots. + r(r+1)(r-1)$$

Solution: We will first obtain a generating function for the numeric function

$$a_r = (r+1)r(r-1)$$

We know that

$$\frac{1}{1-z} = 1 + z + z^2 + z^3 + \ldots + z^r + \ldots$$

Differentiating both sides three times w.r.to z, we have

$$\frac{6}{(1-z)^4} = 3\cdot 2\cdot 1 + 4\cdot 3\cdot 2z + \ldots + (r+3)(r+2)(r+1)z^r + \ldots$$

Multiplying both sides by z^2, we get

$$\frac{6z^2}{(1-z)^4} = 3\cdot 2\cdot 1z^2 + 4\cdot 3\cdot 2z^3 + \ldots + r(r+1)(r-1)z^r$$

In view of Theorem 2, the desired sum is the coefficient of z^r in

$$\frac{1}{1-z}\left[\frac{6z^2}{(1-z)^4}\right] = \frac{6z^2}{(1-z)^5}$$

The coefficient of z^r in $\dfrac{6z^2}{(1-z)^5}$ = coefficient of z^{r-2} in $\dfrac{6}{(1-z)^5}$

$$= 6 \times \frac{5 \cdot 6 \cdot \ldots (r-2+5-1)}{\lfloor\underline{r-2}}$$

$$= 6 \times \frac{1 \cdot 2 \cdot 3 \ldots (r+2)}{1 \cdot 2 \cdot 3 \cdot 4 \lfloor\underline{r-2}}$$

$$= \frac{6 \lfloor\underline{r+2}}{\lfloor\underline{4} \lfloor\underline{r-2}}$$

Thus $3 \cdot 2 \cdot 1 + 4 \cdot 3 \cdot 2 + \ldots + r(r+1)(r-1) = \dfrac{6 \lfloor\underline{r+2}}{\lfloor\underline{4} \lfloor\underline{r-2}}$

§ 4.5. Solution of Combinatorial Problems using Generating Functions

The generating function representation of numeric functions is most useful in solving combinatorial problems. Recall that the number of combinations of n distinct objects taken r at a time is $\dfrac{\lfloor\underline{n}}{\lfloor\underline{r} \lfloor\underline{n-4}}$ which is denoted as ${}^nC_r$. We also know that for a fixed positive integer n.

$$(1+z)^n = {}^nC_0 + {}^nC_1 z + {}^nC_2 z^2 + \ldots + {}^nC_r z^r + \ldots + {}^nC_n z^n \qquad \ldots(1)$$

This shows that $(1+z)^n$ is the generating function for the numeric function $a_r = {}^nC_r$(Note that ${}^nC_r = 1$ and ${}^nC_r = 0$ for $r > n$). We give below a simple combinatorial argument to derive (1). Consider the coefficient of the term z^r in the expansion of $(1+z)^n$. In computing the product $(1+z)^n$ each factor $1+z$ will contribute either a 1 or a z. In particular, to make the term z^r, exactly r of the n factors $1+z$ contribute a z each and remaining $n-r$ of the factors contribute a 1 each. Consequently, the coefficient of z^r is the number of ways of selecting r of the n $(1+z)$ factors to make up the terms z^r which is equal to ${}^nc_r$. Thus we can interpret that each factor in the product $(1+z)^n$ contains two terms 1 and z corresponding to selecting an object zero times (since $z^0 = 1$) or once.

Hence if an object may be selected 0, 1, 2,...., k times or if there are k objects of a given kind, a factor 1+z on the left hand side of (1) is replaced by

$$1+z+z^2+ - - - +z^k$$

We now give some examples to illustrate how to build generating functions for combinational problems.

Example 4.24. Find a generating function for the numeric function a_r, where a_r is equal to the number of ways to select r objects from 10 objects $x_1, x_2, x_3, ..., x_{10}$ among which x_1 can be selected at most twice, x_2 can be selected at most thrice and the others can be selected only once.

Solution: Since we can select the object x_1 zero times, once or twice, the generating function of the given problem will contain the factor $(1+z+z^2)$. Similarly since object x_2 can be selected zero times once, twice, or thrice, the generating function will also have the factor $(1+z+z^2+z^3)$. Now since each of the remaining eight objects can be selected zero times or once, the generating function will have eight factors $(1+z)$. Hence the generating function of a_r is

$$A(z) = (1+z+z^2)(1+z+z^2+z^3)(1+z)^8$$

Example 4.25. Find a generating function for a_r, the number of ways to select r balls from a pile of three green, three white, three blue and three red balls.

Solution: Clearly the generating function will be a multiplication of four factors corresponding to each colour green, white, blue and red. Since there are three balls of each colour, each factor will be $(1+z+z^2+z^3)$. Hence required generating function is

$$(1+z+z^2+z^3)^4$$

Note: Sometimes it is easy to find the right generating function by first converting the given problem into equivalent problem having integer solutions only. For example, the selection problem is Example 4.25 can be equivalently stated as the number of integer solutions to the equation

$$n_1 + n_2 + n_3 + n_4 = r, \qquad 0 \leq n_i \leq 3$$

where n_1 represents the number of green balls chosen, n_2 the number of white balls, n_3 the number of blue balls and n_4 the number of red balls.

Example 4.26. Find the generating function for a_r, the number of ways to selecting r objects from n objects with unlimited repetitions. Hence find a_r.

Solution: Since each object can be selected as many times as we wish the generating function is

$$A(z) = (1+z+z^2+....)^n$$

Observe that the contribution from each of the factors

$$1+z+z^2+....$$

will correspond to the number of times one of the object is selected.

To find a_r, we shall find the coefficient of z^r in A(z). We can write A(z) as

$$A(z) = \left(\frac{1}{1-z}\right)^n = (1-z)^{-n}$$

$$\therefore \text{ Coefficient of } z^r = (-1)^r \frac{(-n)(-n-1)...(-n-r+1)}{\underline{|r}}$$

$$= \frac{\underline{|n+r-1}}{\underline{|r}\,\underline{|n-r}}$$

$$= {}^{n+r-1}C_r$$

Example 4.27. Using generating function, find the number of ways of selecting six objects from three types of objects with repetition of up to four objects of each type.

Solution: Clearly the generating function of a_r, the number of ways to select **r** objects from three types of objects with repetition of up to four objects of each type is

$$A(z) = (1+z+z^2+z^3+z^4)^3$$

Now we are required to find a_6, the coefficient of z^6 in A(z)

We can write $A(z)$ as

$$A(z)=\left(\frac{1-z^5}{1-z}\right)^3$$

$$=(1-z^5)^3\,(1-z)^{-3}$$

$$=(1-3z^5+3z^{10}-z^{15})\,(1-z)^{-3}$$

$\therefore$ Coefficient of z^6 in $A(z)$ = coefficient of z^6 in $(1-z)^{-3}-3\times$coefficient of z in $(1-z)^{-3}$

$$=\frac{3\cdot 4\cdot\ldots 8}{\lfloor 6}-3$$

$$=25$$

Example 4.28. Using generating function, find the number of distributions of 12 identical objects into:

(a) Five different boxes with at most four objects in each box.

(b) Seven different boxes with at least one object in each box.

(c) Three different boxes with at most five objects in the first box.

Solution: (a) The generating function for the problem is

$$A(z) = (1+z+z^2+z^3+z^4)^5$$

Observe that each factor $1+z+z^2+z^3+z^4$ in $A(z)$ corresponds to having none, one, two, three or four objects in a box. Clearly the coefficient of z^{12} in $A(z)$ will be our answer. Since

$$A(z) = (1+z+z^2+z^3+z^4)^5$$

$$=\left(\frac{1-z^5}{1-z}\right)^5$$

$$=(1-5z^5+10z^{10}-10z^5+5z^{20}-z^{25})(1-z)^{-5}$$

$\therefore$ Coefficient of z^{12} = coefficient of z^{12} in $(1-z)^{-5}-5\times$ coefficient of z^7 in $(1-z)^{-5}+10\times$coefficient of z^2 in $(1-z)^{-5}$

$$= {}^{16}C_{12}-5\times{}^{11}C_7+10\times{}^{6}C_2$$

(b) The generating function for the problem is

$$A(z) = (z + z^2 + z^3 + \ldots)^7$$

Observe that each factor $z + z^2 + z^3 + \ldots$ corresponds to having one, two ... objects in a box. Since we can write

$$\begin{aligned} A(z) &= (z + z^2 + \quad \ldots)^7 \\ &= z^7(1 + z + z^2 \ldots)^7 \\ &= z^7(1-z)^{-7} \end{aligned}$$

$\therefore$ Coefficient of z^{12} in $A(z)$ = coefficient of z^5 in $(1-z)^{-7}$

$$= {}^{11}C_5$$

(c) The generating function is

$$A(z) = (1 + z + z^2 + z^3 + z^4 + z^5)(1 + z + z^2 + \ldots)^2$$

Observe that the factor $(1 + z + z^2 + \ldots + z^5)$ corresponds to having at most five objects in first box whereas the factors $(1 + z + z^2 + \ldots)$ correspond to having any number of objects in other boxes. Since

$$A(z) = \left(\frac{1-z^6}{1-z}\right)(1-z)^{-2} = \left(1-z^6\right)(1-z)^{-3}$$

$\therefore$ Coefficient of z^{12} in $A(z)$ = coefficient of z^{12} in $(1-z)^{-3}$

$$\begin{aligned} &- \text{coefficient of } z^6 \text{ in } (1-z)^{-3} \\ &= {}^{14}C_{12} - {}^{8}C_6 \\ &= 63 \end{aligned}$$

Example 4.29. Using generating function, find the number of ways to select 10 balls from a large pile of red, white and blue balls if

(a) the selection has at least two balls of each colour.

(b) the selection has at most two red balls.

(c) the selection has an even number of blue balls.

Solution: (a) This selection problem is equivalent to the number of integer solutions of the equation

$$n_1 + n_2 + n_3 = 10, \qquad n_i \geqslant 2$$

Here n_1 represents the number of red balls chosen, n_2 the number of white and n_3 the number of blue balls.

This problem does not ask for the general solution of the ways to select r objects. However in building a generating function, we automatically consider all values of r, not just 10. The generating functions is

$$\begin{aligned} A(z) &= (z^2+z^3+...)^3 \\ &= z^6(1+z+z^2+...)^3 \\ &= z^6(1-z)^{-3} \end{aligned}$$

$$\therefore \text{Coefficient of } z^{10} \text{ in } A(z) = \text{Coefficient of } z^4 \text{ in } (1-z)^{-3} = {}^6C_4 = 15$$

(b) In this case, the generating functions is given by

$$\begin{aligned} A(z) &= (1+z+z^2)(1+z+z^2+...)^2 \\ &= (1+z+z^2)(1-z)^{-2} \end{aligned}$$

$$\therefore \text{Coefficient of } z^{10} = {}^{11}C_{10}+{}^{10}C_9+{}^9C_8 = 30$$

(c) In this case, the generating function is given by

$$A(z) = (1+z+z^2+...)^2(1+z^2+z^4+z^6+...)$$

Observe that the factor $1+z^2+z^4+z^6+......$ corresponds to having 0, 2, 4, 6,... (even) number of blue balls in the selection. We can write

$$A(z) = (1+z^2+z^4+...)(1-z)^{-2}$$

$$\begin{aligned} \therefore \text{Coefficient of } z^{10} \text{ is } A(z) &= \text{Sum of coefficients of } z^{10}, z^8, z^6, z^4, z^2 \text{ and } 1 \text{ in } (1-z)^{-2} \\ &= {}^{11}C_{10}+{}^9C_8+{}^7C_6+{}^5C_4+{}^3C_2+1 \\ &= 36 \end{aligned}$$

Example 4.30. Find a generating function for the number of ways to select five non–consecutive integers form 1, 2, ..., n. What is your answer for $n = 10$?

Solution: Suppose x_1, x_2, x_3, x_4, x_5 represents any five non-consecutive integers form 1, 2, ..., n such that $x_1 < x_2 < x_3 < x_4 < x_5$. Let n_1 be the number of integers from 1, 2, ..., n which are less than the chosen number x_1. Let n_2, n_3, n_4 and n_5 be the number of integers lying between (x_1, x_2), (x_2, x_3), (x_3, x_4) and (x_4, x_5) respectively. (For example, if x_1, x_2, x_3, x_4, x_5 are 1, 3, 7, 10, 12 then $n_1 = 0$, $n_2=1$, $n_3 = 3$, $n_4 = 2$, $n_5 = 1$.) Moreover, let n_6 be the number of integers from 1, 2, ..., n which are greater than x_5.

Now the given problem is equivalent to find the number of ways to select $n-5$ integers from 1, 2, ...,n other than x_1, x_2, x_3, x_4 and x_5 and satisfying the above conditions. This problem is equivalent to find the number of integer solutions to.

$$n_1+n_2+n_3+ ...+ n_6 = n-5 \text{ with } n_1, n_6 \geqslant 0 \text{ and } n_2, n_3, n_4, n_5 \geqslant 1$$

The generating function is therefore

$$\begin{aligned} A(z) &= (1+z+z^2+...)^2(z+z^2+z^3+...)^4 \\ &= z^4(1-z)^{-6} \end{aligned}$$

Note that we have to obtain the coefficient of z^{n-5} in $A(z)$. For $n = 6$, we have to find the coefficient of z^{10-5} in $A(z)$, that is, coefficient of z^5 in $z^4(1-z)^{-6}$.

$$\therefore \text{ Required answer} = \text{Coefficient of } z \text{ in } (1-z)^{-6}$$

$$= 6$$

Example 4.31. Use generating functions to show that

$$\sum_{k=0}^{r} {}^mC_k \cdot {}^nC_{r-k} = {}^{m+n}C_r$$

Solution: Let $A(z) = (1+z)^n$ and $B(z) = (1+z)^n$.

Then $a_r = {}^nC_r$ and $b_r = {}^mC_r$.

Now $A(z)\cdot B(z) = (1+z)^{n+m}$

Coefficient of z^r in $A(z)\,B(z) = {}^{m+n}C_r$.

We know from theorem 1(iii) of § 4.4 that coefficient of z^r in $A(z)B(z)$ is given by

$$\sum_{k=0}^{r} a_k b_{r-k} = \sum_{k=0}^{r} {}^nC_k \cdot {}^mC_{r-k}$$

Hence $$\sum_{k=0}^{r} {}^nC_k \cdot {}^mC_{r-k} = {}^{m+n}C_r$$

Example 4.32. Prove that

$${}^mC_0 \times {}^nC_0 + {}^mC_1 \times {}^nC_1 + \ldots {}^mC_n \times {}^nC_n = {}^{m+n}C_n \quad \text{where } n \le m$$

Solution: We have

$$A(z) = (1+z)^m$$
$$= {}^mC_0 + {}^mC_1 z + {}^mC_2 z^2 \ldots + {}^mC_m z^m$$
$$B(z) = (1+z)^n$$
$$= {}^nC_0 + {}^nC_1 z + {}^nC_2 z^2 \ldots + {}^nC_n z^n$$
$$= {}^nC_n + {}^nC_{n-1} z + {}^nC_{n-2} z^2 + \ldots + {}^nC_{n-r} z^r + \ldots + {}^nC_0 z^n$$
$$(\because {}^nC_r = {}^nC_{n-r})$$

Now $A(z) \cdot B(z) = (1+z)^{m+n}$ and coefficient of z^n in $(1+z)^{m+n}$ is ${}^{m+n}C_n$. We also know from theorem 1(iii) of §4.4 that coefficient of z^n in $A(z)B(z)$ is given by

$$\sum_{k=0}^{n} a_k b_{n-k}$$
$$= \sum_{k=0}^{n} {}^mC_k \cdot {}^nC_k \quad \left(\because b_{n-k} = {}^nC_{n-k} = {}^nC_k\right)$$

$$\therefore {}^nC_0 + {}^nC_0 + {}^mC_1 \times {}^nC_1 + \ldots + {}^mC_n \times {}^nC_m = {}^{m+n}C_n$$

Exercise

1. Find the coefficient of z^{20} in $(z + z^2 + z^3 + z^4 + z^5)(z^2 + z^3 + \ldots)^5$.
2. Use Generating functions to find the number of ways to distribute **r** apples among eight children if

   (a) each child gets at least one apple.

(b) each child gets an even number of apples.

3. How many ways are there to get a sum 25 when 10 dice are rolled?

   [Hint: Generating function of the problem is $(z+z^2+z^3+...+z^6)^{10}$. Required number is the coefficient of z^{25}.]

4. How many ways are there to select 25 toys from seven types of toys with between two and six of each type?

   [Hint:Generating function of the problem is $(z^2+z^3+...+z^6)^7$.]

5. How many ways are there to distribute 25 identical balls into seven distinct boxes if the first box can have no more than 10 balls but any amount can go in each of the other six boxes.

   [Hint: Generating function of the problem is $(1+z+...+z^{10})(1+z+z^2+...)^6$.]

6. Find the generating function for the number of ways to select an integer between 0 and 999999 whose sum of digits is r.

7. Find the number of ways to select 5 letters from the word MATH in which M and A can appear any number of times but T and H at most once.

   [Hint: Generating function is $(1+z+z^2+....)^2(1+z)^2$. Find coefficient of z^5.]

8. Determine the number of ways in which $2r+1$ marbles can be distributed among three distinct boxes so that no box will contain more than r marbles.

9. Find the number of integer solutions to the equation $n_1+n_2+n_3+n_4=7$ with $0 \leqslant n_i \leqslant 3$

   [Hint: Generating function of the problem is $(1+z+z^2+z^3)^4$ and the desired number is the coefficient of z^7.]

10. Find the number of integer solutions to the equation $n_1+n_2+n_3+n_4=10$ with n_1 odd, n_2 even and $2 \leqslant n_i \leqslant 6$, $i=1, 2, 3, 4$.

[Hint: Generating function is $(z^3+z^5)(z^2+z^4+z^6)(z^2+z^3+z^4+z^5+z^6)^2$. Now find the coefficient of z^{10}.]

11. A set of **r** balls is selected from unlimited supply of red, blue, white and yellow balls. A selection must satisfy the condition that either the number of red balls is even and the number of blue balls is odd or the number of white balls is even and the number of yellow balls is odd. Let a_r denote the total number of such selection. Determine the generating function A(z).

[Hint: Generating function $A(z) = 2(1+z^2+z^4+z^6+....)(z+z^3+z^5+...)(1+z+z^2+...)^2-(1+z^2+z^4+...)^2(z+z^3+z^5+...)^2$.]

12. Prove the relation ${}^nC_r = {}^{n-1}C_r + {}^{n-1}C_{r-1}$

13. Use binomial expansions to evaluate

(a) $\sum_{k=0}^{m} {}^mC_k \cdot {}^nC_{r+k}$ (b) $\sum_{k=0}^{r} (-1)^k\, {}^nC_k \cdot {}^nC_{r-k}$

$\overline{k=0}$

(c) ${}^nC_1 + 2\,{}^nC_2 + ... + i\,{}^nC_i + ... + n \cdot {}^nC_n$

Answers

1. ${}^{14}C_9 - {}^9C_4$ 2. (a) ${}^{r-1}C_{r-8}$ (b) ${}^{(r/2+8-1)}C_{r/2}$

3. ${}^{24}C_{15} - 10 \times {}^{18}C_{10} + {}^{10}C_2 \times {}^{12}C_3$ 4. ${}^{17}C_{11} - {}^{12}C_6 \times {}^7C_1 + {}^7C_1 \times {}^7C_2$

5. ${}^{31}C_{25} - {}^{20}C_{14}$ 6. $(1+z+z^2+...+z^9)^6$

7. 20 8. ${}^{2r+3}C_{2r+1} - 3 \times {}^{r+2}C_r$

9. ${}^{10}C_7 - 4 \times {}^6C_3$ 10. 1

13. (a) ${}^{m+n}C_{r+2k}$

(b) $\begin{cases} 0, & \text{if } r \text{ is odd} \\ (-1)^{r/2} \cdot {}^nC_{r/2} & \text{if } r \text{ is even} \end{cases}$

(c) $n \cdot 2^{n-1}$

§ 4.6. Solution of Recurrence Relations by the Method of Generating Functions

Generating functions are extremely useful in solving and studying the recurrence relations. We first explain the procedure to solve recurrence relation with the help of an example.

Consider the recurrence relation

$$a_r = 3a_{r-1}+2, \qquad r \geqslant 1 \qquad \text{...(1)}$$

with $a_0 = 1$

Observe that the recurrence relation (1) is not valid for $r = 0$ since a_{-1} is not defined. Multiplying both sides of (1) by z^r, we obtain

$$a_r z^r = 3a_{r-1}z^r + 2\,z^r, r \geqslant 1 \qquad \text{...(2)}$$

Summing (2) for all $r \geqslant 1$, we have

$$\sum_{r=1}^{\infty} a_r z^r = 3\sum_{r=1}^{\infty} a_{r-1} z^r + 2\sum_{r=1}^{\infty} z^r \qquad \text{...(3)}$$

Note that we cannot take summation from $r \geqslant 0$ because equation (1) is not valid for $r = 0$

Let $A(z) = a_0+a_1z+a_2z^2+\ldots+a_rz^r+\ldots$

Then (3) can be written as

$$A(z) - a_0 = 3zA(z) + \frac{2z}{(1-z)} \qquad \text{...(4)}$$

because $\displaystyle\sum_{r=1}^{\infty} a_{r-1}z^r = z\sum_{r=1}^{\infty} a_{r-1}z^{r-1} = z\sum_{r=0}^{\infty} a_r z^r$

$$= zA(z)$$

and $\displaystyle\sum_{r=1}^{\infty} z^r = z\sum_{r=1}^{\infty} z^{r-1}$

$$= z\sum_{r=0}^{\infty} z^r = \frac{z}{1-z}$$

From (4), we have

$$
\begin{aligned}
(1-3z)A(z) &= \frac{2z}{1-z} + a_0 \\
&= \frac{2z}{1-z} + 1 \\
&= \frac{1+z}{1-z} \\
\Rightarrow A(z) &= \frac{1+z}{(1-3z)(1-z)} \\
&= \frac{2}{1-3z} - \frac{1}{1-z} \\
\Rightarrow a_r &= 2\cdot 3^r - 1
\end{aligned}
$$

General Procedure

Consider the recurrence relation $c_0\, a_r + c_1\, a_{r-1} + ... c_r a_{r-k} = f(r)$ ...(1) where c_0, c_1, c_r, ..., c_k are constants. Obviously, relation (1) is valid for $r \geq k$. Multiplying both rides of (1) by z^r and taking summation from $r = k$ to $r = \infty$, we get

$$\sum_{r=k}^{\infty}(c_o a_r + c_1 a_{r-1} + c_2 a_{r-2} + . . . + c_k a_{r-k})z^r = \sum_{r=k}^{\infty} f(r) z^r$$

We know that

$$\sum_{r=k}^{\infty}\left(c_0 a_r z^r\right) = c_0\left[\sum_{r=0}^{\infty} a_r z^r - a_0 - a_1 z - a_2 z^2 . . . - a_{k-1} z^{k-1}\right]$$

$$\sum_{r=k}^{\infty}\left(c_1 a_{r-1} z^r\right) = c_1 z\left[\sum_{r=0}^{\infty} a_r z^r - a_0 - a_1 z - a_2 z^2 . . . - a_{k-2} z^{k-2}\right]$$

. . . .

. . . .

$$\sum_{r=k}^{\infty} c_k a_{r-k} z^r = c_k z^k\left[\sum_{r=0}^{\infty} a_r z^r\right]$$

Substituting these values in (1) and writing

$$A(z) = \sum_{r=0}^{\infty} a_r z^r$$

we get

$$A(z) = \frac{1}{c_0 + c_1 z + \ldots + c_k z^k} \left[\begin{array}{l} \sum_{r=k}^{\infty} f(r) z^r + c_0 \left(a_0 + a_1 z + \ldots + a_{k-1} z^{k-1}\right) \\ +c_1 z\left(a_0 + a_1 z + \ldots + a_{k-2} z^{k-2}\right) \\ + \ldots\ldots\ldots\ldots\ldots\ldots\ldots\ldots \\ +c_{k-1} z^{k-1}\left(a_0\right) \end{array} \right]$$

We now obtain a_r from $A(z)$

We give more illustrative examples.

Example 4.33. Use generating functions to solve the recurrence relation

$$a_n = a_{n-1} + n \qquad \text{with} \quad a_0 = 1$$

Solution: Multiply both sides of the given recurrence relation by z^n and summing over $n = 1$ to ∞, we get

$$\sum_{n=1}^{\infty} a_n z^n = \sum_{n=1}^{\infty} a_{n-1} z^n + \sum_{n=1}^{\infty} n z^n$$

$$\Rightarrow A(z) - a_0 = zA(z) + \frac{z}{(1-z)^2}$$

where $A(z) = a_0 + a_1 z + a_2 z^2 + \cdots + a_r z^r + \cdots$

$$\Rightarrow (1-z)A(z) = 1 + \frac{z}{(1-z)^2} \qquad (\because a_0 = 1)$$

$$\Rightarrow \quad A(z) = \frac{1}{(1-z)} + \frac{z}{(1-z)^3}$$

$$\Rightarrow \quad A(z) = (1-z)^{-1} + z(1-z)^{-3}$$

$$\Rightarrow \quad a_n = 1 + {}^{n+1}C_{n-1}$$

because coefficient of z^n in $(1-z)^{-1}$ is 1 and in $z(1-z)^{-3}$ is ${}^{n+1}C_{n-1}$. But ${}^{n+1}C_{n-1} = {}^{n+1}C_2$, therefore, we can write

$$a_n = 1 + {}^{n+1}C_2.$$

Example 4.34. Use generating functions to solve the recurrence relation

$$a_r = a_{r-1} + a_{r-2} \quad \text{with } a_1 = 2 \text{ and } a_2 = 3$$

Solution: Observe that a_0 is not given. Therefore, the recurrence relation is not valid for $r = 2$ and is valid only for $r \geqslant 3$. From the given relation, we have for $r = 2$.

$$a_2 = a_1 + a_0$$

If we choose $a_0 = 1$ then the given recurrence relation is valid for $r \geqslant 2$. In this case, we multiply both sides of the given relation by z^r and summing from $r = 2$ to $r = \infty$, we obtain

$$\sum_{r=2}^{\infty} a_r z^r = \sum_{r=2}^{\infty} a_{r-1} z^r + \sum_{r=2}^{\infty} a_{r-2} z^r$$

$$\Rightarrow \left[A(z) - a_0 - a_1 z\right] = z\left[A(z) - a_0\right] + z^2 A(z)$$

$$\Rightarrow A(z) = \frac{1+z}{1-z-z^2} \qquad (\because a_0 = 1, a_1 = 2)$$

$$= \frac{1+z}{(1-\alpha z)(1-\beta z)}$$

$$\left(\text{where } \alpha = \frac{1}{2}\left(1+\sqrt{5}\right) \text{ and } \beta = \frac{1}{2}\left(1-\sqrt{5}\right)\right)$$

$$\Rightarrow A(z) = \left(\frac{1+\alpha}{\alpha-\beta}\right)\left(\frac{1}{1-\alpha z}\right) + \left(\frac{1+\beta}{\beta-\alpha}\right)\left(\frac{1}{1-\beta z}\right) \text{(by partial fraction)}$$

$$\Rightarrow A(z) = \left(\frac{5+3\sqrt{5}}{10}\right)(1-\alpha z)^{-1} - \left(\frac{-5+3\sqrt{5}}{10}\right)(1-\beta z)^{-1}$$

$$\Rightarrow a_r = \left(\frac{5+3\sqrt{5}}{10}\right)\alpha^r - \left(\frac{-5+3\sqrt{5}}{10}\right)\beta^r, \ r \geq 0$$

$$\therefore a_r = \left(\frac{5+3\sqrt{5}}{10}\right)\left(\frac{1+\sqrt{5}}{2}\right)^r - \left(\frac{-5+3\sqrt{5}}{10}\right)\left(\frac{1-\sqrt{5}}{2}\right)^r, \qquad r \geq 0$$

Example 4.35. Use generating functions to solve the recurrence relation

$$a_r - 5a_{r-1} + 6a_{r-1} = 2^r + r, \qquad r \geqslant 2$$

with $a_0 = 1$ and $a_1 = 1$

Solution: Multiplying both sides of the relation by z^r and summing from $r = 2$ to ∞, we obtain.

$$\sum_{r=2}^{\infty} a_r z^r - 5\sum_{r=2}^{\infty} a_{r-1} z^r + 6\sum_{r=2}^{\infty} a_{r-2} z^r = \sum_{r=2}^{\infty} 2^r z^r + \sum_{r=2}^{\infty} r z^r$$

$$\Rightarrow A(z) - a_0 - a_1 z - 5z\left[A(z) - a_0\right] + 6z^2 A(z)$$

$$= \frac{4z^2}{1-2z} + z\left[\frac{1}{(1-z)^2} - 1\right]$$

because $\sum_{r=2}^{\infty} 2^r z^r = (2z)^2 + (2z)^3 + \ldots$

$$= (2z)^2\left[1 + (2z) + (2z)^2 + \ldots\right] = \frac{4z^2}{1-2z}$$

and $\sum_{r=2}^{\infty} r z^r = \left(2z^2 + 3z^3 + 4z^4 + \ldots\ldots\ldots\right)$

$$= z\left[2z + 3z^2 + \ldots\right] = z\left[\frac{1}{(1-z)^2} - 1\right]$$

Simplifying, we obtain

$$A(z) = \frac{1 - 8z + 27z^2 + 35z^3 + 14z^4}{(1-z)^2(1-2z)^2(1-3z)}$$

$$= \frac{5/4}{(1-z)} + \frac{1/2}{(1-z)^2} - \frac{3}{1-2z} - \frac{2}{(1-2z)^2} + \frac{17/4}{1-3z}$$

$$\Rightarrow a_r = \frac{5}{4} + \frac{1}{2}(r+1) - 3\cdot 2^r - 2(r+1)2^2 + \frac{17}{4}3^r$$

$$= \frac{7}{4} + \frac{r}{2} - r\cdot 2^{r+1} - 5\cdot 2^r + \frac{17}{4}3^r$$

Example 4.36. Solve the recurrence relation

$$a^2_r - 2a^2_{r-1} = 1 \text{ with } a_0 = 2$$

Solution: Let $b^r = a^2_r$. Then given relation becomes

$$b_r - 2b_{r-1} = 1 \text{ with } b_0 = 4$$

Now multiplying both sides by z^r and summing from $r = 1$ to ∞, we obtain

$$\sum_{r=1}^{\infty} b_r z^r - 2\sum_{r=1}^{\infty} b_{r-1} z^r = \sum_{r=1}^{\infty} z^r$$

$$\Rightarrow \left[B(z) - b_0\right] - 2z\,B(z) = \frac{z}{1-z}$$

$$\Rightarrow (1-2z)B(z) = \frac{z}{1-z} + 4 = \frac{4-3z}{1-z}$$

$$\Rightarrow B(z) = \frac{4-3z}{(1-2z)(1-z)}$$

$$= 5\cdot\frac{1}{(1-2z)} - \frac{1}{(1-z)}$$

$$\Rightarrow \quad b_r = 5\cdot 2^r - 1$$

$$\therefore \quad a_r = \sqrt{5\cdot 2^r - 1}$$

Simultaneous Recurrence Relations

Now we consider a set of simultaneous recurrence relations and solve it using generating functions.

Example 4.37. Use generating functions to solve the following simultaneous recurrence relations

$$a_r = 3a_{r-1} + 2b_{r-1}$$

$$b_r = a_{r-1} + b_{r-1} \qquad \text{with } a_0 = 1 \text{ and } b_0 = 0$$

Solution: Multiplying both equations by z^r and summing from $r = 1$ to ∞, we get

$$\sum_{r=1}^{\infty} a_r z^r = 3\sum_{r=1}^{\infty} a_{r-1} z^r + 2\sum_{r=1}^{\infty} b_{r-1} z^r$$

and $$\sum_{r=1}^{\infty} b_r z^r = \sum_{r=1}^{\infty} a_{r-1} z^r + \sum_{r=1}^{\infty} b_{r-1} z^r$$

$$A(z)-1 = 3z\,A(z)+2z\,B(z) \qquad ...(1)$$

and $$B(z) = z\,A(z)+z\,B(z) \qquad ...(2)$$

Solving (1) and (2), we obtain

$$A(z) = \frac{1-z}{1-4z+z^2}$$

$$= \frac{\left(3+\sqrt{3}\right)}{6\left[1-\left(2+\sqrt{3}\right)z\right]} + \frac{\left(3-\sqrt{3}\right)}{6\left[1-\left(2-\sqrt{3}\right)z\right]}$$

and $$B(z) = \frac{z}{1-4z+z^2}$$

$$= \frac{\sqrt{3}}{6\left[1-\left(2+\sqrt{3}\right)z\right]} - \frac{\sqrt{3}}{6\left[1-\left(2-\sqrt{3}\right)z\right]}$$

$$\Rightarrow \quad a_r = \frac{3+\sqrt{3}}{6}\left(2+\sqrt{3}\right)^r + \frac{3-\sqrt{3}}{6}\left(2-\sqrt{3}\right)^r$$

and $$b_r = \frac{\sqrt{3}}{6}\left(2+\sqrt{3}\right)^r - \frac{\sqrt{3}}{6}\left(2-\sqrt{3}\right)^r$$

Exercise

1. Use generating functions to solve the simultaneous recurrence relations

$$a_n = a_{n-1} + b_{n-1} + c_{n-1}$$

$$b_n = 3^{n-1} - c_{n-1}$$

$$c_n = 3^{n-1} - b_{n-1} \text{ with } a_1 = b_1 = c_1 = 1$$

(Hint: First obtain a_0, b_0 and c_0)

2. Solve the recurrence relation

$$a_n = 3a_{n-1} + 4a_{n-2} \text{ with } a_0 = a_1 = 1$$

3. Solve the recurrence relation

$$a_{r+2} - 2a_{r+1} + a_r = 2^r \text{ with } a_0 = 2 \text{ and } a_1 = 1$$

4. Consider the recurrence relation

$$a_{n+1} = a_n + n, n = 0, 1, 2, \dots \text{ with } a_0 = 1$$

Show that the solution of the recurrence relation is

$$a_n \neq \frac{n(n-1)}{2} + 1$$

5. Use the method of generating functions to find the solution of the recurrence relation

$$a_{r+2} - 5a_{r+1} + 6a_r = 2 \quad \text{with } a_0 = 1 \text{ and } a_1 = 2$$

6. Solve the recurrence relation

$$d_r = 2d_{r-1} - d_{r-2} \quad \text{with } d_1 = 1.5 \text{ and } d_2 = 3$$

Answers

1. $$a_r = \begin{cases} \frac{1}{4}(3^r + 3), & \text{r is even} \\ \frac{1}{4}(3^r + 1), & \text{r is odd} \end{cases}$$

$$b_r = c_r = \frac{1}{4}\left(3^r - (-1)^r\right)$$

2. $a_n = \left(\frac{2}{5}\right)4^n + \left(\frac{3}{5}\right)(-1)^n$

3. $a_r = 1 - 2r + 2^r$

5. $a_r = 1 - 2^r + 3^r$

6. $d_r = \frac{3}{2}r$

CHAPTER FIVE

Boolean Algebra

In this chapter, we shall study Boolean algebra as an abstract structure. The definition of a Boolean algebra which will be given now is one given by Huntington in 1904. In fact, Boolean algebra originated in the works of the English Mathematician George Boole (1813-1865). The original purpose of this algebra was to simplify logical statements and solve logic problems. Today it is the backbone of design and analysis of computer and other digital circuits.

§ 5.1. Boolean Algebra

Definition: Let B be a none-empty set with two binary operations $+$ and $*$, a unary operation $'$, and two distinct elements 0 and 1. Then B is called a Boolean algebra if the following axioms hold for any $a, b, c \in B$.

$[B_1]$ **Commutative laws** : The operations $+$ and $*$ are commutative.

In other words,

$$a+b=b+a \quad \text{and} \quad a*b=b*a \quad \forall\ a, b\in B$$

$[B_2]$ **Identity laws** : For any $a\in B$

$$a+0=a \qquad \text{and} \qquad a*1=a$$

That is, both operations + and * have identity elements denoted by 0 and 1 respectively.

$[B_3]$ **Distributive laws** : Each binary operation is distributive over the other. That is, for any $a, b, c \in B$,

$$a + (b*c) = (a+b) * (a+c) \quad \text{and} \quad a*(b+c) = (a*b) + (a*c)$$

$[B_4]$ **Complement laws** : For each a in B, there exists an element a' in B such that

$$a + a' = 1 \text{ and } a * a' = 0$$

We sometimes denote a Boolean algebra by $(B, +, *, ', 0, 1)$. The elements 0 and 1 are called zero element (identity for +) and unit element (identity for *) of B respectively while a' is called complement of a in B.

We will usually drop the symbol * between a and b and write $a * b$ simply as ab. Some authors use the symbols $\vee$ and $\wedge$ in place of the symbols + and * respectively and denote the complement of an element a by the symbol $\bar{a}$ instead of a'.

We mention here that there exist other sets of axioms which can equally well define a Boolean algebra, though, of course, each set is derivable from the other. Moreover, we shall also give an alternative definition of a Boolean algebra in terms of an associated partial ordering. The following example shows that the algebra of sets is a Boolean algebra.

Example 5.1: Let S be a non-empty set and $P(S)$ be the power set of S. Then $P(S)$ is a Boolean algebra with respect to union and intersection as two binary operations + and * respectively and complement of a set with respect to S as unary operation $'$. ϕ and S will act as 0 and 1 respectively.

Solution: We shall show that the power set $P(S)$ of a non-empty set S forms a Boolean algebra with respect to union and intersection as two binary operations + and * respectively and complement of a subset A of S with respect to S, *i.e.*, $S-A$ as unary operation $'$ on A.

1. **Commutative laws** : We know from set theory that

$$A \cup B = B \cup A \text{ and } A \cap B = B \cap A \qquad \forall\ A, B \in P(S)$$

Thus commutative laws are satisfied.

2. **Identity laws** : We know that ϕ and S belong to $P(S)$ such that

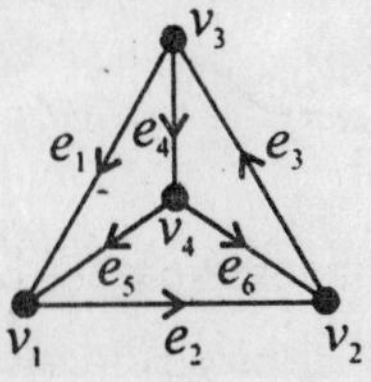

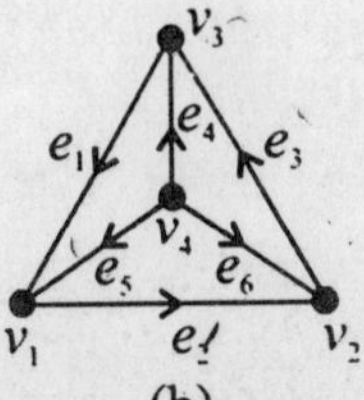

Level of a Vertex

In a rooted tree T vertex v is said to be at **level** (or depth) l if v is at a distance of l from the root. Since there exists only one path between any two vertices, therefore the level of a vertex v in a rooted tree can also be defined as the length of the path from the root to the vertex v.

Clearly the root is at level 0. In Fig. 11.7, level of $x = 1$, level of $z = 4$.

Height of a Binary Tree

The maximum level, l_{max}, of any vertex in a binary tree is called the *height* of the binary tree. The height of binary tree in Fig. 11.7 is 5.

If the *height* of a binary tree is k then the binary tree is called k-level binary tree. In a rooted tree, all the vertices at any particular level are shown on a horizontal line.

Arborescences

Definition: A digraph G is to be an **arborescence** if G contains no circuit (neither directed nor semi-directed) and G has exactly one vertex v of zero in-degree.

The vertex v with zero in-degree is called root of the arborescence. An arborescence is shown below:

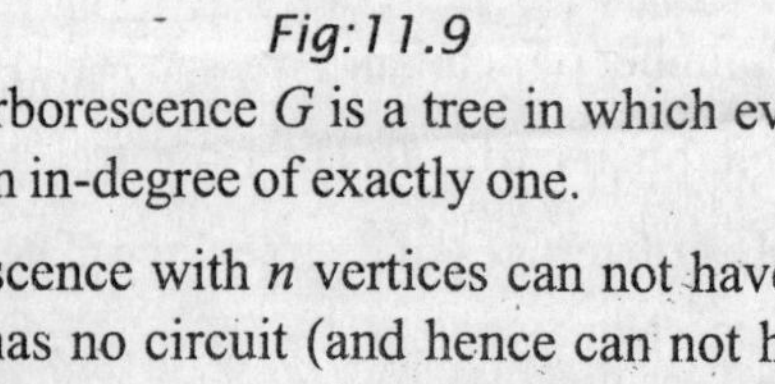

Fig: 11.9

Theorem 1: An arborescence G is a tree in which every vertex other than the root has an in-degree of exactly one.

Proof: An arborescence with n vertices can not have more than $n-1$ edges because it has no circuit (and hence can not have more edges than a tree). Consider the sum of in-degrees of all the vertices in G. Since G has at most $n - 1$ edges, this sum is $\leq n - 1$. That is,

$$d^-(v_1) + d^-(v_2) + \ldots + d^-(v_n) \leq n-1.$$

Now because there is exactly one vertex in G whose in-degree is zero, others $n-1$ terms must be positive integers. Hence, they must all be

Case (iii) If $a' = 1$. Then $a * a' = 0 \Rightarrow a * 1 = 0$

$$\Rightarrow a = 0$$

Thus $a' \neq 1$ because a is not equal to 0.

Therefore B either has only two elements 0 and 1, or B has four elements because if there is an element a in B different from 0 and 1, then B must have another element a' different from 0, 1 and a. Hence no Boolean algebra can have exactly three elements.

In the following examples, * has been dropped between a and b.

Example 5.11: Prove that for any a, b and c in a Boolean algebra the following are equal.

(a) $(a+b)(a'+c)(b+c)$ (b) $ac+a'b+bc$

(c) $(a+b)(a'+c)$ (d) $ac+a'b$

Solution: We have

$(a+b)(a'+c)(b+c) = (a+b)(a'b+c)$ by distributive law

$= a(a'b+c)+b(a'b+c)$ by distributive law

$= aa'b+ac+ba'b+bc$ by distributive law

$= (aa')b+ac+a'bb+bc$ by commutative and associative laws

$= 0b+ac+a'b+bc$ $\because aa' = 0$ and $bb = b$

$= ac+a'b+bc$ $\because 0b=0$

Thus (a) and (b) are equal

Now we show that (c) is equal to (b).

$(a+b)(a'+c) = a(a'+c)+b(a'+c)$ by distributive law

$= aa'+ac+ba'+bc$ by distributive law

$= 0+ac+a'b+bc$ by commutative and complement laws

Thus (b) is equal to (c).

Finally, we show that (b) is equal to (d).

$ac+a'b+bc = ac+a'b+(a+a')bc$ $\because a+a'=1$

$= ac+a'b+abc+a'bc$ by distributive law

$= (ac + abc) + (a'b + a'bc)$ by associative and commutative law

$= (ac + (ac)b) + (a'b + (a'b)c)$ by associative and commutative law

$= ac + a'b$ by absorption law

Hence (b) and (d) are equal.

Thus all the four are equal.

Example 5.12: In any Boolean algebra, show that

(1) $(a + b)(b + c)(c + a) = ab + bc + ca$

(2) $(a + b')(b + c')(c + a') = (a' + b)(b' + c)(c' + a)$

Solution: (1)

L.H.S $= (a + b)(b + c)(c + a)$

$= (a + b)[(b + c)(c + a)]$ by associative law

$= (a + b)[(c + b)(c + a)]$ by commutative law

$= (a + b)[c + ba]$ by distributive law

$= a(c + ba) + b(c + ba)$ by distributive law

$= ac + aba + bc + bba$ by distributive and associative laws

$= ac + (aa)b + bc + (bb)a$ by commutative and associative laws

$= ac + ab + bc + ba$ $\because aa = a$

$= ac + bc + (ab + ab)$ by commutative and associative laws

$= ac + bc + ab$ $\because a + a = a$

$= ab + bc + ac$ by associative and commutative laws

= R.H.S

(2) $(a + b')(b + c')(c + a') = [(a + b')(b + c')](c + a')$

$= [(a + b')b + (a + b')c'](c + a')$

$= (ab + b'b + ac' + b'c')(c + a')$

$$= (ab + 0 + ac' + b'c')(c + a')$$
$$= (ab + ac' + b'c')(c + a')$$
$$= (ab + ac' + b'c')c + (ab + ac' + b'c')a'$$
$$= abc + ac'c + b'c'c + aba' + ac'a' + b'c'a'$$
$$= abc + 0 + 0 + 0 + 0 + b'c'a'$$
$$= abc + a'b'c'$$

Similarly, we can show that

$$(a' + b)(b' + c)(c' + a) = abc + a'b'c'$$

Hence we have

$$(a + b')(b + c')(c + a') = (a' + b)(b' + c)(c' + a)$$

Example 5.13: In a Boolean algebra, if $a + x = b + x$ and $a + x' = b + x'$ then prove that $a = b$

Solution : We are given that

$$a + x = b + x \quad \text{... (1)}$$

and $$a + x' = b + x' \quad \text{... (2)}$$

Now $a = a + 0$ by identity law

$= a + xx'$ $\because xx' = 0$

$= (a + x)(a + x')$ by distributive law

$= (b + x)(b + x')$ by (1) and (2)

$= b + xx'$ by distributive law

$= b + 0$ $xx' = 0$

$= b$ by identity law

Example 5.14: In any Boolean algebra, prove that $b = c$ if and only if both $a + b = a + c$ and $ab = ac$ holds.

Proof : If $b = c$ then we have

$a + b = a + c$ and $ab = ac$ both hold.

Now we show that

$a + b = a + c$ and $ab = ac \Rightarrow b = c$

$$\begin{aligned} b &= b(b+a) && \text{by absorption law} \\ &= b(c+a) && \because a+b=a+c \Rightarrow b+a=c+a \\ &= bc+ba && \text{by distributive law} \\ &= bc+ab && \text{by commutative law} \\ &= bc+ac && \text{by given condition} \\ &= (b+a)c && \text{by distributive law} \\ &= (c+a)c && \because b+a=c+a \\ &= c && \text{by absorption law} \end{aligned}$$

Exercise

1. Write the dual of each of the following

   (a) $(a * 1) * (o + a') = o$

   (b) $a + a'b = a + b$

2. Show that the set $B = \{0, a, b, 1\}$ together with the operations $\vee, \wedge$ and $'$ defined by

∨	0	a	b	1
0	0	a	b	1
a	a	a	1	1
b	b	1	b	1
1	1	1	1	1

∧	0	a	b	1
0	0	0	0	0
a	0	a	0	a
b	0	0	b	b
1	0	a	b	1

′	
0	1
a	b
b	a
1	0

   is a Boolean algebra.

3. Show that algebra of sets is a Boolean algebra with respect to suitable operations.

4. Show that a non-empty subset S of a Boolean algebra is a sub algebra if it is closed under $*$ and $'$.

5. Show that a mapping f from a Boolean algebra B to another Boolean algebra B' which preserves the operations $+$ and $'$ also preserves the operation $*$.

6. If a and b are elements of a Boolean algebra B then show that $a = b$ if and only if $ab' + a'b = 0$

7. Prove that in any Boolean algebra

   (a) $a + a'b = a + b$

   (b) if $ax = bx$ and $ax' = bx'$ then $a = b$

   (c) $ab + ab' + a'b + a'b' = 1$

8. Let B_n be the set of all n-tuples of the form $(a_1, a_2, ..., a_n)$ where each a_i is either 0 or 1. Define suitable operations on B_n so that it becomes a Boolean algebra.

   [Hint : Define $(a_1, a_2,... , a_n) + (b_1, b_2, ... , b_n) = (a_1 + b_1, a_2 + b_2, ... , a_n + b_n)$, $(a_1, a_2,... a_n) * (b_1, b_2 ... , b_n) = (a_1 b_1, a_2 b_2 ..., a_n b_n)$ and $(a_1, a_2 ..., a_n)' = (a'_1, a'_2, ..., a'_n)$]

Answer

(1) (*a*) $(a + 0) + (1 * a') = 1$ (*b*) $a(a' + b) = ab$

§5.3. Boolean Algebra as Lattices

Definition: Let B be a Boolean algebra and $a, b \in B$. We say $a \le b$ if $ab' = 0$

Theorem 1: Let B be a Boolean algebra and $a, b \in B$. Then $a \le b$ if and only if $a + b = b$.

Proof: We know by Theorem 7 of § 5.1 that

$ab' = 0$ is equivalent to $a + b = b$

Thus $a \le b$ if and only if $a + b = b$

Theorem 2: Let B be a Boolean algebra. Then the relation $\le$ defined as $a \le b$ if and only if $ab' = 0$ is a partial order on B.

Proof: $\le$ is reflexive. Since $aa' = 0$ therefore $a \le a$ for all $a \in B$

$\le$ is anti-symmetric. Let $a, b \in B$ such that

$a \le b$ and $b \le a$. Then $ab' = 0$ and $ba' = 0$

Now

$$
\begin{aligned}
a &= a.1 \\
&= a(b+b') \\
&= ab+ab' \\
&= ab+0 \\
&= ab+ba' \\
&= ba+ba' \\
&= b(a+a') \\
&= b.1 \\
&= b
\end{aligned}
$$

Thus ≤ is anti-symmetric

≤ is transitive. Suppose that $a \le b$ and $b \le c$. Then $ab' = 0$ and $bc' = 0$.

Now

$$
\begin{aligned}
ac' &= a.1.c' = a(b+b')c' \\
&= (ab+ab')c' \\
&= a(bc')+(ab')c' \\
&= a.0+0.c' \\
&= 0+0=0
\end{aligned}
$$

Hence $a \le c$.

Thus ≤ is a partial order on B.

Theorem 3: Let B be a Boolean algebra. Then $(B, \le)$ where ≤ is defined as $a \le b$ if and only if $ab' = 0$, is a lattice. Moreover the identities 0 and 1 are the least and the greatest elements of this lattice.

Proof: We have already shown that $(B, \le)$ is a partial ordered set. To show that $(B, \le)$ is a lattice, we will show that for any elements $a, b \in B$, join of a and b is $a + b$ and meet of a and b is ab. That is, we will show that

Sup $\{a, b\} = a \vee b = a + b$ and $\inf\{a, b\} = a \wedge b = ab \;\; \forall a, b \in B$

Since $a.(a+b)' = a(a'b') = (aa')b' = 0b' = 0$

$\therefore \qquad a \le a + b$

Similarly, $\qquad b \le a + b$

$a \leq a+b$ and $b \leq a+b \Rightarrow a+b$ is an upper bound of the set $\{a, b\}$

Let c be any other upper bound of $\{a, b\}$. Then

$$a \leq c \text{ and } b \leq c$$
$$\Rightarrow ac' = 0 \text{ and } bc' = 0$$
$$\Rightarrow ac' + bc' = 0$$
$$\Rightarrow (a+b)c' = 0$$
$$\Rightarrow a+b \leq c$$

Thus $a+\mathrm{b}$ is the least upper bound of the set $\{a, b\}$, which by definition, is the joint of a and b denoted by $a \vee b$. Similarly we can show (or by duality) that ab is the infimum of $\{a, b\}$.

Thus $\qquad \inf\{a, b\} = a \wedge b = ab.$

Hence B is a lattice where + and ′ are join and meet operations.

Finally, if $a \in B$ then $0\,a' = 0$ and hence $0 \leq a$. This shows that 0 is the least element of B. Similarly $a.1' = a0 = 0$ for all $a \in B$ implies that $a \leq 1$ for all $a \in B$. Thus 1 is the greatest element of B. Thus B is bounded lattice.

Theorem 4 Let $(B, +, ., ')$ be a Boolean algebra. Then the lattice $(B, \vee, \wedge)$, where $a \vee b = a+b$ and $a \wedge b = ab$ is bounded, complemented and distributive. Conversely, if $(B, \vee, \wedge)$ is a bounded, complemented and distributive lattice then $(B, +, ., ')$ is a Boolean algebra, where $a+b = a \vee b$, $ab = a \wedge b$ and a' is a complement of a in $(B, \vee, \wedge)$.

Proof: Let $(B, +, \cdot, ')$ be a Boolean algebra. Then $(B, \vee, \wedge)$ is a bounded lattice by Theorem 3 above. Since $\vee$ and $\wedge$ are precisely + and $\cdot$ respectively, axioms B_2 and B_4 in the definition of a Boolean algebra show that $(B, \vee, \wedge)$ is also distributive and complemented.

Conversely suppose $(B, \vee, \wedge)$ is bounded, complemented and distributive lattice with 0 and 1 as the least and the greatest elements. For $a, b \in B$, define

$$a+b = a \vee \mathrm{b}$$

and

$$a \cdot b = a \wedge \mathrm{b}$$

Then the binary operation + and $\cdot$ are commutative with 0 and 1 as their respective identities. The distributive laws follow from the definition of a distributive lattice. Thus the axioms B_1 to B_3 in the definitions for a Boolean algebra are verified. Since $(B, \vee, \wedge)$ is a complemented lattice, we can find a complement of each $a \in B$. We denote this complement of a by a'. Now we have

$$a + a' = 1 \text{ and } aa' = 0$$

Thus axiom B_4 is also satisfied. Thus $(B, +, \cdot, ')$ is a Boolean algebra.

Remark: Many authors define a Boolean algebra as a bounded, complemented and distributive lattice. The preceding theorem shows that this definition is equivalent to ours.

§5.4. Representation Theorem for Finite Boolean Algebras

The partial order structure induced on the set B of a Boolean algebra $(B, +, \cdot, ')$ also enables us to prove the representation theorem for finite Boolean algebras. By a finite Boolean algebra we mean a Boolean algebra with a finite number of elements. We shall show that a finite Boolean algebra has exactly 2^n elements for some $n>0$. Moreover, there is a unique Boolean algebra of 2^n elements for every $n>0$.

Let $(B, +, ., ')$ be a Boolean algebra. Then $(B, \leq)$ is a lattice, where $a \leq b$ if and only if $ab' = 0$. We recall that an element a in B is an atom if it covers 0. In other words, an element a in B is called an atom if $0 < a$ and there is no element b in B such that $0 < b$ and $b < a$. For example, atoms of a power set Boolean algebra $P(S)$, are precisely singleton subsets of S. As another example, consider the Boolean algebra $B = \{1, 2, 5, 7, 10, 14, 35, 70\}$ of factors of 70 under the operations *lcm* (for +) and *gcd* (for .) The atoms of this Boolean algebra are precisely 2, 5 and 7.

In the following lemma, we give some simple properties about atoms in a finite Boolean algebra.

Lemma. Let $(B, +, \cdot, ')$ be a finite Boolean algebra. Then

(i) for every non-zero element b, there exists atleast one atom a such that $a \leq b$.

(ii) if a and b are distinct atoms then $ab = 0$

(iii) if b is any non-zero element in B and $a_1, a_2, a_3 \ldots, a_k$ be all atoms of B such that $a_i \leq b$, $i = 1 \ldots k$, then $b = a_1 + a_2 + \ldots + a_k$ and this representation is unique.

Proof: (i) Let $(B, +, \cdot, ')$ be a finite Boolean algebra with 0 as the least element. Let b be any non-zero element of B. We shall show that there exists atleast one atom a in B such that $a \leq b$. If b itself is an atom then we have nothing to do. If b is not an atom then there exists b_1 in B such that $0 < b_1 < b$. If b_1 is an atom, we are done. Otherwise there exists b_2 in B such that $0 < b_2 < b_1 < b$. Continuing in this manner, since B is finite, there exits an atom b_i for some i such that $0 < b_i < \ldots < b_2 < b_1 < b$. It follows that for every non-zero element b, there exists atleast one atom a such that $a \leq b$.

(ii) Let a and b be two distinct atoms of B. If $ab \neq 0$ then by (i), there exists an atom c such that $c \leq ab$. Note that $ab = a \wedge b = \inf\{a, b\}$ Therefore $c \leq ab \leq a$. Since a itself is an atom it follows that $a = c$. Similarly $b = c$. Hence $a = b$. In other words, if a and b are distinct atoms then $ab = 0$.

(iii) Let $a_1, a_2, \ldots, a_k$ be distinct atoms of B such that $a_1 \leq b$, $a_2 \leq b, \ldots,$ $a_k \leq b$

We claim that

$$b = a_1 + a_2 + \ldots + a_k.$$

Since $\quad a_i \leq b$ for each $i = 1, 2, \ldots, k$

$$\Rightarrow \sup\{a_1, a_2, \ldots, a_k\} \leq b$$

$$\Rightarrow a_1 + a_2 + \ldots + a_k \leq b$$

We now show that

$$b \leq a_1 + a_2 + \ldots + a_k$$

For notational convenience, let $c = a_1 + a_2 + \ldots + a_k$.

To show $b \leq c$, we shall show $bc' = 0$. $b \leq c$ will follow by (i).

If possible, suppose $bc' \neq 0$. Then by (i), there exists an atom a such that

$$a \leq bc'$$

Since $bc' \leq b$ and $bc' \leq c'$ by transitive property of $\leq$, we have

$$a \leq b \text{ and } a \leq c'$$

Since a is an atom and $a \leq b$, therefore a must be equal to one of the atoms $a_1, a_2, \ldots, a_k$. Thus $a \leq a_1 + a_2 + \ldots + a_k = c$

Now $a \leq c'$ and $a \leq c \Rightarrow a \leq c \wedge c' = cc' = 0$

$\Rightarrow a = 0$, which is impossible because a is an atom.

Thus $bc' = 0$ which implies $b \leq c$ by (i)

Now $b \leq c$, $c \leq b$ and $\leq$ is anti-symmetric give

$$b = c = a_1 + a_2 + \ldots + a_k$$

Uniqueness: Suppose $b = b_1 + b_2 + \ldots + b_r$, where each b_i is an atom and $b_i \neq b_j$ be another representation of b.

$\Rightarrow b_i \leq b$ for each i, because b is the supremum of $b_1, b_2, \ldots, b_r$

Now consider an atom b_i, $1 \leq i \leq r$. Since $b_i \leq b$, we have

$$\text{Inf } \{b_i, b\} = b_i$$

$\Rightarrow \quad b_i b = b_i, \quad 1 \leq i \leq r$

$\Rightarrow b_i (a_1 + a_2 + \ldots + a_k) = b_i \qquad \because a_1 + a_2 + \ldots + a_k = b$

$\Rightarrow b_i a_1 + b_i a_2 + \ldots b_i a_k = b_i \qquad$ by distributive law

$\Rightarrow$ for some a_j, $1 \leq j \leq k$, $\quad b_i a_j \neq 0$

$\Rightarrow b_i = a_j \qquad$ by (ii)

This shows that every b_i is equal to some a_j and hence representation of b as sum of atoms is unique (except for order)

Corollary Let $(B, +, ., ', 0, 1)$ be a Boolean algebra. Then sum of all atoms in B equals 1.

Now we state and prove **representation theorem** for finite Boolean algebras.

Theorem: Let $(B, +, \cdot, ')$ be a finite Boolean algebra. Let S be the set of atoms of B. Then $(B, +, \cdot, ')$ is isomorphic to the Boolean algebra $(P(S), \cup, \cap, -)$ of power set of S.

Proof: Let $(B, +, \cdot, ')$ be a finite Boolean algebra and let $S = \{a_1, a_2, ..., a_n\}$ be the set of all distinct atoms of B. By the last lemma, every element $x \neq 0$ has a unique representation as a sum of atoms. That is,

$$x = a_1 + a_2 + ... + a_k, \text{ where } a_i\text{'s are atoms.}$$

Define a mapping $f: B \to P(S)$ as follows

$$f(0) = \phi \text{ and } f(x) = \{a_1, a_2, ..., a_k\} \text{ where}$$

$x = a_1 + a_2 + ... + a_k$ is the unique representation of x as a sum of atoms.

Suppose x, y are any two elements in B. Suppose

$$x = a_1 + a_2 + ... + a_r + b_1 + b_2 + ... + b_s$$

and $$y = b_1 + b_2 + ... + b_s + c_1 + c_2 + ... + c_t$$

where each $a_i, 1 \leq i \leq r$, $b_j, 1 \leq j \leq s$ and $c_k, 1 \leq k \leq t$ are atoms of B.

Then $$x + y = a_1 + a_2 + ... + a_r + b_1 + b_2 + ... b_s + c_1 + c_2 + ... + c_t$$

and $$xy = b_1 + b_2 + ... + b_s$$

because if a_i and a_j are distinct atoms then $a_i a_j = 0$ and

$$a + a = a \text{ and } a.a = a \text{ for any } a \in B.$$

Hence

$$\begin{aligned} f(x + y) &= \{a_1, a_2, ..., a_r, b_1, b_2 ... , b_s, c_1, c_2 ... c_t\} \\ &= \{a_1, a_2 ... a_r, b_1, b_2, ..., b_s\} \cup \{b_1, b_2 ... , b_s, c_1, c_2 ..., c_t\} \\ &= f(x) \cup f(y) \\ f(xy) &= \{b_1, b_2, ..., b_s\} \\ &= \{a_1, a_2 ..., a_r, b_1, b_2, ..., b_s\} \cap \{b_1, b_2 ... , b_s, c_1, c_2 .., c_t\} \\ &= f(x) \cap f(y) \end{aligned}$$

Let $S - \{a_1, a_2, ..., a_r, b_1, b_2, ...b_s\} = \{d_1, d_2, ..., d_p\}$

We claim $z = d_1 + d_2 + ... + d_p$ is the complement of x. For this, we shall show $x + z = 1$ and $xz = 0$

Clearly, $x + z = a_1 + a_2 + ... + a_r + b_1 + b_2 + ... + b_s + d_1 + d_2 + ... + d_p$ is the sum of all atoms of B. Therefore, $x + z = 1$.

Also $xz = 0$ because $a_i a_j = 0$ for distinct atoms a_i and a_j.

Hence $z = x'$.

Now $f(x') = \{d_1, d_2, ..., d_p\} = S - \{a_1, a_2, ..., a_r, b_1, b_2, ..., b_s\}$

$= S - f(x) =$ complement of $f(x)$

Further, by uniqueness of the representation, we see that f is one-one and onto.

(To show f in one-one, consider $x, y \in B$ such that $x \neq y$, we can write

$$x = a_1 + a_2 + ... + a_m$$

and $y = b_1 + b_2 ... + b_k$ as sums of atoms.

$$x \neq y \Rightarrow a_1 + a_2 + ... + a_m \neq b_1 + b_2 + ... + b_k$$

$$\Rightarrow \{a_1, a_2 ...., a_m\} \neq \{b_1, b_2 ..., b_k\}$$

$$\Rightarrow f(x) \neq f(y)$$

Thus f in one-one.

To show that f is onto, let $\{a_1, a_2, ..., a_n\}$ be any subset of S. Then

$$x = a_1 + a_2 + ... + a_n$$

is a unique elements in B and $f(x) = \{a_1, a_2 ..., a_n\}$. Hence f is onto.

Hence f is an isomorphism. Thus Boolean algebra $(B, +, \cdot, ')$ and $(P(S), \cup, \cap, -)$ are isomorphic to each other.

Corollary : Every finite Boolean algebra has 2^n elements for some positive integer n.

Proof: Since, by above theorem, every finite Boolean algebra B is isomorphic to power set Boolean algebra $P(S)$ and power set $P(S)$ has 2^n elements, where n is the number of elements in S, the set of atoms of B. Hence B has 2^n elements for some $n > 0$.

Example 5.15: Consider the Boolean algebra $B = \{1, 2, 5, 7, 10, 14, 35, 70\}$ with $+, \cdot$ and $'$ defined as follows:

$$a + b = \text{lcm}(a, b)\,,\ ab = \gcd(a, b) \text{ and } a' = 70/a$$

In this Boolean algebra, atoms are 2, 5 and 7 and B is isomorphic to the Boolean algebra $(P(S), \cup, \cap, -)$, where $S = \{2, 5, 7\}$.

Example 5.16 In any Boolean algebra B, show that

$$a \leq b \Rightarrow a + bc = b(a + c), \text{ where } a, b, c \in B.$$

Proof: we have

$$\begin{aligned} a + b &= a(b + b') + b \\ &= ab + ab' + b && \text{by distributive law} \\ &= ab + 0 + b && \because a \leq b \Leftrightarrow ab' = 0 \\ &= ab + b && \text{by identity law} \\ &= b && \text{by absorption} \end{aligned}$$

$$\begin{aligned} \text{Therefore,} \quad a + bc &= (a+b)(a+c) && \text{by distributive law} \\ &= b(a + c) && \text{in view of the result just proved.} \end{aligned}$$

Example 5.17: Show that the lattice whose Hasse diagram is given below is not a Boolean algebra.

Solution: Observe that elements a and e are both complements of c. But Theorem 3 of § 5.1 says that such an element is unique. Thus given lattice cannot be a Boolean algebra.

Example 5.18: Consider the lattice $\{1, 2, 4, 5, 10, 20\}$ of all positive divisors of 20 under the divisibility relation. Then this lattice cannot be a Boolean algebra because it has six elements and $6 \neq 2^n$ for any integer $n > 0$. Thus we conclude that divisors of 20 is not a Boolean algebra.

§5.5. Boolean Functions

Let $(B, +, \cdot, ')$ be a Boolean algebra. By a constant, we shall mean any symbol, such as 0 and 1, which represents a specified element of B. By a variable, we mean a symbol, which represents an arbitrary element of B.

A Boolean function or a Boolean polynomial is an expression derived from a finite number of applications of the operations $+$, $\cdot$, and $'$ to the elements of a Boolean algebra.

Expressions such as ab, $(a' + b)' + ab'x + ab$, and $a' + b'$ are Boolean functions.

In any Boolean algebra, we know that $2a = a + a = a$, $3a = a + a + a = a$ and in general $na = a$ where n is any positive integer.

Also

$a^2 = a.\, a = a$, $a^3 = a.\, a.\, a = a$ and in general $a^k = a$, where k is any positive integer. Thus no multiples or powers appear in the Boolean polynomials.

Definition : A Boolean expression of n variables $x_1, x_2, \ldots x_n$ is said to be a **minterm** or a **minimal polynomial** if it is of the form

$$f_1(x_1) \cdot f_2(x_2) \cdot f_3(x_3) \ldots f_n(x_n)$$

where $\quad f_i(x_i) = x_i$ or x_i' for all $i = 1, 2, \ldots n$

For example, $x_1.x_2$, $x_1'.x_2$, $x_1.x_2'$ and $x_1'.x_2'$ are min-terms in two variables x_1 and x_2. Similarly $x_1x_2'x_3$ is an example of minterms in three variables x_1, x_2 and x_3.

Theorem 1: There are exactly 2^n minterms in n variables $x_1, x_2, \ldots x_n$.

Proof: A minterm in n variables $x_1, x_2, \ldots, x_n$ is an expression of the form

$$f_1(x_1) \cdot f_2(x_2) \cdot \ldots, f(x_n),$$

where each $f_i(x_i) = x_i$ or x_i' for all $i = 1, \ldots, n$

Clearly, there are two ways of selecting $f_i(x_i)$ namely x_i or x_i' for each $i = 1, \ldots, n$. Thus there are 2^n different minterms in n variables.

Disjunctive Normal Form

Definition : A Boolean function in n variables $x_1, x_2, \dots, x_n$ is said to be in disjunctive normal form (in short, $D\,N$ form) if it is a sum of min-terms. Also 1 and 0 are said to be in disjunctive normal form.

In other words, a Booleasn function in n variables $x_1, x_2, \dots, x_n$ is said to be in disjunctive normal form if the function is a sum of terms of the type

$$f_1(x_1)f_2(x_2)f_3(x_3)\dots f_n(x_n)$$

where $f_i(x_i) = x_i$ or x_i' for all $i = 1, 2, \dots n$ and no two terms are same. Also 1 and 0 are said to be in disjunctive normal form.

The disjunctive normal form is also called the sum-of-products canonical form.

Observe that in a disjunctive normal form (sum of products canonical form) any particular minterm may or may not be present. Since there are 2^n minterms in n variables, we can have only 2^{2^n} different DN forms. These DN forms include the DN form of 0 in which no minterm is present in the sum and also the DN form of 1 where all the minterms are present in the sum. In any case, every Boolean function given in DN form in n variables is equal to one of the 2^{2^n} Boolean functions.

Example 5.19: Write the functions $f = (xy' + xz)' + x'$ in DN form.

Solution: we have

$$\begin{aligned}
f &= (xy' + xz)' + x' \\
&= (xy')'(xz)' + x' \\
&= (x' + y)(x' + z') + x' \\
&= x' + yz' + x' \\
&= x' + (x' + yz') = x' + yz' \\
&= x'(y + y')(z + z') + yz'(x + x') \\
&= x'yz + x'yz' + x'y'z + x'y'z' + xyz' + x'yz' \\
&= x'yz + xyz' + x'yz' + x'y'z + x'y'z'
\end{aligned}$$

because the minterm $x'yz'$ is appearing twice.

Example 5.20: Write the Boolean function $f = x_1 + x_2$ in sum of products canonical form in three variables x_1, x_2 and x_3.

Solution: we have

$$f = x_1 + x_2$$
$$= x_1 (x_2 + x'_2)(x_3 + x'_3) + x_2 (x_1 + x'_1)(x_3 + x'_3)$$
$$= x_1 x_2 x_3 + x_1 x_2 x'_3 + x_1 x'_2 x_3 + x_1 x'_2 x'_3 + x_1 x_2 x_3$$
$$+ x_1 x_2 x'_3 + x'_1 x_2 x_3 + x'_1 x_2 x'_3$$
$$= x_1 x_2 x_3 + x_1 x_2 x'_3 + x_1 x'_2 x_3 + x_1 x'_2 x'_3 + x'_1 x_2 x_3 + x'_1 x_2 x'_3$$

Complete Disjunctive Normal Form

As seen earlier, if there are n variables, then the total number of min terms will be 2^n. Therefore any *DN* form can have at most 2^n minterms.

Definition: A disjunctive normal form in n variables, which contains all the 2^n minterms is called the complete disjunctive normal form.

For example,

$$f = xy + x'y + xy' + x'y'$$

is the complete disjunctive normal form in two variables

It can be seen by simplification of the complete disjunctive normal form or by the following theorem that the complete disjunctive normal form is identically equal to 1.

Theorem 2: If each of n variables is assigned the value 0 or 1 in an arbitrary, but fixed manner then exactly one minterm of the complete disjunctive normal form in the n variables will have the value 1 and all other minterms will have the value 0.

Proof: Consider the complete disjunctive normal form in n variables $x_1, x_2, ..., x_n$. Then it has all the 2^n minterms of the form

$$f_1(x_1) f_2(x_2) \ldots f_n(x_n)$$

where $f_i(x_i) = x_i$ or x_i' for each i = 1, 2, ..., n.

Now assign the values 0 or 1 to the variables $x_1, x_2, ..., x_n$. Select a minterm from the complete normal form as follows: use x_i if x_i is assigned the value 1 and use x_i' if x_i is assigned the value 0 for each x_i , $i = 1, 2,..., n$. The term so selected is then a product of n ones and hence is equal to 1. All other terms in the complete normal form will contain atleast one factor 0 and hence will be 0.

Corollary1: Two functions are equal if and only if their respective disjunctive normal forms contain the same minterms.

Proof: Two functions with same minterms are obviously equal. Conversely, if two functions are equal, then they must have same value for every choice of value for each variable. In particular, they assume the same value for each set of values 0 and 1, which may be assigned to the variables. By Theorem 2 above, the combinations of values of 0 and 1 which, when assigned to the variables, make the function assume the value 1 uniquely determine the terms which are present in the *DN* form for the function. Hence, both *DN* forms contain the same min-terms.

Corollary2: To establish any identity in Boolean algebra, it is sufficient to check the value of each function (on both sides of the identity) for all combinations of 0 and 1, which may be assigned to the variables.

We have seen in the preceding theorems that a Boolean function is completely determined by the values it takes for each possible assignment of 0 and 1 to the respective variables. This suggests that Boolean functions could be easily specified by giving a table to represent such properties. If such a table has been given, then the function, in disjunctive normal form, may be written down by inspection. We simply look at the conditions where the function takes the value 1 then the sum of corresponding minterms (where function takes the value 1) gives the function, although the function so obtained may not be in simplest form. The following example will explain this method.

Example 5.21: Find and simplify the function specified by the table as given on page 258.

Solution: We observe that function $f(x, y, z)$ takes the value 1 for the combinations represented by rows 2 and 4. Therefore the disjunctive normal form of f will contain two minterms each corresponding to the conditions given in rows 2 and 4. In row 2, the values of x, y, z are given respectively as 1, 1, 0 and so the corresponding minterm will be xyz'. Similary the other minterm will be $xy'z'$ which is taken with respect to row 4. Thus the function

Row	x	y	z	f(x,y,z)
1	1	1	1	0
2	1	1	0	1
3	1	0	1	0
4	1	0	0	1
5	0	1	1	0
6	0	1	0	0
7	0	0	1	0
8	0	0	0	0

$$f(x, y, z) = xyz' + xy'z'$$

We now simplify $f(x, y, z)$ by using laws of Boolean algebra

$$\therefore\ f(x, y, z) = xz'\ (y + y') = xz'$$

It may be verified that this function satisfies all the other rows also.

Rule for finding complement of a function in *DN* form

We can find by inspection the complement of any function given in disjunctive normal form. If a function *f* is given in disjunctive normal form then its complement f′ is obtained by omitting from the complete disjunctive normal form the terms which appear in the function f.

Example 5.22: Find the complement of the following functions: -

(a) $f = a'b + ab'$

(b) $f = a'bc + abc' + a'b'c + a'b'c'$

Solution: (a) The complete disjunctive normal form in two variables *a*, *b* is

$$ab + ab' + a'b + a'b'$$

Therefore the complement of $f = a'b + ab'$ is given by

$$f' = ab + a'b'$$

(*b*) The complete disjunctive normal form in three variables *a*, *b*, *c* is

$$abc + abc' + ab'c + ab'c' + a'bc + a'bc' + a'b'c + a'b'c'$$

Therefore the complement of $f = a'bc + abc' + a'b'c + a'b'c'$ is given by

$$f' = abc + ab'c + ab'c' + a'bc'$$

Solved Examples

Example 5.23: Find the function of three variables x, y and z which is 1 if either $x = y = 1$ and $z = 0$ or if $x = z = 1$ and $y = 0$ and is 0 otherwise

Solution: By given conditions, the required function takes the value 1 at two points namely when $x = 1$, $y = 1$, $z = 0$ and when $x = 1$, $y = 0$, $= 1$. The corresponding minterms are

$$xyz' \text{ and } xy'z.$$

Hence the required function f in *DN* form is equal to sum of these two terms, i.e.,

$$f = xyz' + xy'z$$

Example 5.24: Express each of the following in *DN* form in the smallest possible number of variables

(i) $xy' + xz + xy$

(ii) $(x'y + xyz' + xy'z + x'y'z't + t')'$

Solution:

(i) Let $f = xy' + xz + xy$

$= xy' + xy + xz$ by commutative law

$= x(y' + y) + xz$ by distributive law

$= x + xz$ by complement law

$= x$ by absorption law

Now f contains only one variable. Hence

$$f = x$$

is its disjunctive normal form in smallest possible variables.

(ii) Let $f = [x'y + xyz' + xy'z + x'y'z't + t']'$

We first consider the expression without complement, i.e.,

$$x'y + xyz' + xy'z + x'y'z't + t'$$

and express it in disjunctive normal form in four variables x, y, z and t.

Now

$$
\begin{aligned}
x'y+xyz'+xy'z+x'y'z't+t' &= x'y\,(z+z')\,(t+t')+xyz'\,(t+t')+xy'z \\
&\quad (t+t')+x'y'z't+t'(x+x')\,(y+y')\,(z+z') \\
&= x'y\,(zt+z't+zt'+z't')+xyz'\,(t+t')+ \\
&\quad xy'z\,(t+t')+x'y'z't+t'\,(xyz+x'yz+ \\
&\quad xy'z+xyz'+x'y'z+x'yz'+xy'z'+x'y'z') \\
&= x'yzt+x'yz't+x'yzt'+x'yz't'+xyz't+ \\
&\quad xyz't'+xy'zt+xy'zt'+x'y'z't+ \\
&\quad xyzt'+x'yzt'+xy'zt'+xyz't'+x'y'zt' \\
&\quad +x'yz't'+xy'z't'+x'y'z't'
\end{aligned}
$$

Using the law $a+a=a$, we get,

$$
\begin{aligned}
&x'yzt+x'yz't+x'yzt'+x'yz't'+xyz't+xyz't'+xy'zt+xy'zt'+ \\
&x'y'z't+x'yzt'+x'y'zt'+xy'z't'+x'y'z't'
\end{aligned}
$$

which contains 13 terms.

Now writing the missing terms from the complete *DN* form, we get the required function (which is complement of this function) as

$$f=xyzt+x'y'zt+xy'z't$$

Example 5.25: Express the following Boolean functions

$$f(x,y,z)=(x+y)\,(x+y')\,(x'+z)$$

in *DN* form in three variables

Solution: We can write

$$
\begin{aligned}
f(x,y,z) &= (x+yy')\,(x'+z) \qquad \text{using distributive law} \\
&= x(x'+z) \\
&= xz \\
&= x(y+y')z \\
&= xyz+zy'z
\end{aligned}
$$

Aliter: We can also proceed as follows. We first construct the table for values of f for all possible values x, y and z and then we write the *DN* form from the table corresponding to the rows where f takes value 1.

The table for above function is given on page 261.

Hence *DN* form of the function is

$$xyz+xy'z$$

x	y	z	f(x,y,z)
1	1	1	1
1	1	0	0
1	0	1	1
1	0	0	0
0	1	1	0
0	1	0	0
0	0	1	0
0	0	0	0

Example 5.26: In a Boolean algebra, show that

$$f(x, y) = x f(1, y) + x' f(0, y)$$

Solution: The complete *DN* form of $f(x, y)$ is

$$f(x, y) = xy + xy' + x'y + x'y' = x(y + y') + x'(y + y') \quad \text{... (1)}$$

Putiting $x = 1$ and therefore $x' = 0$, we get

$$f(1, y) = y + y'$$

Again putting $x = 0$ and therefore $x' = 1$, we get

$$f(0, y) = y + y'$$

$$\therefore \quad f(x, y) = x f(1, y) + x' f(0, y) \quad \text{by (1)}$$

Example 5.27: Write all 16 possible functions of two variables x and y

Solution: All possible 16 functions are listed in the following table:

x	y	f_1	f_2	f_3	f_4	f_5	f_6	f_7	f_8	f_9	f_{10}	f_{11}	f_{12}	f_{13}	f_{14}	f_{15}	f_{16}
1	1	1	1	1	1	0	1	1	1	0	0	0	1	0	0	0	0
1	0	1	1	1	0	1	1	0	0	1	0	1	0	1	0	0	0
0	1	1	1	0	1	1	0	1	0	0	1	1	0	0	1	0	0
0	0	1	0	1	1	1	0	0	1	1	1	0	0	0	0	1	0

Therefore, the 16 functions are :

$$f_1 = xy + x'y + xy' + x'y' = 1$$

$$f_2 = xy + x'y + xy' \qquad = x + x'y$$

$$f_3 = xy + x'y + x'y' \qquad = xy + x'$$

$$f_4 = xy + xy' + x'y' \qquad = xy + y'$$

$$f_5 = x'y + xy' + x'y' \qquad = x'y + y'$$

$$f_6 = xy + x'y \;=\; y$$

$$f_7 = xy + xy' \;=\; x$$

$$f_8 = xy + x'y' \;=\; xy + x'y'$$

$$f_9 = x'y + x'y' = x'$$

$$f_{10} = xy' + x'y' = y'$$

$$f_{11} = x'y + xy' = x'y + xy'$$

$$f_{12} = xy \;=\; xy$$

$$f_{13} = x'y \;=\; x'y$$

$$f_{14} = xy' \;=\; xy'$$

$$f_{15} = x'y' \;=\; x'y'$$

$$f_{16} = 0 \;=\; 0$$

Exercise

1. Express each of the following in DN form in the smallest possible number of variables
   (a) $x + x'y$
   (b) $(u + v + w)(uv + u'w)'$
2. Write complete disjunctive normal form in three variables x, y and z. Determine which term equals 1 if
   (a) $x = 1, y = z = 0$
   (b) $x = z = 1$ and $y = 0$
3. Write disjunctive normal form in the three variables x, y and y of the function $f = x + y'$.
4. Write the function f of x, y and z which is 1 if and only if any two or more of the variables are 1.
5. Find, by inspection, the complement of each of the following :
   (a) $xy + x'y$
   (b) $x'y'z' + x'yz + xy'z'$.

6. Prove that there are exactly 2^{2^n} distinct functions of n variables in a Boolean algebra.
7. Write and simplify the two functions f_1 and f_2 specified by the table

x	y	z	f_1	f_2
1	1	1	0	1
1	1	0	1	1
1	0	1	0	0
1	0	0	1	0
0	1	1	0	0
0	1	0	0	0
0	0	1	0	0
0	0	0	0	1

Answers

1. (a) $xy + xy' + x'y$ (b) $uv'w + u'vw' + uv'w'$
3. $xyz + xyz' + xy'z + xy'z' + x'y'z + x'y'z'$
4. $xy + yz + zx$
5. (a) $xy' + x'y'$ (b) $xyz + xyz' + xy'z + x'y'z + x'yz'$
7. $f_1 = xz', f_2 = xy + x'y'z'$

§ 5.6 Conjunctive Normal Form

Conjunctive normal form is a dual of disjunctive normal form. Thus all the results that we proved for *DN* forms can be extended to this form by duality.

Definition : A Boolean function is said to be in **conjunctive normal form** (in short, *CN* form) in n variables $x_1, x_2, \ldots, x_n$, for $n > 0$, if the function is a product of terms of the type

$$f_1(x_1) + f_2(x_2) + \ldots + f_n(x_n)$$

where $f_i(x_i) = x_i$ or x_i' for each $i = 1, 2, ..., n$, and no two terms are same. Moreover 0 and 1 are also said to be in conjunctive normal form.

The terms of the type

$$f_1(x_1) + f_2(x_2) + ... + f_n(x_n),$$

where $f_i(x_i) = x_i$ or x_i' for each $i = 1, 2, ..., n$ are called **max-terms** or maximal polynomials.

Theorem 1: Every function in Boolean algebra, which contains no constants is equal to a function in *CN* form

Proof : Let f be the Boolean function which contains no constant. If f contains an expression of the form $(x + y)'$ or $(xy)'$ for some variables x and y then De Morgan's rule may be applied to get $x'y'$ and $x' + y'$, respectively. This process may be continued until each $'$ which appears applies only to a single variable.

Next, by applying the distribution law, f can be reduced to products. Now suppose some term does not contain either x_i or x_i' for some variable x_i. Then $x_i x_i'$ may be added to this term without changing the function. Continuing this process for each missing variable in each factors in f will give an equivalent function whose factors contain x_i or x_i' for each $i = 1, 2, ..., n$.

Finally, using $aa = a$, we can eliminate the duplicate terms and with this the proof is complete.

Example 5.28: Write the function $(xy' + xz)' + x'$ in *CN* form.

Solution: Let

$$\begin{aligned}
f &= (xy' + xz)' + x' \\
&= (xy')'\,(xz)' + x' \\
&= (x' + y)\,(x' + z') + x' \\
&= x' + (x' + y)\,(x' + z') \\
&= (x' + x' + y)\,(x' + x' + z) \\
&= (x' + y)\,(x' + z') \\
&= (x' + y + zz')\,(x' + z' + yy') \\
&= (x' + y + z)\,(x' + y + z')\,(x' + y + z')\,(x' + y' + z') \\
&= (x' + y + z)\,(x' + y + z')\,(x' + y' + z')
\end{aligned}$$

Definition: (*Complete conjunctive normal form*) : The conjunctive normal form in n variables which contains 2^n factors (max-terms) is called complete conjunctive normal form in n variables.

Theorem 2: Let f be a complete conjunctive normal form in n variables. Then exactly one maxterm (factor) will have the value 0 and all other max-terms (factors) will have the value 1.

Proof: The proof is dual of Th. 2 of § 5.5

Example 5.29: Find the Boolean function f in *CN* form that is given by the following table:

x	y	z	f
1	1	1	1
1	1	0	0
1	0	1	0
1	0	0	0
0	1	1	1
0	1	0	1
0	0	1	1
0	0	0	0

Solution: To get the expression in *CN* form, we look at the values of $f(x, y, z)$ when it is 0. We see from the table that f takes value 0 at 2nd, 3rd, 4th and 8th row. Thus

$$f = (x' + y' + z)\,(x' + y + z')\,(x' + y + z)\,(x + y + z)$$

[Note: the maxterm corresponding to 2nd row is $(x' + y' + z)$]

Rule for finding complement of a function given in *CN* form

As in disjunctive normal form, we can use the conjunctive normal form to find complements of functions written in this form by inspection. The complement of any function written in *CN* form is that function whose factors are exactly those factors of the complete conjunctive normal form, which are missing from the given function.

Example 5.30: Find the complement of $f = (x + y')\,(x' + y)$

Solution : The given function f is written in *CN* form of two variables. The complete conjunctive normal form in two variables is

$$(x + y)\,(x + y')\,(x' + y)\,(x' + y')$$

Therefore the complement of f is

$$(x+y)(x'+y').$$

Example 5.31: Find the conjunctive normal form for the function

$$f = xyz + x'yz + xy'z' + x'yz'$$

Solution: We know that $(f')' = f$.

$$\therefore f = [(xyz + x'yz + xy'z' + x'yz')']'$$

$$= [(xyz)'(x'yz)'(xy'z')'(x'yz')']' \quad \text{by De Morgan's law}$$

$$= [(x'+y'+z')(x+y'+z')(x'+y+z)(x+y'+z)]'$$

by De Morgan's law

$$= (x+y+z)(x'+y+z')(x+y+z')(x'+y'+z)$$

Exercise

(1) Express each of the following in the *CN* form in the smallest possible number of variables

(a) $x + x'y$

(b) $(u+v+w)(uv+u'w)'$

(c) $(x+y)(x+y')(x'+z)$

(2) Write the *CN* form in thiee variables x, y and z

(a) $x + y'$

(b) $(x+y)(x'+y')$

(3) Write the function of x, y and z which is 0 when any two or more of the variables are 0 otherwise it is 1.

(4) Find by inspection the complement of each of the following

(a) $(x+y)(x'+y)(x'+y')$

(b) $(x+y+z)(x'+y'+z)(x'+y'+z')$

(5) Change the function $f = uv + u'v + u'v'$ from *DN* form to *CN* form

(6) Change the function $f = (x+y')(x'+y)(x'+y')$ from *CN* form to *DN* form.

(7) Let $f(x_1, x_2, x_3) = [(x_1 + x_2)' + x_1'x_3]'$ be a Boolean expression (function) over two valued Boolean algebra. Write $f(x_1, x_2, x_3)$ in both *DN* and *CN* forms.

(8) Write the expression $f(x_1, x_2, x_3) = x_1x_2 + x_1x_3 + x_2'x_3$ in both *DN* and *CN* forms.

(9) Express the function given by the table below in both *DN* form and *CN* form.

(x_1, x_2, x_3)	$f(x_1, x_2, x_3)$
(0, 0, 0)	1
(0, 0, 1)	0
(0, 1, 0)	1
(0, 1, 1)	0
(1, 0, 0)	0
(1, 0, 1)	1
(1, 1, 0)	0
(1, 1, 1)	1

(10) For any Boolean function $f(x_1, x_2)$, show that

$$f(x_1, x_2) = [x_1 + f(0, x_2)]\,[x_1' + f(1, x_2)].$$

(11) Simplify the following Boolean functions:

(a) $ab + abc + bc$

(b) $(ab' + c)(a+b')c$

Answers

(1) (a) $x + y$ (b) $(u + v + w)(u + v + w')(u + v' + w')(u' + v' + w)(u' + v' + w')$ (c) $(x + z)(x + z')(x' + z)$

(2) (a) $(x + y' + z)(x + y' + z')$ (b) $(x + y + z)(x + y + z')(x' + y' + z)(x' + y' + z')$

(3) $(x + y + z)(x + y + z')(x + y' + z)(x' + y + z)$

(4) (a) $(x + y')$ (b) $(x + y + z')(x + y' + z)(x + y' + z')(x' + y + z)(x' + y + z')$

(5) $u' + v$ (6) $x'y'$

(7) *DN* form of $f = x_1x_2x_3 + x_1x_2x'_3 + x_1x'_2x_3 + x_1x'_2x'_3 + x'_1x_2x'_3$

CN form of $f = (x_1+x_2+x_3)\,(x_1+x_2+x'_3)\,(x_1+x'_2+x'_3)$

(8) *DN* form of $f = x_1\,x_2\,x_3 + x_1\,x_2\,x'_3 + x_1\,x'_2\,x_3 + x'_1\,x'_2\,x_3$

CN form of $f = (x_1+x_2+x_3)\,(x_1+x'_2+x_3)\,(x_1+x'_2+x'_3)\,(x'_1+x_2+x_3)$

(9) $f(x_1, x_2, x_3) = (x_1+x_2+x'_3)\,(x_1+x'_2+x'_3)\,(x'_1+x_2+x_3)\,(x'_1+x'_2+x_3)$

(11) (a) $ab+ac$ (b) $ac+b'c$

§ 5.7 Minimization of Boolean Functions: Karnaugh Map

In this section, we shall concern with the problem of obtaining a minimal Boolean function equivalent to a given Boolean function. These problems arise in the design of switching circuits because the cost of the circuit, to some extent, depends on the number of switches in the circuit. The goal of minimization is to reduce to a minimum number of switches or gates required by a circuit.

A general method of simplifying a Boolean function to obtain a minimal form is to use basic laws and identities such as $a + ab = a$ of a Boolean algebra. Consider the following example:

Example 5.32: Simplifying the Boolean function

$$f = ab'cd + cb + cd' + ac' + a'bc' + b'c'd'$$

Solution: Here

$$\begin{aligned}
f &= ab'cd + cb + cd' + ac' + a'bc' + b'c'd' \\
&= (ab'd + b + d')c + (a + a'b + b'd')c' \\
&= c\,(b + d' + ab'd) + c'\,[(a + a')\,(a + b) + b'd'] \\
&\qquad \text{using } a + a'\,b = (a + a')\,(a+b) \\
&= c\,[\,(b + d' + b')\,(b + d' + ad)] + c'\,[a + b + b'd'] \\
&\qquad \text{using distributive law \& } a + a' = 1 \\
&= c\,(b+d' + a)\,(b + d' + d) + c'(a + b + b')\,(a + b + d') \\
&= c\,(a + b + d') + c'\,(a+b+d') \quad \because\ d + d' = 1,\ b + b' = 1 \\
&= (a + b + d')\,(c + c') \\
&= a + b + d'
\end{aligned}$$

From the above, it is clear that the choice of which Boolean laws to use in any particular simplification operation is primarily determined by the skill of the person performing Boolean manipulations and this skill is partly a matter of experience. Because of this difficulty in reducing a Boolean functions to its simplest (minimal) form, a method based on Karnaugh maps has been developed.

Karnaugh Maps: The Karnaugh map is a pictorial representation of truth table of the Boolean functions. This method is easy to use when Boolean function has six or fewer variables. Since function of one variable can be simplified easily, there is no need to illustrate it. We illustrate the method when number of variables in a function is 2, 3 and 4.

Case of two Variables

We consider the case when the Boolean function f is of two variables, say x and y. In the first figure below, we have constructed a 2×2 matrix of squares with each square containing one possible input combination of variables x and y. The Karnaugh map of the function is the 2×2 matrix obtained by placing 0s and 1s in the square according to whether the functional value is 0 or 1 for the input combination associated with that square. For example, the Boolean function

$$f = xy + x'y$$

is represented by the Karnaugh map as shown in second figure below.

	x	x'
y	xy	$x'y$
y'	xy'	$x'y'$

	x	x'
y	1	1
y'	0	0

Karnaugh map of $f = xy + x'y$

We now consider the method to obtain minimal form of the function by using Karnaugh map. The application of the Boolean law $xy + x'y = x$, when seen in the context of a Karnaugh map, becomes the replacement of two adjacent squares (squares having one side in common) containing 1s by a rectangle containing two squares. The absorption law $x + xy = x$ has its counterpart on a Karnaugh map as well. It is simply the grouping of adjacent squares into the largest possible rectangle of such squares and we will use the largest rectangle instead of individual squares. To find minimal form of the function, we first consider all largest rectangles composed of the adjacent squares with 1s in them. From the set of these largest rectangles, the minimum number of rectangles are taken such that every square with 1 is part of atleast one such rectangle.

Example 5.33: Use the Karnaugh map method to find a minimal *DN* form (sum-of-products form) of the following functions

(a) $f(x, y) = xy + xy'$

(b) $f(x, y) = xy + x'y + x'y'$

(c) $f(x, y) = xy + x'y'$

Solution:

(a) We first represent $f(x, y)$ by a Karnaugh map. The Karnaugh map representation of $f(x, y) = xy + xy'$ is the following :

	x	x'
y	1	0
y'	1	0

We have represented two adjacent squares with 1s in them by a rectangle. This rectangle represents x. Hence

$$f(x, y) = x.$$

(b) The representation of $f(x, y) = xy + x'y + x'y'$ by Karnaugh map is as follows :

	x	x'
y	1	1
y'	0	1

The function $f(x, y)$ contains two pairs of adjacent squares with 1 (indicated by two rectangles) which includes all the squares of $f(x, y)$ which contain 1. The horizontal pair (rectangle) represents y and vertical pair (rectangle) represents x'. Hence

$$f(x, y) = y + x' = x' + y$$

is its minimal form.

(c) The Karnaugh map representation of $f(x, y) = xy + x'y'$ is given below:

	x	x'
y	1	0
y'	0	1

Observe that $f(x, y)$ consists of two rectangles as shown in the figure. Thus

$$f(x, y) = xy + x'y'$$

is the minimal form.

Case of three Variables

We now turn to the case of a function of three variables, say x, y and z. The Karnaugh map corresponding to Boolean functions $f(x, y, z)$ is shown in figure 3(a).

	xy	xy'	$x'y'$	$x'y$
z				
z'				

Fig 3(a)

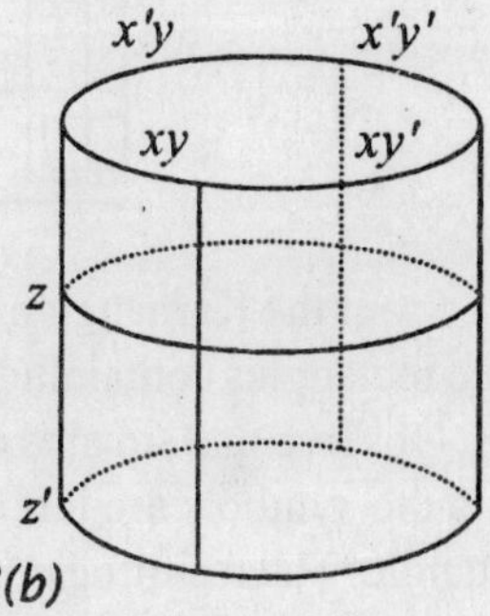

Fig 3(b)

In figure 3(a), each square represents the minterm corresponding to the column and row intersecting in that square. In order that every pair of adjacent products in figure 3(a) are geometrically adjacent, the right and left edges of the must be identified. This is equivalent to cutting out, bending and gluing the map along the identified edges to obtain the cylindrical figure as shown in figure 3(b).

By a *basic rectangle* in Karnaugh map with three variables, we mean a square, two adjacent squares or four squares which form a 1 x 4 or a 2 x 2 rectangle.

Suppose that the Boolean function $f(x, y, z)$ has been represented in the Karnaugh map by placing 0s and 1s in the appropriate squares. A minimal form of $f(x, y, z)$ will consist of the least number of *maximal basic rectangles* (a basic rectangle which is not contained in any larger basic rectangle) of f which together include all the squares with 1 (in them) of f.

Example 5.34: Find using Karnaugh maps a minimal form for each of the following Boolean fucntions

(a) $f(x, y, z) = x\,y\,z + x\,y\,z' + x'\,y\,z' + x'\,y'\,z'$

(b) $f(x, y, z) = x\,y\,z + x\,y\,z' + xy'z + x'yz + x'\,y'\,z$

(c) $f(x, y, z) = x\,y\,z + x\,y\,z' + x'\,y\,z' + x'\,y'\,z' + x'\,y'\,z$

Solution

(a) The Karnaugh map corresponding to the given function is given below

	xy	xy'	$x'y'$	$x'y$
z	1	0	1	0
z'	1	0	0	1

From the Karnaugh map, we see that $f(x, y, z)$ has three maximal basic rectangles containing squares with 1 which are shown by rectangles. Observe that squares corresponding to xyz' and $x'yz'$ are adjacent. Thus the symbols are left open ended to signify that they join in one rectangle. The resulting minimal Boolean function is

$$xy + yz' + x'y'z$$

(b) The Karnaugh map corresponding to the function

$$f = xyz + xyz' + xy'z + x'yz + x'y'z$$

is given below which has five squares with 1s in them corresponding to the five minterms of f.

	xy	xy'	$x'y'$	$x'y$
z	1	1	1	1
z'	1	0	0	0

From the Karnaugh map, we see that $f(x, y, z)$ has two maximal basic rectangles containing all the squares with 1, which are shown by rectangles. One of the maximal basic rectangle is the two adjacent squares which represents xy and the other is the 1x 4 square which represents z. Both are needed to cover all the squares with 1. So, the minimal form of $f(x, y, z)$ is given by

$$f(x, y, z) = xy + z$$

(c) The Karnaugh map corresponding to the function

$$f(x, y, z) = xyz + xyz' + x'yz' + x'y'z' + x'y'z$$

is given below which has five squares with 1s in them corresponding to the five minterms of f.

	xy	xy'	$x'y'$	$x'y$
z	1	0	1	0
z'	1	0	1	1

As shown by the rectangles, $f(x, y, z)$ has four maximal basic rectangles. To cover all squares with 1s in them, it is necessary here to include basic rectangles which represent xy and $x'y'$ and only one of two rectangles which correspond to $x'z'$ and yz'. Thus $f(x, y, z)$ has two minimal forms :

$$f(x, y, z) = xy + x'y' + x'z'$$

and

$$f(x, y, z) = xy + x'y' + yz'.$$

Case of four Variables

The Karnaugh map corresponding to Boolean functions $f(x, y, z, w)$ with four variables x, y, z and w is shown below. Each of the 16 squares corresponds to one of the 16 min-terms with four variables

$$xyzw, xyzw', \ldots, x'yz'w$$

as indicated by the labels of the row and column of the square.

	xy	xy'	$x'y'$	$x'y$
zw				
zw'				
$z'w'$				
$z'w$				

Here again, we consider the first and last column to be adjacent and the first and last rows to be adjacent, both by wrap around.

A basic rectangle in a four-variable Karnaugh map is a square, two adjacent squares, four squares which form a 1 × 4 or 2 × 2 rectangle or eight squares which form a 2 × 4 rectangle. The minimization technique for a Boolean function $f(x, y, z, w)$ is the same as for three variables function.

Example 5.35: Use Karnaugh maps to find a minimal form for the following Boolean functions

(a) $f(x, y, x, w) = x'\,y\,z\,w + x\,y'\,z\,w' + x'\,y'\,z\,w' + x\,y\,z'\,w' + xy'\,z'w'$

(b) $f(x, y, z, w) = xy' + x\,y\,z + x'\,y'\,z' + x'\,y\,z\,w'$

Solution:

(a) The karnaugh map representation of the given function is shown below which has five squares with 1s.

	xy	xy'	$x'y'$	$x'y$
zw	0	0	0	1
zw'	0	1	1	0
$z'w'$	1	1	0	0
$z'w$	0	0	0	0

in them corresponding to the five min-terms of f. A minimal cover of all 1s of the map consists of the three maximal basic rectangles as shown in the figure. Thus the minimal form is

$$f(x, y, z, w) = y'zw' + xz'w' + x'yzw$$

(b) The Karnaugh map representation of the given function is shown below. Observe that there are four squares with 1s in them representing xy'. Similarly, there are two squares with 1 representing xyz and so on.

	xy	xy'	$x'y'$	$x'y$
zw	1	1		
zw'	1	1		1
$z'w'$		1	1	
$z'w$		1	1	

The minimum number of maximal basic rectangles to cover all 1s of the map is 3 as shown in the figure. Thus the minimal form is

$$f(x, y, z) = xz + y'z' + yzw'.$$

Observe that the upper left 2 ×2 rectangle represents xz while the other 2 ×2 rectangle represents $y'z'$.

Example 5.36: Use a Karnaugh map to find a minimal form of the function

$$f(x, y, z, w) = xyzw + xyzw' + xy'zw' + x'y'zw + x'y'zw'$$

Solution: The Karnaugh map of the given function is shown below:

	xy	xy'	$x'y'$	$x'y$
zw	1	0	1	0
zw'	1	1	1	0
$z'w'$	0	0	0	0
$z'w$	0	0	0	0

As shown by the rectangles, $f(x, y, z, w)$ has four maximal basic rectangles of 1 × 2 size. To cover all 1s, it is necessary to include basic rectangles which represent xyz and $x'y'z$ and only one of the two dotted rectangles which cover 1 at the square corresponding to $xy'zw'$.

Hence we obtain two minimal forms, namely,

$$f(x, y, z, w) = xyz + x'y'z + xzw'$$

and

$$f(x, y, z, w) = xyz + x'y'z + y'zw'.$$

Exercise

Q.1 Describe the Karnaugh maps for three and four variables.

Q.2 Use the Karnaugh map representation to find a minimal form of each of the following functions:

(a) $f(x, y) = x'y + xy$

(b) $f(x, y, z) = xyz + xy'z + x'yz + x'y'z$

(c) $f(x, y, z) = xyz' + xy'z' + x'y'z' + x'y'z + x'yz + x'yz'$

Q.3 Use the Karnaugh map to find a minimal form of each of the following functions:

(a) $f = xyz'w' + xyz'w + xy'zw + xy'zw' + x'y'zw + x'y'zw' + x'yw'z'$

(b) $f = xyzw' + xy'zw' + xy'z'w' + xy'z'w + x'y'zw + x'y'zw' + x'y'z'w' + x'yz'w'$

Q.4 Find the minimal form of the Boolean function of four variables represented by the Karnaugh map given below:

	xy	xy'	$x'y'$	$x'y$
zw	1	0	0	1
zw'	0	0	0	0
$z'w'$	0	0	0	0
$z'w$	1	0	0	1

Answers

2 (a) $f = y$ (b) $f = z$ (c) $f = z' + x'z$

3 (a) $f = y'z + xyz' + yz'w'$

(b) $f = xzw' + xy'z' + x'y'z + x'z'w$

4 $f = yw$

CHAPTER SIX

Switching Circuits and Logic Circuits

One of the major applications of Boolean algebra is to the switching circuits (an electrical network consisting of switches) that involve two-state devices. The simplest example of such a device is a switch or contact. The theory introduced here holds equally well for such two-state devices as rectifying diodes, magnetic cores, transistors, various type of electron tubes etc. With the advent of computers, the algebra of circuits is receiving more attention because of the significant use of Boolean algebra in the design and simplification of complex circuits which are involved in electronic computers, dial telephone switching systems etc.

§ 6.1. Switching Circuits

By a switch we mean a contact or a device which permits or stops the flow of electric current. The switch can assume two states 'closed' or 'open' (ON or OFF). When the switch is closed the current flows in the circuit. When the switch is open current does not flow. We will use $a, b, c \ldots x, y, z, \ldots$etc. to denote switches in a circuit. If two switches operate so that they open and close simultaneously we denote them by the same letter. Again, if two switches be such that one is open if and only if the other is closed, we represent them by a and a'.

There are two basic ways in which switches are generally interconnected. These are referred to as 'in series' and in 'parallel'.

Definition 1: Two switches *a* and *b* are said to be connected 'in series' if the current flows only when both are closed and current does not flow if any one or both are open.

Two switches *a* and *b* connected in series a circuit is denoted by *ab* and is represented as shown in the following diagram:

Definition 2: Two switches *a* and *b* are said to be connected 'in parallel' if current flows when any one or both are closed and current does not flow when both are open.

Two switches *a* and *b* connected in parallel in a circuit is denoted by $a + b$ and is represented as shown in the following diagram:

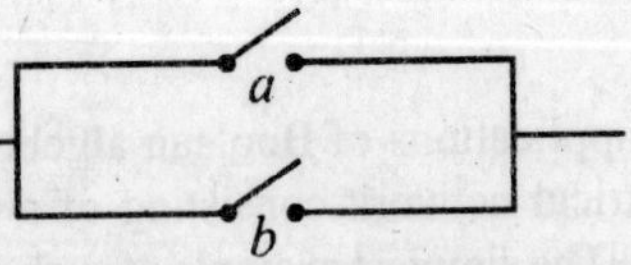

We assign the value 1 to a switch which is always closed and the value 0 to a switch which is always open. If two switches *a* and *a′* both appear then *a* is 1 if and only if $a'=0$.

Theorem: The algebra of switches is a Boolean algebra.

Proof: We know that the set B = {0, 1} with operations +, · and ′ defined by the following tables forms a Boolean algebra of two elements

+	0	1
0	0	1
1	1	1

·	0	1
0	0	0
1	0	1

a	a′
0	1
1	0

Consider the following correspondence between the elements and operations of the switching algebra and the Boolean algebra B of two elements described above.

Boolean algebra	x	x'	$+$	$\cdot$	0	1
	$\updownarrow$	$\updownarrow$	$\updownarrow$	$\updownarrow$	$\updownarrow$	$\updownarrow$
Switching algebra	x	x'	connected in parallel (+)	connected in series (·)	open	closed

With the help of this correspondence, every series-parallel circuit corresponds to a Boolean function and conversely every Boolean function corresponds to a circuit. Two circuits S_1 and S_2 are defined to be equivalent if both are open (current does not pass through either) or both are closed (current passes through both) for any given position of switches involved. We now verify that switching algebra satisfies all the axioms of Boolean algebra:

[B_1] **Commutative laws**: Since circuits

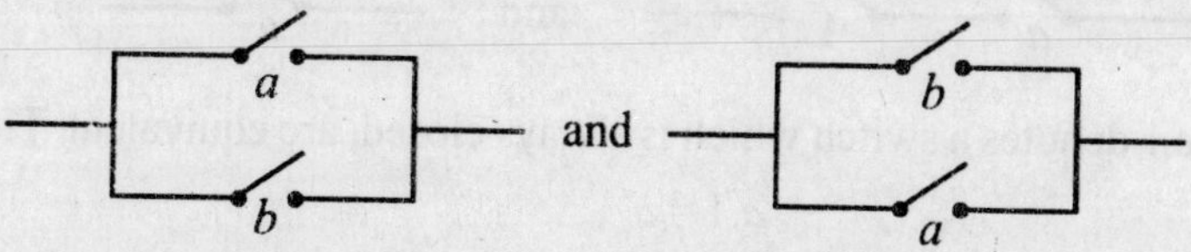

are equivalent, we have $a+b=b+a$

Similarly, circuits

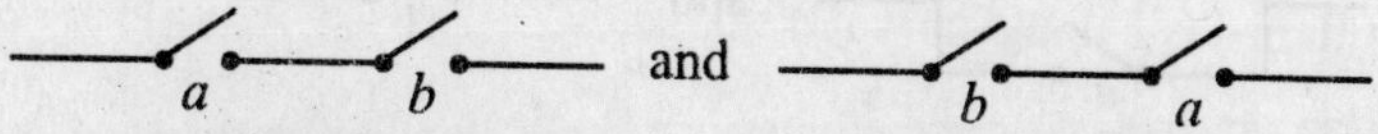

are equivalent, we have $a\cdot b=b\cdot a$

[B_2] **Distributive laws**: Since circuits

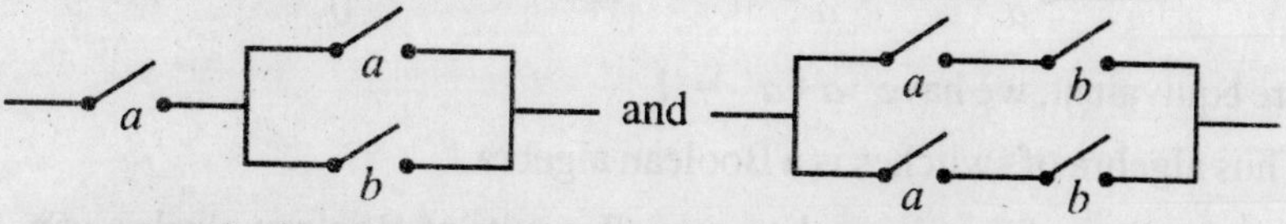

are equivalents, we have $a.(b+c) = ab+bc$

Similarly, we have,

$$a+b\cdot c = (a+b)\cdot(a+c)$$

because circuits

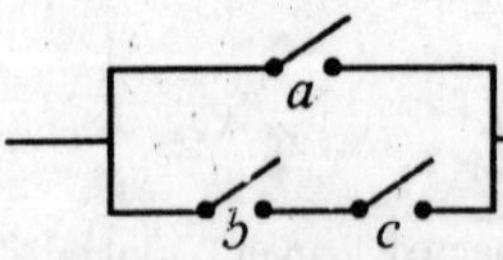 and 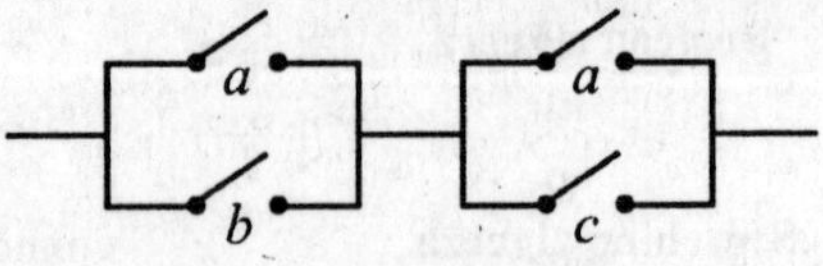

are equivalent

[B_3] **Identity laws**: Since circuits

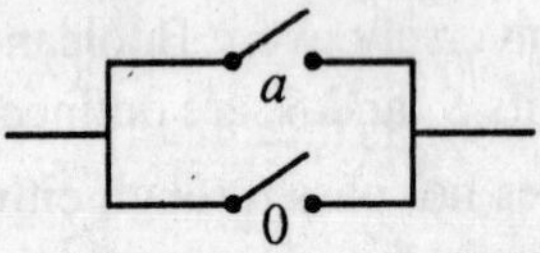 and 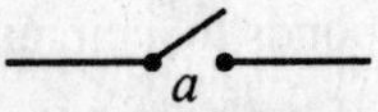 ,

where 0 represents a switch which is always open, are equivalent, therefore, we have, $a+0=a$

Similarly, circuits

 and 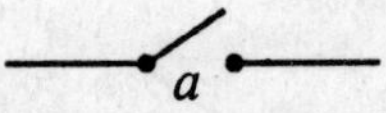

where 1 denotes a switch which is always closed, are equivalent. Thus

$$a.1=a$$

[B_4] **Complement laws**: Since the circuits

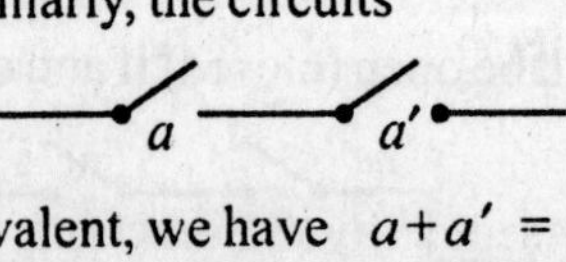 and 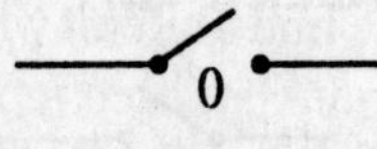

are equivalent, we have $a+a' = 1$

Similarly, the circuits

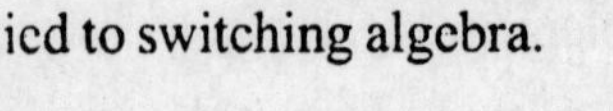 and  ,

are equivalent, we have $a+a' = 1$

Thus algebra of switches is a Boolean algebra.

In view of the above theorem, all results of Boolean algebra can be applied to switching algebra.

Example 6.1. Draw a circuit which realizes (represents) the Boolean function

$$f=abc'+a'(b+c')$$

Solution: The circuit is given by the diagram

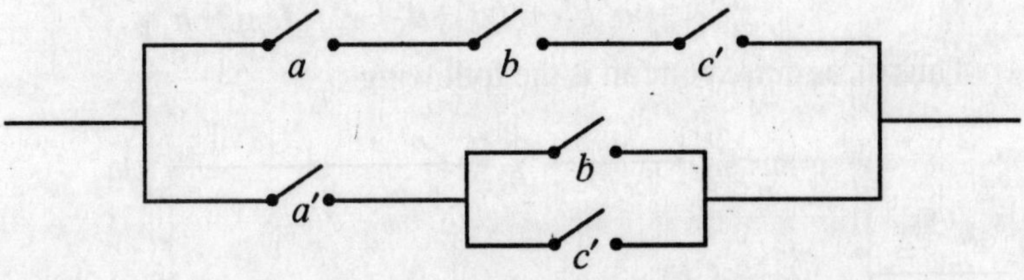

Example 6.2. Find the function that represents the circuit

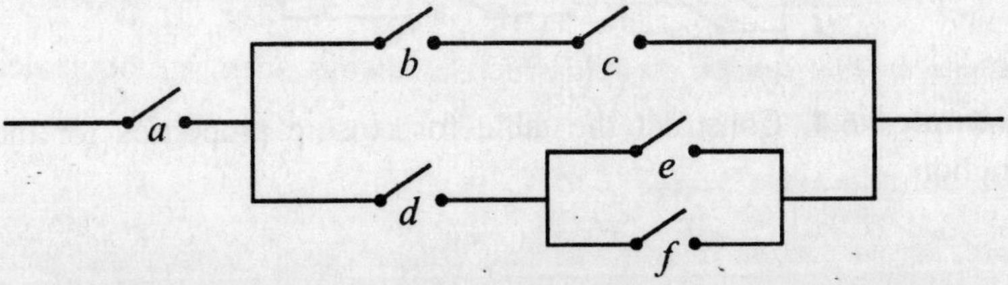

Solution: The function f which represents the circuit is given by

$$f = a(bc+d(e+f))$$

Example 6.3. Find the function f that represents the circuit

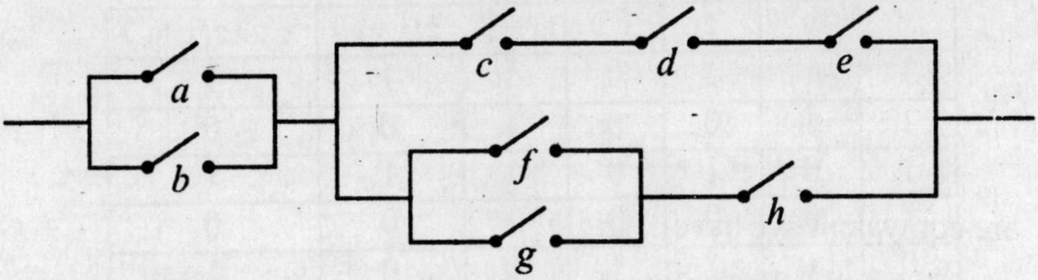

and hence find a circuit which would be open (closed) if and only if the above circuit is closed (open).

Solution: The given circuit is represented by the function

$$f = (a+b)(cde+(f+g)h)$$

The required circuit which will be open if and only if the given circuit is closed will be given by the complement of f. Now,

$$\begin{aligned} f' &= [(a+b)(cde+(f+g)h)]' \\ &= (a+b)' + [cde + (f+g)h]' \end{aligned}$$

$$= a'b' + (c'+d'+e')[(f+g)'+h']$$

$$= a'b' + (c'+d'+e')[f'g'+h']$$

Thus the required circuit is the following

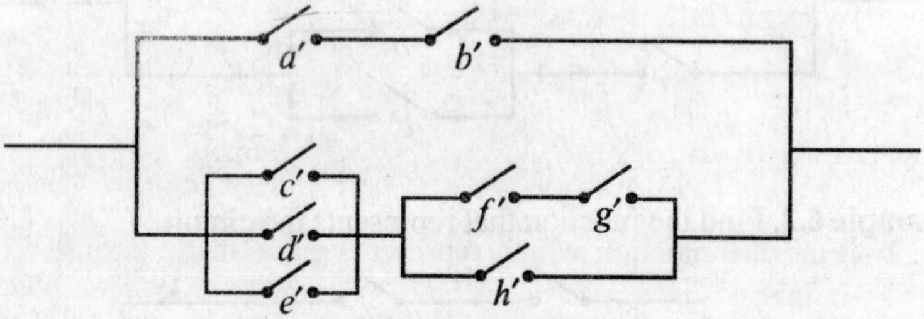

Example 6.4. Construct the table for closure properties for the function

$$f = x'z+z(x+y')$$

Solution: A table of closure properties for a function is identical to a truth table for a propositional function. Therefore, the table of closure properties for the function

$f = x'z+z(x+y')$ is given below:

x	y	z	$x'y$	$x+y'$	$z(x+y')$	$x'y+z(x+y')$
1	1	1	0	1	1	1
1	1	0	0	1	0	0
1	0	1	0	1	1	1
1	0	0	0	1	0	0
0	1	1	1	0	0	1
0	1	0	1	0	0	1
0	0	1	0	1	1	1
0	0	0	0	1	0	0

Exercise

1. Draw circuits which realize each of the following expressions without simplifying the expressions:

   (a) $abc+ab(dc+ef)$

   (b) $(a+b'+c)(a+bc') + c'd + d(b'+c)$

   (c) $x[y(z+w) + z(u+v)$

2. Find the function which represents each of the circuits given below:

(a) 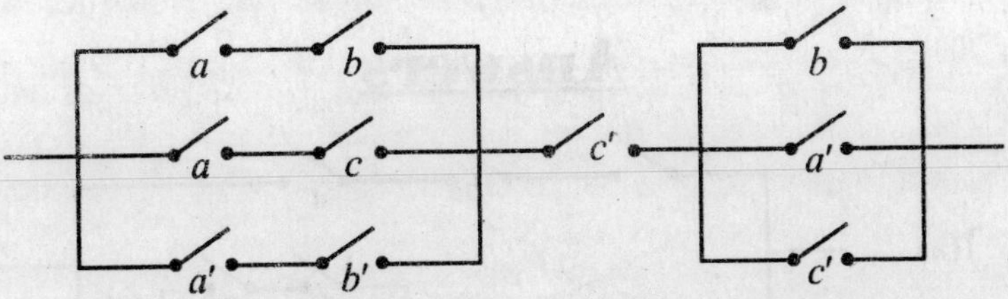

(b) 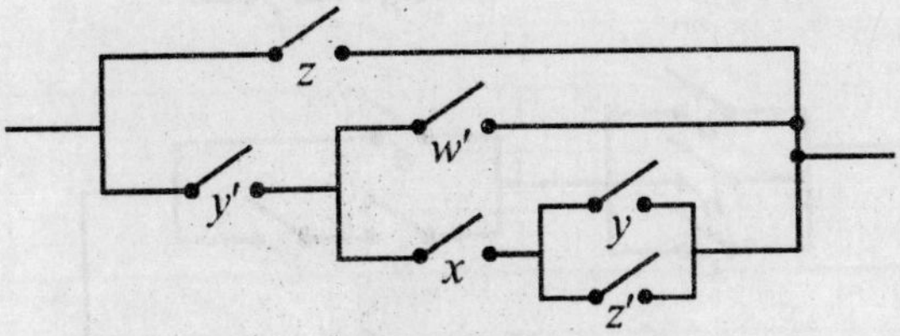

3. Find a circuit which is closed when the following circuit is open and open when the given circuit is closed.

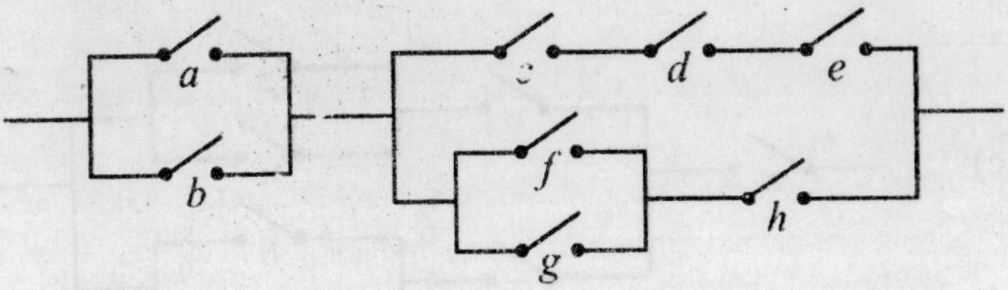

4. Find circuits which realize each of the functions given in the following table:

x	y	z	f_1	f_2	f_3
1	1	1	0	1	1
1	1	0	1	0	1
1	0	1	1	0	0
1	0	0	0	1	1
0	1	1	0	1	1
0	1	0	0	1	0
0	0	1	0	0	0
0	0	0	1	1	0

Answers

1(a)

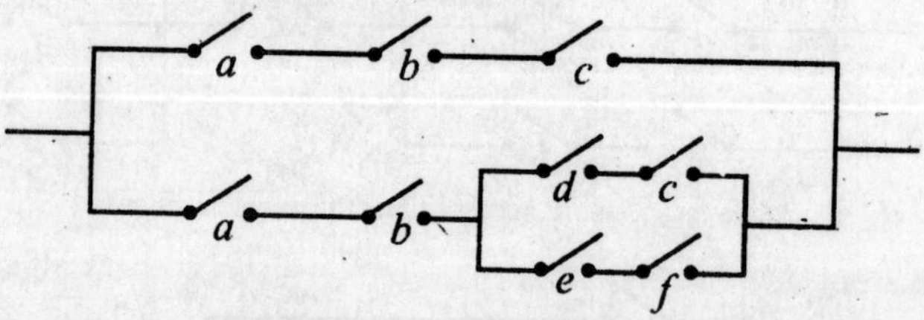

(b)

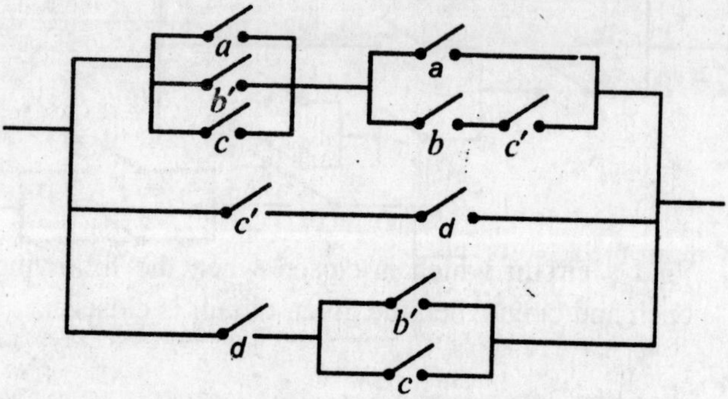

(c)

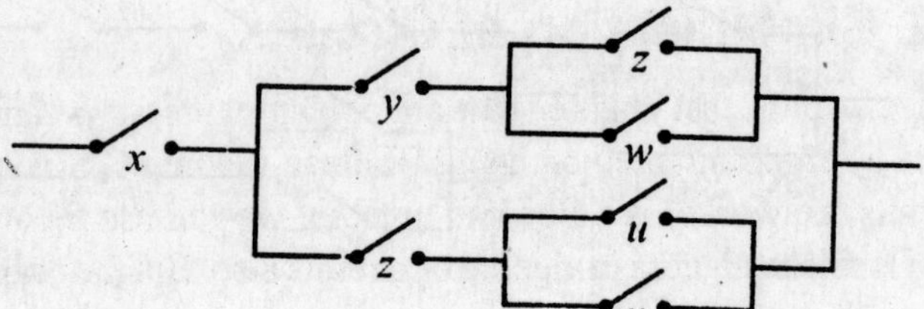

2 (a) $(ab + ac + a'b')\, c'\, (b + a' + c')$

(b) $z + y'\, [w' + x\, (y+z')]$

3

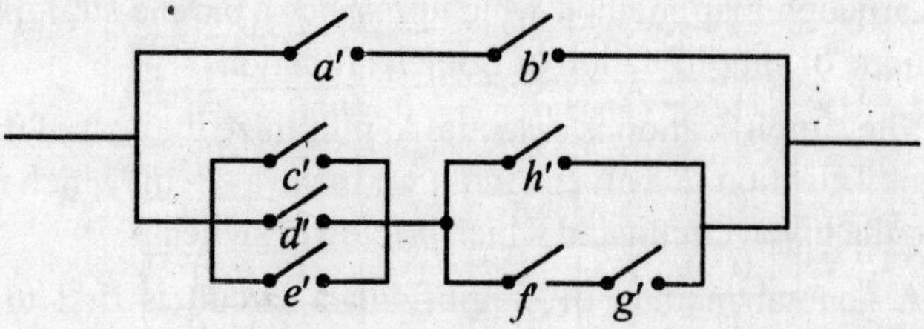

4 Circuit for f_1 is

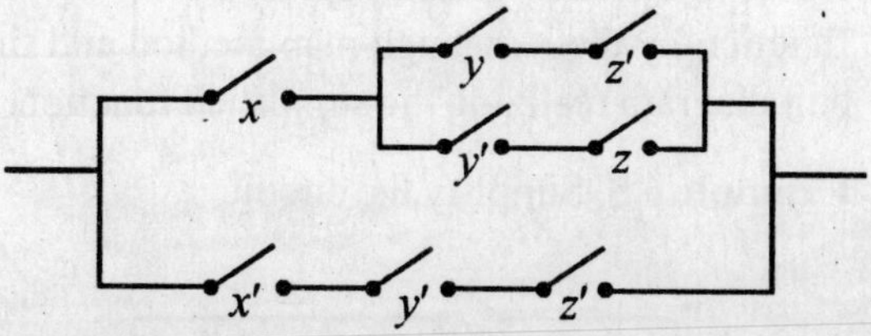

Circuit for f_2 is

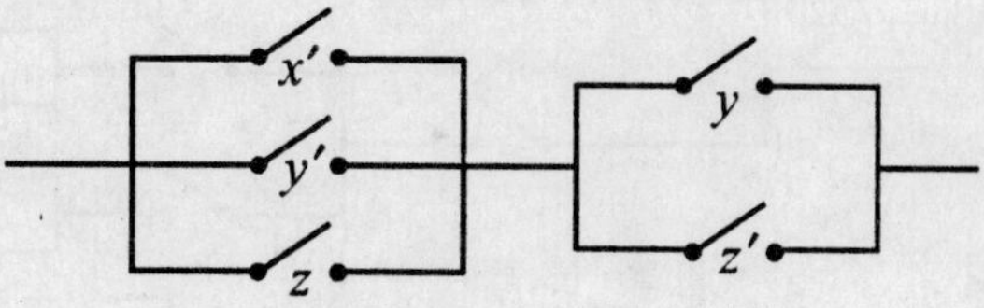

Circuit for f_3 is

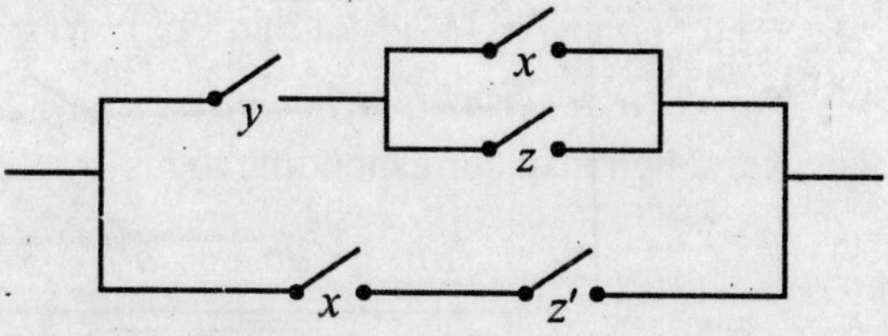

§ 6.2. Simplification of Circuits

We have seen in last chapter that a Boolean expression can be simplified by algebraic method using Boolean theorems. Since algebra of circuits (networks) is a Boolean algebra, we can use theorems and rules of Boolean algebra in algebra of circuits also. But the drawback of this technique is that there are no specific rules for proceeding step by step to manipulate the process of simplification. Even if a minimum is obtained, one may not be sure that it is minimum. Another method is the Karnaugh map method which provides a simple straight forward technique of simplification of Boolean functions.

The simplification is essential to minimize the cost of the circuit. By simplification of a given circuit we mean a circuit which is equivalent to the given circuit and which has fewer switches.

A general method of simplifying a circuit is first to find the Boolean function which represents the circuit, then to simplify the function by the Karnaugh map method and finally to draw a new circuit diagram realizing the simplified function.

Example 6.5. Simplify the circuit

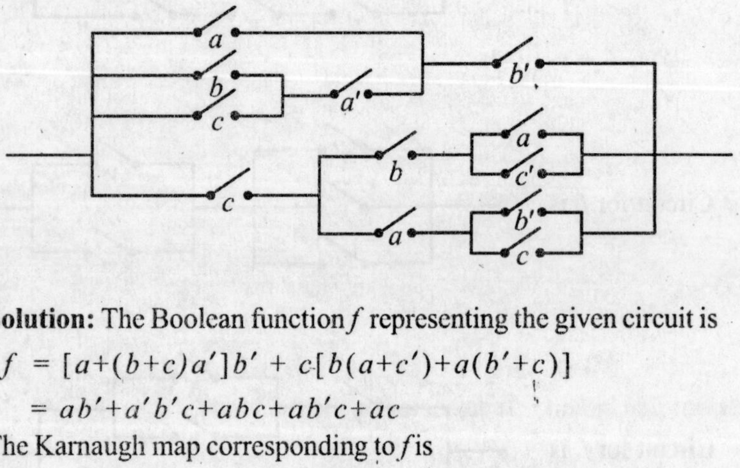

Solution: The Boolean function f representing the given circuit is

$$f = [a+(b+c)a']b' + c[b(a+c')+a(b'+c)]$$
$$= ab'+a'b'c+abc+ab'c+ac$$

The Karnaugh map corresponding to f is

	ab	ab'	$a'b'$	$a'b$
c	1	1	1	
c'		1		

The simplified function is

$$f = ab' + ac + b'c$$

This expression is the minimal sum of products. The circuit corresponding to this form will contain 6 switches. But if it is expressed as

$$f = ab' + (a+b')c$$

then the corresponding circuit will contain five switches. Thus we see that we have to see both forms of the simplified expression before constructing the circuit. Hence we get the following circuit.

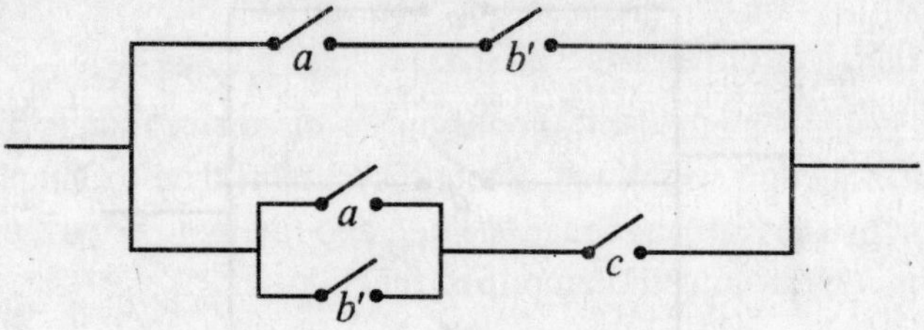

Example 6.6. Simplifying the circuit given below:

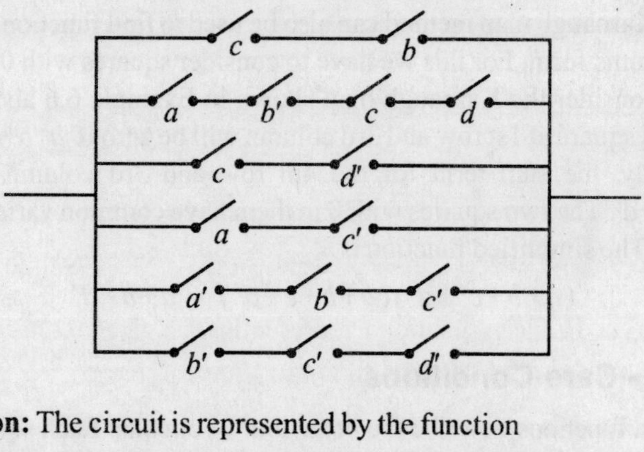

Solution: The circuit is represented by the function

$$f = bc + ab'cd + cd' + ac' + a'bc' + b'c'd'$$

	ab	ab'	$a'b'$	$a'b$
cd	1	1	0	1
cd'	1	1	1	1
$c'd'$	1	1	1	1
$c'd$	1	1	0	1

The function contains four variable. The Karnaugh map of f is given on the last page (to find simplified form of f) .

There are three basic rectangles each of the type 2×4. The simplified function is $f = a+d'+b$

Hence the simplified circuit is

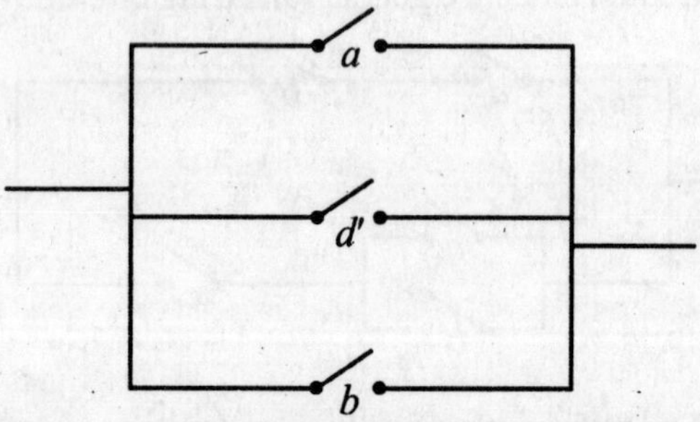

Note: Karnaugh map method can also be used to find function in product of sums form. For this we have to consider squares with 0 in them only. Consider the Karnaugh map shown in Example 6.6 above. The entry in square at 1st row and 3rd column will be zero if $a+b+c'+d'=0$. Similarly, the sum term for the 4th row and 3rd column will be $a+b+c+d'$. The two squares with 0 in them have common variables a,b and d'. The simplified function is

$$(a+b+c'+d')(a+b+c+d') = a+b+d'$$

Don't - Care Conditions

Boolean functions describe the behaviour of circuits. Each square of a Karnaugh map represents the output of the circuit corresponding to a combination of values of the input variables. Some times it happens that certain input combinations never occur. In such situations the output of the circuit network, is not specified. These situations are referred to as don't-care conditions. The square on the Karnaugh map corresponding to a don't care conditions is indicated by – or × and such square is known as don't care square. A don't care square may be assumed either as a square with 1 or a square with 0 as desired while forming the basic rectangles for simplification. Any one of such squares or some of them may be included or may not be included while forming basic rectangles.

Example 6.7. Suppose a circuit is defined by the following table where × denotes don't-care condition. Draw the circuit as simplified as possible.

x	y	z	$f(x,y,z)$
0	0	0	1
1	0	0	1
0	1	0	0
0	0	1	×
1	1	0	0
1	0	1	1
0	1	1	0
1	1	1	0

Solution: we can directly represent the function f given by the table in Karnaugh map as follows:

	xy	xy'	$x'y'$	$x'y$
z	0	1	×	0
z'	0	1	1	0

It is clear that if we take 0 in the don't care square then we have two basic rectangles of the type 1× 2 and the corresponding Boolean function will be

$$f = xy' + z'z'$$

But if we take 1 in the don't care square then we have one basic rectangle of the type 2×2 and the corresponding Boolean function will be

$$f = y'$$

Hence simplified circuit is

Thus the addition of don't care condition has made our circuit simpler.

Exercise

1. Simplify the following circuits:

(a)

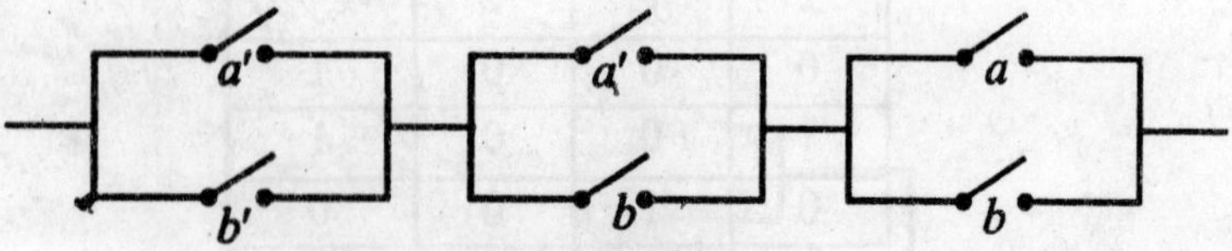

(b)

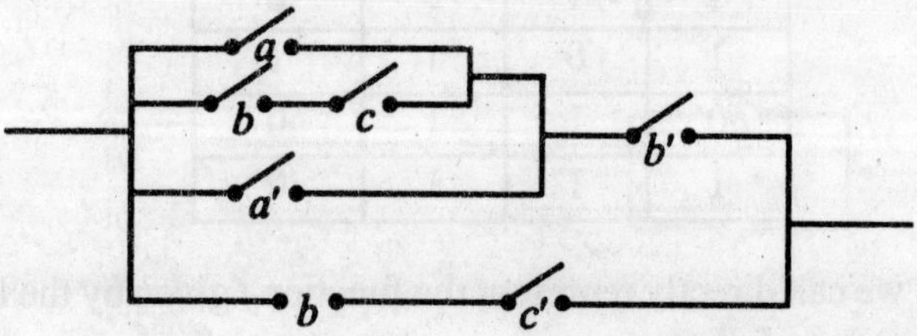

(c)

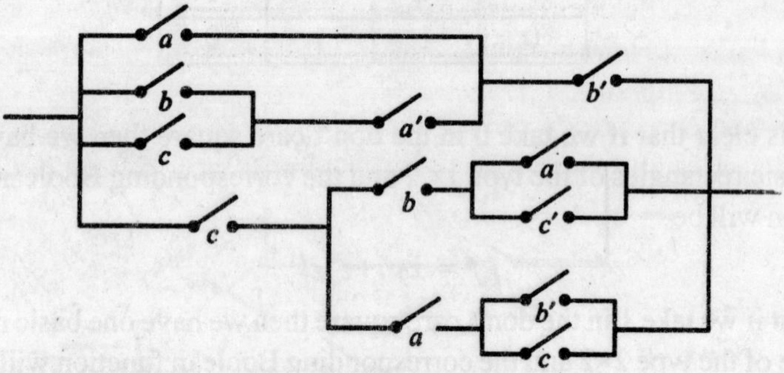

2. Suppose that we have a circuit defined by the function

$$f = a(b'+c)+ a'b'c'$$

Suppose further that abc′ and $a'b'c$ are impossible to occur. Draw the circuit with least switches.

Answers

1(a) 

(b) 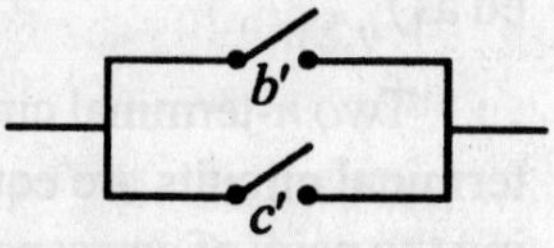

(c) 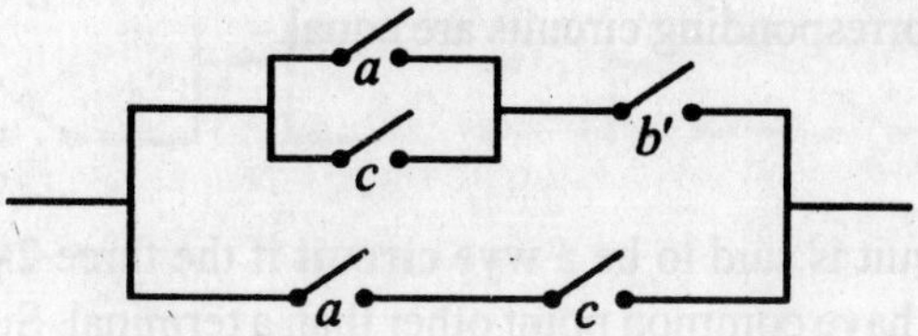

2 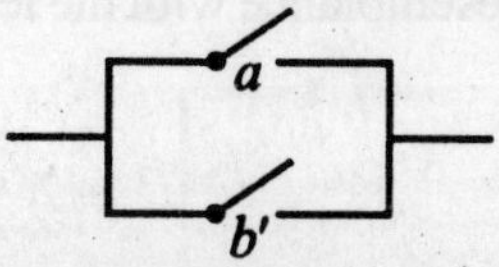

§ 6.3. Non-Series Parallel Circuits

In the previous sections, we have discussed circuits which have two terminals. Such circuits are called 2-terminal circuits. We now study the concept of *n*-terminal circuits. The following diagram shows a 3-terminal circuit.

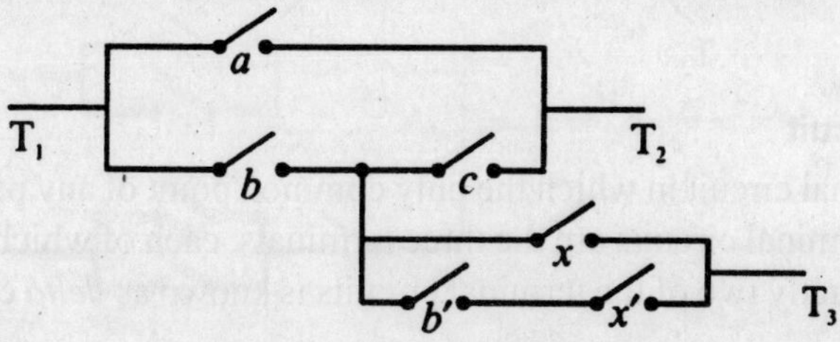

This circuit is a combination of three 2-terminal circuits joining terminal points T_1 and T_2, T_2 and T_3, and T_1 and T_3. The Boolean function corresponding to the 2-terminal circuit joining terminals T_i and T_j will be denoted by f_{ij} for each i and j, where $i \neq j$.

In general, an *n*-terminal circuit can be defined as a configuration of switches connected by wire in which *n* points are designated as terminals. The $n(n-1)/2$ possible Boolean functions corresponding to the 2-terminal circuit joining T_i to T_j for each i and j, $i \neq j$, will be denoted as f_{ij}.

Two *n*-terminal circuits are said to be *equivalent* if each pair of 2-terminal circuits are equivalent or, as well, if the functions representing the pairs of corresponding circuits are equal.

Wye Circuit

A 3-terminal circuit is said to be a **wye circuit** if the three 2-terminal circuits involved have common point other than a terminal. Such a circuit is called wye circuit because of its resemblance with the letter Y.

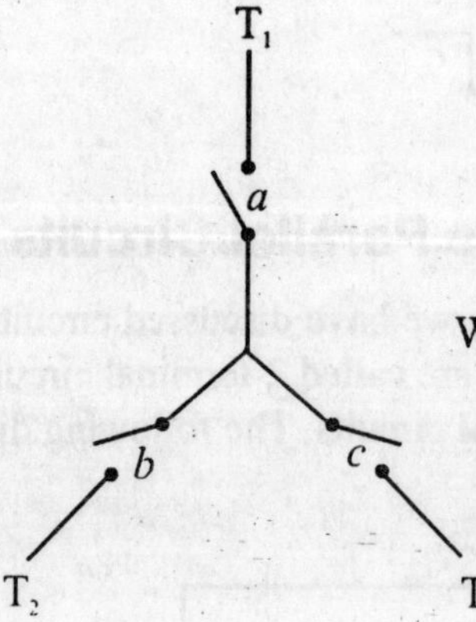

Wye circuit

Delta Circuit

A 3-terminal circuit in which the only common point of any pair of the three 2-terminal circuits are the three terminals, each of which is common to exactly two of the terminal circuits is known as *delta* circuit.

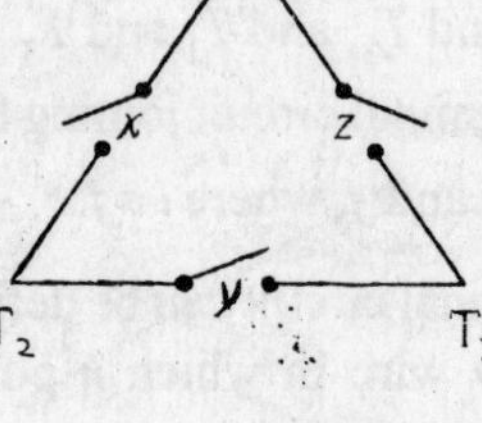

Delta circuit

Bridge circuit

A non-series-parallel circuit of the type shown below is known as ***bridge circuit.***

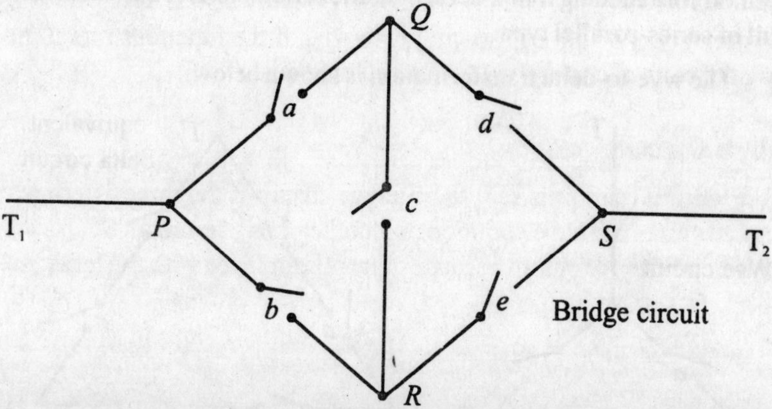

Bridge circuit

The vertices of the circuit are labeled with capital letters for reference. The point Q is the central point of a wye circuit with terminals P, R, and S.

Star Circuit

An n-terminal circuit is said to be a *star* if it has a common central point other than the terminal points. Clearly star circuit is a generalization of wye circuit.

Mesh Circuit

An n-terminal circuit is called a mesh if in this circuit only common points of any pair of $\frac{1}{2}n(n\text{-}1)$, 2-terminal circuits are the terminals only, each of which is common to exactly (n-1) 2- terminal circuits. If n=3, then mesh circuit is simply a delta circuit.

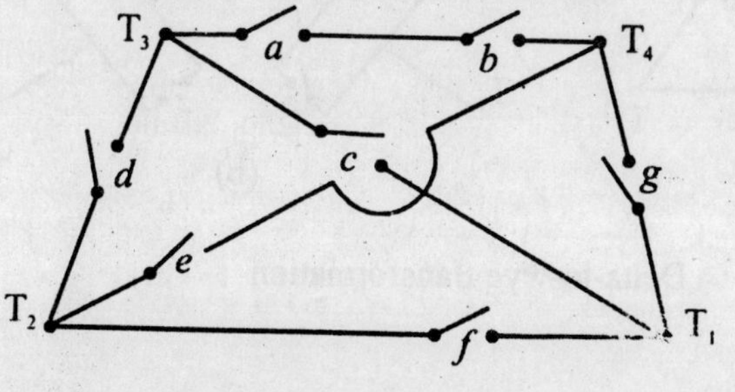

4-terminal mesh circuit

Wye-to-delta and delta-to-wye Transformations

As in ordinary circuit theory, there exist wye-to-data and delta-to-wye transformations. We introduce these transformations to develop a method for reducing a non-series-parallel circuit to an equivalent circuit of series-parallel type.

The wye-to-delta transformation is shown below:

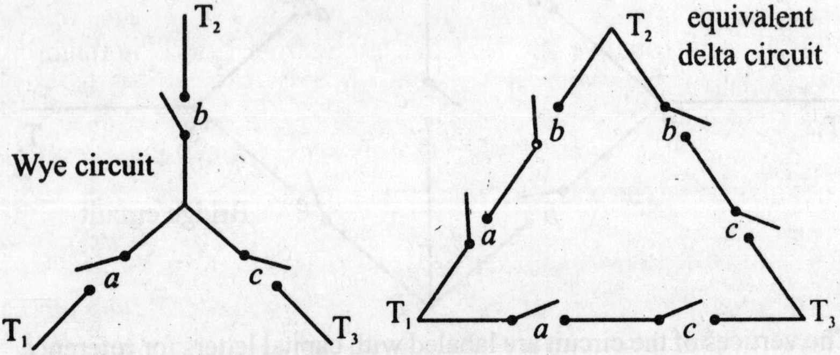

Wye-to-delta transformation

Clearly, wye-to-delta transormation gives an equivalent 3- terminal circuit since 2- terminal circuits formed in each case are series connections of the same pair of switches.

A delta-to-wye transformation is given below:

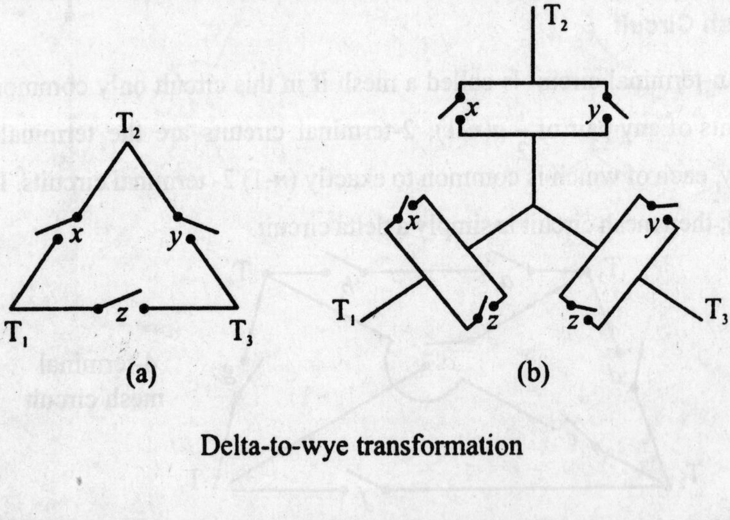

Delta-to-wye transformation

Now to see that the circuits in figure (b) above are equivalent to the circuits in (a), we note that the 2-terminal circuit from T_1 to T_2 in (b) is nothing but the circuit given by the Boolean functions.

$$(x+z)(x+y)$$

which is equivalent to the function

$$x + yz$$

in view of the distributive law. Note that circuit from T_1 to T_2 in figure (a) is given by

$$x+yz$$

Similarly, we can check the other two circuits.

Star-To-Mesh Transformations

Star and mesh circuits are generalizations of wye and delta circuits respectively. Therefore wye-to-delta and delta-to-wye transformations may be generalized to a star-to-mesh and mesh-to-star transformations respectively. Figure below suggest the method employed.

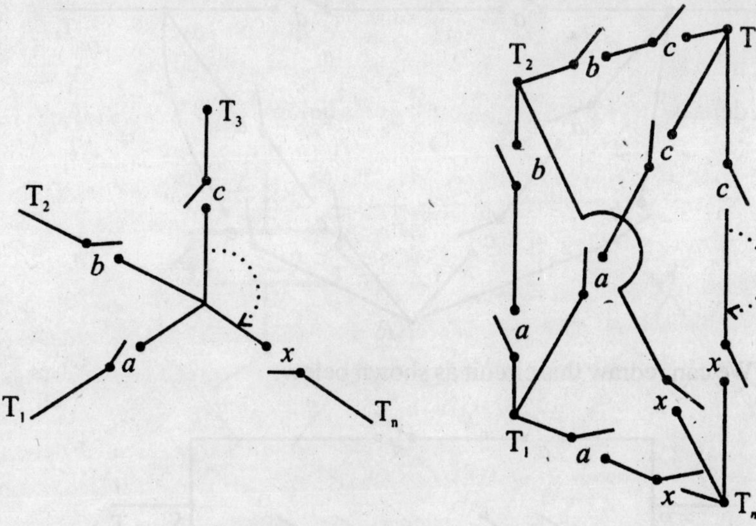

Star-to-mesh transformation

To find series-parallel circuit equivalent to a bridge circuit

We can use wye-to-delta transformation to obtain a Boolean function to represent the bridge circuit (which is not a series-parallel circuit) given below and thus obtain a series-parallel circuit which is equivalent

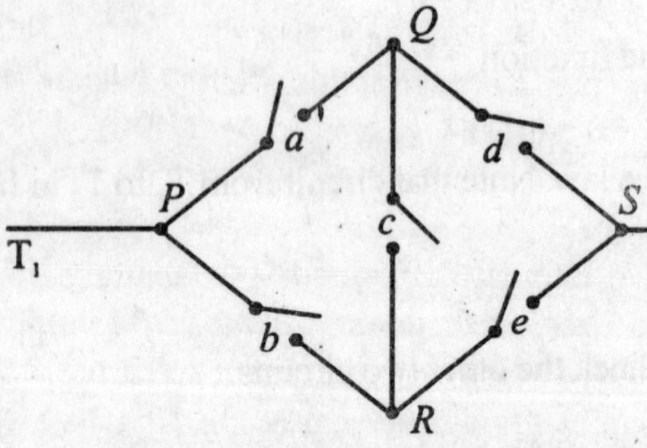

Bridge circuit

to the given bridge circuit. Consider the bridge circuit given above. The point Q can be taken as central point of a wye circuit with terminals P, R and S. Applying the wye-to-delta transformation we get an equivalent circuit shown below:

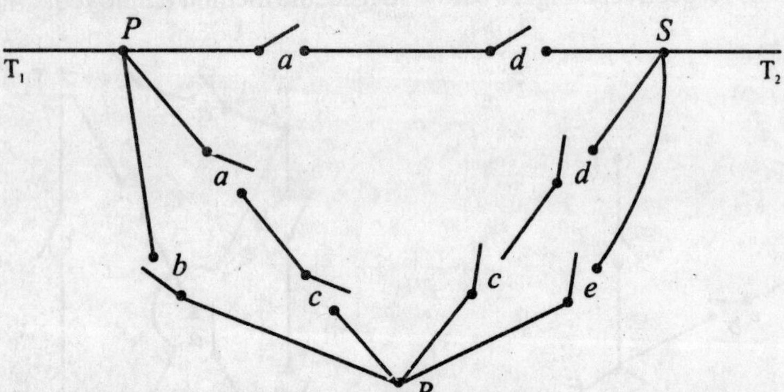

We can redraw this circuit as shown below:

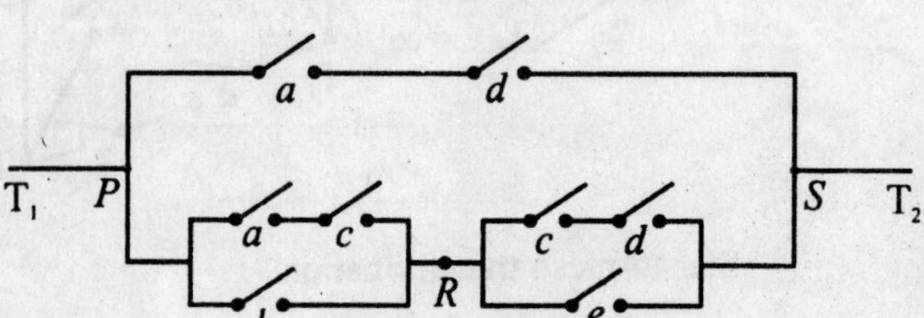

From this figure, it follows that the given bridge circuit is represented by the Boolean function

$$f = ad+(ac+b)\ (cd+e)$$

Conversely, any Boolean function of the form

$$f = ad+(ac+b)\ (cd+e)$$

can be realized by a bridge circuit. For this, we first locate the delta circuit and then transform it into a wye circuit.

Alternative Methods

We now explain two alternative methods of obtaining the Boolean function for a circuit. These methods are easier to use in simple cases but have the disadvantage of being essentially trial and error methods and hence error may occur in complicated cases.

The first of these methods consists of examining the circuit for all possible combinations of closed switches which allow a current through the circuit. To illustrate the method we consider again the bridge circuit.

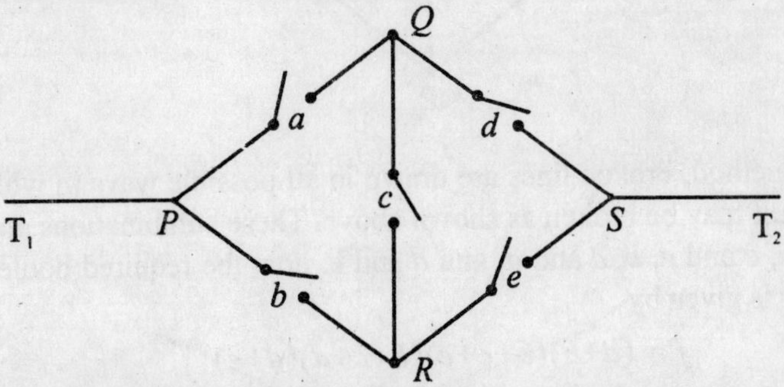

From the circuit it is clear that there are only 4 possible paths from T_1 to T_2, namely

$$T_1—P—Q—S—T_2,$$

$$T_1—P—Q—R—S—T_2,$$

$$T_1—P—R—Q—S—T_2,$$

and $$T_1—P—R—S—T_2,$$

through which current can flow from T_1 to T_2. These paths corresponds to the combinations $a\,d$, $a\,c\,e$, $b\,c\,d$ and $b\,e$. Hence the required Boolean function is

$$f = ad + be + ace + bcd$$

The second alternative method is to consider all ways in which the circuit may be broken (by a combination of open switches)

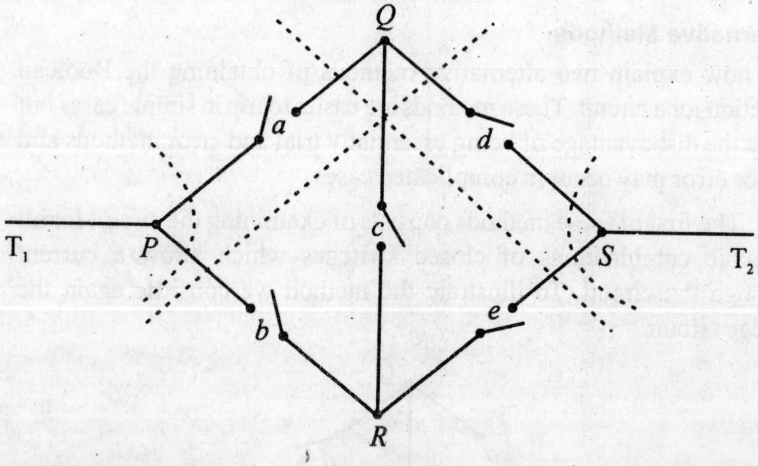

In this method, broken lines are drawn in all possible ways in which the circuit may be broken as shown above. These combinations are a and b, a, c and e, b, c and d, and d and e. now the required boolean function is given by

$$f = (a+b)(a+c+e)(b+c+d)(d+e)$$

which has the value 0 if any of the four sets of switches is open. It can be easily seen that the three Boolean functions

$$ad + (ac+b)(cd+e)$$

$$ad + be + ace + bcd$$

and $(a+b)(a+c+e)(b+c+d)(d+e)$ are equivalent.

Solved Examples

Example 6.8. Simplify the circuit given below:

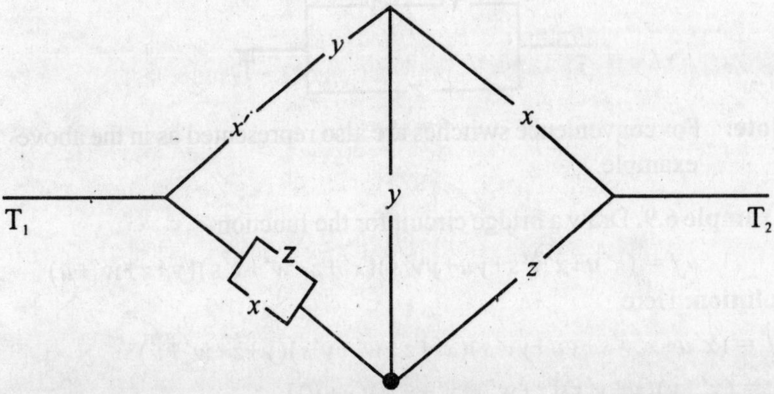

Solution: we draw broken lines through the circuit in all possible ways in which the circuit could be broken by a combination to open switches as shown below:

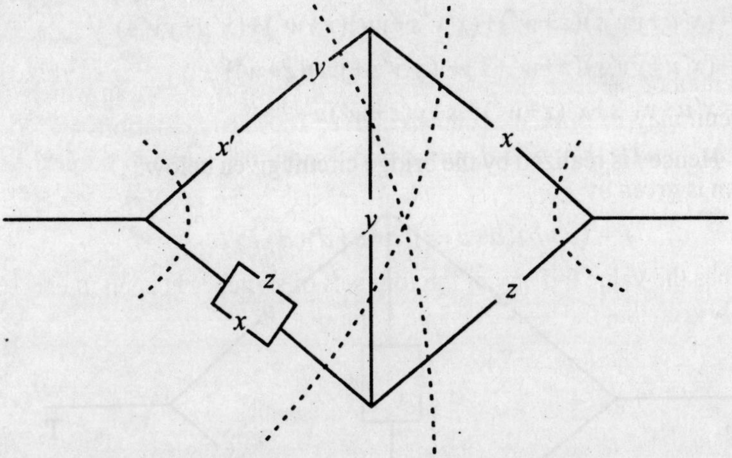

Now we can write the Boolean function *f* representing this circuit:

$$f = (x'y+x+z)(x+z)\ (x'y+y+z)\ (x+z+y+x)$$
$$= (x'y+z+xy)(x+z)$$

$$= (y+z)(x+z)$$
$$= xy+z$$

Therefore, the circuit equivalent to the given circuit is

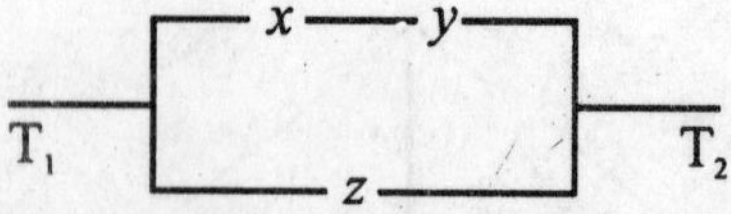

Note: For convenience switches are also represented as in the above example.

Example 6.9. Draw a bridge circuit for the function

$$f = (x'u+x'v's+yu+yv's)(x'+z+w'+v's)(y+z+w'+u)$$

Solution: Here,

$$\begin{aligned} f &= (x'u+x'v's+yu+yv's)(x'+z+w'+v's)(y+z+w'+u) \\ &= (x'+y)(u+v's)[z+w'+(x'+v's)(y+u)] \\ &= (x'+y)\ (u+v's)\ (z+w')+(x'+z)(x'+v's)(u+v's)(y+u) \\ &= (x'+y)\ (u+v's)\ (z+w')+(x'+yv's)(u+yv's) \\ &= (x'+y)\ (u+v's)\ (z+w')+(x'u+yv's) \\ &= (x'u+yv's)(z+w')+(x'v's+yu)(z+w')+(x'u+yv's) \\ &= (x'u+yv's)(z+w'+1)+\ (x'v's+yu)(z+w') \\ &= x'u+yv's+x'(z+w')v's+y(z+w')u \end{aligned}$$

Hence f is realized by the bridge circuit given below:

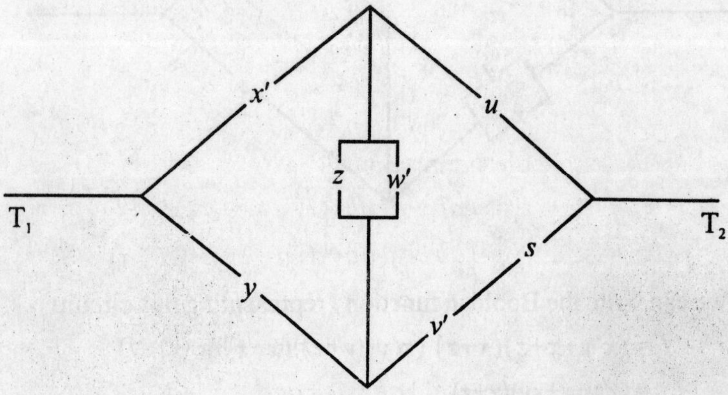

Example 6.10. Find the Boolean function which represents bridge circuit given below. Simplify, if possible.

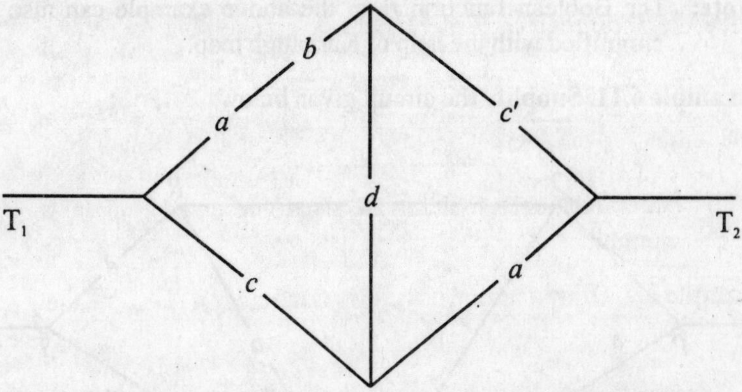

Solution: We draw broken lines as shown below through the circuit in all possible ways in which the circuit could be broken by a combinations of open switches.

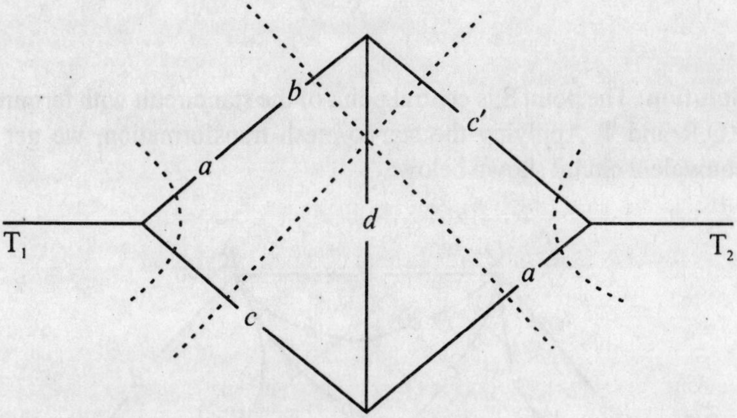

Hence the required Boolean function will be

$$\begin{aligned} f &= (ab+c)(ab+d+a)(c+d+c')(c'+a) \\ &= (ab+c)(ab+d+a)(1+d)(c'+a) \\ &= [ab+c(d+a)](c'+a) \\ &= abc'+cc'(d+a)+ab+ac(d+a) \\ &= abc'+ab+ac+acd \end{aligned}$$

$$= ab+ac$$

$$= a(b+c)$$

Note: The Boolean function f in the above example can also be simplified with the help of Karnaugh map.

Example 6.11. Simplify the circuit given below:

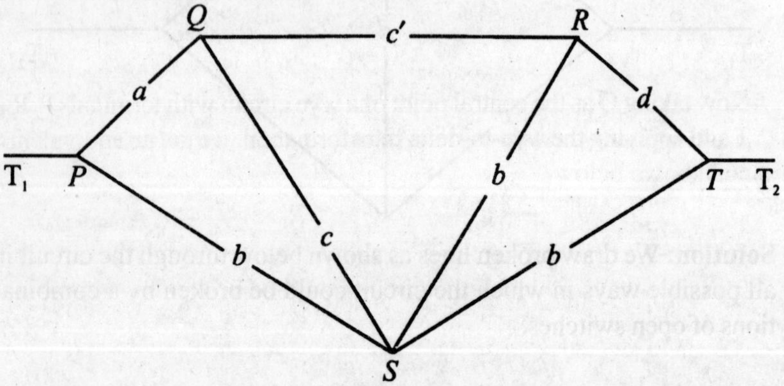

Solution: The point S is central point of the star circuit with terminals P,Q,R and T. Applying the star-to-mesh transformation, we get an equivalent circuit shown below:

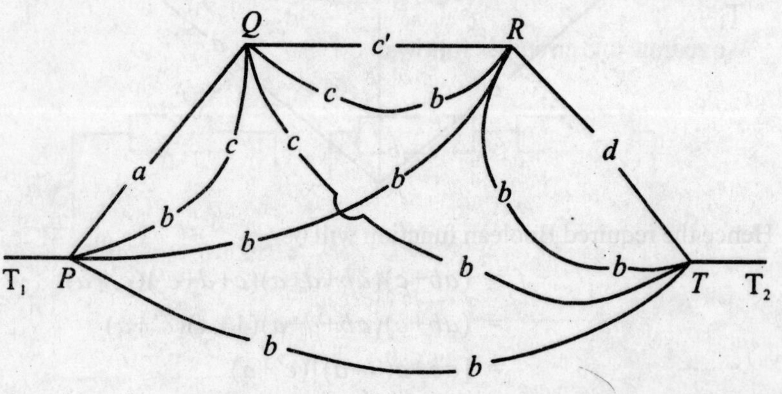

which is equivalent to

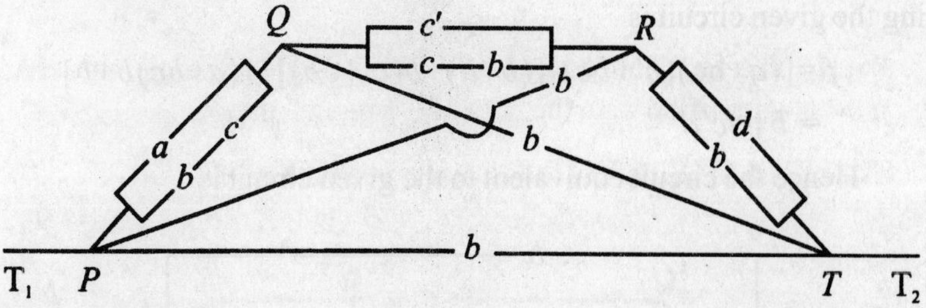

Now taking Q as the central point of a wye circuit with terminals P, R and T and applying the wye-to-delta transformation, we get an equivalent circuit shown below

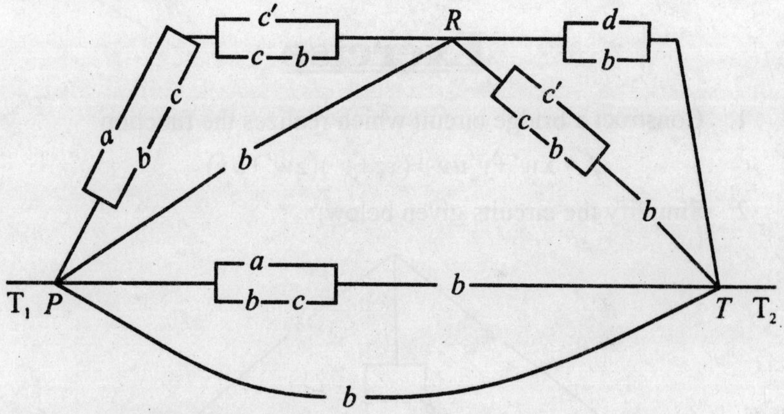

We redraw this circuit as follows:

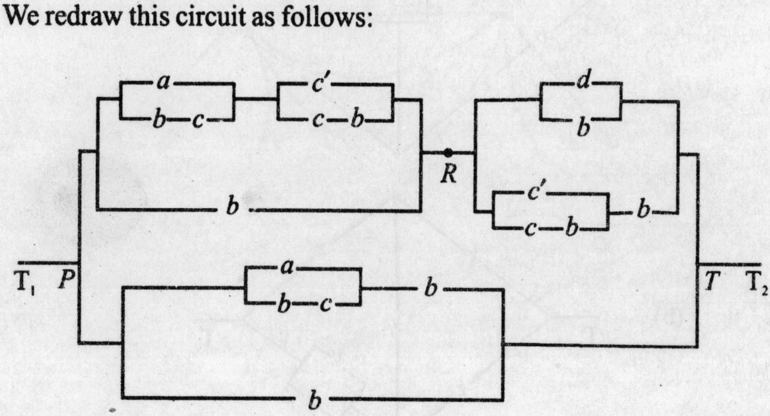

From this figure, it follows that the Boolean function representing the given circuit is

$$f=[(a+bc)(c'+bc)][(b+d)+b(c'+cb)]+[(a+bc)b+b]$$
$$= b+ac'd$$

Hence the circuit equivalent to the given circuit is

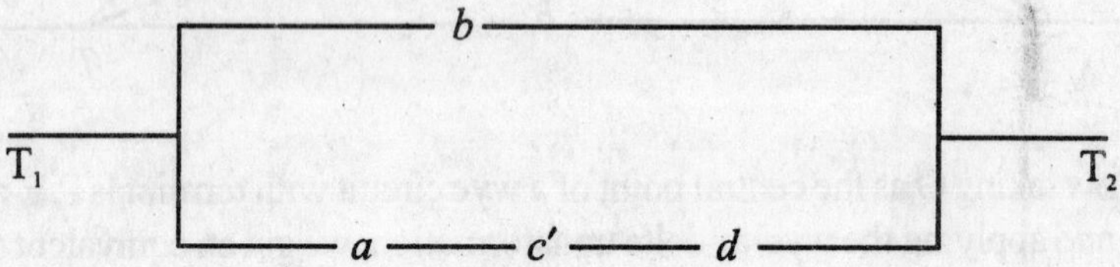

Exercise

1. Construct a bridge circuit which realizes the function
$$f= xw'+y'uv +(xz+y')(zw'+uv)$$
2. Simplify the circuits given below:

(a)

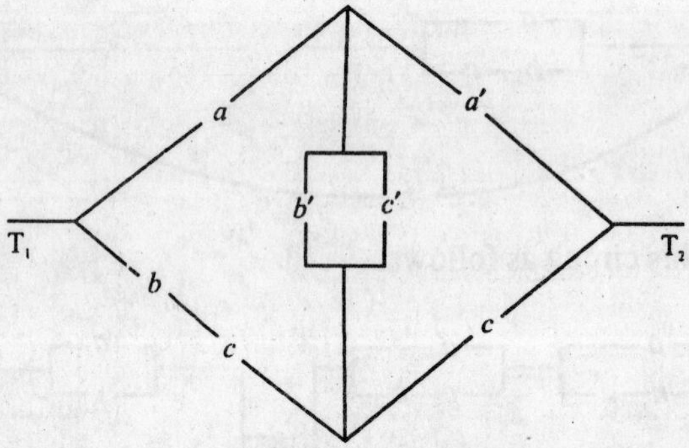

(b)

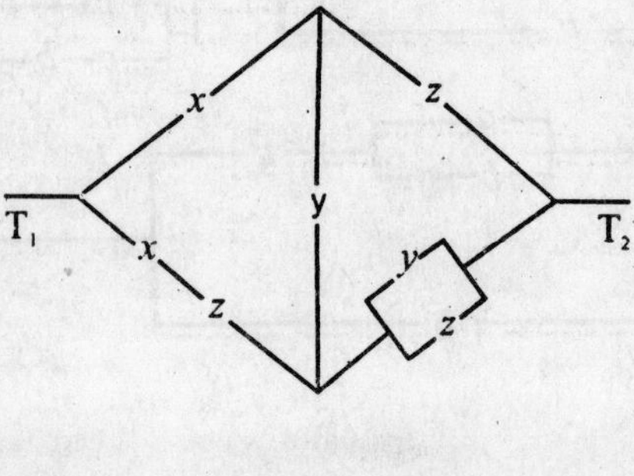

3. Find series-parallel circuit equivalent to the circuit given below:

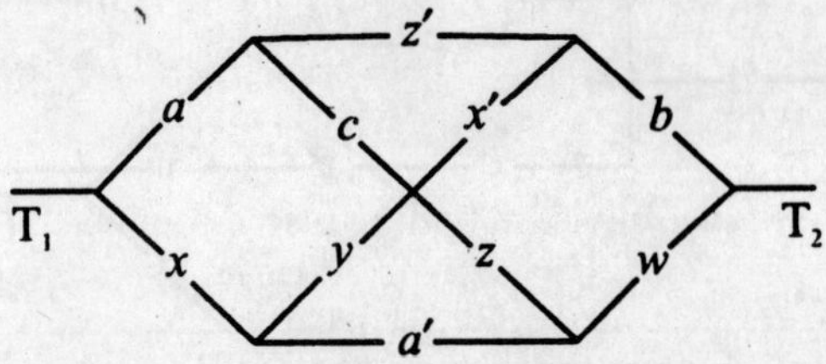

Answers

1.

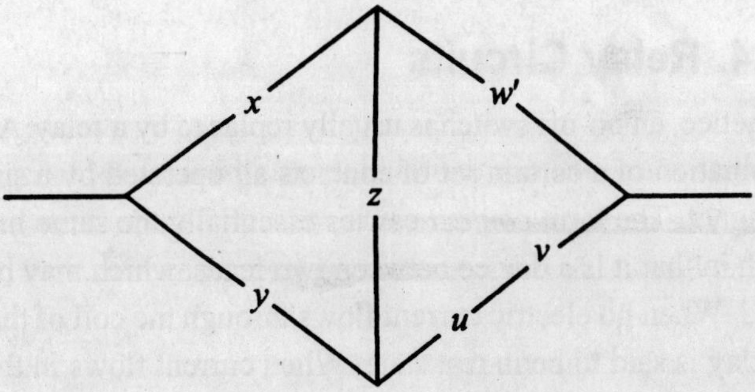

2. (a)

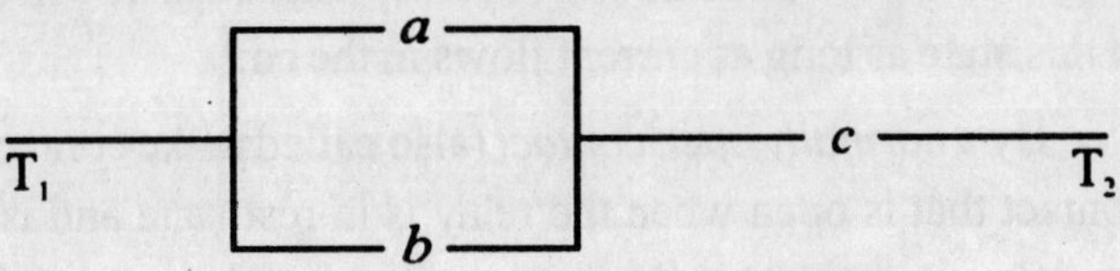

(b)

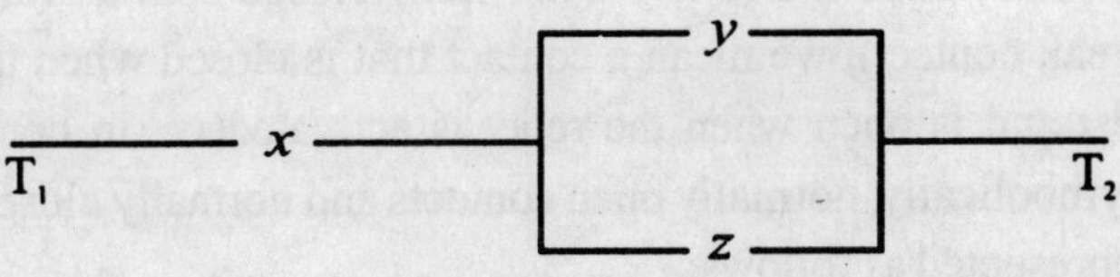

3.

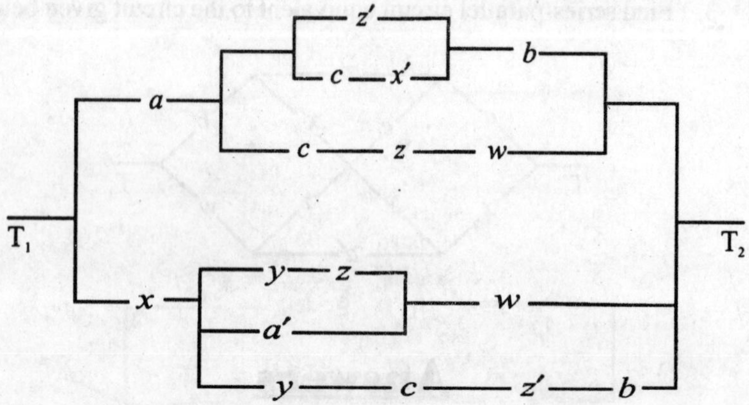

§ 6.4. Relay Circuits

In practice, an on-off switch is usually replaced by a relay. A *relay* is a combination of a certain set of contacts all operated by a single electromagnet. The term *contact* carries essentially the same meaning as switch in that it is a device between two leads, which may be open or closed. When no electric current flows through the coil of the magnet, the relay is said to be in rest state. When current flows in the coil the relay is activated and said to be in its operate state. It will continue to be in this state as long as current flows in the coil.

By a *normally open contact* (also called make contact), we mean a contact that is open when the relay is in rest state and is closed when the relay is activated. By a *normally closed contact* (also known as break contact), we mean a contact that is closed when the relay is at rest and is open when the relay is activated(*i.e.* in operative state). Symbolically, normally open contacts and normally closed contact are represented as follows:

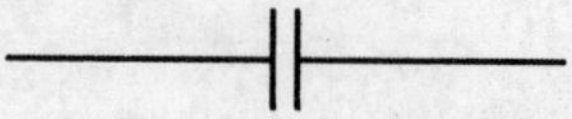

Normally open contact
(a)

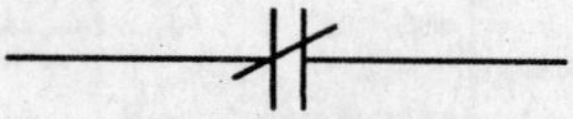

Normally closed contact
(b)

The following figure indicates a typical relay. The on-off switch

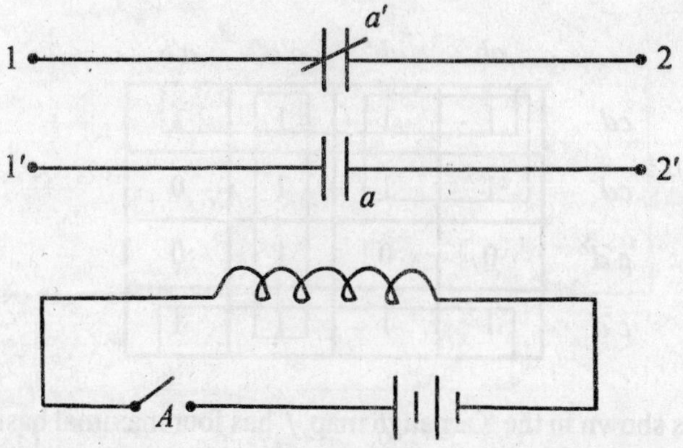

controlling the electromagnetic coil will be denoted by a capital letter, say A, and all normally open contacts by a and all normally closed contacts by a'. It may be noted that a normally open contact and a normally closed contact on the same relay will always be in opposite states, whether the relay is at rest or is activated. We can connect relay contacts in parallel and in series in exactly the same manner as on-off switches are interconnected and all the results of Boolean algebra can be applied to the analysis and synthesis of relay networks.

Example 6.12. Simplify the relay networks given below.

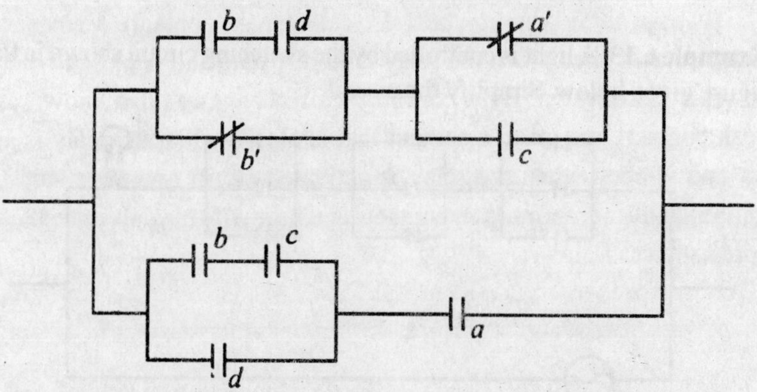

Solution: The boolean function corresponding to the given circuit is

$$f = (bd+b')(a'+c)+(bc+d)a$$

Representing f by a Karnaugh map, we have

	ab	ab'	$a'b'$	$a'b$
cd	1	1	1	1
cd'	1	1	1	0
$c'd'$	0	0	1	0
$c'd$	1	1	1	1

As shown in the Karnaugh map, f has four maximal basic rectangles. Thus the minimal form of f is given by

$$f = d + a'b' + ac$$

The simplified network is given below:

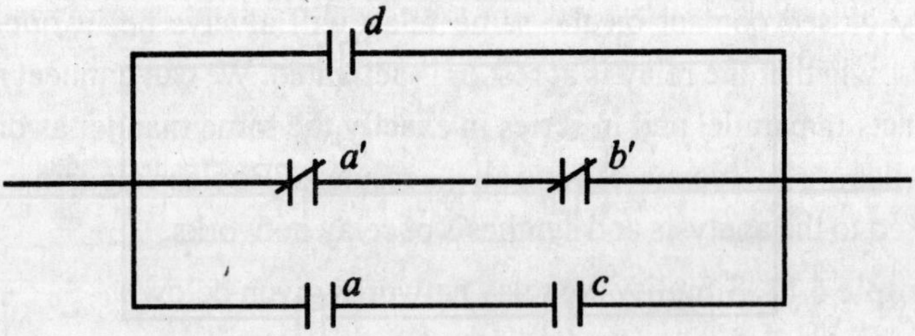

Example 6.13. A light is controlled by the switching circuit shown in the figure given below. Simplify the circuit.

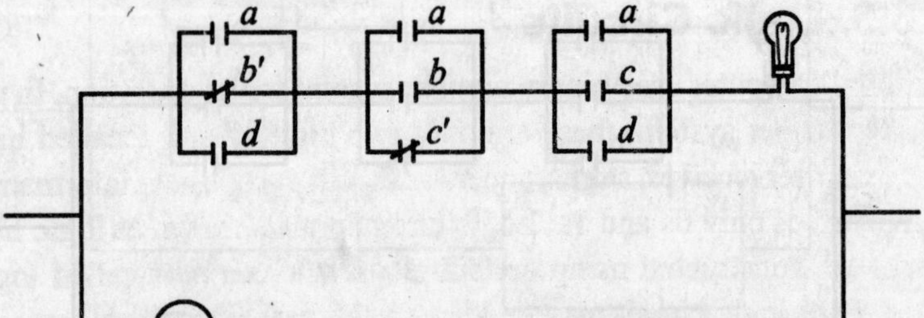

Solution: The above circuit can be described by the boolean function

$$f = (a+b'+d)(a+b+c')(a+c+d)$$

The simplified form of f is obtained from the Karnaugh map of f given below:

	ab	ab'	$a'b'$	$a'b$
cd	1	1	0	1
cd'	1	1	0	0
$c'd'$	1	1	0	0
$c'd$	1	1	1	1

As shown in the Karnaugh map, there are three maximal basic rectangles, namely, 2×4, 1×4 and 2×2 rectangles.

Hence $$f = a + c'd + bd$$

$$= a + (b + c')d$$

Thus the equivalent circuit which will perform the same control function is given below:

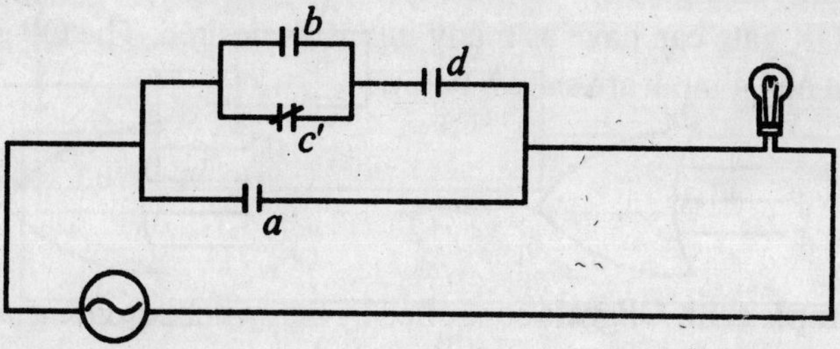

§ 6.5. Logic Circuits

A digital computer used binary number system for its operation. In the binary number system, there are only two digits 0 and 1 called bits. The computer receives, stores, understands and manipulates information composed of only 0s and 1s. Logic circuits (also known as logic network) are constructed using certain elementary circuits called logic gates. Each logic circuit may be viewed as a machine which contains one or more input devices and exactly one output device. There are three basic logic gates namely OR gate, AND gate, and NOT gate. By connecting these gates in different ways, we can build circuit called logic circuits that perform arithmetic and other operations associated with the human brain. We first define logic gates.

1. **OR gate:** An OR gate has two or more inputs but it has only one output. Let x and y be two inputs. The output of OR gate is denoted by $x+y$, where $x+y$ is defined by the table (b) given below:

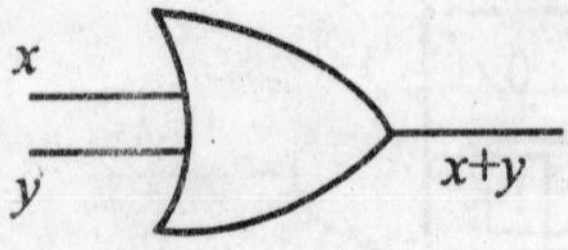

OR gate

(a)

x	y	$x+y$
1	1	1
1	0	1
0	1	1
0	0	0

(b)

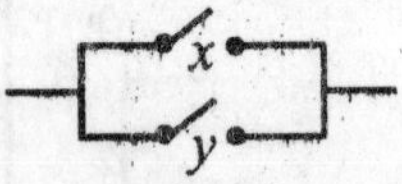

(c)

Thus output of OR gate is 1 if any one of the input is 1 otherwise it is zero. The logical operation of an OR gate can easily be explained with the help of two switches connected in parallel as shown in figure (c) above. The figure (a) shows standard symbol for two input OR gate. An OR gate can have as many inputs as desired. The OR gate with 3 input and 4 input are shown below:

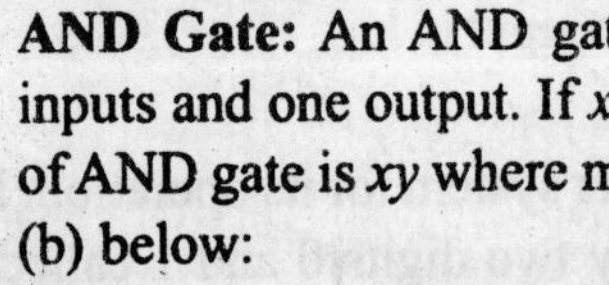

3-input OR gate

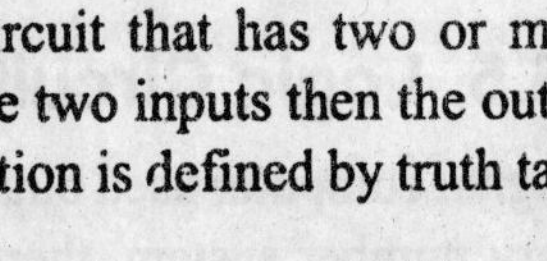

4-input OR gate

2. **AND Gate:** An AND gate is a circuit that has two or more inputs and one output. If x and y are two inputs then the output of AND gate is xy where multiplication is defined by truth table (b) below:

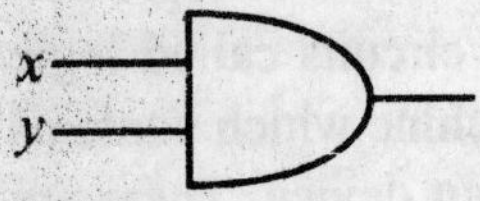

AND gate

(a)

x	y	xy
1	1	1
1	0	0
0	1	0
0	0	0

(b)

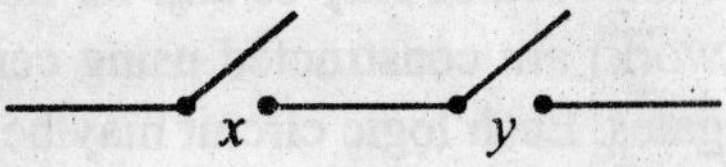

(c)

Thus output of AND gate is 1 if both the inputs are 1 otherwise it is zero. The logical operation of an AND gate can easily be explained with the help of two switches connected in series as shown in figure (c) above. The figure (a) shows standard symbol for two input and AND gate. An AND gate can have as many inputs as desired. The AND gate with 3-input and 4-input are shown below:

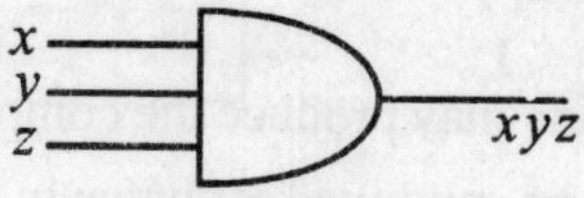

3-input AND gate

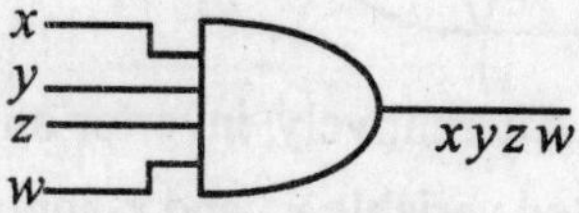

4-input AND gate

3. **NOT gate:** A NOT gate is a circuit that has one input and one output. If x is the input then output of NOT gate is denoted by x' where x' is defined by the truth table (b) below. A NOT Gate is also called *inverter*.

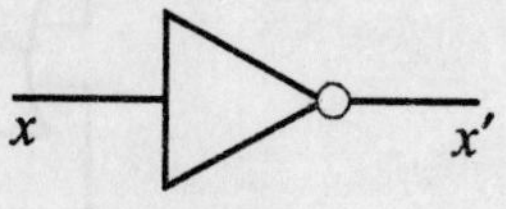

(a)

x	x'
1	0
0	1

(b)

Thus the output of NOT gate is complement of the input. The output is 1 if the input is 0, and output is 0 if the input is 1.

Logic Networks

It is quite obvious that the above gates can be interconnected to form an electronic circuit that realizes any given Boolean expression

Example 6.14. Construct a circuit using gates to realize the Boolean expression

$$f = (x_1+x_2)(x_1'+x_3)+(x_3+x_4)'$$

Solution: The logic network is given on page 312.

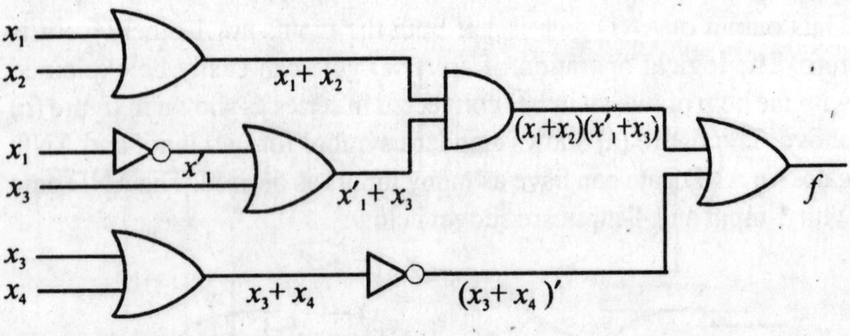

Alternatively, invertor on x_1 input line may produce the complemented variable x_1' and x_3 input line may be combined as shown in the figure below. Observe that this network has 4 input corresponding to four variable x_1, x_2, x_3, and x_4.

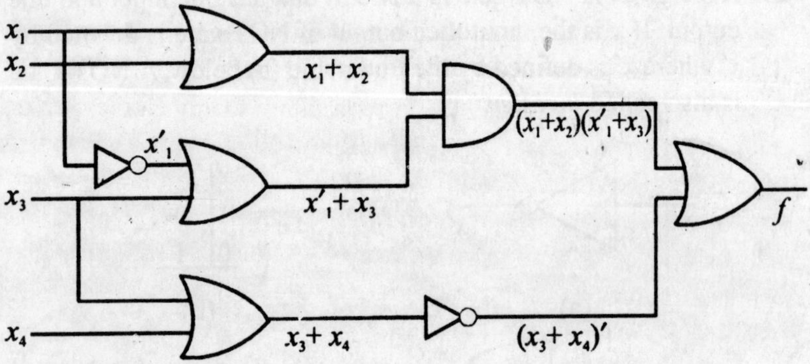

4. **NAND and NOR gates:** There are two additional gates frequently used in computer known as NAND gates and NOR gates. Figure (a) below represents a NAND gate and its associated truth table. A NAND gate is equivalent to an AND gate followed by a NOT gate.

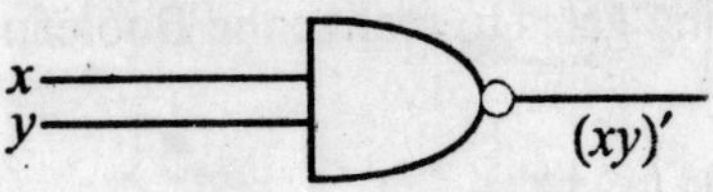

NAND gate

x	y	NAND
1	1	0
1	0	1
0	1	1
0	0	1

(a)

A NAND Gate has two or more inputs and one output. Further the output of a NAND gate is 0 if and only if all inputs are 1.

A NOR gate is equivalent to an OR gate followed by a NOT gate. Figure (b) below represents a NOR gate and its associated truth table.

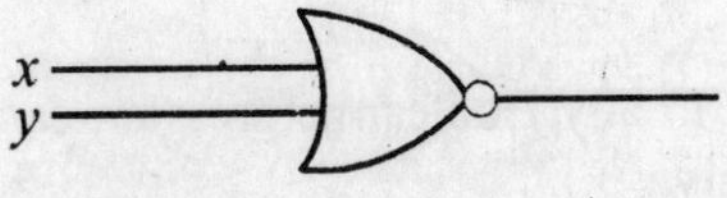

NOR gate

x	y	NOR
1	1	0
1	0	0
0	1	0
0	0	1

(b)

A NOR gate also has two or more inputs and one output. The output of a NOR gate is 1 if and only if all the inputs are 0.

Example 6.14. Show that logic circuit corresponding to any Boolean expression can be realized by using NAND gates alone.

Solution: Since logic circuits corresponding to any Boolean expression are constructed using AND, OR and NOT gates, therefore it is enough to show that AND, OR and NOT gates can be replaced by NAND gates only. The following figures show how OR gates, AND gates and NOT gates can be replaced by suitable interconnections of NAND gates.

(i) 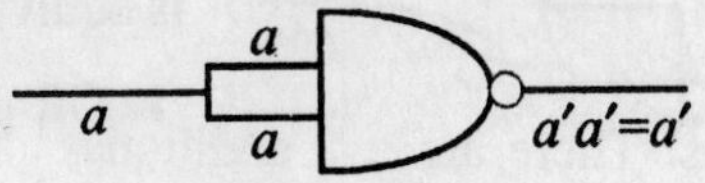

is equivalent to NOT gate

(ii) 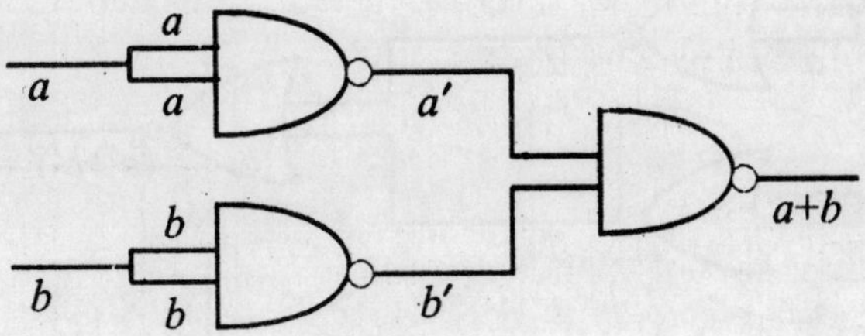

is equivalent to OR gate

(iii)
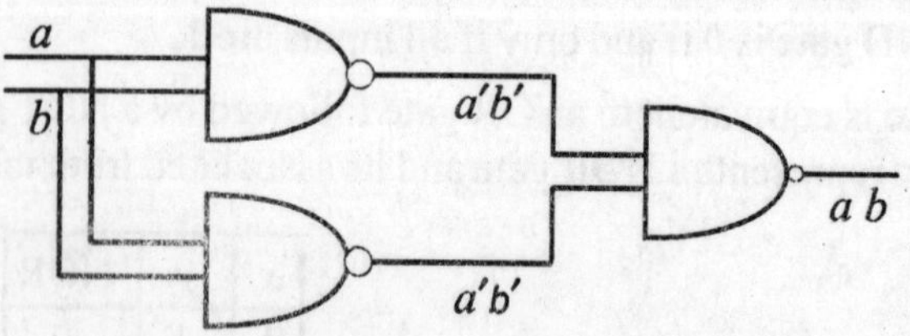

is equivalent to NAND gate

Thus logic circuit corresponding to any Boolean expression can be constructed using NAND gates only.

Example 6.15. Show that logic circuit to realize any Boolean expression can be constructed using NOR gates alone.

Solution: As in above example, we shall show that OR gate, AND gate and NOT gate can be replaced by suitable interconnections of NOR gates.

(i)
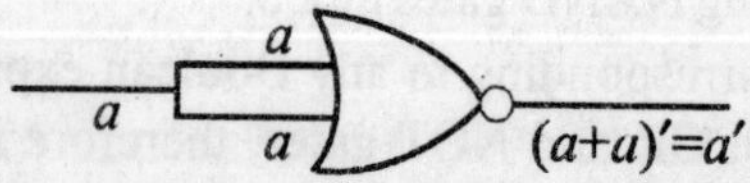

is equivalent to NOT gate

(ii)
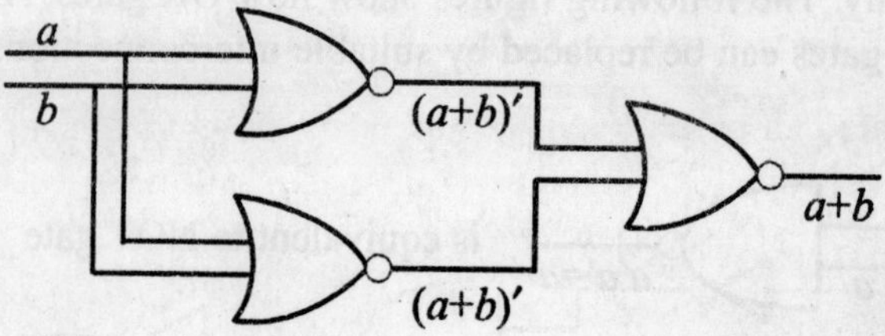

is equivalent to OR gate

(iii)
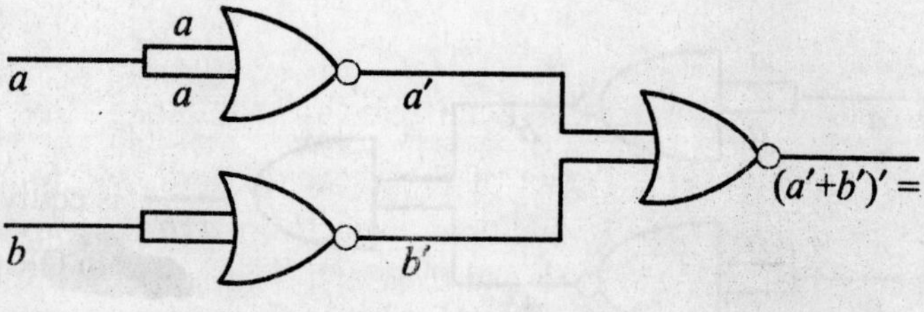

Thus logic circuit corresponding to any Boolean expression can be constructed using NOR gates only.

Some More Solved Examples

Example 6.16. Show that the set of gates {OR, NOT} is functionally complete

Or

Show that an AND gate can be replaced by a suitable interconnections of OR gate and NOT gates.

Solution: Since by DeMorgan's law,

$$xy = (x'+y')'$$

therefore an AND gate can be replaced by one OR gate and three NOT gates. The logic circuit is given below:

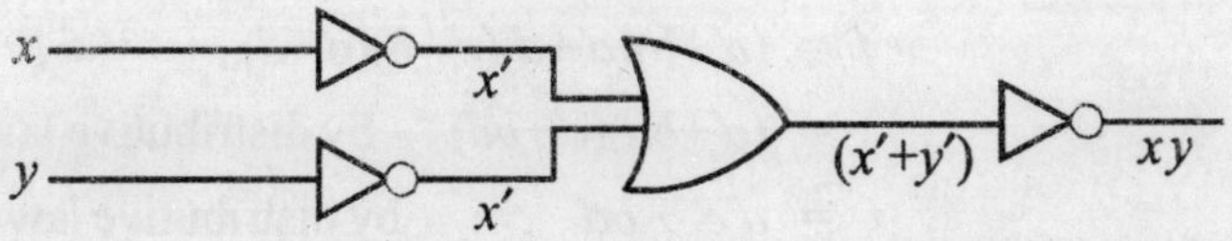

Example 6.17. Show that an OR gate can be replaced by a suitable interconnection of AND gates and NOT gates.

Solution: By DeMorgan's law, we have

$$x+y = (x'\ y')'$$

Thus an OR gate can be replaced by one AND gate and three NOT gates as shown below:

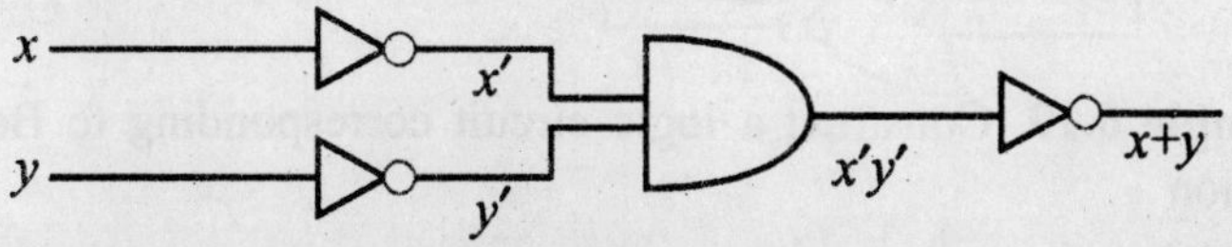

Example 6.18. Simplify the circuit given on page 316.

Solution: Clearly the output of first, second, third, and fourth OR gates are $a'+b$, $a'+d$, $c'+b$ and $c'+d$ respectively. Thus the inputs first (upper) AND gate are $a'+b$ and $a'+d$ and inputs of second (lower) AND gate are $c'+b$ and $c'+d$. Hence inputs of last AND gate are $(a'+b)(a'+d)$ and $(c'+b)(c'+d)$. Therefore, the boolean expression corresponding to given circuit is

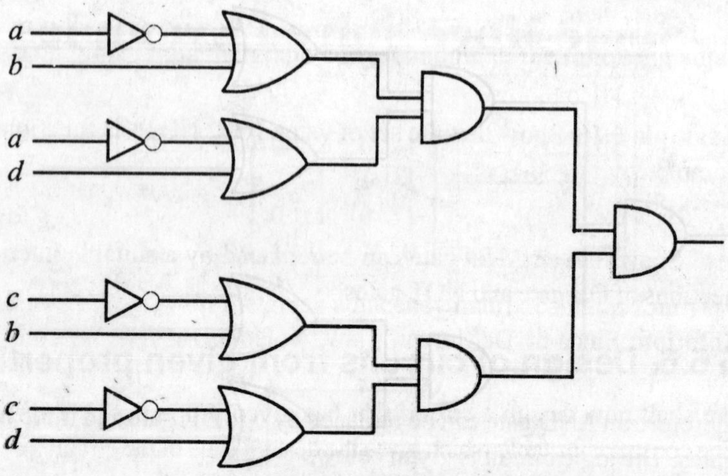

$$f = (a'+b)(a'+d)(c'+b)(c'+d)$$
$$= (a'+bd)(c'+bd) \quad \text{by distributive law}$$
$$= a'c' + bd \quad \text{by distributive law}$$

The simplified circuit corresponding to $f=a'c'+bd$ is given below:

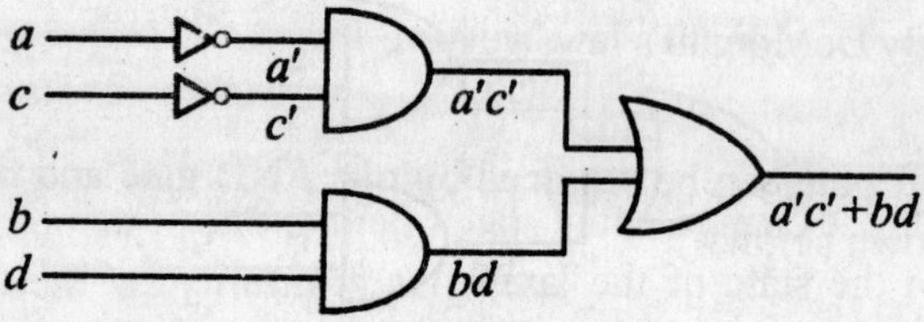

Example 6.19. Construct a logic circuit corresponding to Boolean function

$$f = xyz' + yz' + x'y$$

Solution: Clearly the logic circuit will contain 3 AND gate and one OR gate. The inputs of the first AND gate wil be x, y and z', inputs of second AND gate will by y and z' and inputs of last AND gate will be x' and y. The logic circuit is given on page 317.

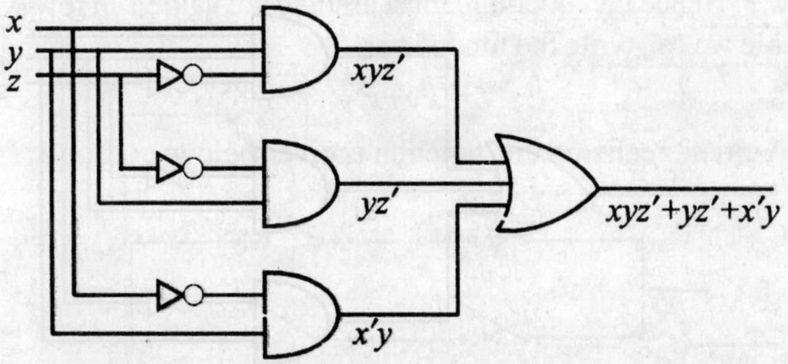

§ 6.6. Design of circuits from given properties

We shall now design a circuit that has given properties. For this, we first construct the table which gives the desired state of the circuit for each possible combinations of states for the separate switches. The Boolean function corresponding to the table is then written and if possible simplified. From the simplified expression a circuit is drawn. We explain the procedure with the help of examples.

Example 6.20. Design a circuit connecting two switches and a light bulb in such a way that either switch may be used to control the light independently of the state of the other.

Solution: Let x and y denote the two switches in the circuit. Each of x and y is to control the state of the lamp independently of the other. It means that the change in the state of either one of them must cause a change in the state of the lamp. We arbitrarily set the lamp 'on' when x and y are both closed. We then get the table of values for the closure function, say f of the lamp shown in table below:

Row	x	y	f
1	1	1	1
2	1	0	0
3	0	1	0
4	0	0	1

Note that either row 2 or row 3 represents a change of state of a single switch and thus is required to change the state of the light. Hence function must take the value 0 for rows 2 and 3. Finally, row 4 represents a change of state of a single switch from the state of either row 2 or row 3. Hence the function f must assume the value 1 in row 4. From the table we can write the function as

$$f = xy + x'y'$$

A circuit realizing this function is given below:

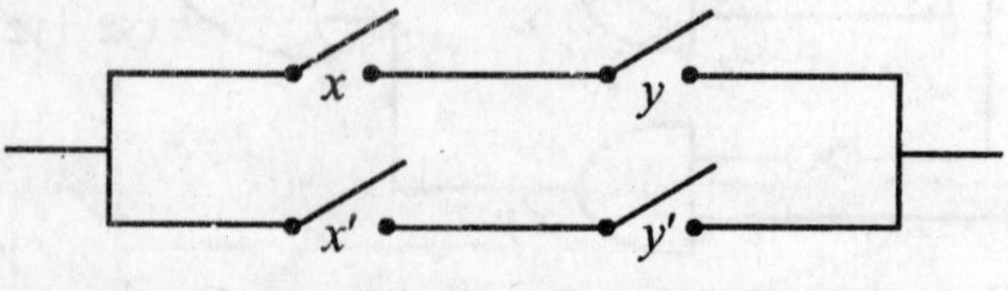

Alternatively, the diagram may also be given as below:

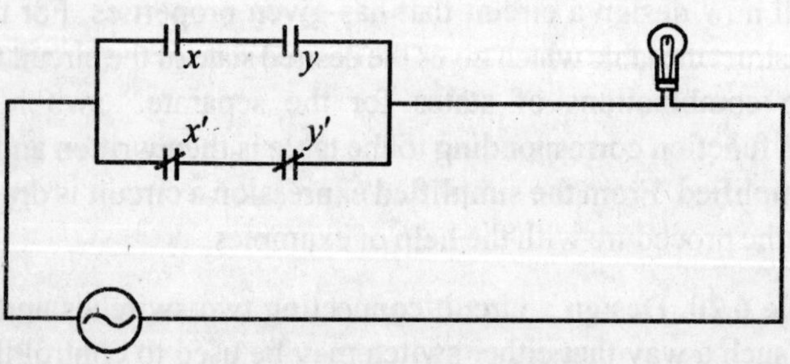

Example 6.21. Design as simply as possible a series-parallel circuit for the operation of a light independently from three switches x, y and z.

Solution: As discussed in Example 6.20 above, we arbitrarily set the lamp 'on' when x,y and z are closed. We then get the table of values for the closure function, say f of the lamp as given below:

row	x	y	z	f
1	1	1	1	1
2	1	1	0	0
3	1	0	1	0
4	1	0	0	1
5	0	1	1	0
6	0	1	0	1
7	0	0	1	1
8	0	0	0	0

From this table, we get the function f as follows

$$f = xyz+xy'z'+x'yz'+x'y'z$$
$$= x(yz+y'z')+x'(yz'+y'z)$$

Hence the circuit corresponding to f is

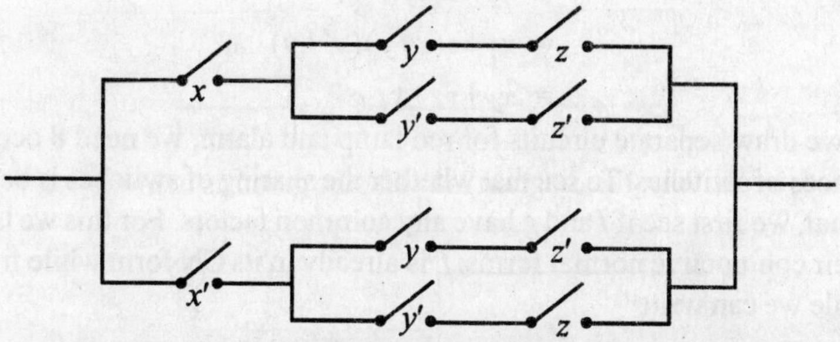

Example 6.22. An aircraft has three engines. Each engine is provided with a switch which closes as soon as there is any mechanical fault in that engine. Although the aircraft can run with just one engine operating, it is desired to have a red lamp appear when there is a fault in any one of the engines and an alarm to ring when there is a fault in any two of them. Design a two terminal circuit for this, sharing switches whenever possible.

Solution: we denote the three switches by x, y and z. Let f and g denote the closure functions for the red lamp (R) and the alarm(A) respectively. The table of value for the closure function f and g of lamp and alarm is given below:

row	x	y	z	f	g
1	1	1	1	1	1
2	1	1	0	1	1
3	1	0	1	1	1
4	1	0	0	1	0
5	0	1	1	1	1
6	0	1	0	1	0
7	0	0	1	1	0
8	0	0	0	0	0

From the table, we see that

$$f = x+y+z$$

and
$$\begin{aligned} g &= xyz+xyz'+xy'z+x'yz \\ &= xy+xy'z+yz \\ &= xy+z(xy'+y) \\ &= xy+z(x+y)(y'+y) \\ &= xy+yz+xz \end{aligned}$$

If we draw separate circuits for red lamp and alarm, we need **8** occurrences of switches. To see that whether the sharing of switches is beneficial, we first see if f and g have any common factors. For this we take their conjunctive normal forms. f is already in its CN form while from table we can write

$$g = (x+y+z)(x+y+z')(x+y'+z)(x'+y+z)$$

Here although $x+y+z$ is a common factor in f and g, it will not result in any saving to share it because the remaining three factors of g, even upon simplification would require at least five occurrences of switches. So we write f as f_1+f_2 where $f_1 = x$ and $f_2 = y+z$. Then

$$\begin{aligned} g &= (y+z+xx')[x+(y+z')(y'+z)] \\ &= (y+z)(x+yz+y'z') \\ &= (y+z)(x+yz) \end{aligned}$$

Now we may try to use the factor $y+z$ common in f_2 and g. The resulting circuit is as follows:

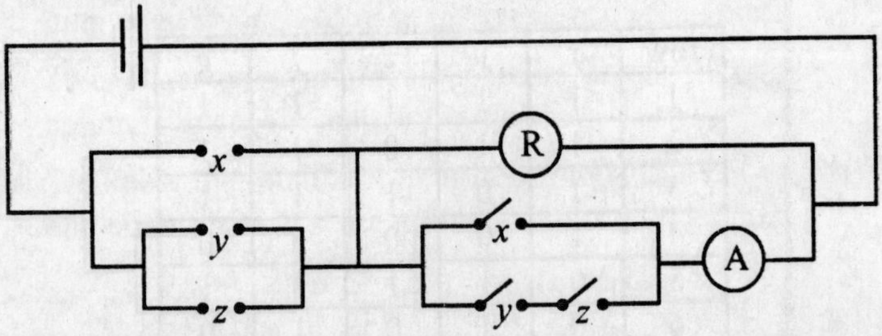

But in this circuit the alarm will also ring even when only x is closed which is not desired. To avoid this we insert switch x' as shown in figure below which is a correct solution.

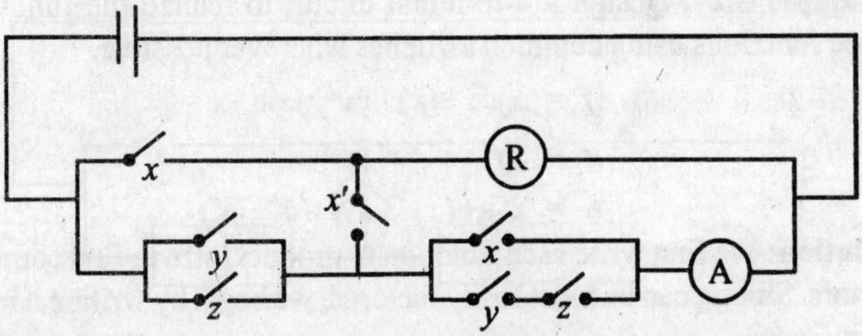

Note that there is a saving of switches because 7 occurrences are needed.

Example 6.23. Suppose there is a committee of four persons A,B,C and D. Any motion is taken as passed by the committee if and only if either A and any of B, C, D say yes or at least three out of four members say yes. Any member who says yes press a button. A light is to shine only when a motion is passed. Draw the circuit.

Solution: Let a, b, c, and d denote the four switches (buttons) given to members A, B, C, and D respectively. Now the light should shine when a and any of b, c, d are closed. That is when $a(b+c+d)$ is 1. Similarly light should shine when any three of a,b,c,d are 1. That is, when any of abc, bcd, acd, abd is 1. Thus the closure function for the lamp is given by

$$\begin{aligned} f &= a(b+c+d) + abc+bcd+acd+abd \\ &= ab+ac+ad+bcd \\ &= a(b+c+d)+bcd \end{aligned}$$

Hence the required circuit is given by the following figure.

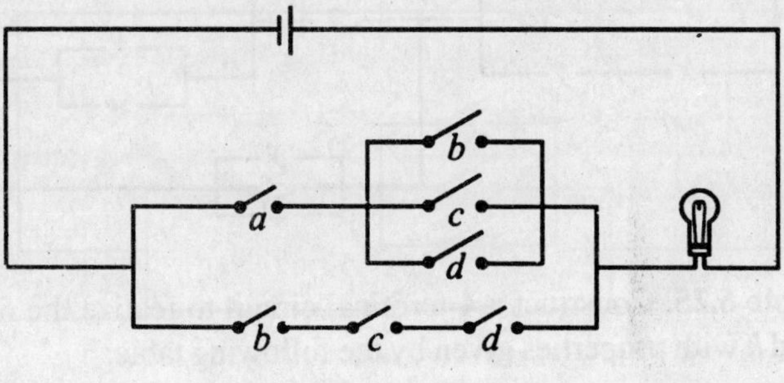

Example 6.24. Design a 4-terminal circuit to realize the following three functions using common switches wherever possible.

$$f = xy'z+(xy'+x'y)zw$$
$$g = xy'zw'+x'yzw'$$
$$h = x'y+(xy'+x'y)(z'+w')$$

Solution: We first write each function in product form to find common factors. Since g can be most easily factored, we begin by writing it in the form

$$g = (xy'+x'y)zw'$$

Examining f and h for possible factors in common with g, we find

$$\begin{aligned} f &= z[xy'+(xy'+x'y)w] \\ &= z[xy'y'+x'yy'+(xy'+x'y)w] \\ &= z[(xy'+x'z)y'+(xy'+x'y)w] \\ &= z(xy'+x'y)(y'+w) \end{aligned}$$

Similarly,

$$\begin{aligned} h &= x'y+(xy'+x'y)(z'+w') \\ &= x'yy+xy'y+(xy'+x'y)(z'+w') \\ &= (x'y+xy')y+(xy'+x'y)(z'+w') \\ &= (x'y+xy')(y+z'+w') \end{aligned}$$

Hence the required circuit is given below where circuits between T_0 and T_1, T_0 and T_2 and T_0 and T_3 realize functions g, f and h respectively.

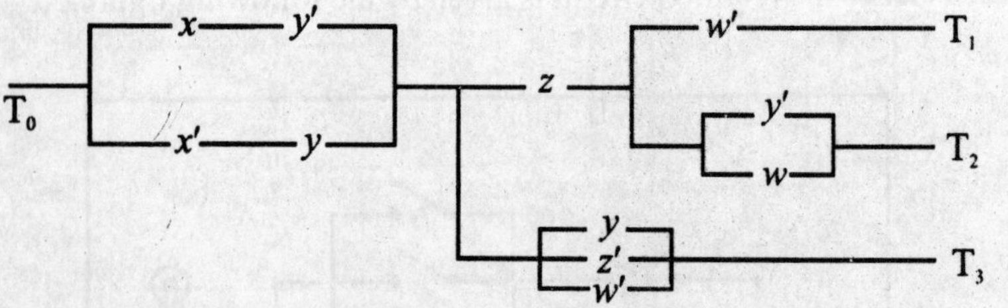

Example 6.25. Construct a 4-terminal circuit to realize the functions f, g, and h with properties given by the following table:

(x,y,z)	f	g	h
(0,0,0)	1	0	1
(0,0,1)	1	0	1
(0,1,0)	0	1	1
(0,1,1)	1	0	0
(1,0,0)	0	0	0
(1,0,1)	1	1	1
(1,1,0)	0	1	1
(1,1,1)	1	0	0

Solution: We first write each function in conjunctive normal form:

$$f = (x+y'+z)(x'+y+z)(x'+y'+z)$$
$$g = (x+y+z)(x+y+z')(x+y'+z')(x'+y+z)(x'+y'+z')$$
$$h = (x+y'+z')(x'+y+z)(x'+y'+z')$$

The common factors in f, g and h are noted and are indicated below with dotted vertical lines. Simplifications are made where possible

$$f = (x'+y'+z) \;\vdots\; (y'+z)$$
$$g = (x'+y'+z) \;\vdots\; (y'+z') \;\vdots\; (x+y)$$
$$h = (x'+y'+z) \;\vdots\; (y'+z') \;\vdots$$

The required circuit is given below where circuits between To and T_1, T_0 and T_2, T_0 and T_3 are corresponding to functions f, g and h respectively.

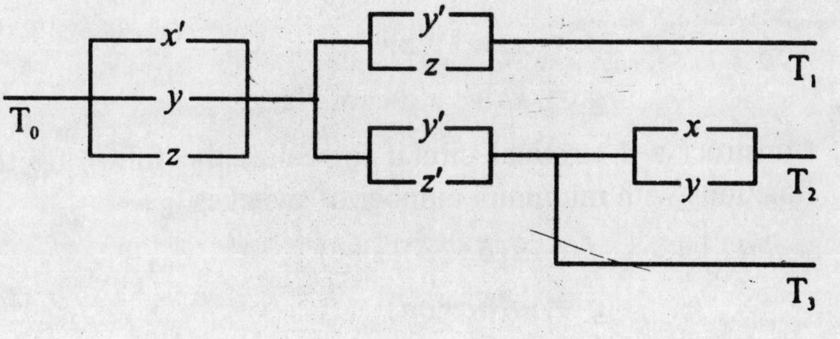

Exercise

1. Simplify the relay network given below:

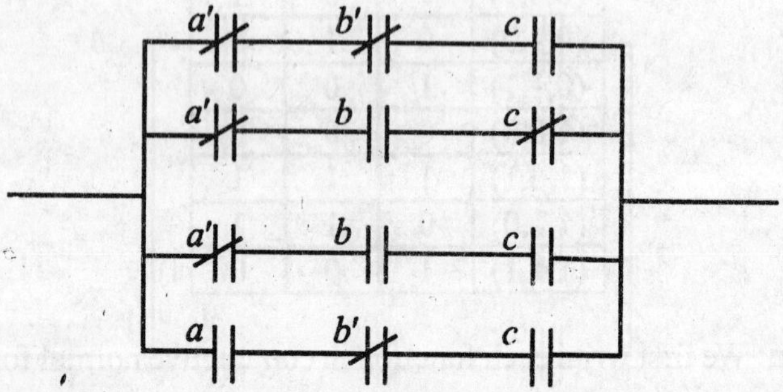

2. Use Karnaugh maps to redesign logic circuit given below so that it becomes a minimal AND-OR circuit

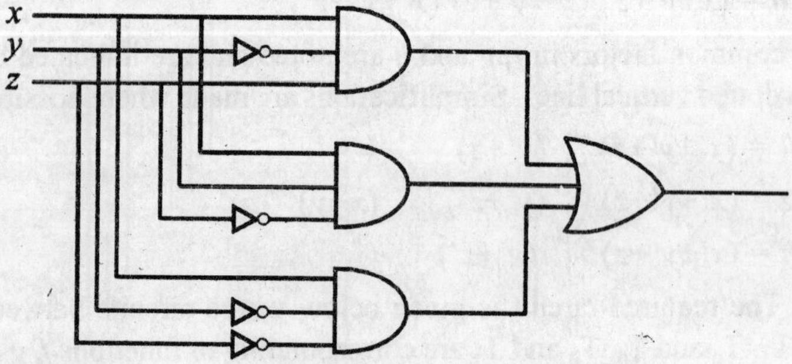

3. Construct a 3-terminal circuit to realize both the following functions

$$f = xzw + y'zw$$

$$g = xzw + y'zw + x'y'z$$

4. Construct a 4-terminal circuit to realize the following three functions with minimum number of switches

$$f = a(b+cd)(x+y)$$

$$g = a(bc+cd)$$

$$h = a(bc'+b'cd)$$

(Hint: Ten switches are sufficient for this circuit)

5. Design a circuit which connects two switches at the two entrances of a room to light a bulb in the room. The switches work independently so that any of the switch may be used to operate the bulb.

6. A digital system has 4-bit input from 0000 to 1111. Design a logic circuit that produces an output 1 whenever the equivalent decimal input is greater than 13.

Answers

1

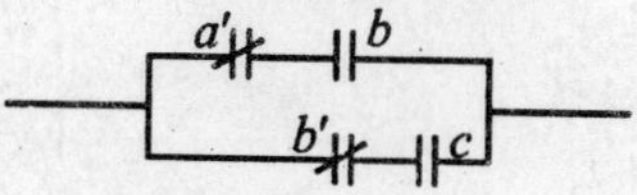

2

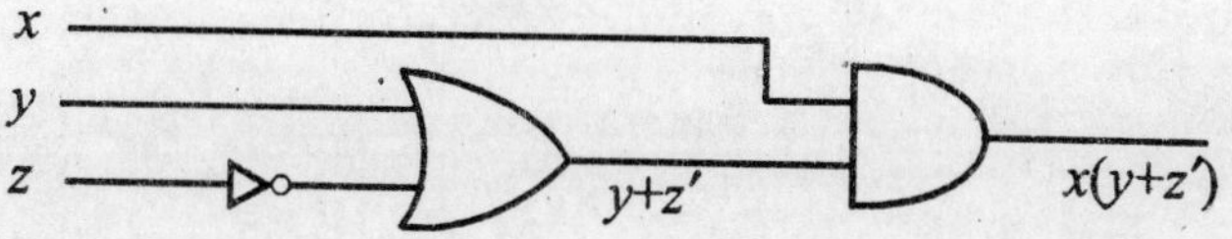

3

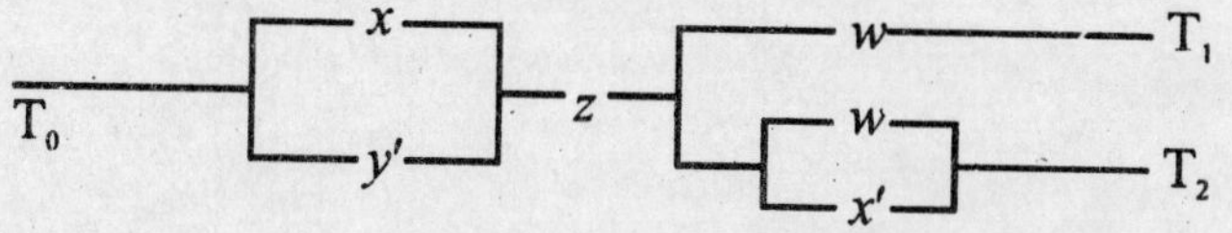

4

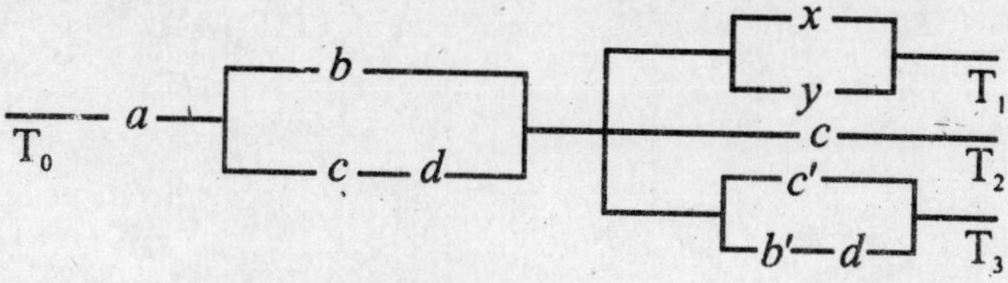

5

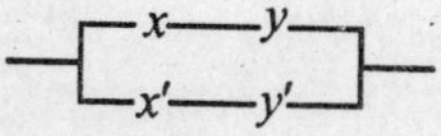

6. The function is given by $f = xyzw + xyzw$

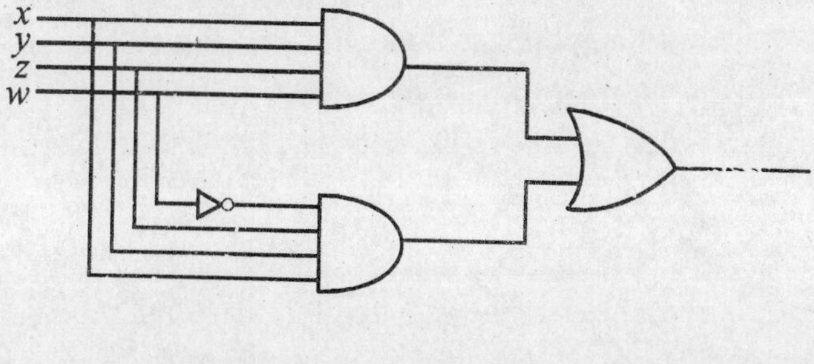

[illegible]

[illegible] digital system has 4-bit input from 0000 to 1111 [illegible]

[illegible]

[illegible]

CHAPTER SEVEN

Basic Concepts in Graph Theory

In chapter 2, we discussed binary relations and mentioned how they may be represented diagramatically. In fact, in many problems, a graphical representation of the binary relation on a finite set of objects is a very convenient form of representation. A relation on a finite set may be represented by joining x to y by an edge with arrow, provided (x, y) belongs to the given relation. We might also require to consider situations where there may be more than one edge joining two vertices. This gives rise to the study of graphs and directed graphs.

Graph theory begins with very simple geometric ideas and has many powerful applications to solve problems in electrical networks, computer sciences, operations research and social sciences. In this chapter, we will define graph as an abstract mathematical system. However in order to provide some intuitive feelings, we shall also represent graphs diagramatically. Any such diagram will also be called a graph. We begin with some of the basic terminology of graph theory.

§ 7.1 Graphs

Definition: A graph G is a pair (V, E), where $V = \{v_1, v_2, ...\}$ is a non-empty set whose elements are called vertices (or nodes) and $E = (e_1, e_2, ...\}$ is a set such that each element e_k of E is identified with an unordered pair (v_i, v_j) of vertices. The elements of E are called edges of G. The vertices v_i, v_j associated with edge ek are called the end vertices of e_k and the edge e_k is then denoted as $e_k = (v_i, v_j)$.

Graphs are usually represented by diagrams in which a vertex is represented by a dot or a small circle and an edge is represented by a line segment joining its end vertices.

Example 7.1: Let $V = \{v_1, v_2, v_3, v_4\}$ and $E = \{e_1, e_2, e_3, e_4, e_5\}$ be such that

$$e_1 = (v_1, v_2), e_2 = (v_4, v_3), e_3 = (v_1, v_3), e_4 = (v_2, v_4) \text{ and } e_5 = (v_1, v_2).$$

Then $G = (V, E)$ is a graph.

This graph is represented as in Fig 7.1

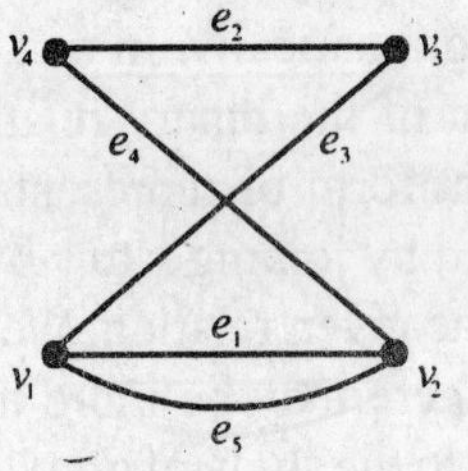

Fig: 7.1

Parallel edges in a graph

We observe from the definition of a graph $G = (V, E)$ that while the elements of E are distinct, more than one edge in E may have the same pair of end vertices. All edges having the same pair of end vertices are called *parallel* edges. For example, the edges e_1 and e_5 in Fig 7.1 are parallel edges.

Self-loop in a graph

We note that the definition of a graph allows an edge to be of the form

(v_i, v_i). Such an edge having the same vertex as both its end vertices is called a self-loop. Edge E_5 in the graph of Fig. 7.2 is a self-loop.

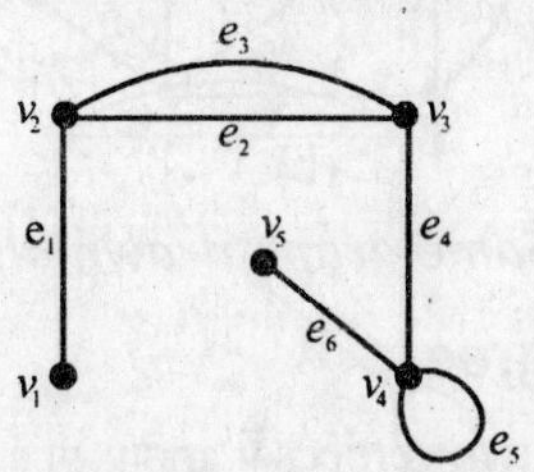

Fig: 7.2 Graph with five vertices and six edges

Simple graph

A graph that has neither self-loops nor parallel edges is called a *simple* graph.

Graphs in Fig. 7.1 and Fig. 7.2 are not simple. The graph in Fig. 7.3 below represents a simple graph whose vertex set V is the set $\{v_1, v_2, v_3, v_4\}$ and whose set of edges E consists of $e_1 = (v_1, v_2)$, $e_2 = (v_2, v_3)$, $e_3 = (v_1, v_3)$ and $e_4 = (v_3, v_4)$.

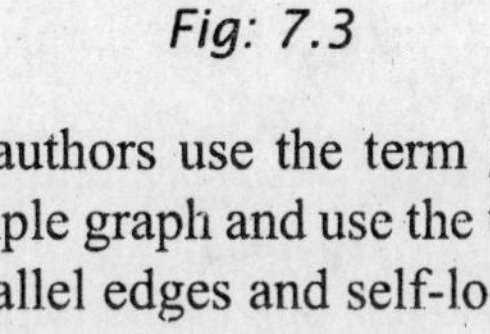

Fig: 7.3

Remarks : (1) Some authors use the term graph to mean what we have called a simple graph and use the term multigraph to mean a graph with parallel edges and self-loop.

(2) It should be noted that, in drawing a graph, it is immaterial whether the edges are drawn straight or curved, long or short. What is important is how the vertices are joined up. For example, the three graphs drawn in Figs. 7.4 (a), (b) and (c) are the same.

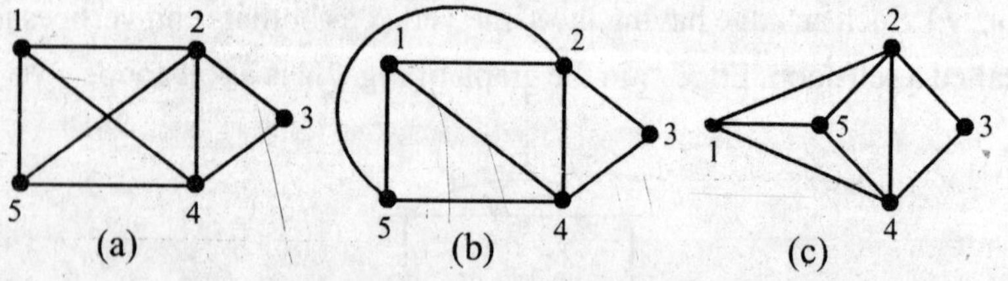

Fig: 7.4 Same graph drawn differently

Incidence and Degree

Let e_k be an edge joining two vertices v_i and v_j of a graph G. Then the edge e_k is said to be **incident** on each of its end vertices v_i and v_j. For example, in the graph of Fig. 7.3, edge e_3 is incident on vertices v_1 and v_3. Two vertices in a graph are said to be **adjacent** if there exists an edge joining the vertices. For example, in the graph of Fig. 7.3, vertices v_1 and v_2 are adjacent while vertices v_1 and v_4 are not adjacent.

Two non-parallel edges are said to be adjacent if they are incident on a common vertex. For example, edges e_1 and e_2 in the graph of Fig. 7.3 are adjacent because they are incident on a common vertex v_2, while edges e_1 and e_4 are not adjacent.

The **degree** of a vertex v in a graph G, written as $d(v)$, is equal to the number of edges which are incident on v with self-loop counted twice, In the graph of Fig. 7.2,

$$d(v_1) = 1$$

$$d(v_2) = 3$$

$$d(v_3) = 3$$

$$d(v_4) = 4$$

$$d(v_5) = 1$$

The degree of a vertex is sometimes also referred to as its valency.

Finite and Infinite Graphs

A graph $G = (V, E)$ is said to be **finite** if both sets V and E are finite; otherwise it is called infinite graph. The graphs in Figs. 7.1 and 7.2 are finite. In this book, we shall study only finite graphs and unless otherwise stated the term 'graph' will always mean a finite graph.

Isolated Vertex, Pendant Vextex and Null Graphs

A vertex having no edge incident on it is called an **isolated vertex**. In other words, a vertex v is said to be isolated vertex if degree $d(v)$ of v is zero. For example, the vertex v_6 in the graph of Fig. 7.5 is isolated vertex. A vertex is said to be **pendant** vertex if its degree is one. Vertex v_5 in the graph of Fig. 7.5 is a pendant vertex.

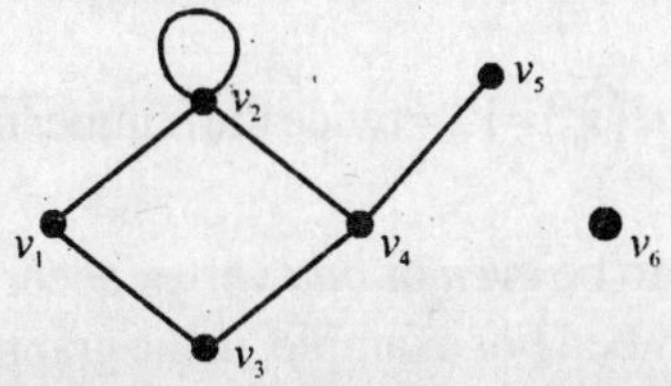

Fig: 7.5

A graph $G = (V, E)$ is said to be *null* graph if the set of vertices V is non-empty but the set of edges E is empty. A null graph is thus a graph in which every vertex is an isolated vertex. A null graph of five vertices is shown in Fig 7.6. A null graph having one vertex only is called *trivial* graph.

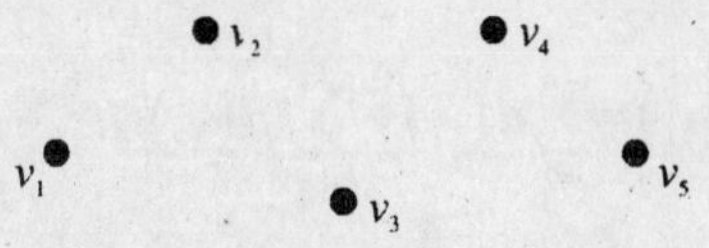

Fig: 7.6 Null graph of five vertices

We now derive two simple but important results.

Theorem 7.1: The sum of the degrees of all vertices in a graph G is equal to twice the number of edges in G.

Proof: Let G be a graph with e edges and n vertices $v_1, v_2, ..., v_n$. Since each edge is incident on two vertices, it contributes 2 to the sum of the degrees of the graph. Thus the sum of degrees of all the vertices in G is twice the number of edges in G.

That is,

$$\sum_{i=1}^{n} d(v_i) = 2e$$

This result is also known as **handshaking lemma.**

To verify the above theorem, consider the graph of Fig. 7.5. In this graph, we have

$d(v_1) = 2$, $d(v_2) = 4$, $d(v_3) = 2$, $d(v_4) = 3$ and $d(v_5) = 1$, while $d(v_6) = 0$. Therefore

$$\sum_{i=1}^{6} d(v_i) = 12 = \text{twice the number of edges.}$$

A vertex is said to be *even* or *odd* vertex according as its degree is an even or an odd number. For example, in the graph of Fig. 7.5, vertices v_1, v_2, v_3 and v_6 are even vertices whereas v_4 and v_5 are odd vertices.

Theorem 7.2: The number of odd vertices (*i.e.*, vertices of odd degree) in a graph is always even.

Proof: Let the number of vertices in a graph G be n. Without loss of generality, we can assume that the degrees of the first r vertices $v_1, v_2, \ldots, v_r$ be even and those of the remaining $(n - r)$ vertices be odd. Then

$$\sum_{i=1}^{n} d(v_i) = \sum_{i=1}^{r} d(v_i) + \sum_{i=r+1}^{n} d(v_i). \qquad \text{..................(1)}$$

By Theorem 7.1, the sum on the left-hand side of (1) is even. The first sum on the right hand side is also even, because each term in this sum is even. Hence the second sum on the right hand side must also be even:

$$\sum_{i=r+1}^{6} d(v_i) = \text{an even number..} \qquad \text{......................(2)}$$

Since each term $d(v_i)$ in (2) is odd, the total number of terms in the sum must be even (to make the sum an even number). In other words, $n - r$, the number of vertices of odd degree, should be even. Hence the theorem

§ 7.2 Isomorphic Graphs

Definition: Two graphs $G_1 = (V_1, E_1)$ and $G_2 = (V_2, E_2)$ are said to be **isomorphic** if there exists a one-to-one correspondence between their vertex sets (V_1 and V_2) and a one-to-one correspondence between their edge sets (E_1 and E_2) so that the corresponding edges of G_1 and G_2 are incident on the corresponding vertices of G_1 and G_2.

In other words, if vertices v_1 and v_2 in G_1 correspond respectively to vertices v_1' and v_2' in G_2, then an edge in G_1 with end vertices v_1 and v_2 must correspond to an edge in G_2 with v_1' and v_2' as end vertices and vice versa.

Example 7.2: Show that the graphs given below are isomorphic

Fig: 7.7

Solution: We can establish the correspondence between the two graphs as follows:

Vertex Correspondence: The vertices u_1, u_2, u_3, and u_4 correspond to v_1, v_2, v_3 and v_4 respectively.

Edge Correspondence: The edges 1, 2, 3, 4, 5 and 6 correspond to e_1, e_2, e_3, e_4, e_5, and e_6 respectively.

Example 7.3: Show that the graphs given below are not isomorphic

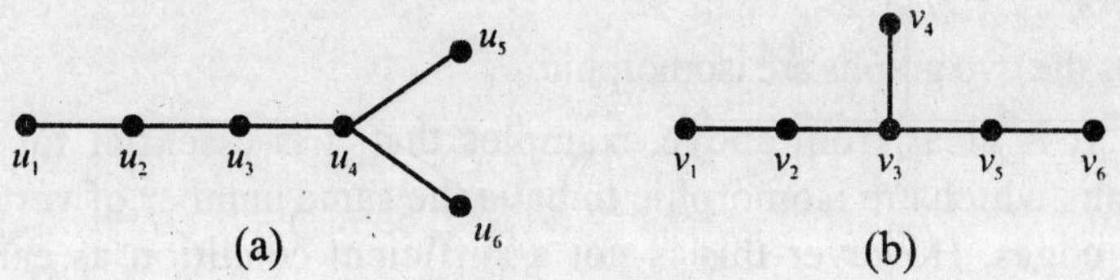

Fig: 7.8

Solution: The two graphs in Fig 7.8 are not isomorphic can be shown as follows: If the graph in Fig 7.8(a) were to be isomorphic to the graph in Fig 7.8(b), then vertex u_4 in (a) must correspond to vertex v_3 in (b), because there are no other vertices of degree three. Now in (a) there are two pendant vertices u_5 and u_6 adjacent to u_4 while in (b) there is only one pendant vertex v_4, adjacent to v_3. Hence these two graphs are not isomorphic.

Example 7.4: Show that the graphs given below are isomorphic.

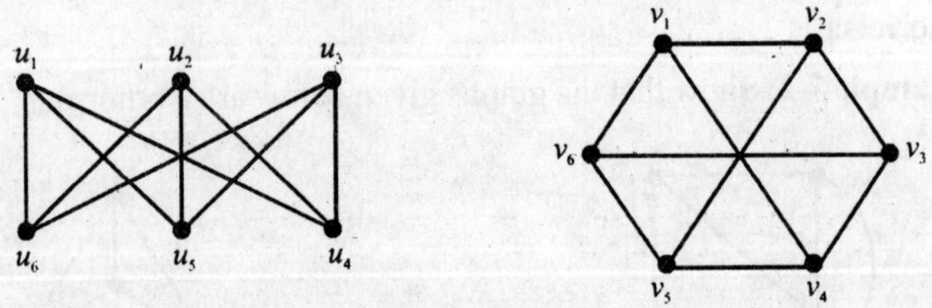

Fig: 7.9

Solution: The above two graphs are isomorphic can be seen as follows: We define the correspondence between vertex sets and edge sets as given below.

Vertex Correspondence : $u_1 \leftrightarrow v_1$, $u_2 \leftrightarrow v_3$, $u_3 \leftrightarrow v_5$, $u_4 \leftrightarrow v_2$, $u_5 \leftrightarrow v_4$ and $u_6 \leftrightarrow v_6$.

Edge Correspondence : $(u_1, u_4) \leftrightarrow (v_1, v_2)$, $(u_1, u_5) \leftrightarrow (v_1, v_4)$, $(u_1, u_6) \leftrightarrow (v_1, v_6)$, $(u_2, u_4) \leftrightarrow (v_3, v_2)$, $(u_2, u_5) \leftrightarrow (v_3, v_4)$, $(u_2, u_6) \leftrightarrow (v_3, v_6)$, $(u_3, u_4) \leftrightarrow (v_5, v_2)$, $(u_3, u_5) \leftrightarrow (v_5, v_4)$, $(u_3, u_6) \leftrightarrow (v_5, v_6)$.

Thus the two graphs are isomorphic.

It is clear from above examples that it is essential for two graphs which are isomorphic to have the same number of vertices and edges. However this is not a sufficient condition as can be seen from Example-7.3.

§ 7.3 Subgraphs and Complements

Definition: Let $G=(V,E)$ be a graph. A graph $H=(V',E')$ is said to be a subgraph of G if E' is a subset of E and V' is a subset of V such that an edge (v_i, v_j) is in E' only if v_i and v_j are in V'. For example, the graph in Fig. 7.10(b) is a subgraph of the graph in Fig. 7.10(a).

(a) (b)

Fig: 7.10 Graph (a) and one of its subgraph (b)

Spanning Subgraph

A subgraph H of a graph G is said to be spanning subgraph if all the vertices of G are present in the subgraph H.

It is easy to observe that every graph is its own subgraph. Also, if H is a subgraph of a graph G and K is a subgraph of H then K is a subgraph of G.

Edge-disjoint Subgraphs

Two subgraphs H_1 and H_2 of a graph G are said to be *edge-disjoint* if H_1 and H_2 do not have any edge in common. For example, the two graphs in Figs. 7.11(a) and (b) are edge-disjoint subgraphs of the graph in Fig 7.10(a). It may be noted that although edge-disjoint graphs do not have any edge in common, they may have vertices in common.

(a) (b)

Fig: 7.11

Vertex-disjoint Subgraphs

Two subgraphs H_1 and H_2 of a graph G are said to be *vertex-disjoint* if H_1 and H_2 do not have any vertices in common. Obviously, every vertex-disjoint subgraphs are edge-disjoint.

Complement of a Subgraph

Definition: Let $H = (V', E')$ be a subgraph of a graph $G = (V, E)$. The complement of subgraph H with respect to the graph G is the subgraph $\overline{H}(G) = (V, E - E')$. For example, Fig 7.12(c) shows the complement of the subgraph in Fig. 7.12(b) with respect to the graph in Fig. 7.12(a)

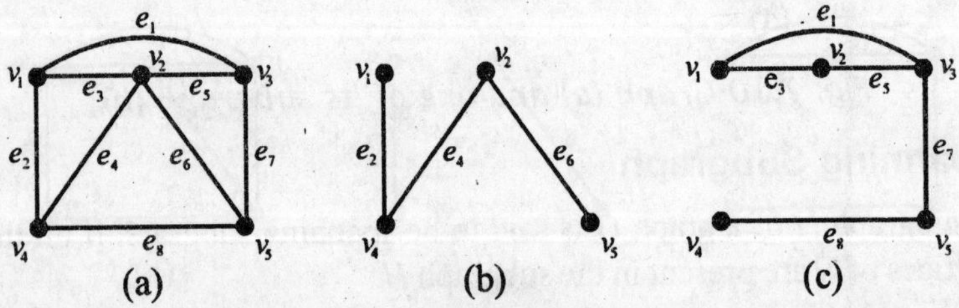

Fig: 7.12

Thus, the complement of a subgraph H with respect to graph G is obtained from G by removing the edges of H.

Complement of a Simple Graph

Definition: Let $G = (V, E)$ be a simple graph. The complement of G is the graph $\overline{G} = (V, E')$ such that on edge e is in E' if and only if it is not in E. In other words, two vertices v_i and v_j are adjacent in $\overline{G}$ if and only if they are not adjacent in G. For example, Fig. 7.13 shows a graph and its complement.

(a) Graph *(b) Complement of graph in (a)*

Fig: 7.13

A simple graph which is isomorphic to its complement is called *self-complementary* graph. For example, the graph given below is self-complementary.

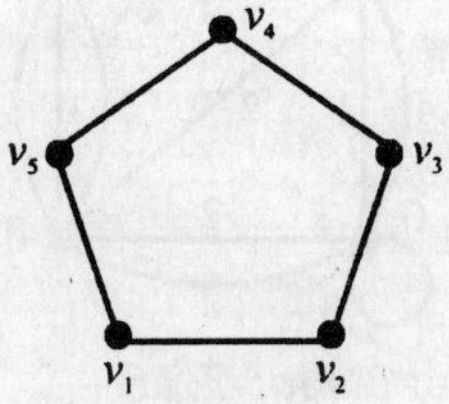

Fig: 7.14

The Four Multicoloured Cubes Problem

We conclude this section with a puzzle which is also known as **Instant Insanity.**

Problem: We are given four cubes. The six faces of each cube are coloured red, blue, green and white. Is it possible to stack the cubes one on the top of another to form a column such that all four colours appear on each side of the resulting column.

Solution: **Step 1**: We solve the problem by drawing a graph G with four vertices R, B, G and W, one for each colour. Pick a cube and call it cube 1. Then draw an edge between two vertices if and only if the cube in question has the corresponding colours on opposite faces. In

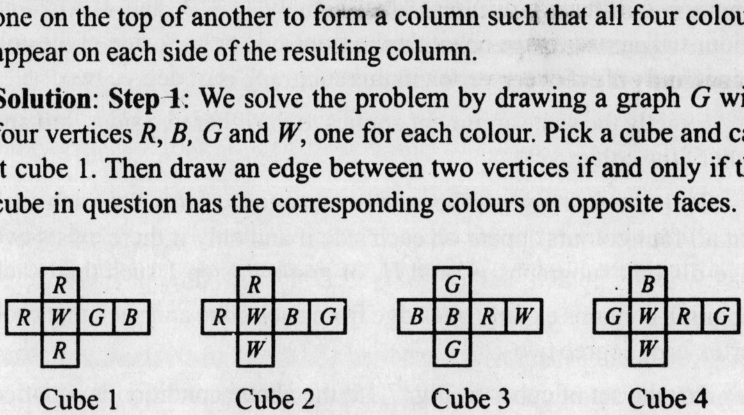

Fig: 7.15 Four unfolded cubes

other words, if a red face in cube 1 has a green face opposite to it then draw an edge between vertices R and G in the graph. Do the same for the remaining two pairs of faces in cube 1. Put label 1 on all three edges resulting from cube 1. Repeat the procedure for the other three cubes one by one on the same graph.

The graph corresponding to the cubes given in Fig. 7.15 is shown in Fig. 7.16.

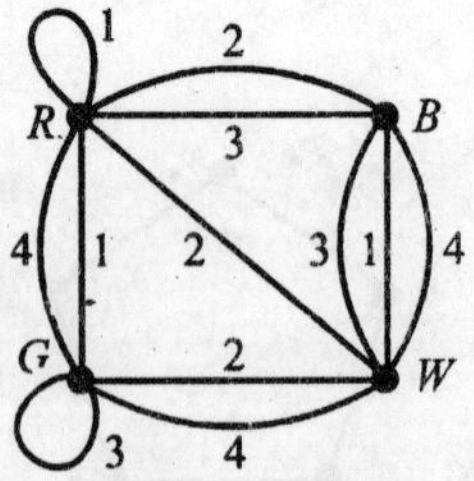

Fig: 7.16

Step 2: Consider the graph resulting from this representation. The degree of each vertex is the total number of faces with the corresponding colour.

Consider two opposite vertical sides of the desired column of four cubes, say, facing front and back. A subgraph will represent these eight faces, four facing front and four back. Each of the four edges in this subgraph will have a different label namely 1, 2, 3, and 4. Also, no colour will appear twice on either the front side or back side of the column if and only if every vertex in this subgraph is of degree two.

Exactly the same argument applies to the other two sides, left and right of the column.

Thus the four given cubes can be arranged to form a column such that all four colours appear on each side if and only if there exists two edge-disjoint subgraphs H_1 and H_2 of graph in step 1 such that each subgraph contains exactly one edge from each cube and such that each vertex is of degree two.

For the set of cubes in Fig. 7.15, the above condition is satisfied and the two subgraphs are shown in Fig. 7.17

Fig: 7.17

The desired column is shown below in Fig 7.18

Fig: 7.18

Solved Examples

Example 7.5: What is the maximum number of vertices in a graph with 35 edges and all vertices are of degree at least 3.

Solution: Let n be the number of vertices. Since the graph has 35 edges and each edge contribute 2 to the sum of degrees of all vertices, therefore sum of degrees of all vertices is $2 \times 35 = 70$. Also, each vertex is of degree at least 3, therefore

$$70 \geqslant 3n$$

The largest integer n satisfying this inequality is 23. Hence maximum number of vertices in a graph with 35 edges such that each vertex is of degree at least 3 is 23.

Example 7.6: Explain why the two graphs in Fig. 7.19 are not isomorphic

Fig: 7.19

Solution: In the graph (a), no vertices of degree two are adjacent while in the graph (b) vertices of degree two are adjacent. Since isomorphism preserves adjacency of vertices, the graphs are not isomorphic.

Example 7.7: Prove that if G is self complementary then G has $4k$ or $4k+1$ vertices, where k is an integer.

Solution: Let the number of vertices in a self-complementary graph G be n. We know that maximum number of edges in a simple graph with n vertices is $\frac{n(n-1)}{2}$. Since G is self-complementary and isomorphic graphs have equal number of edges, therefore, G and its complement each must have $\frac{n(n-1)}{4}$ edges. Hence n or n-1 must be a multiple of 4. Thus

$$n = 4k \text{ or } n\text{-}1 = 4k, \text{ where } k \text{ is an integer.}$$

$$\Rightarrow \quad n = 4k \text{ or } 4k+1, \text{ where } k \text{ is an integer.}$$

Example 7.8: Show that the four cubes problem in Fig. 7.20 has no solution

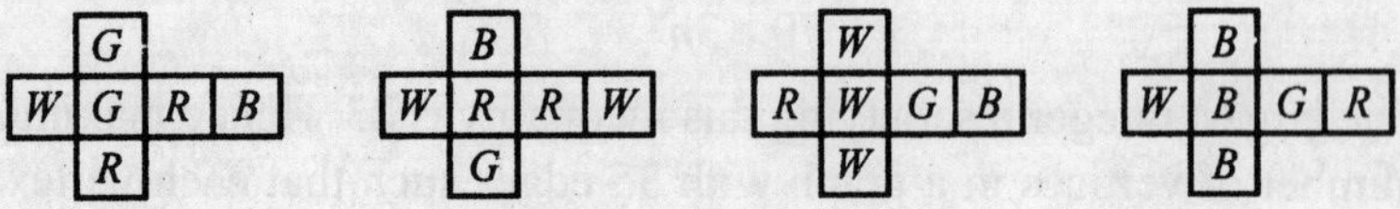

Cube 1 Cube 2 Cube 3 Cube 4

Fig: 7.20

Solution: Using the method described in the four cubes problem, we obtain the following graph G for the set of cubes in Fig. 7.20.

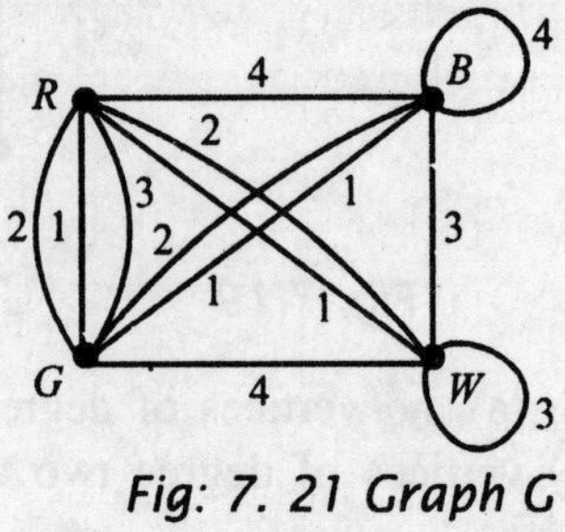

Fig: 7. 21 Graph G

Since in G, we can not find two edge-disjoint subgraph H_1 and H_2 such that each subgraph contains exactly one edge from each cube and degree of each vertex in each subgraph is two, therefore we can not stake the given four cubes one on the top of another to form a column such that all four colours appear on each side of the resulting column.

Example 7.9: Is it possible to have a group of nine people such that each person is acquainted with exactly five of the other people?

Solution: No such group of nine people can exist. For, if we try to draw a graph with a vertex for each person and an edge between each pair of people who know each other, then we would have a graph with nine vertices all of degree 5. But then sum of degrees of all vertices in this graph would be 45 which is not possible because sum of degrees of all vertices in any graph is always even.

Exercise

1. Draw all simple graphs of three vertices.

2. Draw the graph $G = (V, E)$, where

   (a) $V = \{A, B, C, D, E\}$ and $E = \{(A, B), (A, C), (A, D), (B, C), (B, E), (C, E), (C, D)\}$

   (b) $V = \{v_1, v_2, v_3, v_4, v_5\}$ and $E = \{(v_1, v_2), (v_1, v_3), (v_3, v_1), (v_2, v_2), (v_2, v_4), (v_3, v_4), (v_4, v_5)\}$

3. Describe formally the graphs in Fig. 7.22(a) and Fig. 7.22(b)

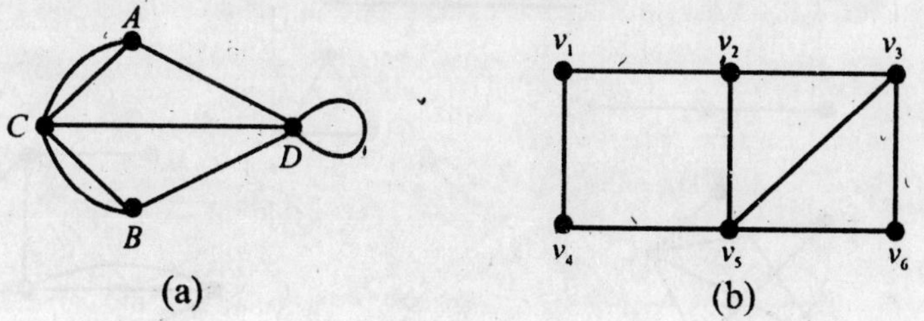

Fig: 7.22

4. Show that the maximum number of edges in a simple graph with n vertices is $\frac{n(n-1)}{2}$

5. Can a graph with seven vertices be isomorphic to its complement?

6. Show that the two graphs in Fig. 7.23(a) and 7.23(b) are isomorphic.

(a) (b)

Fig: 7.23

7. Suppose a simple graph has 15 edges, 3 vertices of degree 4 and all other vertices of degree 3. Find the number of edges in the graph.

8. Find a solution of the four cubes problem for the set of cubes given below:

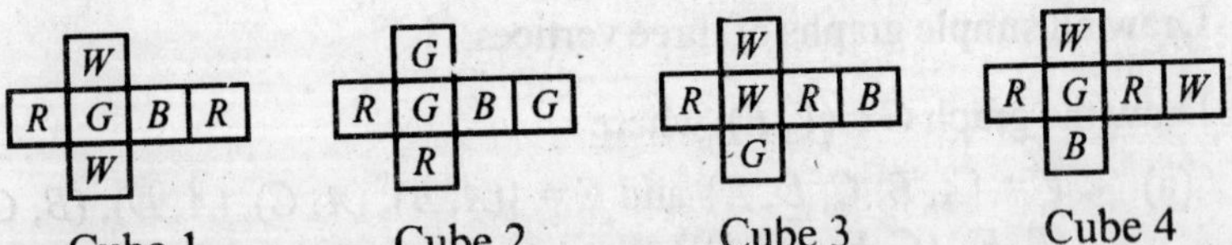

Fig: 7.24

9. Show that a simple graph G with at least two vertices contains two vertices of the same degree.

Answers

1. 

2. 

(a) (b)

3.(a) $G = (V, E)$, where $V = \{A, B, C, D\}$ and $E = \{(A, C), (A, C), (B, C), (B, C), (A, D), (B, D), (C, D), (D, D)\}$

(b) $G = (V, E)$, where $V = \{v_1, v_2, v_3, v_4, v_5, v_6\}$ and $E = \{(v_1, v_2), (v_1, v_4), (v_2, v_3), (v_2, v_5), (v_3, v_5), (v_3, v_6), (v_4, v_5), (v_5, v_6)\}$

5. No

7. 9

8.

§ 7.4 Walk, Paths, and Circuits

In this section, we shall study some additional terminology associated with a graph.

Walk

Definition: A **walk** in a graph $G = (V, E)$ is a finite alternating sequence of vertices and edges $v_0, e_1, v_1, e_2, v_2, ..., v_{k-1}, e_k, v_k$ begining and ending with vertices such that v_{i-1} and v_i are the end vertices of the edge e_i, $1 \le i \le k$ and all the edges are distinct. Vertices v_0 and v_k with which a walk begins and ends are called its **end** or **terminal** vertices. All other vertices are called internal vertices of the walk. Note that in a walk vertices can appear more than once.

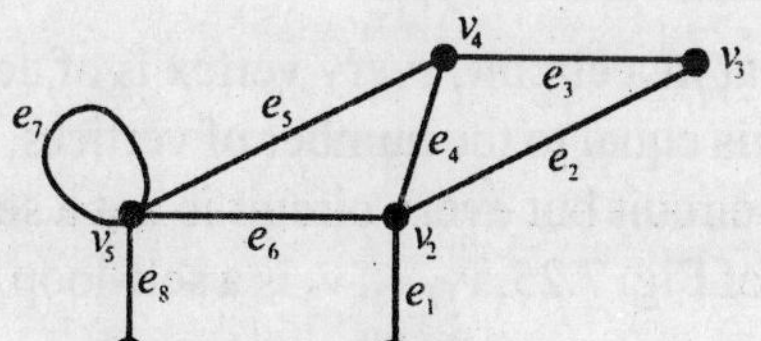

Fig: 7.25

A walk is said to be an **open walk** if its end (or terminal) vertices are distinct. A walk which is not open is called **closed**. Thus a closed walk is a walk which begins and ends at the same vertex. For example, in the graph of Fig. 7.25, the sequence $v_1, e_1, v_2, e_4, v_4, e_3, v_3, e_2, v_2, e_6, v_5$ is an open walk while the sequence $v_2, e_2, v_3, e_3, v_4, e_5, v_5, e_7, v_5, e_6, v_2$ is a closed walk.

Path

Definition: An open walk in which no vertex appears more than once is called a *path*. For example, in the graph of Fig. 7.25, $v_1, e_1, v_2, e_2, v_3, e_3, v_4$ is a path whereas $v_1, e_1, v_2, e_2, v_3, e_3, v_4, e_4, v_2$ is not a path. It is clear from the definition that a path does not intersect itself.

Length of a Path

The number of edges in a path is called the length of the path. For example, the length of the path v_1, e_1, v_2, e_2, v_3 in the graph of Fig. 7.25 is two. It is clear that an edge which is not a self-loop is a path of length one. Observe that a self-loop can be included in a walk but not in a path.

It should be noted that in a path, the terminal vertices are of degree one and rest of the vertices are of degree two.

Circuit

Definition: A closed walk is called a **circuit** if all its vertices except the end vertices are distinct. For example, in the graph of Fig. 7.25, $v_1, e_1, v_2, e_6, v_5, e_8, v_6, e_9, v_1$ is a circuit.

A circuit is also called a **cycle**.

It is clear that in a circuit, every vertex is of degree 2 and that the number of edges is equal to the number of vertices. Every self- loop in a graph is also a circuit but every circuit is not a self-loop. For example, in the graph of Fig. 7.25, v_5, e_7, v_5 is a self-loop and therefore a circuit whereas $v_2, e_2, v_3, e_3, v_4, e_4, v_2$ is a circuit but not a self-loop.

§ 7.5 Connected Graphs and Components

An important concept in graph theory is that of connectedness of vertices in a graph.

Definition: A graph G is said to be **connected** if there exists a path between every pair of vertices in G. A graph which is not connected is called a **disconnected** graph. Thus a graph G is disconnected if we can find a pair v_i, v_j of vertices in G such that there is no path between v_i and v_j. For example the graph in Fig 7.26(a) is connected whereas the graph in Fig 7.26(b) is not connected.

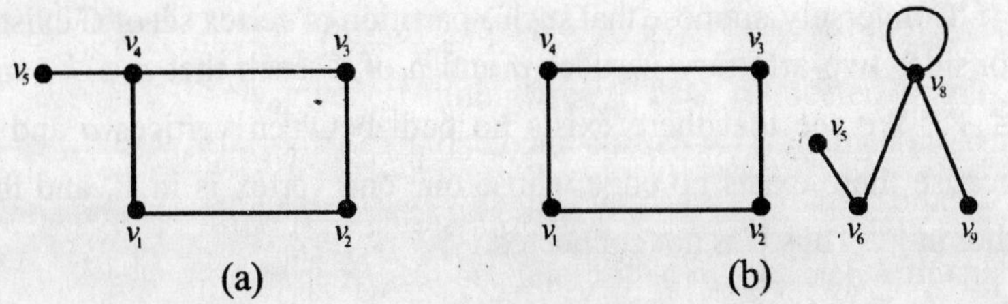

Fig: 7.26

It is easy to see that a disconnected graph consists of two or more connected subgraphs. A **component** of a graph G is a maximal connected subgraph of G. In other words, a component of G is not a proper subgraph of any other connected subgraph of G. For example, the graph G of Fig. 7.27 is not connected. This disconnected graph has

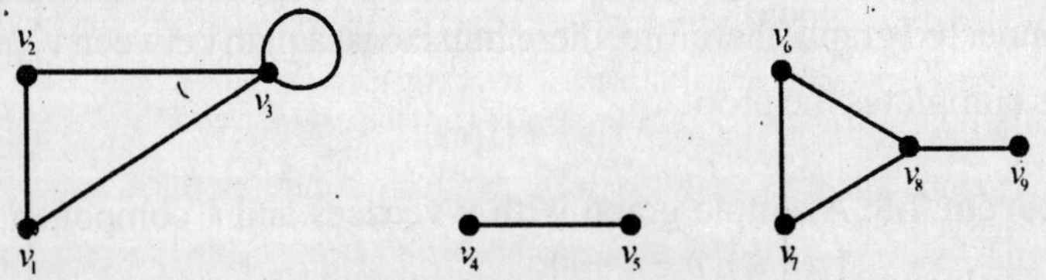

Fig: 7.27 A disconnected graph with three components

three components whose vertex sets are $\{v_1, v_2, v_3\}$, $\{v_4, v_5\}$ and $\{v_6, v_7, v_8, v_9\}$ respectively. Note that if a graph G is connected then it has only one component, namely G itself. Moreover an isolated vertex in a graph is also a component of the graph.

Theorem 7.3 : A graph G is disconnected if and only if the vertex set V of G can be partitioned into two non-empty, disjoint subsets V_1 and V_2 such that there exists no edge in G whose one end vertex is in V_1 and the other in V_2.

Proof : Let G be a disconnected graph. Choose any vertex a in G. Let V_1 be the set of all those vertices which are joined by paths to a. Since G is not connected, V_1 does not contain all vertices of G. Let V_2 be the set of vertices of G which are not in V_1. Then V_2 is non-empty and no vertex in V_1 is joined to any vertex in V_2. Hence V_1, V_2 is the required partition.

Conversely, suppose that such a partition of vertex set of G exists. Consider two arbitrary vertices a and b of G such that $a \in V_1$ and $b \in V_2$. We see that there exists no path between vertices a and b because there exists no edge whose one end vertex is in V_1 and the other in V_2. Thus G is not connected.

Theorem 7.4: If a graph (conected or disconnected) has exactly two vertices of odd degree, then there exists a path joining these two vertices.

Proof : Let G be a graph with all even vertices except vertices v_1 and v_2, which are odd. From Theorem 7.2, which holds for every graph and therefore for every component of a disconnected graph, no graph can have an odd number of odd vertices. Therefore, in graph G, v_1 and v_2 must belong to the same component. Since component of any graph is a connected graph, therefore, there must exist a path between v_1 and v_2. This completes the proof

Theorem 7.5: A simple graph with n vertices and k components can have at most $\dfrac{(n-k)(n-k+1)}{2}$ edges.

Proof: Let G be a simple graph with n vertices and k components. Let $n_1, n_2, n_3, \ldots n_k$, be the number of vertices in each of the k components of G. Then we have

$$n_1 + n_2 + \ldots + n_k = n, \text{ and } \quad n_i \geq 1.$$

We shall use the following inequality to prove the theorem

$$\sum_{i=1}^{k} n_i^2 \le n^2 - (k-1)(2n-k) \quad \text{............... (1)}$$

[Proof of the above inequality:

Since $n_1 + n_2 + \ldots\ldots + n_k = n$, we have

$$\sum_{i=1}^{k} (n_i - 1) = n_1 + n_2 + \ldots\ldots + n_k - k = n - k$$

Squaring both sides, we get

$$\left(\sum_{i=1}^{k} (n_i - 1)\right)^2 = n^2 + k^2 - 2nk$$

$$\Rightarrow \quad \sum_{i=1}^{k} (n_i - 1)^2 + \text{sum of terms of the type } 2(n_i - 1)(n_j - 1)$$

$$= n^2 + k^2 - 2nk$$

$$\Rightarrow \quad \sum_{i=1}^{k} (n_i^2 - 2n_i) + k \le n^2 + k^2 - 2nk \text{ because } (\mathrm{n_i} - 1) \ge 0 \,\forall i$$

$$\Rightarrow \quad \sum_{i=1}^{k} n_i^2 - 2\sum_{i=1}^{k} n_i \le n^2 + k^2 - 2nk - k$$

$$\Rightarrow \quad \sum_{i=1}^{k} n_i^2 \le n^2 + k^2 - 2nk - k + 2n \quad \text{because } \sum_{i=1}^{k} \mathrm{n_i} = n$$

$$\Rightarrow \quad \sum_{i=1}^{k} n_i^2 \le n^2 - (k-1)(2n-k)]$$

Since we know that the maximum number of edges in a simple graph with n vertices is $\frac{n(n-1)}{2}$, therefore, the maximum number of edges in the ith component of G (which is a simple connected graph having n_i vertices) is $\frac{1}{2} n_i (n_i - 1)$. Thus, the maximum number of edges in G is

$$\sum_{i=1}^{k} \frac{1}{2} n_i (n_i - 1) = \frac{1}{2}\left[\sum_{i=1}^{k} n_i^2 - \sum_{i=1}^{k} n_i\right] = \frac{1}{2}\left[n^2 - (k-1)(2n-k) - n\right]$$

$$= \frac{1}{2}(n-k)(n-k+1)$$

This completes the proof.

Corollary: If a simple graph G with n vertices has more than $\frac{(n-1)(n-2)}{2}$ edges then G is connected.

Proof : Let G be a simple graph with n vertices and more than $\frac{(n-1)(n-2)}{2}$ edges. Suppose G has k components. Then by Theorem 7.5, G can have at most $(n-k)(n-k+2)/2$ edges. But since G has more than $(n-1)(n-2)/2$ edges. we must have

$$\frac{(n-k)(n-k+1)}{2} > \frac{(n-1)(n-2)}{2}$$

which is possible only when $k = 1$. Thus G is connected.

Example 7.10: Let G be a simple graph with 6 vertices and 11 edges. Then by above corollary, the graph G is connected.

§ 7.6 Operations on Graphs

In this section, we introduce a few operations involving graphs.

Union of two Graphs

Let $G_1 = (V_1, E_1)$ and $G_2 = (V_2, E_2)$ be two graphs. The **union** of G_1 and G_2, denoted as $G_1 \cup G_2$ is the graph $G_3 = (V_1 \cup V_2, E_1 \cup E_2)$.

For example, two graphs G_1 and G_2 and their union are shown in Fig. 7.28

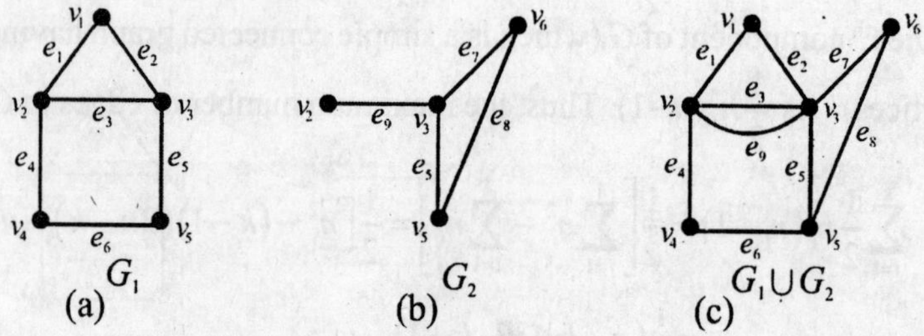

Fig: 7.28 Union of two graphs

Intersection of two Graphs

The intersection of two graphs $G_1 = (V_1, E_1)$ and $G_2 = (V_2, E_2)$, denoted as $G_1 \cap G_2$, is the graph $G_3 = (V_1 \cap V_2, E_1 \cap E_2)$. That is, the vertex set of G_3 consists of only those vertices which are present in both G_1 and G_2, and the edge set of G_3 consists of only those edges which are present in both G_1 and G_2.

The intersection of the graph G_1 and G_2 of Fig. 7.28(a) and (b) is shown in Fig. 7.29.

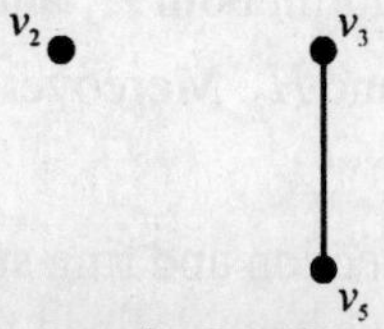

graph $G_1 \cap G_2$

Fig: 7.29

Ring Sum of two Graphs

The ring sum of two graphs $G_1 = (V_1, E_1)$ and $G_2 = (V_2, E_2)$ denoted as $G_1 \oplus G_2$, is the graph $G_3 = (V_1 \cup V_2, E_1 \cup E_2 - E_1 \cap E_2)$. That is, the ring sum of two graphs G_1 and G_2 is a graph consisting of the vertex set $V_1 \cup V_2$ and of edges that are either in G_1 or G_2 but not in both.
For example, the ring sum of of the graphs G_1 and G_2 of Fig. 7.28(a) and (b) is the graph in Fig. 7.30.

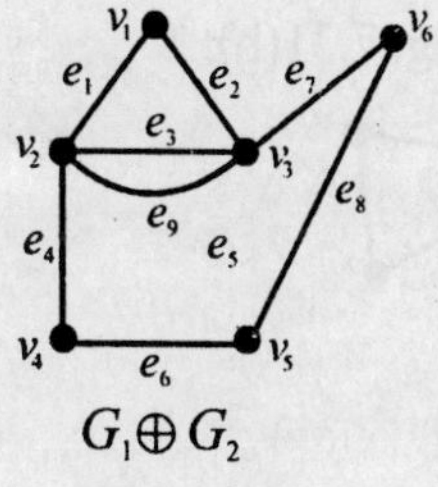

$G_1 \oplus G_2$

Fig: 7.30

It is clear that if graphs G_1 and G_2 are edge disjoint then $G_1 \oplus G_2 = G_1 \cup G_2$ and if H is a subgraph of a graph G then $G \oplus$ H is that subgraph of

G which remains after all the edges in H have been removed from G. Therefore $G \oplus H = \overline{H}(G)$, the complement of H with respect to G.

Decomposition of a Graph

A graph G is said to be decomposed into two subgraphs H_1 and H_2 if

$$H_1 \cup H_2 = G$$

and $H_1 \cap H_2 =$ a null graph.

In other words, every edge of graph G occurs either in subgraph H_1 or in subgraph H_2, but not in both H_1 and H_2. Some vertices may, however, occur in both H_1 and H_2. Moreover in decomposition, isolated vertices are disregarded.

Note that union, intersection and ring sum are binary operations. That is, these operations have been defined for a pair of graphs. These definitions can be extended in an obvious way to include any finite number of graphs.

Next we discuss unary operations on a graph

Vertex Deletion

If v_i is a vertex in a graph $G = (V, E)$, then $G - v_i$ denotes a subgraph of G obtained by deleting (*i.e.* removing) v_i from G. Deletion of a vertex always implies the deletion of all edges incident on that vertex.

For example, deletion of vertex v_1 in the graph of Fig. 7.31(a) results in the graph in Fig. 7.31(b).

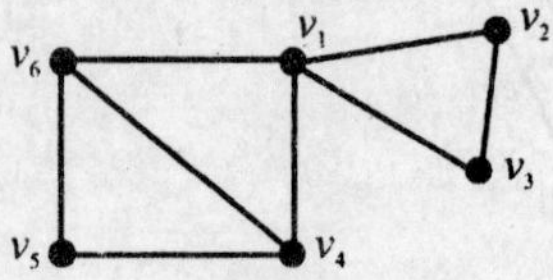

(a) graph G

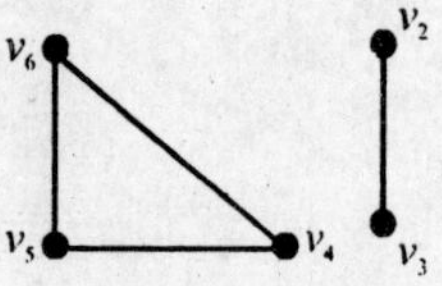

(b) graph $G_1 - V_2$

Fig: 7.31

Edge Deletion

If e_i is an edge of a graph $G = (V, E)$, then $G - e_i$ is the subgraph of G

obtained by removing (deleting) e_i from G. Note that deletion of an edge e_i does not imply deletion of the end vertices of e_i. Thus $G - e_i = G \oplus e_j$.

For example, deletion of edge e_1 in the graph of Fig 7.32(a) results in the graph in Fig. 7.32(b)

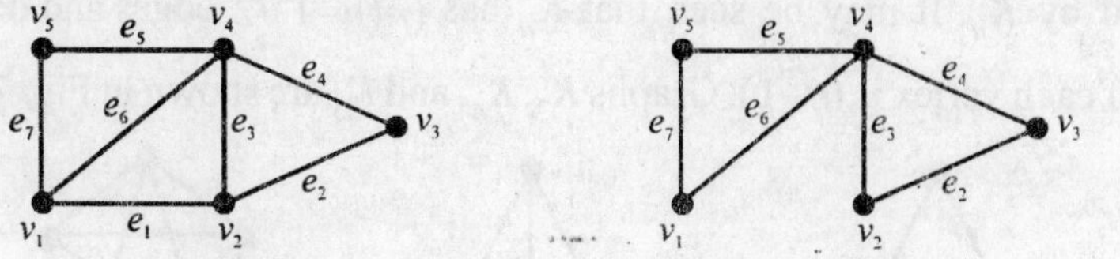

(a) graph G (b) graph $G - e_1$

Fig: 7.32

Fusion of a Pair of Vertices

A pair of vertices x, y in a graph G are said to be fused (merged, identified or short circuited) if the two vertices x and y are replaced by a new vertex such that all the edges in G incident on x and y are now incident on the new vertex. Thus the fusion of two vertices does not change the number of edges in the graph but it reduces the number of vertices by one. For example, fusion of vertices x and y in the graph of Fig. 7.33(a) results in the graph shown in Fig. 7.33(b).

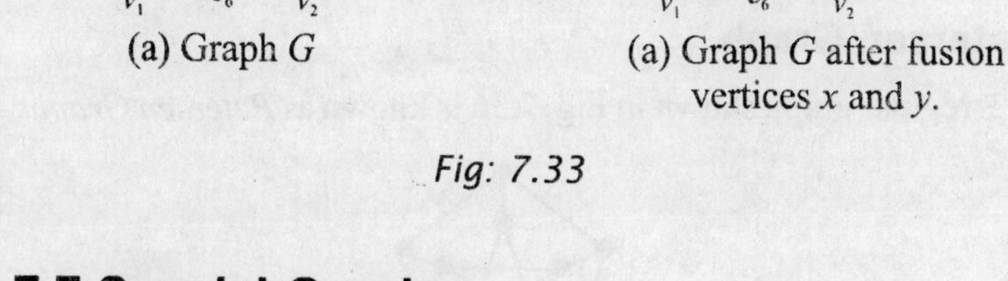

(a) Graph G (a) Graph G after fusion vertices x and y.

Fig: 7.33

§ 7.7 Special Graphs

In this section, we examine some important types of graphs which appear frequently in the theory of graphs.

Complete Graphs

A *complete graph* G is a simple graph in which each pair of vertices are adjacent. If a complete graph G has n vertices, then it will be denoted by K_n. It may be seen that K_n has $(n(n-1)/2$ edges and degree of each vertex is $(n-1)$. Graphs K_3, K_4, and K_5 are shown in Fig. 7.34

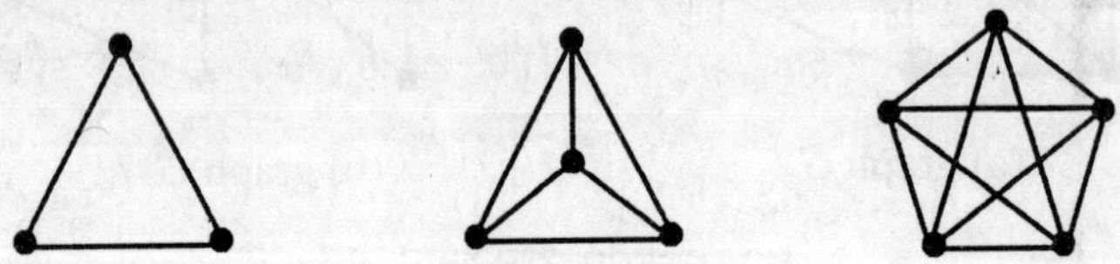

(a) Graph K_3 (b) Graph K_4 (c) Graph K_5

Fig: 7.34

Regular Graphs

A graph G is called *regular* if each vertex in G has the same degree. If a graph G is regular with $d(v) = r$ for each vertex v in G, then G is called r-regular. For example, the graphs in Fig. 7.35(a) and (b) are 3 - regular and 4-regular

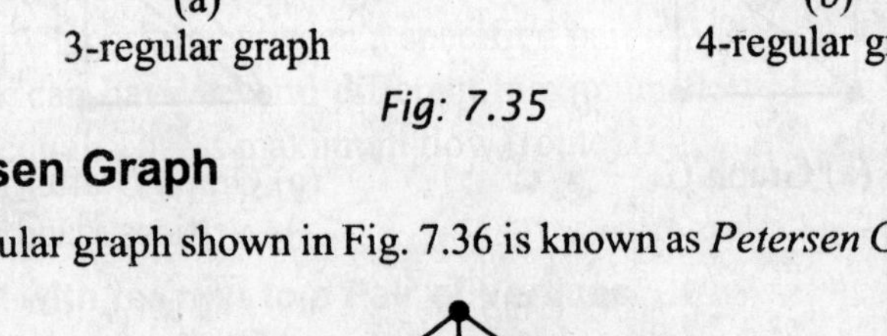

(a) 3-regular graph (b) 4-regular graph

Fig: 7.35

Petersen Graph

A 3-regular graph shown in Fig. 7.36 is known as *Petersen Graphs*.

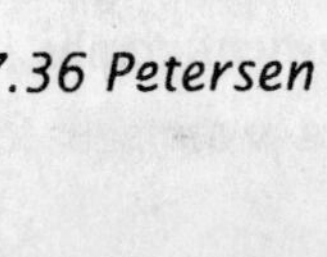

Fig: 7.36 Petersen Graph

Cycle Graphs, Path Graphs and Wheels

A connected 2-regular graph is called a *cycle* graph. We denote the cycle graph with n-vertices by C_n. Cycle graph is also called circuit graph.

The graph obtained from C_n (cycle graph) by removing an edge is called **path graph**. Observe that the number of vertices in path graph is same as the number of vertices in the cycle graph from which it is obtained. The path graph with n vertices is denoted by P_n.

The graph obtained from C_{n-1} (cycle graph with $n-1$ vertices) by joining each vertex to a new vertex v is called the *wheel* with n vertices. The wheel with n vertices is denoted by W_n.

For example, the graphs in Fig. 3.37 are cycle graph, path graph and wheel with 6 vertices.

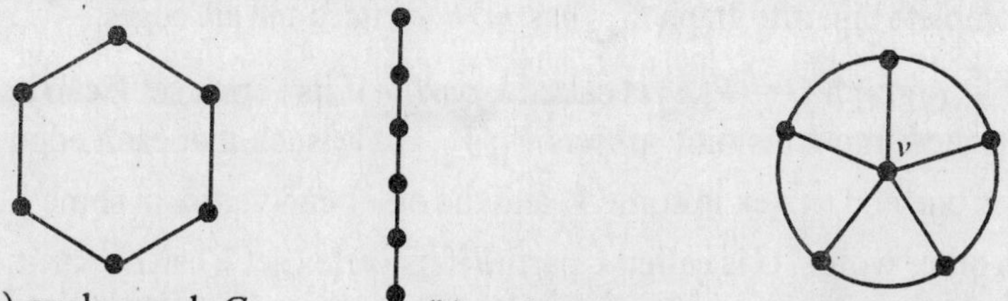

(a) cycle graph C_6 (b) path graph P_6 (c) wheel W_6

Fig: 7.37

Bipartite Graphs

A graph $G = (V, E)$ is called *bipartite* if its vertex set V can be decomposed into two disjoint subsets V_1 and V_2 such that every edge in G joins a vertex in V_1 with a vertex in V_2. In other words, a graph $G = (V, E)$ is bipartite if its vertex set V can be partitioned into two subsets V_1 and V_2 such that each edge in G has one end vertex in V_1 and another in V_2. For example the graph in Fig. 7.38 (a) is a bipartite graph.

In other words, $G = (V, E)$ is bipartite if V can be written as the union of two disjoint sets V_1 and V_2 so that no two members of V_1 or V_2 are adjacent.

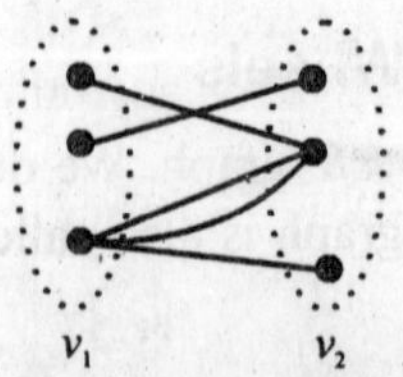

(a) Bipartite graph

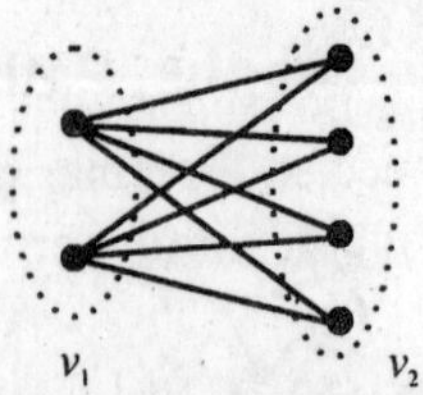

(b) The complete bipartite graph $K_{2,4}$

Fig: 7.38

Complete Bipartite Graph

Let $G = (V, E)$ be a bipartite graph and let V_1 and V_2 be the partition of the vertex set V of G. The bipartite graph G is called *complete bipartite* if each vertex in V_1 is joined to each vertex in V_2 by just one edge. This graph is denoted by $K_{m,n}$ if V_1 has m vertices and V_2 has n vertices. For example, the complete bipartite graph $K_{2,4}$ is shown in Fig. 7.38(b). Note that the complete bipartite graph $K_{m,n}$ has $m+n$ vertices and mn edges.

A graph $G = (V, E)$ is called *k-partite* if its vertex set V can be partitioned into k disjoint subsets $V_1, V_2, ..., V_k$ such that each edge of G has one end vertex in some V_i and the other end vertex in some $V_j \neq V_i$. In other words, G is called k-partite if its vertex set V can be partitioned into k disjoint subsets $V_1, V_2, ..., V_k$ such that no edge in G joins the vertices in the same subset. A complete k-partite graph G is a simple k-partite graph with vertex set partition $\{V_1, V_2, ..., V_k\}$ such that for every vertex x in V_i and every vertex y in V_j, $i \neq j$, there is an edge joining x and y in G.

Platonic Graphs

Consider the following five regular solids (also known as platonic solids, because of plato's interest in them). A platonic graph is a graph

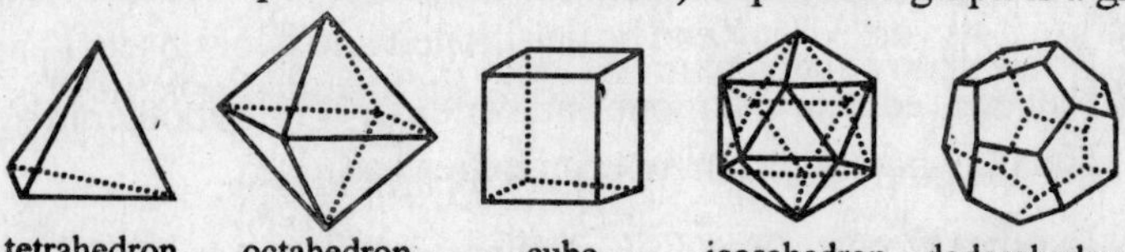

Fig: 7.39

formed from the vertices and edges of a platonic solid. In Fig. 7.40, we have five platonic graphs corresponding to above five platonic solids

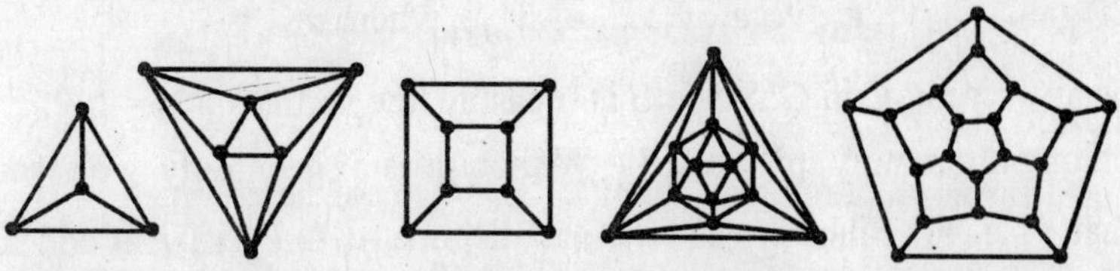

Fig: 7.40 Platonic Graphs

It is clear that in a platonic graph, all vertices have the same degree. Thus they are regular graphs. Moreover each face has equal number of edges in a platonic graph and number of edges in a platonic graph $= \frac{1}{2} d.V$ where d is the degree of any vertex and V is the number of vertices in the graph. It will be seen later on that these graphs are planar also.

Solved Examples

Example 7.11: Show that the maximum number of edges in a complete bipartite graph of n vertices is $\frac{n^2}{4}$

Solution: Let G be a complete bipartite graph with n vertices. Let n_1 and n_2 be the number of vertices in the partitions V_1 and V_2 of vertex set of G Since G is complete bipartite, each vertex in V_1 is joined to each vertex in V_2 by exactly one edge. Thus G has $n_1\ n_2$ edges where $n_1 + n_2 = n$.

But we know that maximum value of $n_1 n_2$ subject to $n_1 + n_2 = n$ is $\frac{n^2}{4}$ .Thus the maximum number of edges in G is $\frac{n^2}{4}$.

Example 7.12: Prove that a graph is bipartite iff all its circuits are of even length.

Solution: Let G be a bipartite graph and let V_1 and V_2 be the partition of the vertex set of G. Let

$$v_1, e_1, v_2, e_2, v_3, ..., e_n, v_{n+1}, \text{ where } v_{n+1} = v_1$$

be any circuit C in G. Since G is bipartite, the vertices v_i, $i = 1, 2, ...,n$, belong alternately to V_1 and V_2. Without loss of generality, we assume that $v_1 \in V_1$. Since V_1 and V_2 are disjoint, $v_i \in V_1$ iff i is odd and $v_j \in V_2$ iff j is even. That is, $v_1, v_3, v_5, ...$ belong to V_1 and $v_2, v_4, v_6, ...$ belong to V_2. Now since C is a circuit and $v_1 \in V_1$, therefore $V_{n+1} \in V_1$. Hence n+1 is odd. Thus n is even and so the length of circuit C is even.

Conversely, suppose every circuit of a graph G has even length. We arbitrarily select a vertex in each component of G and paint each of them with colour 1. We now sucsessively paint all the vertices of G as follows. If a vertex has colour 1 then any vertex adjacent to it will be painted with colour 2. Next, if a vertex has colour 2, then any vertex adjacent to it will be painted with colour 1. Continue this process till every vertex in G has been painted. Since every circuit in G is of even length, no adjacent vertices will have the same colour. Let V_1 be the set of all those vertices in G which have colour1 and V_2 be the set of all those vertices which have colour. 2. Then V_1 and V_2 are a partition of the vertex set of G such that no two vertices in V_1 or V_2 are adjacent. Hence G is bipartite.

Example 7.13 If the intersection of two paths in a graph is a disconnected graph, show that the union of the two paths has at least one circuit.

Solution: Let P_1 and P_2 be two paths in a graph G such that $P_1 \cap P_2$ is disconnected graph. We want to show $P_1 \cup P_2$ contains a circuit. Since $P_1 \cap P_2$ is disconnected, it has a pair of vertices u,v such that there is no path between u and v in $P_1 \cap P_2$. Since u, v are in $P_1 \cap P_2$, both u and v are in P_1 as well as in P_2. Since P_1 and P_2 are paths, there exists a path in P_1 as well as in P_2 connecting u and v. Since as there is no path in $P_1 \cap P_4$ connecting u and v, we may assume that these paths in P_1 and P_2 connecting u and v are edge-disjoint. Now union of these two paths connecting u and v forms a circuit in $P_1 \cup P_2$ as shown in the Fig. 7.41.

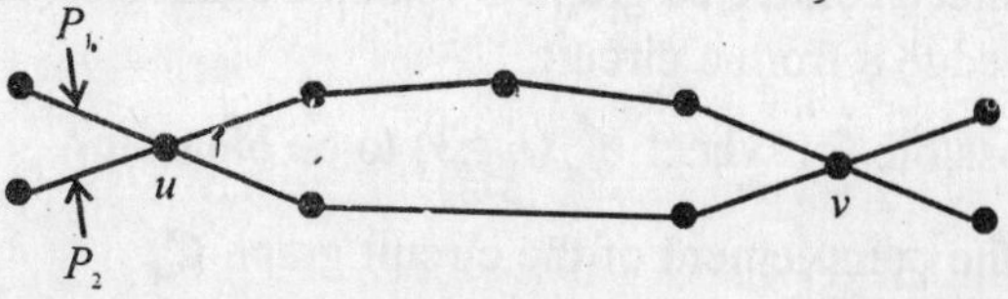

Fig: 7.41

Example 7.14: Is it possible for a bi-partite graph to contain a triangle as a subgraph

Solution: A graph G is said to contain a triangle if G has a subgraph isomorphic with K_3, the complete graph of three vertices. It is clear that triangle is a circuit to length 3. But we know from Example 7.12 that a bipartite graph cannot contain a circuit of odd length. Hence a bipartite graph cannot contain triangle as a subgraph.

Exercise

(1) Let G be a 3-regular graph with n vertices. What is the sum of the degree of the vertices? Show that in such a graph n must be even.

(2) For what values of n is the complete graph K_n a subgraph of K_m?

(3) Is it possible for a bipartite graph to contain a circuit C_n with n odd as subgraph? Give a reason for your answer. For what values of n, the circuit graph C_n is bipartite.

(4) Prove that a graph is bipartite if and only if it contains no circuit of odd length.

(5) Prove that an n-vertex simple graph is not bipartite if it has more than $\frac{n^2}{4}$ edges.

(6) Construct a simple n-regular graph with $2n$ $(n \geq 3)$ vertices having no triangles.

(7) Prove that the union of any two distinct paths (Paths having no edge in common) joining two vertices contains a circuit.

(8) Prove that a connected graph G remains connected after removing an edge e from a circuit.

(9) Is it possible for wheel W_n ($n \geq 3$) to be bipartite?

(10) Draw the complement of the circuit graph C_4.

(11) How many edges has each of the following graphs
(a) K_{10} (b) $K_{5,7}$ (c) W_8 (d) The Petersen graph (e) The octahedron.

(12) Give an example of each of the following:
(a) a bipartite platonic graph
(b) a complete graph that is a wheel
(c) a complete platonic graph.

(13) Which of the following are bipartite graphs?

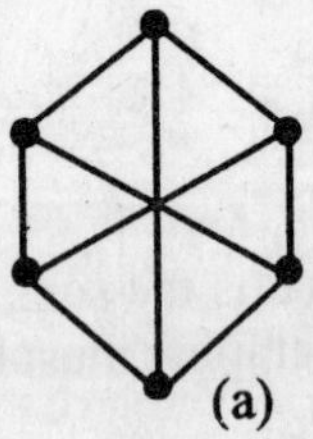

(a)

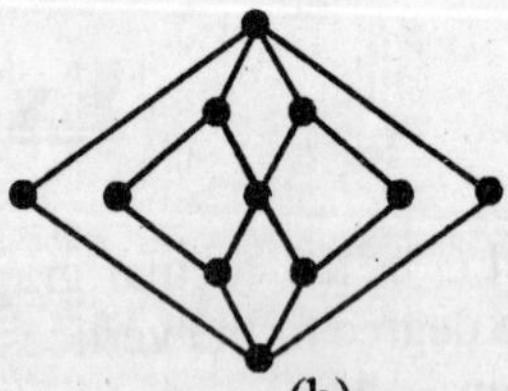

(b)

(Hint: Use Example 7.12)

Answers

(1) $3n/2$ (2) $n \leq m$ (3) No, C_n is bipartite iff n is even

(6) $K_{n,n}$ (9) No, because W_n ($n \geq 3$) contains triangles

(10)

(11) (a) 45 (b) 35 (c) 14 (d) 15 (e) 12

(12) (a) Cube (b) K_4 (c) Tetrahedron

(13) Both are bipartite.

§ 7.8 Eulerian Graphs

The first paper in graph theory was written by *L.* Euler's in 1736. In this paper, Euler gave a solution to what is now called the *Konigsberg Bridge* Problem. His method of proof laid the foundation of the theory of graphs. Konigsberg bridge problem may be stated as follows:
Two istands, *A* and *B*, lying in the Pregel River in Konigsberg (now Kaliningrad in Russia) were connected to each other and to the river banks *C* and *D* with seven bridges as shown in Figure 7.42.

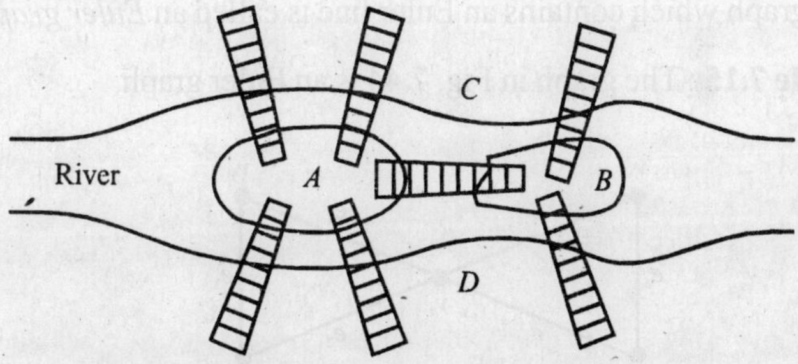

Fig: 7.42 Konigsberg bridge problem

The problem is to start at any of the four land areas — *A*, *B*, *C* or *D* ; walk over each of the seven bridges exactly once, and return to the starting point (without swimming, of course).

Euler represented this situation by means of a graph in which vertices represent the land areas and the edges represent the bridges connecting the land areas. The graph of konigsberg bridge problem is given below:

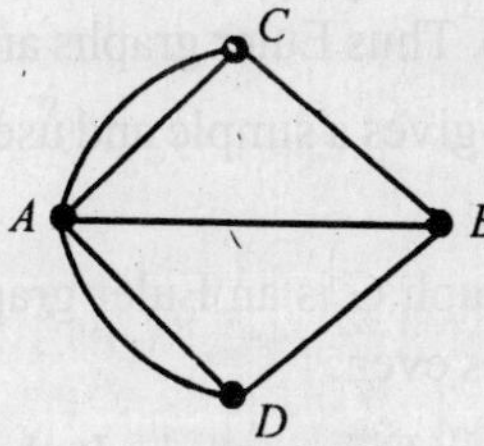

Fig: 7.43 Graph of Konigsberg bridge problem

It is clear that the problem of walking each of the seven bridges exactly once and returning to the starting point is equivalent to finding a closed walk in graph 7.43 that traverses each of the edges exactly once. Euler discovered a very simple criterion for determining whether such a closed walk exists in a graph.

Euler Graphs

Definition: A closed walk in a graph which contains all the edges of the graph is called on *Euler line*. Euler line is also sometimes called Euler circuit.

A graph which contains an Euler line is called an *Euler graph*.

Example 7.15 : The graph in Fig. 7.44 is an Euler graph.

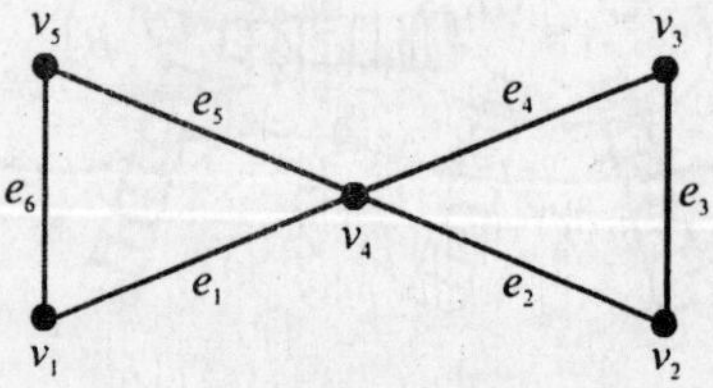

Fig: 7.44

Solution: The closed walk $v_1, e_1, v_4, e_2, v_2, e_3, v_3, e_4, v_4, e_5, v_5, e_6, v_1$ is an Euler line in this graph. Hence it is an Euler graph.

By definition, a walk is always connected and since Euler line is a closed walk which contains all the edges of the graph therefore an Euler graph is always connected except for any isolated vertices which the graph may have. Since isolated vertices do not contribute anything to the understanding of an Euler graph, we assume that Euler graphs do not have isolated vertices. Thus Euler graphs are connected.

The following theorem gives a simple and useful characterization of Euler graphs

Theorem 1: A connected graph G is an Euler graph if and only if the degree of every vertex in G is even.

Proof: Suppose that G is an Euler graph. It therefore contains an

Euler line starting from a vertex, say v_1, in G. Let this Euler line, traversing all the edges of G, be $v_1, e_1, v_2, e_2, v_3, e_3, ..., e_{n-1}, v_n, e_n, v_{n+1}$, where $v_{n+1} = v_1$.

In this Euler line, the vertices $v_1, v_2, ..., v_n$ may not all be distinct and some of these vertices may be repeated but all the edges are distinct. It is then clear that the pair of successive edges e_i and e_{i+1}, $1 \leq i \leq n$-1, contributes 2 to the degree of the vertex v_{i+1}. Thus, vertices $v_2, v_3, ..., v_n$ are of even degree. Besides this vertex v_1 gets a contribution of 2 to its degree from the initial and the final edges e_1 and e_n. Thus all the vertices are of even degree.

Conversely we assume that all vertices of the graph G are of even degree. Let v be any vertex in G. We now construct a walk starting at v and going through the edges of G such that no edge is traced more than once. Since every vertex is of even degree, we can exit from every vertex we enter. Since v is also of even degree, we shall eventually arrive at v when the tracing comes to an end. If this closed walk H,we just traced contains all the edges of G, then G is an Euler graph. If this closed walk does not contain all the edges of G then we remove from G all the edges in H and obtain a subgraph $H' = G - H$ of G formed by the remaining edges. Since both G and H have all their vertices of even degree, therefore, the degrees of vertices of H' are also even. Now since G is connected, H' must touch H at least at one vertex a. Starting from a, we again construct a new walk in graph H'. Since all the vertices of H' are of even degree, this walk in H' must terminate at vertex a. Now this walk in H' can be combined with H to form a new walk, which starts and ends at vertex v. This argument can be repeated until we obtain an Euler line which traverses all the edges of G. Hence G is an Euler graph.

With the help of this theorem, we can immediately tell which graph is an Euler graph and which graph is not an Euler graph.

Example 7.16: Which of the platonic graphs in an Euler graph?

Solution: Since all vertices in the graph of octahedron are of even degree, it is an Euler graph by Theorem 1 above, It may be seen easily that no other platonic graphs is Euler graph because all these graphs contain vertices of odd degree.

Solution of Konigsberg Bridges Problem

Consider the graph of Konigsberg bridge problem given in Fig. 7.43. In this graph, we find that not all vertices are of even degree. Hence it is not an Euler graph. Thus it is not possible to walk over each of the seven bridges exactly once and return to the starting point.

Theorem 2: A connected graph G is an Euler graph if and only if G is the union of some edges-disjoint circuits.

Proof: Suppose that the graph G is a union of some edge-disjoint circuits. Since the degree of each vertex in a circuit is two, the degree of every vertex in G is even. By Theorem 1, G is an Euler graph.

Conversely, let G be an Euler graph. Then by Theorem 1, every vertex in G has even degree. Let v be any vertex of G. Then $d(v) \geq 2$. Thus there are at least two edges incident at v. Let v_1 be any vertex adjacent to v. Since degree of the vertex v_1 is also even, there is a vertex v_2 which is adjacent to v_1 and is different from v. Continuing in this way, we finally arrive at a vertex which has previously been traversed. If v_i is the first such vertex (that is, a vertex which has been previously traversed), then the part of the walk which begins and ends at v_i is a circuit C. We remove C from G and consider the graph $G_1 = G - C$. Since every vertex in C is of degree 2, it follows that every vertex in G_1 must have even degree (the graph G_1 may be disconnected). From G_1, remove another circuit in eaxctly the same way as we removed C from G. Continue this process until no edges are left. Thus G is the union of circuits C, C_1, ..., no two of which have common edges. This completes the proof.

Corollary: Every vertex of a connected Euler graph is contained in some circuit.

Unicursal Graph

In defining an Euler graph, some authors drop the requirement that the Euler line be closed. For Example, the walk $v_1, e_7, v_5, e_6, v_4, e_5, v_1, e_1, v_2, e_4, v_4, e_3, v_3, e_2, v_2$ in the graph of Fig. 7.45, which includes all the edges of the graph and does not retrace any edge, is not closed. We shall call such an open walk that traces (or covers) all edges of a graph without retracing any edge **a unicursal line** or an **open Euler line**. A graph that has a unicursal line will be called **unicursal graph.**

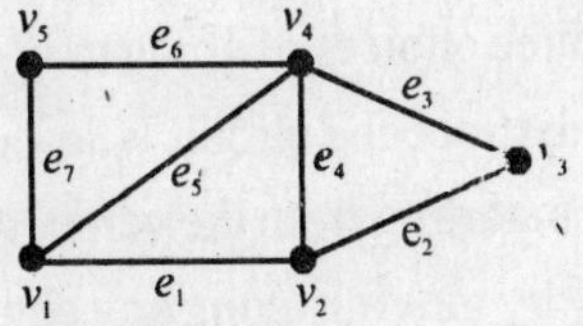

Fig: 7.45

It may be noted that by adding an edge between the initial and final vertices of an open Euler line (*i.e*, unicursal line) we shall get an Euler line. Thus a connected graph is unicursal iff it has exactly two vertices of odd degree.

Theorem 3: Let $G = (V, E)$ be a connected graph with $2k$ odd vertices ($k \geq 1$). Then there exist k edge-disjoint subgraphs such that they together contain all edges of G and that each is a unicursal graph

Proof: Let G be a connected graph with $2k$ odd vertices. Let the odd vertices of G be named $v_1, v_2, v_3, \ldots v_k$; $w_1, w_2, \ldots w_k$. Add k edges in the graph G between the vertex pairs $(v_1, w_1), (v_2, w_2), \ldots, (v_k, w_k)$ to form a new graph G^*. Observe that in G^*, every vertex is of even degree and hence G^* is Euler graph. Let ρ be any Euler line in G^*. Now if we remove from ρ the k edges we added in G to get G^* then ρ will be split into k walks, each of which is a unicursal line and each edge of G is present in exactly one of these unicursal lines. This completes the proof.

Randomly Eulerian Graphs

Consider the graph in Fig. 7.46, which is an Euler graph.

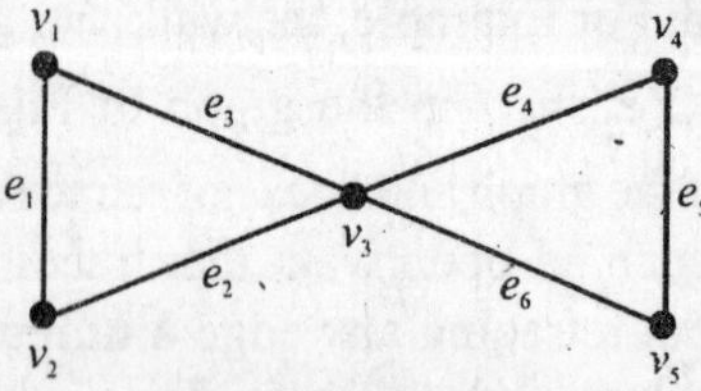

Fig: 7.46

Suppose that we start from vertex v_1 and trace the path v_1, e_1, v_2, e_2, v_3. Now at v_3 we have three choices of going to v_1, v_4 or v_5. If we take the first choice, we will only trace the circuit $v_1, e_1, v_2, e_2, v_3, e_3, v_1$, which is not an Euler line. Thus starting from the vertex v_1, we cannot cover the entire Euler line simply by moving along any edge that has not already been traversed.

We say a graph G is *randomly Eulerion from a vertex* v if whenever we start from vertex v and traverse along the edges of G in an arbitrary way never using any edge twice, we eventually obtain an Euler line.

For example, the graph in Fig. 7.46 is randomly Eulerian from vertex v_3 but not from any other vertex. The graph in Fig 7.47 (a) is not randomly Eulerian from any vertex although it is Euler graph. The graph in Fig 7.47(b) is randomly Eulerian from all its vertices.

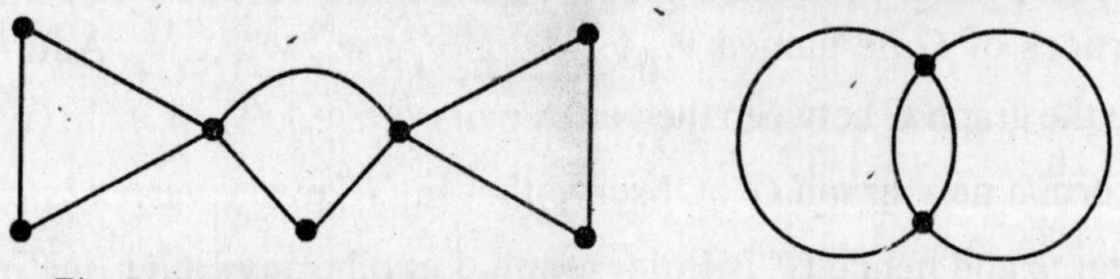

Fig. 7.43(a) Euler graph which is not randomly Eulerian from any vertex

Fig. 7.43(b) Randomly Eulerian from all vertices

Fig: 7.47

The following theorem gives a characterization of a graph which is randomly Eulerian from a vertex v.

Theorem 4: An Euler graph G is randomly Eulerian from a vertex v if and only if every circuit in G contains v.

Proof: We first suppose that G is randomly Eulerian from a vertex v. If possible, we assume that there is a circuit C in G which does not contain v. Consider the subgraph $H = G - C$. Every vertex in H (not necessarily connected) is of even degree. Consider the component of H which contains v. This component is an Euler graph and it contains all the edges incident on v. Thus in this component, there exists an Euler line T starting and ending at vertex v. This Euler line contains all the edges incident on v. Therefore, it cannot be extended to include the edges of C. This contradicts the fact that G is randomly Eulerian from v.

Conversely, suppose vertex v in an Euler graph G is present in every circuit of G. If possible, suppose that G is not randomly Eulerian from v. Then there exists a closed walk T, starting and ending at v, containing all the edges incident on v, but not containing all the edges of G. Moreover, since G is connected, there exists a vertex u different from v such that it is the end vertex of an edge not in T.

Let H be the subgraph of G obtained by removing edges of T from G. Then v is an isolated vertex in H. In H, every vertex is of even degree. So the component of H which contains u is an Euler graph. By corollary of Theorem 2, there is a circuit containing u. This circuit obviously does not contain vertex v. This contradicts the fact that v is in every circuit of G. Thus G must be randomly Eulerian from v. This completes the proof.

We next give a method for constructing an Euler line in a given Euler graph. This method is known as **Fleury's Algorithm**. We require an additional definition before stating the algorithm. An edge in a connected graph G is said to be a *bridge* if the removal of e from G leaves the graph G disconnected. For example, the edge joining the vertices v_4 and v_5 in the graph of Fig. 7.48 is a bridge.

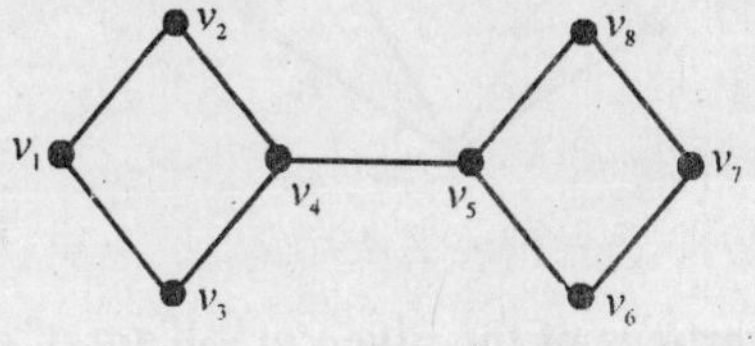

Fig: 7.48

Fleury's Algorithm: Let $G = (V, E)$ be an Euler graph. Then the following algorithm produces an Euler line in G.

Step 1: Choose any vertex v_0 in G as the starting vertex for the Euler line in G and set $T_0 = v_0$

Step 2: Having constructed the walk

$$T_i : v_0, e_1, v_1, e_2, v_2, ..., v_{i-1}, e_i, v_i$$

delete the edge e_i from G and form the graph

$$G_i = G_{i-1} - e_i, i \geq 1 \text{ with } G_0 = G. \text{ Thus}$$

$$G_i = G - \{e_1, e_2, ... e_i\}$$

Step 3: (i) If there is no edge incident at v_i in graph G_i then stop. T_i is the required Euler line

(ii) If there is only one edge incident at v_i in G_i then choose this edge as e_{i+1}. If there are several edges incident on v_i then select any one which is not a bridge on the graph G_i and call it e_{i+1}. Now go to Step 2 with the walk

$$T_{i+1} : v_0, e_1, v_1, e_2, v_2, ... v_{i-1}, e_i, v_i, e_{i+1}, v_{i+1}$$

Example 7.17: Use Fleury's algorithm to construct an Euler line for the graph in Fig. 7.49

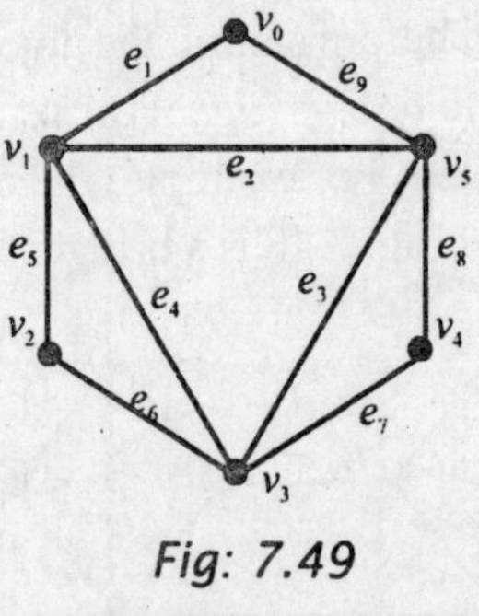

Fig: 7.49

Solution: Since all vertices of the given graph are of even degree, an Euler line exists. According to step 1, we may start from any vertex.

We select vertex v_0. We summarize the results of applying steps 2 and 3 repeatedly in a Table:

Current walk	G_i	Next edge	Reasoning
$T_0 : v_0$	$G_0 = G$	e_1	No edge from v_0 is a bridge. Choose any one.
$T_1 : v_0, e_1, v_1$	$G_1 = G_0 - e_1$	e_2	No edge from v_1 is a bridge. Choose any one
$T_2 : v_0, e_1, v_1, e_2, v_5,$	$G_2 = G_1 - e_2$	e_3	e_9 is a bridge on G_2 at v_5. Choose anyone from e_3, e_8.
$T_3 : v_0, e_1, v_1, e_2, v_5, e_3, v_3$	$G_3 = G_2 - e_3$	e_4	e_7 is a bridge on G_3 at v_3. Choose anyone from e_4 and e_6
$T_4 : v_0, e_1, v_1, e_2, v_5, e_3, v_3, e_4, v_1$	$G_4 = G_3 - e_4$	e_5	only one edge from v_1 remains in G_4.
$T_5 : v_0, e_1, v_1, e_2, v_5, e_3, v_3, e_4, v_1, e_5, v_{2150}$	$G_5 = G_4 - e_5$	e_6	only one edge from v_2 remains in G_5.
$T_6 : v_0, e_1, v_1, e_2, v_5, e_3, v_3, e_4, v_1, e_5, v_2, e_6, v_3$	$G_6 = G_5 - e_6$	e_7	only one edge from v_3 remains in G_6.
$T_7 : v_0, e_1, v_1, e_2, v_5, e_3, v_3, e_4, v_1, e_5, v_2, e_6, v_3, e_7, v_4$	$G_7 = G_6 - e_7$	e_8	only one edge from v_4 remains in G_7.
$T_8 : v_0, e_1, v_1, e_2, v_5, e_3, v_3, e_4, v_1, e_5, v_2, e_6, v_3, e_7, v_4, e_8, v_5$	$G_8 = G_7 - e_8$	e_9	only one edge from v_5 remains in G_8.
$T_9 : v_0, e_1, v_1, e_2, v_5, e_3, v_3, e_4, v_1, e_5, v_2, e_6, v_3, e_7, v_4, e_8, v_5, e_9, v_0$	$G_9 = G_8 - e_9$		

Thus walk T_9 is an Euler line in the graph G. There are many other Euler lines in this graph.

Note: The students are advised to draw graphs $G_1, G_2, \ldots G_9$ for understanding the algorithm thoroughly.

Solved Examples

Example 7.18 : which of the following graphs are Eulerian?

(i) the complete graph K_5

(ii) the complete bipartite graph $K_{1,3}$

(iii) the graph of the octahedron

(iv) the Petersen graph

Solution:

(i) We know that the degree of each vertex in the complete graph K_5 is 5–1 = 4 which is even. Hence degree of each vertex in K_5 is even. Since K_5 is also a connected graph, therefore by Theorem 1, K_5 is an Euler graph.

(ii) The graph $K_{2,3}$ is given below:

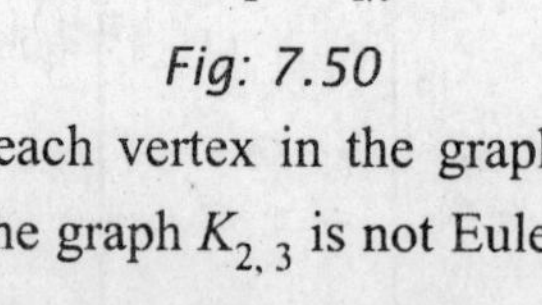

Graph $K_{2,3}$

Fig: 7.50

It is clear that each vertex in the graph $K_{2,3}$ is not of even degree. Hence the graph $K_{2,3}$ is not Euler.

(iii) We know from Fig. 7.40 that each vertex in the graph of the octahedron is of degree 4 which is even. Hence the graph of the octahedron is an Euler graph.

(iv) From Fig. 7.36, it is clear that the Petersen graph is not an Euler graph because each vertex of this graph is of degree 3 which is not even.

Example 7.19 : For what values of n is the graph K_n Eulerian?

Solution: We know that K_n, the complete graph of n vertices is a connected graph in which degree of each vertex is n-1. Since a graph is Eulerian (*i.e.* an Euler graph) if and only if it is connected and degree of each vertex is even, we conclude that K_n is an Euler graph if and only if n is odd.

Example 7.20 : Which complete bipartite graphs $K_{m,n}$ are Euler graphs?

Solution : It is clear from the definition of $K_{m,n}$ graph that it is an Euler graph if and only if both m and n are even.

Example 7.21 : Use Fleury's algorithm to produce an Euler line for the graph in Fig. 7.51.

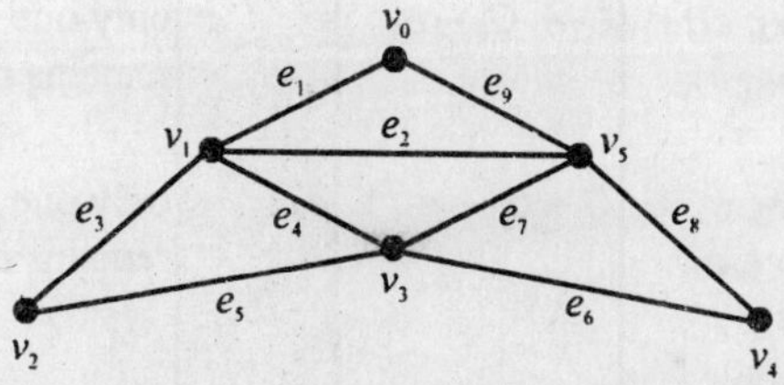

Fig: 7.51

Solution: Since degree of each vertex in the graph of Fig. 7.51 is of even degree, it has an Euler line, According to step 1 of Fleury's algorithm, we can begin from any vertex. Suppose we choose the vertex v_0 as the starting vertex. We summarize the results of applying steps 2 and 3 repeatedly in the following Table:

Current walk	G_i	Next edge	Reasoning
$T_0 : v_0$	$G_0 = G$	e_1	No edge from v_0 is a bridge. Choose any one.
$T_1 : v_0, e_1, v_1$	$G_1 = G_0 - e_1$	e_3	No edge from v_1 is a bridge in G_1. Choose any one

Current walk	G_i	Next edge	Reasoning
$T_2 : v_0, e_1, v_1, e_3, v_2,$	$G_2 = G_1 - e_3$	e_5	Only one edge from v_2 remains in G_2
$T_3 : v_0, e_1, v_1, e_3, v_2, e_5, v_3$	$G_3 = G_2 - e_5$	e_6	No edge from v_3 is a bridge in G_3. Choose any one.
$T_4 : v_0, e_1, v_1, e_3, v_2, e_5, v_3, e_6, v_4$	$G_4 = G_3 - e_6$	e_8	only one edge from v_4 remains in G_4.
$T_5 : v_0, e_1, v_1, e_3, v_2, e_5, v_3, e_6, v_4, e_8, v_5$	$G_5 = G_4 - e_8$	e_2	Edge e_9 is a bridge in G_5 at v_5. Choose any one from e_2 and e_7.
$T_6 : v_0, e_1, v_1, e_3, v_2, e_5, v_3, e_6, v_4, e_8, v_5, e_2, v_1$	$G_6 = G_5 - e_2$	e_4	only one edge from v_1 remains in G_6.
$T_7 : v_0, e_1, v_1, e_3, v_2, e_5, v_3, e_6, v_4, e_8, v_5, e_2, v_1, e_4, v_3$	$G_7 = G_6 - e_4$	e_7	only one edge from v_3 remains in G_7.
$T_8 : v_0, e_1, v_1, e_3, v_2, e_5, v_3, e_6, v_4, e_8, v_5, e_2, v_1, e_4, v_3, e_7, v_5$	$G = G_7 - e_7$	e_9	only one edge from v_5 remains in G_8.
$T_9 : v_0, e_1, v_1, e_3, v_2, e_5, v_3, e_6, v_4, e_8, v_5, e_2, v_1, e_4, v_3, e_7, v_5, e_9, v_0$			

Thus walk T_9 is an Euler line in this graph. Since at several places, there were more than one choices for the edges which could be deleted from the graph, therefore, there are many other Euler lines in this graph.

Exercise

(1) Which platonic graphs are Eulerian?

(2) For what values of n is the wheel W_n an Euler graph?

(3) Give an example of an Euler graph that is not randomly Eulerian.

(4) Use Fleury's algorithm to construct an Euler circuit for the graph in Fig. 7.52.

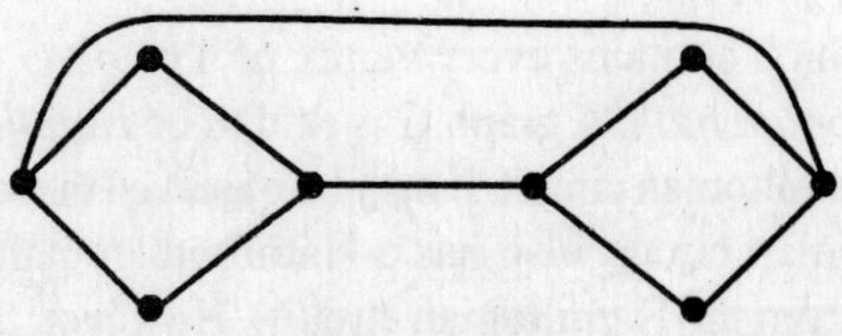

Fig: 7.52

(5) The *line graph* $L(G)$ of a simple graph G is the graph whose vertices are in one-one correspondence with the edges of G and two vertices of $L(G)$ are adjacent if and only if the corresponding edges of G are adjacent. Show that the line graph of a simple Euler graph is an Euler graph. (Hint: degree of a vertex in $L(G)$ is equal to the sum of the degrees of the end vertices of the corresponding edge in G minus two. Since G is Euler, each vertex is of even degree. Hence degree of each vertex in $L(G)$ is also even).

Answers

(1) The graph of the octahedron

(2) $n = 3$

(3) Graph of Fig 7.47(a)

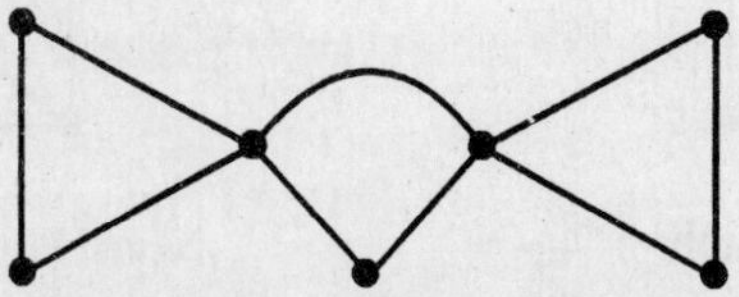

Fig. 7.47(a) Euler graph which is not randomly Eulerian from any vertex

§ 7.9 Hamiltonian Graphs

A problem similar to the problem of finding an Euler line in a graph is finding a Hamiltonian circuit in a graph.

Definition : A *Hamiltonian circuit* in a graph G is a closed walk that traverses every vertex of G exactly once except the starting vertex.

A path which contains every vertex of a graph G exactly once is called *Hamiltonian path*. A graph G is said to be *Hamiltonian graph* if it contains a Hamiltonian circuit. It may be observed that every graph that has a Hamiltonian circuit also has a Hamiltonian path (by removing any one edge from the Hamiltonian circuit). However, there are graphs with Hamiltonian paths that have no Hamiltonian circuits.

It may be noted that an Euler line visits each edge of the graph exactly once whereas a Hamiltonian circuit visits each vertex of the graph exactly once. Unlike the situation for Euler graphs, no characterizations are known for the existence of a Hamiltonian circuit in a graph.

In considering the existence of a Hamiltonian circuit, we shall consider only simple graphs. This is because a Hamiltonian circuit or a Hamiltonian path traverses every vertex exactly once. Hence it cannot contain a self-loop or a set of parallel edges. Thus a general graph must be made simple by removing parallel edges and self-loops before looking for a Hamiltonian circuit in it.

It may be seen that every complete graph of three or more vertices is Hamiltonian. Graphs in Figure 7.53(a) and 7.53(b) are Hamiltonian nd non-Hamiltonian respectively.

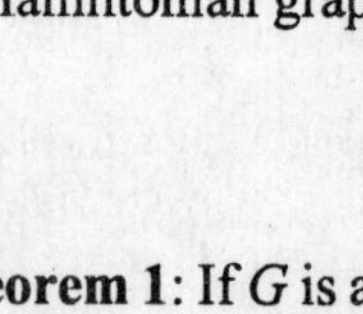

(a)
Hamiltonian graph

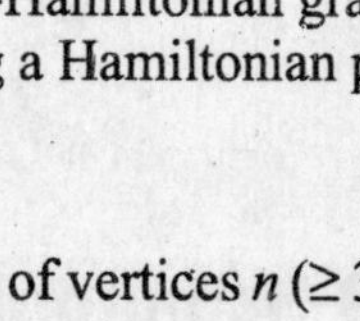

(b)
Non-Hamiltonian graph having a Hamiltonian path.

Fig: 7.53

Theorem 1: If G is a simple graph with number of vertices n (≥ 3) and if

$$\deg(v) + \deg(w) \geqslant n \qquad \text{...} (1)$$

for every pair of non-adjacent vertices v and w, then G is Hamiltonian.

Proof : We shall prove the theorem by contradiction. We assume that there exists a non-Hamiltonian graph with n vertices satisfying the given condition (1) for every pair of non-adjacent vertices v and w. Among all such non-Hamiltonian graphs, let G be a non-Hamiltonian graph with maximum number of edges. Since G is maximal non-Hamiltonian, it follows that there exist two non-adjacent vertices v and w in G such that addition of an edge joining v and w will result in a Hamiltonian graph. Thus in G, there is a Hamiltonian path

$v = v_1, v_2, v_3 ..., v_n = w$ with v and w as the end vertices as shown in figure 7.54.

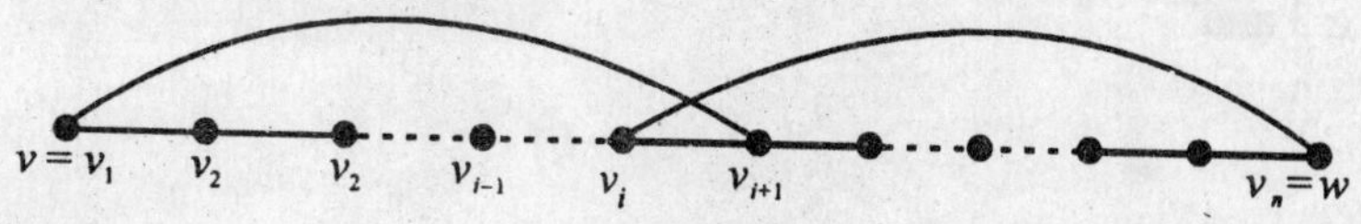

Fig: 7.54

Define

$$S = \{i : v_i \text{ is adjacent to vertex } v \text{ in } G\}$$

$$T = \{ i : v_i \text{ is adjacent to vertex } w \text{ in } G\}$$

Clearly, $|S| = \deg(v)$ and $|T| = \deg(w)$ in G where $|X|$ denotes the number of elements in a set X. We assert that $S \cap T = \phi$ and $|S \cup T| \leq n-1$. For if, $i \in S \cap T$, then the edges (v, v_i) and (v_i, w) would be in G and then

$v = v_1, v_2, ... v_{i-1}, v_i, v_n, v_{n-1}, ... , v_{i+1}, v_1$ would form a Hamiltonian circuit in G, which is a contradiction. Further, $S \cup T \subseteq \{1, 2 ..., n\}$. But since vertex $v_1 = v$ is neither adjacent to v nor adjacent to w, $1 \notin S \cup T$. Therefore

$$|S \cup T| \leq n-1$$

Now, we have

$$\deg(v) + \deg(w) = |S| + |T|$$

$$= |S \cup T| \qquad \because S \cap T = \phi$$

$$\leq n-1$$

But this is a contradiction to our hypothesis. Hence our assumption is wrong. So, G is Hamiltonian.

Corollary 1: If G is a simple graph with n (≥ 3) vertices and if $\deg(v) \geq n/2$ for each vertex v, then G is Hamiltonian.

Proof: The result follows immediately from Theorem 1, since $\deg(v) + \deg(w) \geq n$ for every pair of vertices v and w, whether v and w are adjacent or not.

Corollary 2 : If G is a simple graph with n vertices and e edges with $n \geq 3$ and

$$e \geq \frac{1}{2}\left(n^2 - 3n + 6\right)$$

then G is Hamiltonian

Proof: If possible, let G be non-Hamiltonian. Then by Theorem 1, there exists a pair of non-adjacent vertices v and w such that $\deg(v) + \deg(w) \leq n-1$. Let H be the subgraph of G obtained by deleting vertices v and w from G. Thus the graph H has $n-2$ vertices and $e - [\deg(v) + \deg(w)]$ edges. Since H has $n-2$ vertices, maximum number of edges in H is ${}^{n-2}C_2$. Hence

$$e - \left[\deg(v) + \deg(w)\right] \leq {}^{n-2}C_2 = \frac{(n-2)(n-3)}{2} = \frac{n^2 - 5n + 6}{2}$$

Thus, we have,

$$e \leq \frac{n^2 - 5n + 6}{2} + (n-1) = \frac{n^2 - 3n + 4}{2}$$

which is less than $\frac{1}{2}$ (n^2-3n+6), a contradiction. This completes the proof.

Number of Hamiltonian Circuits in a Complete Graph

In a complete graph with n vertices, there exist $\frac{(n-1)!}{2}$ Hamiltonian circuits. We are interested in finding all the edge-disjoint Hamiltonian circuits in complete graphs. The determination of the exact number of edge-disjoint Hamiltonian circuits in a graph in general is an unsolved problem. However, the following theorem gives the number of edge-disjoint Hamiltonian circuits in a complete graph with odd number of vertices.

Theorem 2 : In a complete graph with n vertices there are $(n-1)/2$ edge-disjoint Hamiltonian circuits, if n is odd number ≥ 3.

Proof : Let G be a complete graph of n vertices. Then G has $n(n-1)/2$ edges. Since a Hamiltonian circuit in G contains n edges, therefore the number of edge-disjoint Hamiltonian circuits in G cannot exceed $(n-1)/2$. We shall show that there are $(n-1)/2$ edge-disjoint Hamiltonian circuits when n is odd. The subgraph of a complete graph of n vertices as shown in Fig. 7.55 is a Hamiltonian circuit.

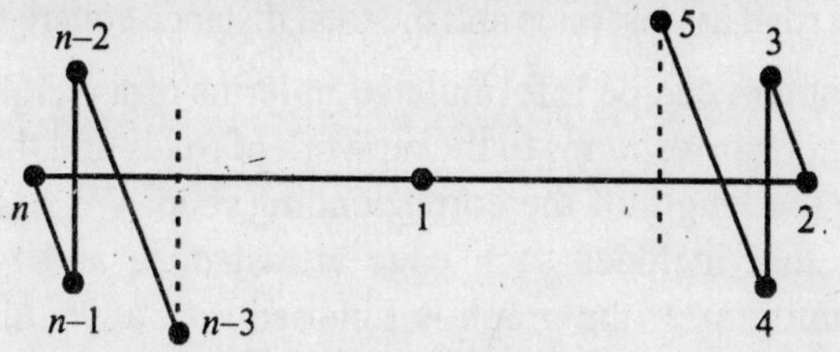

Fig: 7.55

If we keep the vertices fixed on a circle and rotate the polygonal pattern anti-clockwise by

$$\frac{360}{(n-1)}, \quad 2.\frac{(360)}{(n-1)}, \quad 3.\frac{(360)}{(n-1)}, \ldots, \ldots, \frac{(n-3)}{2}.\frac{(360)}{(n-1)}$$

degrees, we find that each rotation results in producing a Hamiltonian circuit that has no edge in common with any of the previous ones. Thus we have obtained $(n-3)/3$ new Hamiltonian circuits, all edge-disjoint from the one in Fig. 7.55 and also edge-disjoint among

Hence we have $\frac{n-3}{2}+1 = \frac{n-1}{2}$ edge disjoint Hamiltonian circuits.

7.10 Weighted Graphs

There are some practical problems associated with Euler graphs and Hamiltonian graphs. To describe these we need the following concepts

Definition : A *weighted graph* is a graph in which a non-negative real number is assigned to each edge. The number assigned to each edge e in called weight of e and is denoted by $w(e)$.

For example, consider a graph that represents the highway connection among cities. If we assign a number to each edge to indicate the distance between the two cities connected by the edge then it represents a weighted graph.

The Chinese Postman Problem

The Chinese Postman problem is as follows :

A postman starts from his post-office in a town to deliver his letters and returns to his starting point every day. What route should he take so that he visits each road at least once and the total distance covered is least?

This problem can be reformulated in terms of a weighted graph, where the graph corresponds to the network of roads and the weight of each edge is the length of the corresponding road. We wish to find a closed walk that includes each edge atleast once and whose total weight is minimum. If the graph is Eulerian then any Euler line is a closed walk of least total weight since each road is visited exactly once. We can find such an Eulerian line by Fleury's algorithm.

If the graph is not Eulerian then the problem is difficult, although there do exist efficient algorithms to find a closed walk of least weight. To illustrate the ideas involved, we take a particular graph in which exactly two vertices have odd degree (Fig. 7.56)

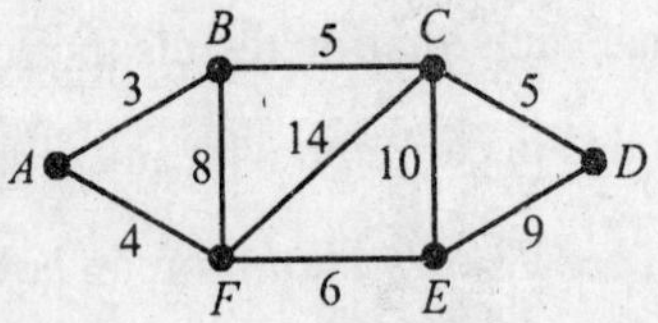

Fig: 7.56

Since this graph has only two vertices of odd degree (vertices B and E), we can find an open Euler line from B to E passing through each edge exactly once. In order to return to the starting point covering least possible distance, we now find the shortest path (path having least weight) from E to B. Among the several paths from E to B, the path $E \to F \to A \to B$ is shortest. The solution of the Chinese postman problem is obtained by taking this shortest path from E to B together with the open Euler line from B to E. Observe that if we combine the shortest path and the open Euler line, we get an Euler graph as shown in Figure 7.57.

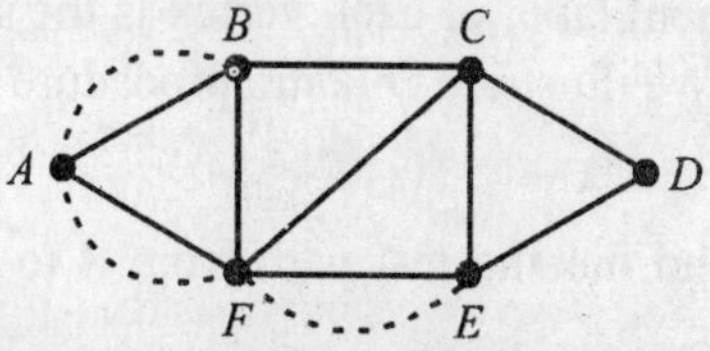

Fig: 7.57

Shortest Path in Weighted Graphs

Let G be a simple weighted graph. The length of an edge from vertex i to a vertex j is denoted by d_{ij}. If there is no edge from vertex i to vertex j then $d_{ij} = \infty$. The shortest path problem is to find the shortest possible path from a specified vertex A to another specified vertex L. There are several well-known procedures to solve this problem. Here we shall discuss an algorithm due to Dijkstra.

Dijkstra Algorithm : This algorithm labels the vertices of the given graph. The algorithm starts by assigning a permanent label O to the starting vertex A and temporary label ∞ to the remaining $n-1$ vertices. At each iteration in the algorithm, another vertex gets a permanent label according to the following rules:

[1] Every vertex j which is not yet permanently labelled gets a new temporary label whose value is given by

$$\min [\text{old label of } j, \text{ old label of } i + d_{ij}],$$

where i is the latest vertex permanently labelled, in the previous iteration and d_{ij} is the length of the edge between vertices i and j. If i and j are not joined by edge then $d_{ij} = \infty$.

[2] The smallest value among all the temporary labels is found and this becomes the permanent label of the corresponding vertex. In case of a tie, select any one of the vertices for permanent labelling.

Steps 1 and 2 stated above are repeated alternately until the destination vertex L gets a permanent label.

The first vertex to be permanently labelled is the starting vertex A. The second vertex to get a permanent label is the vertex nearest to A. The next vertex to be permanently labelled is the next nearest vertex to A. Thus the permanent label of each vertex is the shortest distance of that vertex from A. We illustrate Dijkstra procedure with the help of the following example :

Example 7.22 : Find the shortest path from A to L in the following weighted graph:

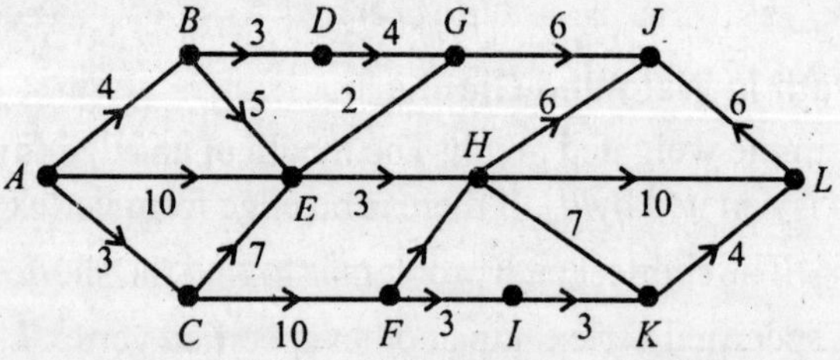

Fig: 7.58

Solution : We shall use an array of length eleven (no. of vertices) to show the temporary and permanent labels of the vertices as we go through the solution. The permanent labels will be shown enclosed in a square and the latest vertex assigned permanent label in the array is indicated by a mark □*. The labeling proceeds as follows :

i	A	B	C	D	E	F	G	H	I	J	K	L
0	[0]*	∞	∞	∞	∞	∞	∞	∞	∞	∞	∞	∞
1.	[0]	4	[3]*	∞	10	∞	∞	∞	∞	∞	∞	∞
2.	[0]	[4]*	[3]	∞	10	13	∞	∞	∞	∞	∞	∞
3.	[0]	[4]	[3]	[7]*	9	13	[illegible]	∞	∞	∞	∞	∞
4.	[0]	[4]	[3]	[7]	[9]*	13	11	∞	∞	∞	∞	∞

i	A	B	C	D	E	F	G	H	I	J	K	L
5.	$\boxed{0}$	$\boxed{4}$	$\boxed{3}$	$\boxed{7}$	$\boxed{9}$	13	$\boxed{11}^*$	12	∞	∞	∞	∞
6.	$\boxed{0}$	$\boxed{4}$	$\boxed{3}$	$\boxed{7}$	$\boxed{9}$	13	$\boxed{11}$	$\boxed{12}^*$	∞	17	∞	∞
7.	$\boxed{0}$	$\boxed{4}$	$\boxed{3}$	$\boxed{7}$	$\boxed{9}$	$\boxed{13}^*$	$\boxed{11}$	$\boxed{12}$	∞	17	19	22
8	$\boxed{0}$	$\boxed{4}$	$\boxed{3}$	$\boxed{7}$	$\boxed{9}$	$\boxed{13}$	$\boxed{11}$	$\boxed{12}$	$\boxed{16}^*$	17	19	22
9.	$\boxed{0}$	$\boxed{4}$	$\boxed{3}$	$\boxed{7}$	$\boxed{9}$	$\boxed{13}$	$\boxed{11}$	$\boxed{12}$	$\boxed{16}$	$\boxed{17}^*$	19	22
10.	$\boxed{0}$	$\boxed{4}$	$\boxed{3}$	$\boxed{7}$	$\boxed{9}$	$\boxed{13}$	$\boxed{11}$	$\boxed{12}$	$\boxed{16}$	$\boxed{17}$	$\boxed{19}^*$	22
11.	$\boxed{0}$	$\boxed{4}$	$\boxed{3}$	$\boxed{7}$	$\boxed{9}$	$\boxed{13}$	$\boxed{11}$	$\boxed{12}$	$\boxed{16}$	$\boxed{17}$	$\boxed{19}$	$\boxed{22}^*$

Thus the shortest distance from A to L is 22. Note that this method gives only the shortest distance. The shortest path can be easily obtained by going backword from the terminal vertex such that we go to that predecessor (vertex) whose label differs exactly by the length of the connecting edge. A tie indicates more than one shortest path. We can also determine the shortest path by keeping a record of the vertices from which each vertex was labelled permanently. This record can be stored in another array of length n, such that whenever a new permanent label is assigned to vertex j, the vertex from which j is directly reached is recorded in the jth position of this array. In the above example, the shortest path is $A \to B \to E \to H \to L$.

The Travelling Salesman Problem

The travelling salesman problem is as follows:

A salesman is required to visit a number of cities. What is the route he should take if he has to start at his home city, visit each city exactly once and then return home travelling the minimum distance?

Suppose we represent cities by the vertices in a graph and roads by edges connecting the cities. The length of the road may be represented as a weight associated with the corresponding edge. In graph theoretic terminology, the travelling salesman problem is equivalent to finding a shortest Hamiltonian circuit.

In a complete weighted graph of n vertices there exist $\dfrac{(n-1)!}{2}$

Hamiltonian circuits. One possible approach to solve the travelling salesman problem is to calculate the weights of all the $(n-1)!/2$ Hamiltonian circuits and then select the shortest one but the labour in this approach is too great even for values of n as small as 20. For arbitrary values of n, no efficient algorithm for solving this problem exists.

Solved Examples

Example 7.23 : Show that the graph in Fig. 7.59 is Hamiltonian

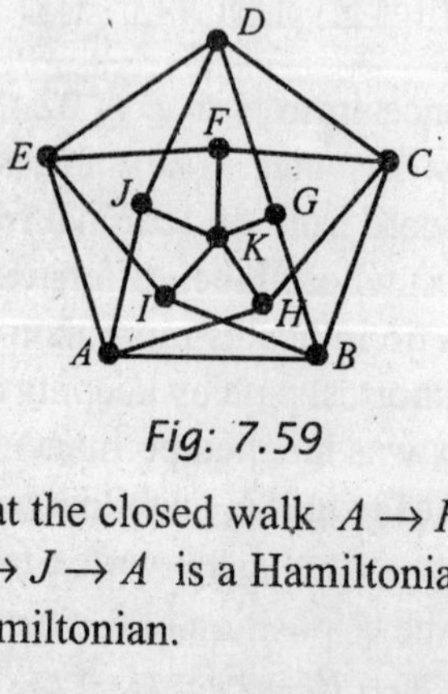

Fig: 7.59

Solution: Observe that the closed walk $A \rightarrow H \rightarrow C \rightarrow F \rightarrow E \rightarrow D \rightarrow G \rightarrow B \rightarrow \text{I} \rightarrow K \rightarrow J \rightarrow A$ is a Hamiltonian circuit.
Hence the graph is Hamiltonian.

Example 7.24 : Show that the following graph is not Hamiltonian.

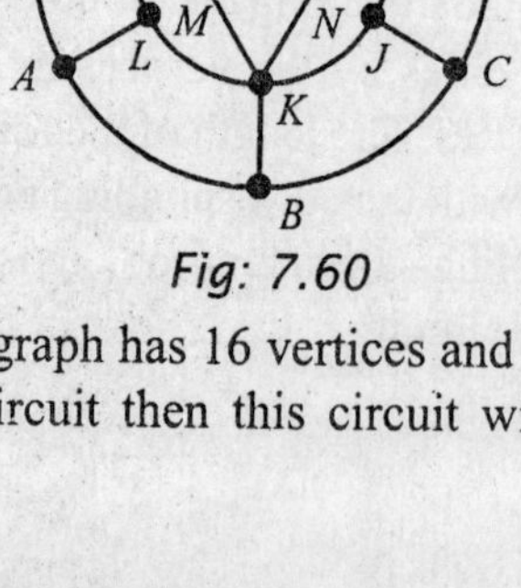

Fig: 7.60

Solution: The given graph has 16 vertices and 27 edges. If the graph has a Hamiltonian circuit then this circuit will contain exactly 16

edges. We know that among all the edges incident on any vertex exactly two can be included in any Hamiltonian circuit. In the above graph, we would have to eliminate one edge at A, three edges at K, three edges at G, one edge at E, three edges at I, one edge at C and one edge at Q. Therefore at least 13 of the 27 edges of G cannot be included in any Hamiltonian circuit. Thus we are left with 14 edges which cannot form a Hamiltonian circuit on the 16 vertices of G. Thus G has no Hamiltonian circuit.

Example 7.25 : Show that the following graph is not Hamiltonian

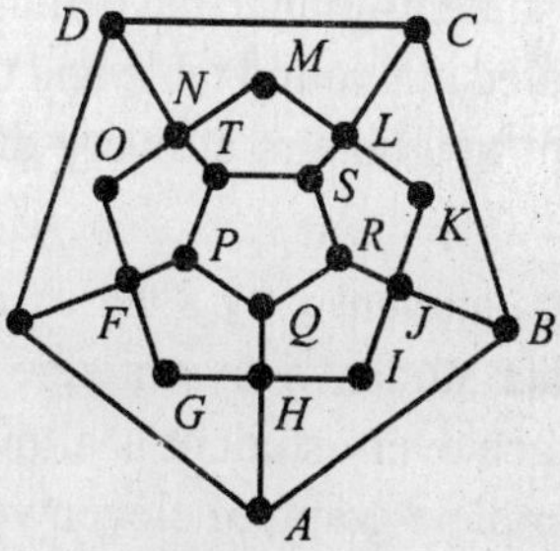

Fig: 7.61

Solution: In this graph, vertices G, I, K M and O are of degree 2. Therefore, we must include all the edges incident on these vertices in the desired Hamiltonian circuit. Since among all the edges incident on a vertex exactly two can be included in any Hamiltonian circuit, we would have to eliminate edges joining D to N and T to N at the vertex N. Similarly edges joining L to C and S, J to B and R, H to A and Q and F to P and E have to be deleted. But then graph becomes disconnected. Hence there exists no Hamiltonian circuit.

Example 7.26 : Under what conditions on r and s does the complete bipartite graph $K_{r,s}$ have a Hamiltonian circuit?

Solution: The vertex set of $K_{r,s}$ can be decomposed into two disjoint subsets V_1 and V_2 containing r and s vertices respectively such that each edge in $K_{r,s}$ joins a vertex in V_1 to a vertex in V_2. Since Hamiltonian circuit is a closed walk that traverses every vertex of the graph exactly once, therefore Hamiltonian circuit in $K_{r,s}$ exists only when $r = s$ and both are ≥ 2 (See Fig 7.62)

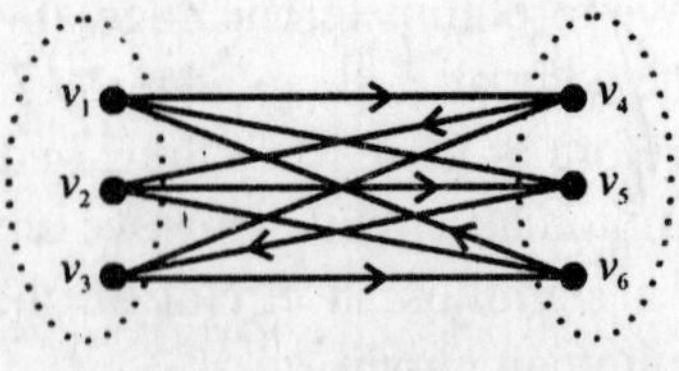

Fig: 7.62

Example 7.28 : Eleven students plan to have dinner together for several days. They will be seated at a round table and the plan calls for each student to have different neighbours at every dinner. For how many days can this be done?

Solution: Representing a student x by a vertex and the possibility of his sitting next to another student y by an edge between x and y, we construct a graph G. Since every student is allowed to sit next to any other student, G is a complete graph of eleven vertices. Every seating arrangement around the table is clearly a Hamiltonian circuit.

The first evening of their dinner they can sit in any order and it will be a hamiltonian circuit. The second evening, if they are to sit in such a manner that every student have different neighbours, we have to find another Hamiltonian circuit in G with an entirely different set of edges from those of first seating. That is, these Hamiltonian circuits are edge disjoint. From Theorem 2, the number of edge-disjoint Hamiltonian circuits in G is $(11-1)/2$ Therefore only five days such arrangements can be done.

Example 7.29 : Solve the travelling salesman problem for the graph in the following Figure

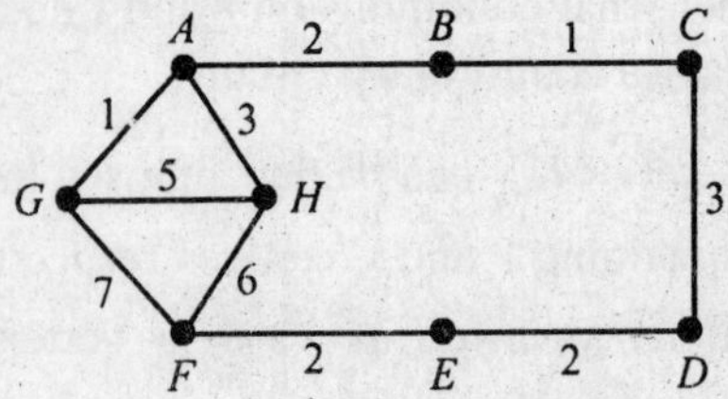

Fig: 7.63

Solution : This graph has the following two Hamiltonian circuits

(a) $A \to B \to C \to D \to E \to F \to H \to G \to A$

(b) $A \to B \to C \to D \to E \to F \to G \to H \to A$

The total weights of Hamiltonian circuits described in (a) and (b) are 22 and 25 respectively. Since the total weight of Hamiltonian circuit in (a) is minimum, salesman should travel according to circuit in (a).

Exercise

1.(a) Which of the platonic graphs have Hamiltonian circuits?

(b) Which complete bipartite graphs are Hamiltonian as well as Eulerian?

(c) For which values of n is the wheel W_n Hamiltonian?

2. Prove that if G is a bipartite graph with an odd number of vertices then G is not Hamiltonian.

3. Show that the graph in Fig. 7.64 is not Hamiltonian.

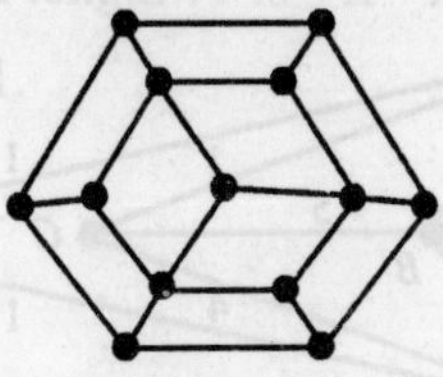

Fig: 7.64

[Hint:Show that it is bipartite graph with odd number of vertices and use problem2 above]

4. Show that the graph in Fig. 7.65 is not Hamiltonian.

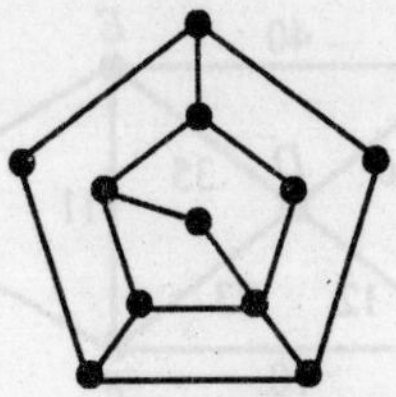

Fig: 7.65

11. Draw a graph that has a Hamiltonian path but does not have a Hamiltonian circuit.

Answers

1.(a) All the five platonic graphs are Hamiltonian.

(b) Complete bipartite graphs $K_{m,n}$ are Hamiltonian as well as Eulerian iff $m = n$ and m, n are even.

(c) For all values of $n \geq 3$.

3.(a) Complete bipartite graph $K_{n,n}$ where n is even.

(b) Complete bipartite graph $K_{m,n}$ where m, n are even but $m \neq n$.

(c) Complete bipartite graph $K_{n,n}$ where n is odd number ≥ 3.

(d) $K_{2,3}$

9. Drawing parallel edges along the path $B \to D \to E \to A \to C$ yields a solution with total weight 24.

10. The shortest path of length 77 is $A \to B \to D \to F \to E \to G$

11. $K_{n,n+1}$ for any $n \geq 1$.

Chapter Eight

Trees and Cut-Sets

The graphs that we come across in most of the applications are connected. Among the connected graphs, trees are probably the most important ones. In this chapter, we shall study trees and its properties. The relationships among circuits, trees and other associated concepts in a graph are also explored.

§ 8.1 Trees

Definition: A connected graph without any circuits is called a *tree*.

Example 8.1: The graph G shown in Fig. 8.1(a) is a tree while the graph shown in Fig. 8.1(b) is not a tree.

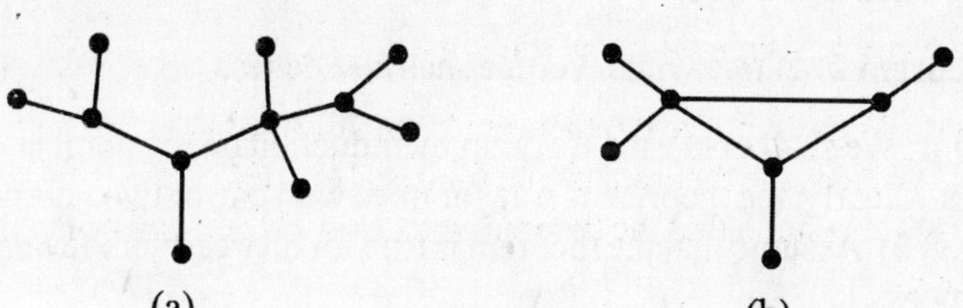

(a) (b)

Fig: 8.1

Example 8.2: Trees with one, two, three and four vertices are shown in Fig. 8.2 below.

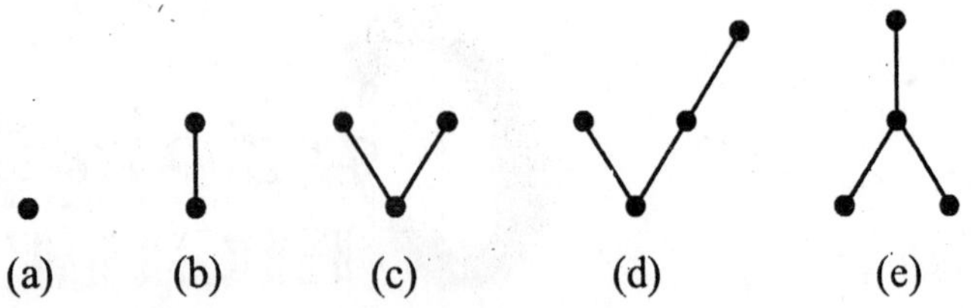

Fig: 8.2

Since parallel edges and self-loops both form circuits, a tree can not have parallel edges and self-loops. Thus a tree has to be a simple graph.

Theorem1: A graph G is a tree iff there is one and only one path between any two vertices of G.

Proof : First suppose that the graph G is a tree. Then by definition of a tree, G is a connected graph. Therefore, there must exist at least one path between any two vertices in G. Now suppose that there are two distinct paths between vertices a and b of G. Then the union of these two paths will contain a circuit and G can not be a tree. Thus there is one and only one path between any two vertices of G.

Conversely, suppose that there is one and only one path between any two vertices of G. We shall show G is a tree. Since there exists a path between any two vertices of G, therefore G is connected. A circuit in a graph with two or more vertices implies that there exists a pair of vertices a, b such that there are two distinct paths between a and b. Since G has one and only one path between any two vertices, G can have no circuit. Thus G is a tree.

Theorem 2 : A tree with n vertices has $n-1$ edges.

Proof: We shall prove the theorem by induction on the number of vertices. Clearly, the theorem is true for trees with one or two vertices (see Fig. 8.2). Assume that the theorem is true for all trees with fewer than n vertices.

Let us consider a tree G with n vertices. Let e_k be any edge in G with end vertices v_i and v_j. According to Theorem 1 above, the edge e_k is the only path between v_i and v_j. Hence, deletion of e_k from G will disconnect the graph. Thus $G-e_k$ is not connected. Further, $G-e_k$ will contain exactly two components, for otherwise the graph G will not be connected. Let these two components of $G-e_k$ be G_1 and G_2. Each of these components is a tree because there were no circuit in G. Let n_1 and n_2 be the number of vertices in G_1 and G_2 respectively. Since $n_1 < n$ and $n_2 < n$, we have by the induction hypothesis

$$\text{number of edges in } G_1 = n_1 - 1$$

and

$$\text{number of edges in } G_2 = n_2 - 1$$

Thus, number of edges in $G - e_k$ is equal to $(n_1-1) + (n_2-1) = (n_1+n_2)-2 = n-2$. Hence G has exactly $n-1$ edges.

Theorem 3: Every connected graph with n vertices and $n-1$ edges is a tree.

Proof: Let G be a connected graph with n vertices and $n-1$ edges. The theorem will be proved if we show that G has no circuit. Suppose that G contains at least one circuit. Since removing an edge from a circuit does not disconnect a graph, we may remove edges, but no vertices from circuits in G until the resulting graph G^* is circuit-free. Now G^* is a connected graph with n vertices and contains no circuit. Thus G^* is a tree with n vertices. Hence G^* has $n-1$ edges (by Theorem 2). But now the graph G has more than $n-1$ edges, a contradiction. Hence G has no circuit. This completes the proof.

Theorem 4: A graph G with n vertices, $n-1$ edges and no circuits is tree.

Proof: Let G be a graph with n vertices, $n-1$ edges and has no circuit. It will be a tree if we show that it is connected. If possible, suppose that G is disconnected. Then G will consist of two or more circuitless components. Without loss of generality let G consist of two components G_1 and G_2. We add an edge e between a vertex v_1 in G_1 and v_2 in G_2. Since v_1 and v_2 are in different components of G, there is no path between v_1 and v_2 in G. Thus addition of edge e will not create a circuit. Thus $G \cup e$ is a circuitless, connected graph (and therefore a tree) of n

vertices and n edges, which is not possible because of Theorem 2. This completes the proof.

Minimally Connected Graph

A connected graph G is said to be *minimally connected* if removal of any edge from G disconnects the graph G.

We now have the following theorem:

Theorem 5: A graph G is a tree iff it is minimally connected.

Proof: Suppose that G is a tree. We show G is minimally connected. Since G is tree, it is connected. If G is not minimally connected then there must exist an edge e in G such that $G-e$ is connected. Therefore, e is in some circuit, which implies that G is not a tree, a contradiction. Thus G is minimally connected.

Conversely, suppose that G is a minimally connected graph. Then G is connected and cannot have a circuit; otherwise, we could remove one of the edges in the circuit and still leave the graph connected. Thus a minimally connected graph is a tree.

Minimum Number of Pendant Vertices in a Tree

Recall that a pendant vertex in a graph is that vertex whose degree is one. In general, trees have several pendant vertices. The minimum number of pendant vertices in a tree is given by the following theorem:

Theorem 6: In any tree (with two or more vertices), there are at least two pendant vertices.

Proof: Let G be any tree having n vertices. Then G has $n-1$ edges. Since each edge contributes two degrees, the sum of the degrees of all vertices in G is $2(n-1)$. Now $2(n-1)$ degrees are to be divided among n vertices in G. Let the number of vertices of degree one in G be x. Since no vertex in a tree can be of zero degree, we have

$$\frac{2(n-1)-x}{n-x} \geq 2$$

$$\Rightarrow \qquad x \geq 2$$

Thus, we must have at least two vertices of degree one in a tree.

§ 8.2 Distance and Centres in a Tree

Definition: Let G be a connected graph and v_i, v_j be any two vertices in G. The distance $d(v_i, v_j)$ between the vertices v_i and v_j is the length of the shortest path (that is, the number of edges in the shortest path) between v_i and v_j. The definition of distance between any two vertices holds in any connected graph and therefore in tree also. If the graph is not a tree, there are generally several paths between a pair of vertices. The shortest path in a graph may not be unique. But in the case of tree, there exists only one path between any pair of vertices.

Example 8.3: Find the distance between the vertices v_1 and v_2 in Fig. 8.3

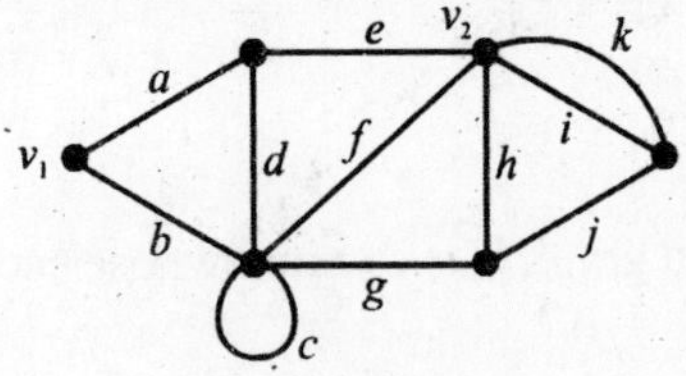

Fig: 8.3

Solution: Some of the paths between vertices v_1 and v_2 in Fig. 8.3 are

$$(a, e), (a, d, g, h), (b, f), (a, d, f), (b, g, h).$$

There are two shortest paths (a,e) and (b,f), each of length 2.

Hence $d(v_1, v_2) = 2$.

Metric: Let $G = (V, E)$ be a graph. A function $f: V \times V \rightarrow R$ is said to be a metric if it satisfies the following conditions:

(i) *Non-negativity* : $f(x, y) \geq 0$ and $f(x, y) = 0$ iff $x = y$

(ii) *Symmetry*: $f(x, y) = f(y, x)\ \forall\ x, y \in V$

(iii) *Triangle inequality*: $f(x, y) \leq f(x, z) + f(z, y)\ \forall\ x, y, z \in V$.

Theorem 1: Prove that the distance between vertices of a connected graph is a metric.

Proof: It is clear from the definition of distance between vertices that it satisfies conditions (i) and (ii) of a metric. Since the distance $d(v_i, v_j)$ is the length of the shortest path between v_i and v_j, this path cannot be longer than a path between v_i and v_j which goes through another vertex v_k. Hence

$$d(v_i, v_j) \leq d(v_i, v_k) + d(v_k, v_j)$$

Thus $d(v_i, v_j)$ is a metric.

Eccentricity of a Vertex

Definition: The eccentricity $E(v)$ of a vertex v in a graph G is the distance between v and the vertex v_i farthest from v in G. Symbolically,

$$E(v) = \max_{v_i \in G} d(v, v_i)$$

Centre of a Graph

Definition: A centre of a graph G is a vertex whose eccentricity in G is minimum.

Example 8.4 : Find the eccentricity of each of the vertices of the graph in Fig. 8.4. Also find the centre of this graph.

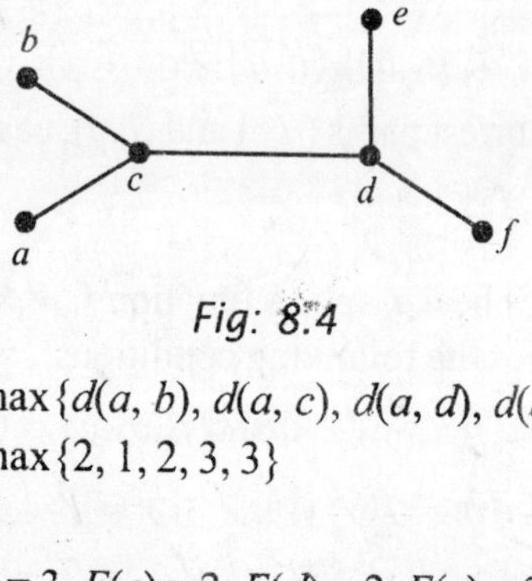

Fig: 8.4

Solution: $E(a) = \max\{d(a, b), d(a, c), d(a, d), d(a, e), d(a, f)\}$
$= \max\{2, 1, 2, 3, 3\}$
$= 3$

Similarly, $E(b) = 3$, $E(c) = 2$, $E(d) = 2$, $E(e) = 3$, $E(f)=3$.

Now there are two vertices c and d having the same minimum eccentricity, namely, 2. Hence this graph has two centres.

In general, one can verify that a graph can have many centres. For example, in the cycle graph C_6 (Fig. 7.39(a)), every vertex is a centre.

However, in the case of a tree, we prove the following theorem:

Theorem 2: Every tree has either one or two centres.

Proof: From the definition of eccentricity of a vertex in a graph it is clear that the maximum distance, max $d(v, v_i)$ from a given vertex v to any other vertex v_i occurs only when v_i is a pendant vertex. Let T be a tree having more than two vertices. By Theorem 6 of § 8.1, tree T must have at least two pendant vertices. Remove all the pendant vertices from T to obtain another graph T'. Then T' is a tree. Further observe that the eccentricities of the vertices in T' get reduced by one from T. Therefore, all vertices that are centres in T will still remain centres in T'. Remove again all pendant vertices from T' to obtain another tree T''. Continue this process of removal of pendant vertices until we are left with either a vertex (which is the centre of T) or an edge whose end vertices are the centres of T. The above process is illustrated in Fig. 8.5. The eccentricity of each of the vertices is shown next to the vertex in the Figure.

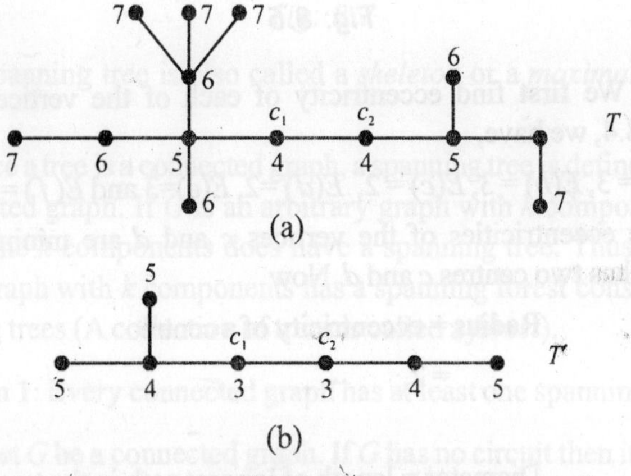

Graph T′ obtained after removing pendant vertices from T

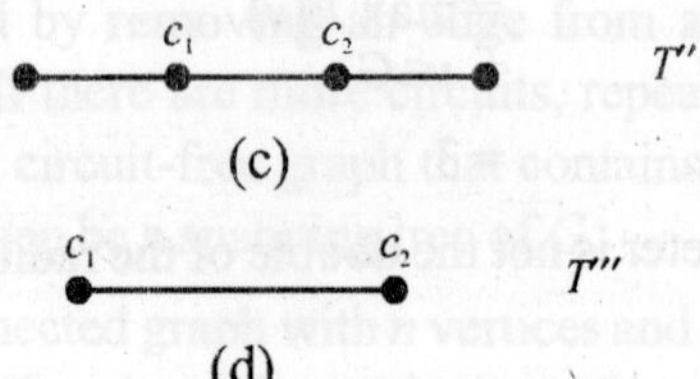

Fig: 8.5 Process of finding centres in a tree.

The following theorem gives a necessary and sufficient condition for a spanning tree to be shortest.

Theorem 1: A spanning tree T of a weighted graph G is a minimal spanning tree iff there exists no other spanning tree of G at a distance of one from T whose weight is smaller than that of T.

Proof: The only if condition is obvious because otherwise we shall get another tree smaller than T by adding a chord and deleting an appropriate branch.

To prove the converse, let T_1 be a spanning tree of G satisfying the hypothesis of the theorem that there is no spanning tree at a distance of one from T_1 whose weight is smaller than that of T_1. We will show that if T_2 is a minimal spanning tree in G then the weights of T_1 and T_2 are equal. So, we assume that T_2 is a minimal spanning tree in G. Since T_2 is a minimal spanning tree, no spanning tree at a distance of one from T_2 can be of smaller weight than that of T_2. Hence T_2 also satisfies the hypothesis of the theorem. Now consider an edge e in T_2 which is not in T_1. Adding e to T_1 forms a fundamental circuit with branches in T_1. Some, but not all, of the branches in T_1 that form the fundamental circuit with e may also be in T_2. Each of these branches in T_1 has a weight smaller than or equal to the weight of e because, by assumption on T_1, there is no spanning tree at a distance of one from T_1 which is smaller than T_1. Thus

Weight of each branches in the circuit $\leq$ weight of edge e.......(1)

Consider all those edges in the fundamental circuit which are not in T_2 (these edges are chords of G with respect to spanning tree T_2). At least one of these edges must form some fundamental circuit with respect to T_2 containing e. Let this edge be b_j. Since T_2 is minimal spanning tree, the weight of b_j cannot be less than that of e. Thus

weight of $e \leq$ weight of b_j ....(2)

From (1) and (2), we have

weight of b_j = weight of e.

Hence the spanning tree,

$$T_1' = (T_1 \cup e) - b_j$$

has the same weight as that of T_1. But T_1' has one edge more in common with T_2 than T_1 and it satisfies the condition of the theorem. This argument can be repeated, producing a finite number of trees of equal weight, T_1', T_1'', ..., each a unit distance closer to T_2, until we get T_2 itself. This proves that if none of the spanning trees at a distance of one from T is smaller than T then no spanning tree smaller than T exists in the graph. This completes the proof.

Algorithms for Minimal Spanning Tree

Here we shall discuss two methods for finding a minimal spanning tree in a weighted graph.

1. Kruskal's Algorithm: In this method, we first list all edges of the weighted graph G in order of non-decreasing weight. We first select a smallest edge of G. Then for each successive step, select a new edge of smallest possible weight that forms no circuit with the previously selected edges. Continue this process until n–1 edges have been selected and these n–1 edges will form the desired minimal spanning tree.

Another algorithm, which does not require listing of all edges in order of non-decreasing weight or checking at each step if a newly chosen edge makes a circuit is due to R.C. Prim. The algorithm is described as follows:

2. Prim's Algorithm: Let G be a connected weighted graph with vertices $v_1, v_2, ..., v_n$. For conveniencc, arrange the given weights of the edges of G in an $n \times n$ table. The weights of non-existent edges are taken as very large, say, ∞ and diagonal elements as empty.

Start from vertex v_1 and connect it to the vertex which has the smallest entry in row 1 of the table. Let this vertex be v_k. Now consider the subgraph with vertices v_1 and v_k and an edge between them. Connect this subgraph to a vertex (other than v_1 and v_k) that has the smallest entry among all entries in rows 1 and k. Let this new vertex be v_i. Next consider the tree with vertices v_1, v_k, and v_i as a subgraph and continue the process until all n vertices have been connected by n–1 edges.

If there is a tie for the smallest entry then tie may be broken arbitrarily.

We now illustrate both the methods of finding minimal spanning tree through an example.

Example 8.8 : Find the minimal spanning tree for the graph in Fig. 8.11 using both Kruskal's and Prim's methods

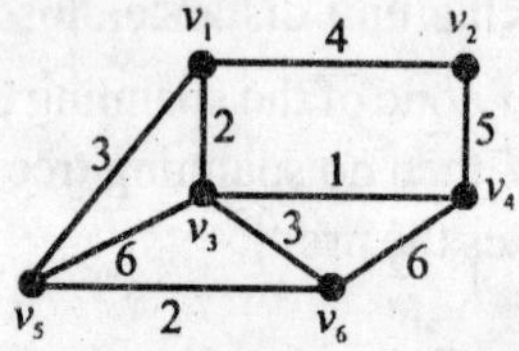

Fig: 8.11

Solution: Minimal spanning tree using Kruskal's algorithm is obtained as follows:

We first list all the edges of the given graph G in order of non-decreasing weight.

$(v_3, v_4), (v_1, v_3), (v_5, v_6), (v_3, v_6), (v_1, v_5), (v_1, v_2), (v_2, v_4), (v_4, v_6), (v_3, v_5)$

We first select the edge (v_3, v_4) whose weight is minimum. Next we select the second edge (v_1, v_3) from the list since it does not form a circuit with the previously selected edge. We successively select the edges $(v_5, v_6), (v_3, v_6)$ and (v_1, v_2). The minimal spanning tree is shown below in Fig. 8.12. Note that at the fourth stage, although the edge (v_1, v_5) is of minimum weight, it cannot be selected because this edge forms a circuit with the three edges already selected. The total of the spanning tree is 12 units.

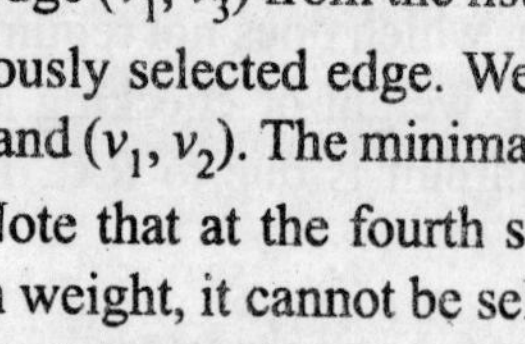

Fig: 8.12

Following Prim's algorithm, we first tabulate the weight of each edge in a table.

	v_1	v_2	v_3	v_4	v_5	v_6
v_1	–	4	2	∞	3	∞
v_2	4	–	∞	5	∞	∞
v_3	2	∞	–	1	6	3
v_4	∞	5	1	–	∞	6
v_5	3	∞	6	∞	–	2
v_6	∞	∞	3	4	2	–

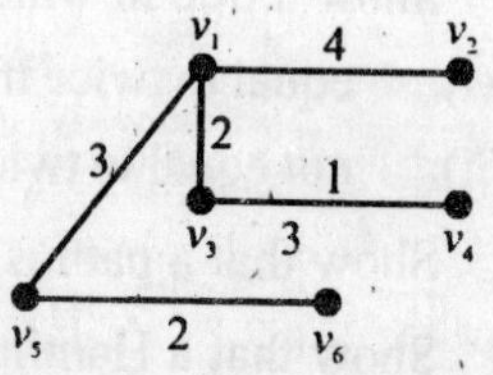

Fig: 8.13

We start with vertex v_1 and pick the smallest entry in row 1 which is (v_1, v_3). We now examine first and third rows of the table and pick the smallest entry which is (v_3, v_4). Thus we have a subgraph with vertices v_1, v_3, and v_4. We now examine rows 1, 3 and 4 and we have to select the vertex other than v_1, v_3 and v_4 whose corresponding entry is smallest. So ruling out the entries (v_3, v_4), (v_4, v_3), (v_1, v_3) and (v_3, v_1) we have a tie between (v_1, v_5) and (v_3, v_6). We arbitrarily select (v_1, v_5). Continuing in this way, the two remaining edges to be selected are (v_5, v_6) and (v_1, v_2). The resulting minimal spanning tree is shown in Fig. 8.13. Although this spanning tree is different from that in Fig. 8.12, but the weight is still 12 units.

Exercise

Q.1 Prove that every tree is a bipartite graph. Find the conditions under which complete bipartite graph $K_{m,n}$ becomes tree.

Q2. A tree has p vertices of degree 1, 2 vertices of degree 2, 4 vertices of degree 3 and 3 vertices of degree 4. Find the value of p.

Q3. A tree has $2n$ vertices of degree 1, $3n$ vertices of degree 2, and n vertices of degree 3. Determine the number of vertices and edges in the tree.

Q4. Give an example of a tree which has:

(i) one centre

(ii) two centres.

Q5. Show a tree in which its diameter is:

(i) equal to twice the radius

(ii) not equal to twice the radius.

Q6. Show that a path is its own spanning tree.

Q7. Show that a Hamiltonian path is a spanning tree.

Q8. Prove that any given edge of a connected graph G is a branch of some spanning tree of G.

Q9. Prove that any circuit in a graph G must have at least one edge in common with a chord set.

Q10. What is the nullity of a complete graph of 7 vertices. Give an example of a graph with nullity equal to 0.

Q11. Determine a minimal spanning tree for the graph in Fig. 8.14 using (i) Kruskal's algorithm (ii) Prim's algorithm.

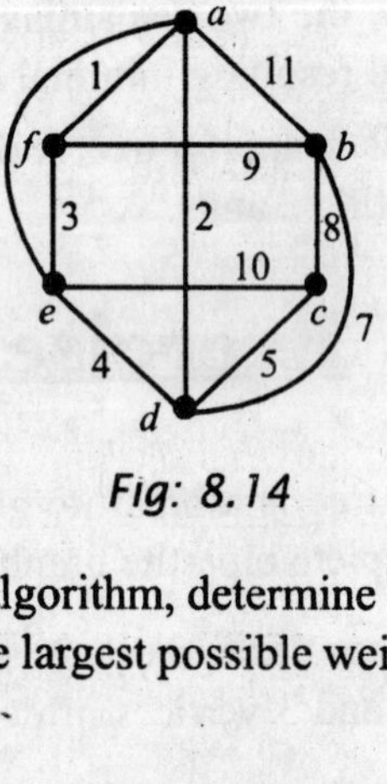

Fig: 8.14

Q12. Using Kruskal's algorithm, determine a maximal spanning tree (*i.e.* a tree with the largest possible weight) for the graph in Fig. 8.15.

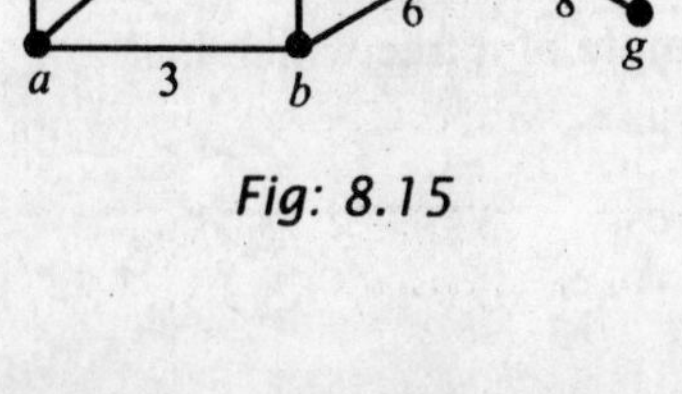

Fig: 8.15

Answers

1. Either $m=1$ or $n=1$
2. $p=12$
3. Number of vertices=12, number of edges=11

4 (i) 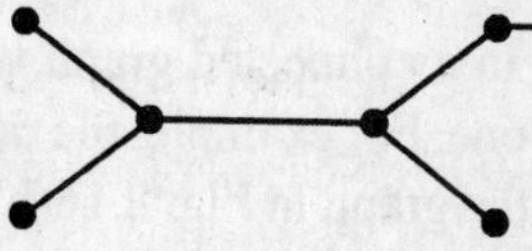 (ii) 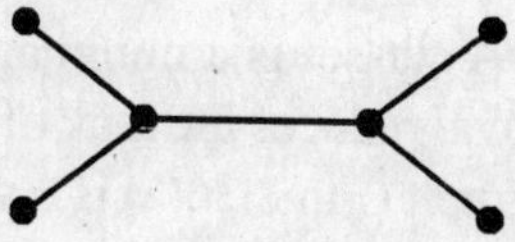

5 (i) 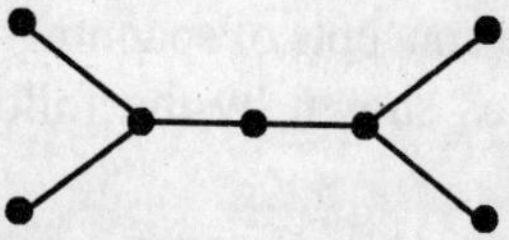 (ii) 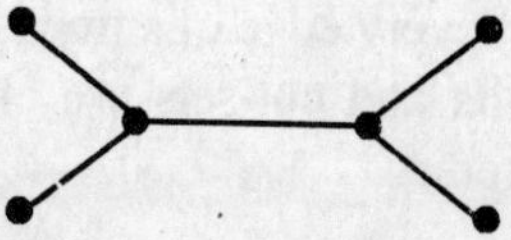

10. 21, Nullity of any tree = 0
11. Total weight of minimal spanning tree = 18 units
12. Total weight of maximal spanning tree = 36 units

§ 8.6 Cut-Sets

Definition: A *cut-set* S of a connected graph G is a minimal set of edges of G whose removal from G disconnects G. Equivalently, a cut-set of a connected graph G is a set of edges whose removal from G leaves G disconnected, and removal of no proper subset of these edges disconnects G.

For example, in Fig. 8.16, the set of edges $\{a, c, d, f\}$ is a cut-set.

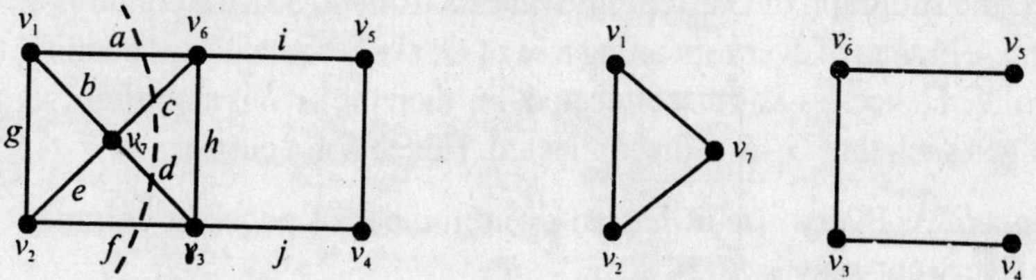

(a) Graph G (b) Graph $G - \{a, c, d, f\}$

Fig: 8.16

Thus removal of a cut-set will divide the graph into two parts. A given graph may have several cut-sets. The sets {*a*, *b*, g}, {*d*, *f*, *h*} are also cut-sets of the graph in Fig. 8.16. The set of edges {*a*, *c*, *d*, *h*) is not a cut-set because one of its proper subsets, {*a*, *c*, *h*} is a cut-set.

A cut-set always cuts a graph into two. Thus the removal of a cut-set from a connected graph *G* reduces the rank of the graph by one. Therefore, we can also define cut-set as follows:

A cut-set is a minimal set of edges in a connected graph whose removal reduces the rank of the graph by one. For example, the rank of the graph *G* in 8.16(a) is 6 while rank of the graph in Fig. 8.16(b) is 5. Since the removal of any edge from a tree breaks the tree into two parts, every edge of a tree is a cut-set. The concepts of spanning trees, circuits and cut-sets are closely related as shown by the following theorems.

Theorem 1: Every cut-set in a connected graph *G* contains at least one branch of every spanning tree of *G*.

Proof: Let *G* be a connected graph and let T be any spanning tree of *G*. Let S be an arbitrary cut-set in *G*. If possible, suppose S does not have any edge of T. Then removal of the set S from *G* (*i.e*, the graph *G–S*) would not disconnect the graph because the graph *G–S* contains T which is connected subgraph of *G*. Obviously this contradicts the definition of the cut-set.

Theorem 2: Any minimal set of edges containing at least one branch of every spanning tree of a connected graph *G* is a cut-set.

Proof: Let *G* be a connected graph and let S be a minimal set of edges containing at least one branch of every spanning tree of *G*. Consider $G-S$, the subgraph of *G* after removal of *S* from *G*. Since *S* contains at least one branch of every spanning tree of *G*, *G*–S contains no spanning tree of *G*. Hence $G-S$ is disconnected. Further, the set *S* is a minimal set of edges such that $G-S$ is disconnected. Hence *S* is a cut-set.

Theorem 3: Every circuit has an even number of edges in common with every cut-set.

Proof: Let *S* be a cut-set and *C* a circuit of a connected graph *G*. Then $G-S$ will be disconnected and has two components. Let V_1 and V_2 be the vertex sets of the two components of $G–S$. Clearly, $V_1 \cap V_2 = \phi$ and $V_1 \cup V_2$ = vertex set of G. If all the vertices of the circuit C are entirely within vertex set V_1 or V_2, then the number of edges common to S and C is zero, an even number.

If some vertices in C are in V_1 and some in V_2, then in traversing the circuit C, we traverse back and forth between the set V_1 and V_2 as illustrated in Fig. 8.17.

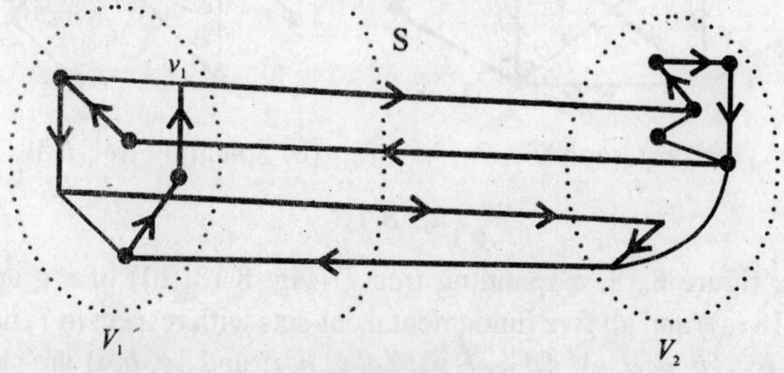

Fig: 8.17

Suppose the circuit C starts from a vertex v_1 in V_1. Since the traversing should end at v_1, it is necessary that every time we meet with an edge of S taking us from a vertex in V_1 to a vertex in V_2, there must be an edge of S taking us from a vertex in V_2 to a vertex in V_1, (Since every edge in S has one end in V_1 and the other in V_2 and no other edge of G has this property). This is possible only if C and S have an even number of edges in common.

Fundamental Cut-Sets

Consider a spanning tree T of a connected graph G. Let b be any branch of T. Now removal of the branch b disconnects T. That is, $\{b\}$ is a cut-set of T. Therefore $\{b\}$ will partitions all vertices of T into two disjoint sets V_1 and V_2. Consider the same partition of vertices of G and the cut-set S in G which corresponds to this partition. The cut-set S will contain only one branch b of T and other edges in S are chords with respect to T. Such a cut-set is called a *fundamental cut-set* with respect to T. Thus we have following:

Definition: Let T be a spanning tree of a connected graph G. A cut-set S of G which contains exactly one branch of T is called a *fundamental cut-set* of G with respect to T.

Since there are n–1 branches in a spanning tree of a graph with n vertices, we have n–1 fundamental cut-sets with respect to a given spanning tree T.

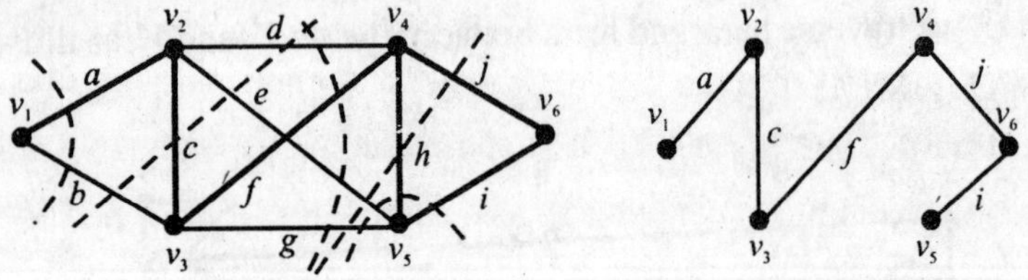

(a) Graph G (b) Spanning tree T of G

Fig: 8.18

In figure 8.18, a spanning tree T (Fig. 8.18 (b)) of a graph G (Fig8.18 (a)) and all five fundamental cut-sets with respect to T, namely, $\{a, b\}$, $\{b, c, d, e\}$, $\{d, e, f, g\}$, $\{e, g, h, i\}$ and $\{g, h, j\}$ are shown through dotted lines.

It should be kept in mind that the term fundamental cut-set has meaning only with respect to a given spanning tree.

The following results give a close relationship between the fundamental circuits and fundamental cut-sets relative to a spanning tree of a graph.

Theorem 4: For a given spanning tree T of a connected graph, a chord c_i that determines a fundamental circuit C occurs in every fundamental cut-set associated with the branches of C and in no other.

Proof: Let C be the fundamental circuit of a connected graph G with respect to a chord c_i of a spanning tree T. Let C contain the branches b_1, b_2, ...,b_k of T, in addition to the chord c_i. That is

$C = \{c_i, b_1, b_2, ..., b_k\}$ is a fundamental circuit with respect to T.

Let S_1 be the fundamental cut-set of G associated with b_1, consisting of q chords in addition to the branch b_1. That is,

$S_1 = \{b_1, c_1, c_2, ...c_q\}$ is a fundamental cut-set with respect to T.

In view of Theorem 3, C and S_1 must have an even number of common edges. Edge b_1 is in both C and S_1, and C_i is only other edge that can possibly also be in S_1. Therefore, we must have two (an even number) edges b_1 and c_1 common to S_1 and C. Thus the chord c_i is one of the chords $c_1, c_2, ...c_q$. Exactly the same argument holds for fundamental cut-sets of G associated with $b_2, b_3, ...b_k$. Thus, the chord c_i is contained in every fundamental cut-set associated with branches in C. Next we show that no other fundamental cut-set of T contains c_i (besides those associated with $b_1, b_2, ...,b_k$). Suppose that the fundamental cut-set S' with respect to some branch b_{k+1} of T contains c_i. Then c_i will be the only common edge in S' and C. This will contradict Theorem 3. Thus the chord c_i is present only in those cut-sets defined by the branches $b_1, b_2, ...b_k$.

Theorem 5: For a given spanning tree T of a connected graph, a branch b_i that determines a fundamental cut set S is contained in every fundamental circuit associated with the chords in S and in no others.

Proof: Let the fundamental cut-set determined by a branch b_i of spanning tree T be

$S = \{b_i, c_1, c_2, ...c_p\}$, where $c_1, c_2, ...c_p$ are chords of T and let C_1 be the fundamental circuit determined by chord c_1. That is,

$C_1 = \{c_i, b_1, b_2,...,b_q\}$, where $b_1, b_2, ...,b_q$ are branches of T.

By Theorem 3, the number of edges common to S and C_1 must be even. Therefore b_i must be in C_1. Exactly the same argument holds for the fundamental circuits associated with chords $c_2, c_3..., c_p$.

Next, we show that no other fundamental circuit of T contains b_i (besides those associated with $c_1, c_2, .., c_p$). Suppose that b_i occurs in a fundamental circuit C_{p+1} made by a chord other than $c_1, c_2,..., c_p$. Then b_i is the only edge common is C_{p+1} and S, which is not possible. Hence the theorem.

We shall illustrate the above theorems with the help of an example.

Example 8.9 : Consider the graph of Fig. 8.18(a) and its spanning tree T in Fig 8.18(b).

(i) Find all fundamental cut-sets associated with the branches in the fundamental circuit made by chord g and verify Theorem 4.

(ii) Find fundamental circuits associated with the chords in the fundamental cut-set determined by f and verify Theorem 5.

Solution:

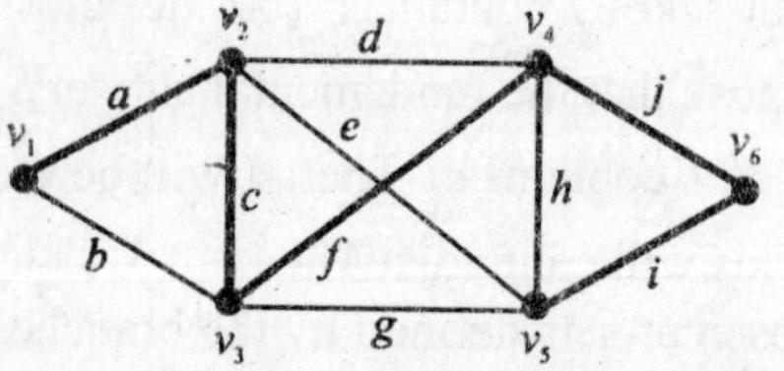

Graph G

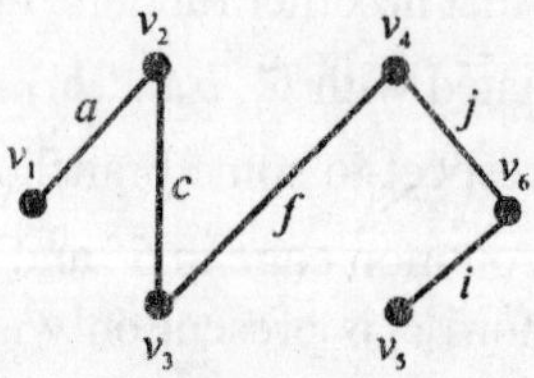

Spanning tree T

(i) The fundamental circuit made by the chord g is

$$\{g, f, j, i\}$$

The three fundamental cut-sets associated with the three branches f, j, i are as follows:

Fundamental cut-set determined by branch $f = \{d, e, f, g\}$

Fundamental cut-set determined by branch $j = \{g, h, j\}$

Fundamental cut-set determined by branch $i = \{g, h, i\}$

Chord g occurs in each of these three fundamental cut-sets and there is no other fundamental cut-set that contains g. This verifies Theorem 4.

(ii) The fundamental cut-set determined by f is

$$\{f, e, d, g\}.$$

The three fundamental circuits associated with the chords e, d, and g are as follows:

Fundamental circuit determined by chord $e = \{e, c, f, i, j\}$

Fundamental circuit determined by chord $d = \{d, c, f\}$

Fundamental circuit determined by chord $g = \{g, f, i, j\}$

Branch f is contained in each of these three fundamental circuits and there is no other fundamental circuit that contains f. This verifies Theorem 5.

Example 8.10 : Show that a Euler graph G can not have a cut-set of odd number of edges.

Solution: Suppose S is a cut-set having an odd number of edges. For convenience, we assume S contains 3 edges. Removal of S from G will disconnect the graph G. Let G_1 and G_2 be the components of $G-S$. Each edge in cut-set S of G joins a vertex in G_1 to a vertex in G_2. Let $S = \{(u_1, v_1), (u_2, v_2), (u_3, v_3)\}$, where u_i are vertices in G_1 and V_i are vertices in G_2. Since G is an Eulerian graph, degree of each vertex in G is even. After removing 3 edges of S, each u_i and v_i would have an odd number of degrees in G_1 and G_2 respectively. But we know that the number of odd vertices in any graph is always even while both graphs G_1 and G_2 have an odd number of odd vertices. Hence G cannot have a cut-set of odd number of edges.

§ 8.7 Edge-Connectivity, Vertex Connectivity and Separability

Edge Connectivity: The number of edges in the smallest cut-set is defined as the edge-connectivity of the graph. Here by smallest cut-set we mean that cut-set which has fewest number of edges. Recall that rank of a graph is equal to $n-k$, where n is the number of vertices and k is the number of components. Since removal of a cut-set from a connected graphs partitions the graph into two components, we can define the edge-connectivity of a connected graph as follows:

The edge connectivity of a connected graph is defined as the minimum number of edges whose removal reduces the rank of the graph by one.

Example 8.11: The edge connectivity of a tree is one because removal of an edge from the tree disconnects the graph.

Example 8.12: The edge connectivity of the graph in Fig 8.18(a) is two.

Example 8.13: The edge connectivity of the complete bipartite graph $K_{2,3}$ is 2. In general, edge connectivity of $K_{m,n}$ is p, where $p = \min\{m, n\}$.

Example 8.14: The edge connectivity of a complete graph of n vertices is $n-1$.

Note: In literature of graph theory, edge connectivity and vertex connectivity of a disconnected graph are defined as zero.

Vertex Connectivity : The *vertex connectivity* of a connected graph G is defined as the minimum number of vertices whose removal from G leaves the remaining graph disconnected. Observe that vertex connectivity is meaningful only for graphs that have three or more vertices and are not complete.

Example 8.15: The vertex connectivity of any tree is one.

Example 8.16: The vertex connectivity of the graph in Fig. 8.18(a) is two.

Example 8.17: The vertex connectivity of complete bipartite graph $K_{m,n}$ is minimum of m and n.

Example 8.18: Find the vertex connectivity of the graph in Fig. 8.19

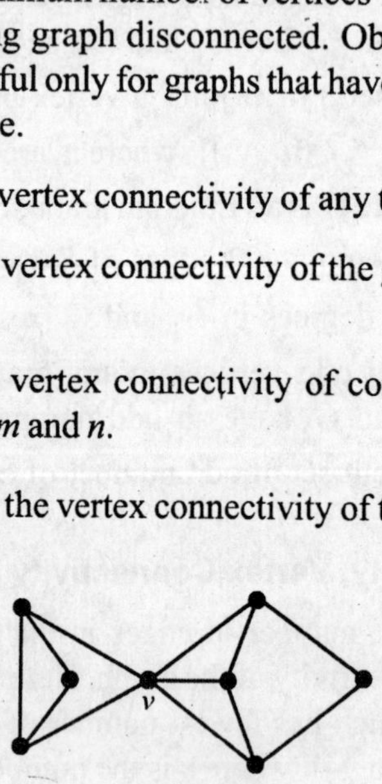

Fig: 8.19

Solution: The vertex connectivity of this graph is one because removal of the vertex v disconnects the graph.

Separable graph: A connected graph whose vertex connectivity is one is called a *separable* graph. For example the graph in figure 8.19 is separable.

Non-separable graph: A connected graph whose vertex connectivity is not one is called a non-separable. Equivalently, a connected graph which is not separable is called non-separable. For example, the graph in Fig. 8.18(a) is not separable.

Cut-Vertex

A *cut-vertex* is a vertex in a separable graph whose removal disconnects the graph. Cut-vertex is also called an *articulation* point. A given separable graph can have more than one cut-vertex. For example, the graph in Fig. 8.20 is separable and both v_1 and v_2 are cut-vertices.

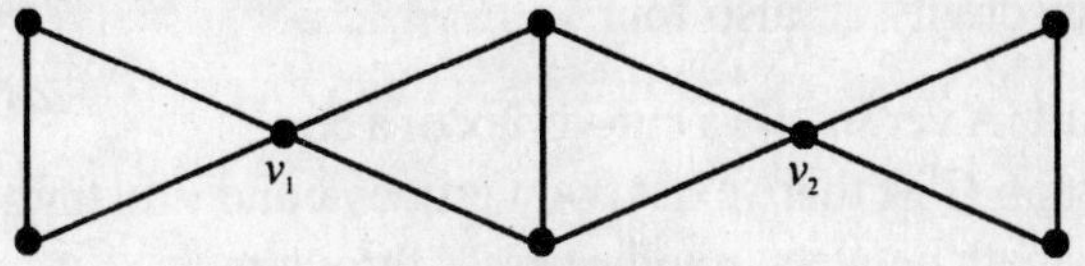

Fig: 8.20

Example 8.19: Prove that in a tree every vertex of degree greater than one is a cut-vertex.

Solution: Let G be a tree and let v be a vertex in G whose degree is greater than one (See Fig. 8.21).

Since degree of v is greater than one, it means there are at least two vertices v_1 and v_2 which are adjacent to v. Clearly, removal of v will put v_1 and v_2 in different components because there is no other path between v_1 and v_2. Hence removal of v disconnects the graph. Thus v is a cut-vertex.

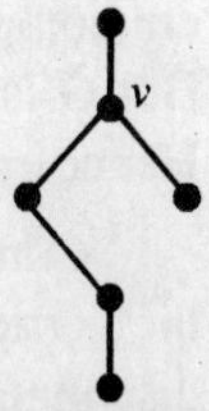

Fig: 8.21

Example 8.20: Find the edge connectivity and vertex connectivity of the graph in Fig. 8.22

Solution: It can be seen that removal of any two vertices does not disconnect the graph. However removal of three vertices a, b and f disconnects the graph. Hence vertex connectivity is three.

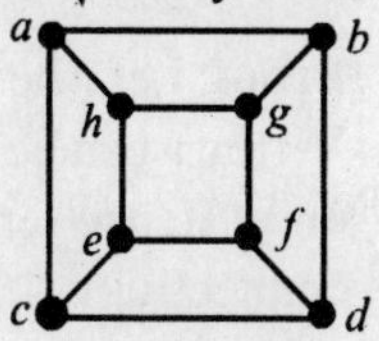

Fig: 8.22

Similarly, we see that removal of any two edges does not disconnect the graph but the set of any three edges incident on a vertex forms a cut-set. Hence edge connectivity is three.

Remark: In the graph of Fig. 8.22, we have

Edge connectivity = Vertex connectivity = degree of any vertex.

Example 8.21: What are the edge connectivity and vertex connectivity of the graph in Fig. 8.23?

Solution: It can be easily verified that removal of three vertices does not disconnect the graph. But the removal of four vertices v_1, v_5, v_6, v_8 disconnect the graph. Hence vertex connectivity is four. It can be easily seen that edge connectivity is also four.

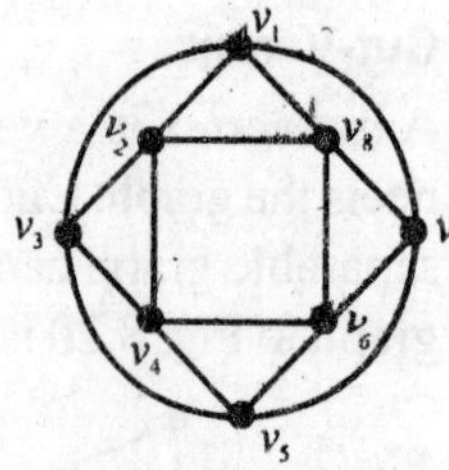

Fig: 8.23

Theorem 1: A vertex v is a cut-vertex of a connected graph G iff there exist two vertices x and y distinct from v such that every path between x and y passes through v.

Proof: First suppose that v is a cut-vertex of a connected graph G. Then $G–v$, the graph obtained after removal of vertex v from G, is disconnected. Let x be a vertex in one of the component G_1 of $G - v$ and y be a vertex in the complement of G_1 in $G–v$. Since x and y are in different components of $G–v$, there exist no path between x and y in $G–v$. But G is connected implies there exist paths between x and y in G. Hence every path between x and y passes through v.

Conversely, suppose v lies on every path between x and y. Then the vertices x and y are not connected in $G–v$. Thus the graph $G–v$ is disconnected. Hence v is a cut-vertex.

Theorem 2: Prove that the edge connectivity of a graph G cannot exceed the degree of the vertex with smallest degree in G.

Proof: Let v be the vertex with the smallest degree in a given graph G. Vertex v can be separated from G by removing the $d(v)$ edges incident on v. Hence removal of $d(v)$ edges disconnects the graph. Thus edge connectivity $\leq d(v)$. This completes the proof.

Theorem 3: The vertex connectivity of any graph G is always less than or equal to edge connectivity of G.

Proof: Let p denote the edge connectivity of G. Therefore, there exists a cut-set S in G containing p edges. Then S partitions the vertices G into two subsets V_1 and V_2 such that every edge in S joins a vertex in V_1 to a vertex in V_2. By removing at most p vertices from V_1 or V_2 on which the edges of S are incident, we will be able to delete S from G

(together with all other edges incident on these vertices). Thus removal of at most p vertices from G will disconnect the graph. Hence vertex connectivity is less than or equal to p.

Corollary: Every cut-set in a non-separable graph with more than two vertices contains at least two edges.

Proof: By definition, a graph G is non-separable if its vertex connectivity is at least two. In view of above theorem, edge connectivity $\geq$ vertex connectivity. Hence edge connectivity of a non-separable graph is at least two which is possible if the graph has at least two edges.

Theorem 4: The maximum vertex connectivity of a connected graph with n vertices and e edges is the integral part of the number $2e/n$.

(The integral part of a number x is denoted as $[x]$)

Proof: We know that every edge of a graph contributes two degrees. Thus sum of degrees of all the vertices in G is $2e$. Since this sum ($2e$) is divided among n vertices, therefore there must be at least one vertex in G whose degree is less than or equal to $2e/n$.
Therefore

edge connectivity of $G \leq 2e/n$ [By Theorem 2]

$\Rightarrow$ vertex connectivity $\leq$ edge connectivity $\leq 2e/n$ [By Theorem 3]

$\Rightarrow$ maximum vertex connectivity possible $= [2e/n]$, the integral part of $2e/n$.

Remark: We can construct a graph with n vertices and e edges having maximum vertex connectivity as follows:
We first draw an n-regular graph in which degree of each vertex is $[2e/n]$ and then add the remaining edges arbitrarily.

Network Flows

Nowadays, we are largely governed by networks-such as network of telephone lines, network of transportation (highways, rails, etc). Thus the mathematical analysis of such networks has become very important. In this section, we will see that network analysis is essentially a study of graphs.

A *network* is represented by a weighted connected graph that contains no self-loops. The vertices denote the stations (or places) and the edges are links through which the given commodity (such as oil, gas, number of messages, etc) flows.

The weight associated with each edge represents the capacity of the edge. The capacity of an edge is the maximum amount of material which can flow through the edge.

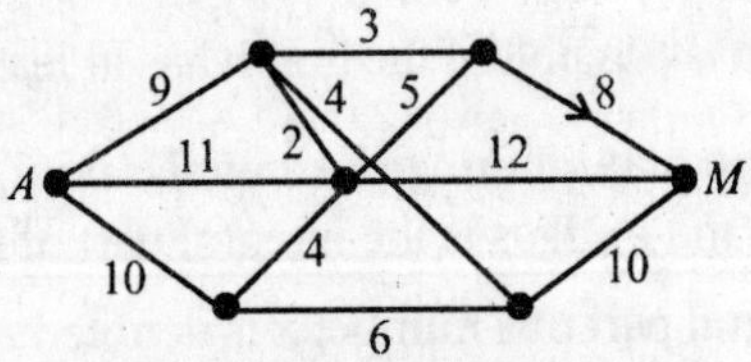

Fig: 8.24 A network of flow

The following assumptions are made for network problems:

(i) At any intermediate vertex, the amount of material entering is equal to the amount of material leaving.

(ii) The flow in any edge cannot exceed the capacity of that edge.

(iii) There is no loss of material during the flow.

In network flows, we are mainly interested in finding the maximum flow possible between a specified pair of vertices. In general, a network can have several different maximum flows but their values will be equal. That is maximum flow (route) is not unique but its value is unique.

Cut-set with respect to a Pair of Vertices

A *cut-set with respect to a pair of vertices a* and *b* in a connected graph *G* is a cut-set of *G* that puts *a* and *b* into two different components. For example, in Graph of Fig. 8.18(a), cut-set $\{b, c, e, d\}$ is a cut-set with respect to v_1 and v_4. The cut-set $\{a, b\}$ is also a cut-set respect to v_1 and v_4. But the cut-set $\{g, h, j\}$ is not a cut-set with respect to v_1 and v_4 because this cut-set does not separate v_1 and v_4.

Capacity of a Cut-set

The capacity of a cut-set S in a weighted connected graph G is defined as the sum of the weights of all the edges in S.

Theorem 4: (Max-flow min-cut theorem). The maximum flow possible between two given vertices a and b in a network is equal to the minimum of the capacities of all cut-sets with respect to a and b.

Proof: Let S be a cut-set with respect to the vertices a and b in a graph G. Since by definition of cut-set, after removal of S from G, there is no path between a and b. Therefore, we conclude that every path in G between a and b must contain at least one edge from S. Thus every flow from a to b or from b to a must pass through one or more edges of S. Therefore,

Value of any flow between a and $b \leq$ Capacity of any cut-set with respect to a and b.

$\Rightarrow$ maximum flow between a and $b \leq$ minimum of capacity of all cut-sets with respect to a and b.

Example 8.22: Find the maximum flow between a and b in the diagram below. In this figure arrows indicate the direction of flows from the node (vertex).

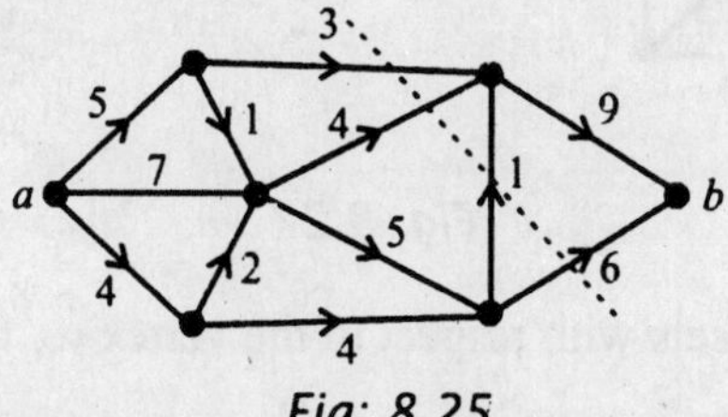

Fig: 8.25

Solution: The minimal cut-set with respect to a and b is shown by dotted line in the network. The capacity of this cut-set is 3+4+1+6 = 14. Thus maximum flow possible between a and b is 14.

Example 8.23: Construct a graph with the following properties:

Edge connectivity of $G = 4$, vertex connectivity = 3, and degree of every vertex of $G \geq 5$.

Solution: Consider the graph as given in fig. 8.26:

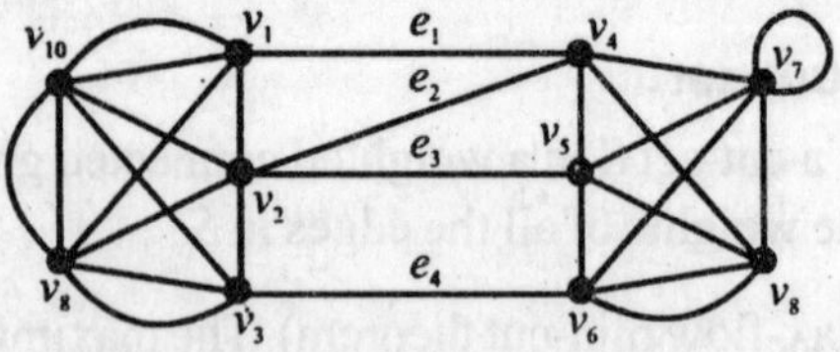

Fig: 8.26

The vertex connectivity of this graph is three because removal of v_1, v_2, v_3 or v_4, v_5, v_6 disconnects the graph. The edge connectivity of this graph is four. $S = \{e_1, e_2, e_3, e_4\}$ is one such cut-set. Observe that degree of any vertex can be increased by drawing parallel edges or by making a self-loop.

Exercise

(1) Show that the edge connectivity and vertex connectivity of the graphs in the following figure is each equal to three.

(a) (b)

Fig: 8.27

(2) List all cut-sets with respect to the vertex v_1, v_2 in the graph

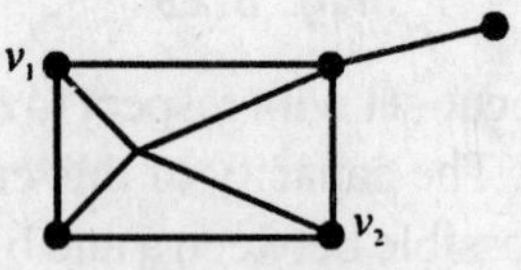

Fig: 8.28

(3) (a) Give an example of a graph whose edge connectivity is one.

(b) Give an example of a graph whose vertex connectivity is one.

(c) Give an example of a graph whose edge connectivity and vertex connectivity both are equal to one.

(4) Prove that in a connected graph G the complement of a cut-set in G does not contain a spanning tree and the complement of a spanning tree does not contain a cut-set.

(5) Give an example of a simple regular graph of degree three that is separable.

(6) Obtain the maximum flow in the network of Fig. 8.29

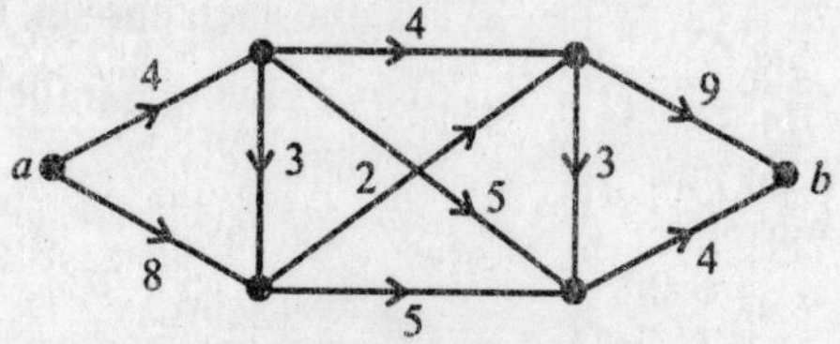

Fig: 8.29

(Arrow represents direction of the flow).

(7) Prove that a graph G will be non-separable iff every pair of its vertices could be present in some circuit in G.

Answers

(3) Consider the adjacent graph. This graph has edge connectivity and vertex connectivity both equal to one.

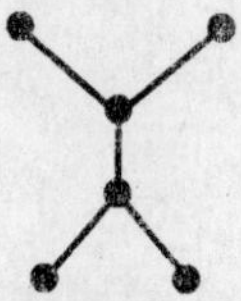

(5)

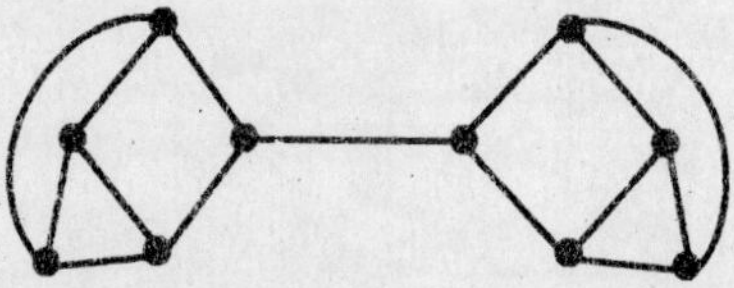

A regular graph of degree 3 that is separable.

(6) 12

CHAPTER NINE

Planar Graphs and Colourings

In this chapter we investigate some more important concepts in graph theory such as planarity, duality and colouring of graphs etc. First we consider planar graphs and give some properties of these ;raphs.

When we colour the map of a given country, we want that its adja-ent states are coloured differently, for easy identification of the states. natural question is: what is the minimum number of colours required paint any map so that adjacent states receive different colours? This uestion was first raised by F.G. Guthrie, in a letter to De Morgan in 852. It has been shown that five colours are sufficient for colouring a ap. However in 1976, Appel and Haken proved that a planar graph an be coloured with four colours only.

9.1. Planar Graphs

graph G is said to be **planar** if the graph can be drawn in a plane so that o edges cross except at vertices. For example, the complete graph ith four vertices is usually represented as in Fig. 9.1(a). But this raph can also be drawn with non crossing of edges as in Fig. 9.1(b) or ig.9.1(c).

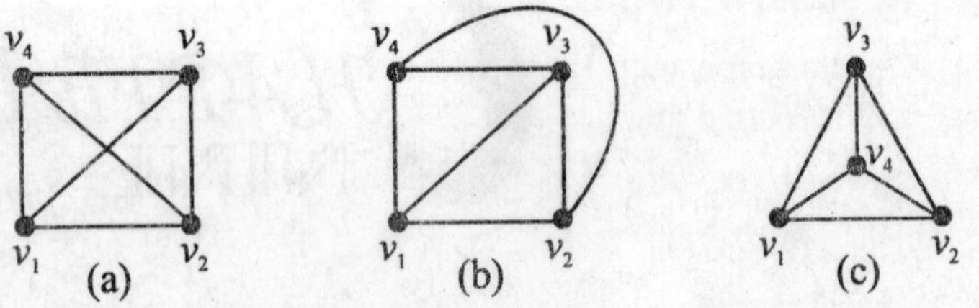

Fig: 9.1 A graph drawn in three different ways

Hence the complete graph with four vertices is planar.

A graph G that cannot be drawn on a plane without crossings of its edges is called **non-planar**. A drawing of a planar graph G on a plane or surface such that no edges intersect is called **embedding** of G. In order to declare that a graph is non-planar, we have to show that it is not possible to draw the graph on a plane without crossings of edges.

As mentioned in Chapter 7, a graph G is a pair (V, E) where the set V consists of objects called vertices and the set E consists of objects called edges and each edge is identified with a pair of vertices. Graph given in this form is called *combinatorial graph* or *abstract graph*. We have represented the given abstract graph by means of a figure. This representation is called **geometric graph**. Hence a given combinatorial graph can have several different geometric graphs. For example, the graph in fig. 9.1 has three different geometric graphs. A particular planar representation of a finite planar graph is called **map**.

If we look at the two planar embeddings of a graph in Fig. 9.1(b) and (c) we find that in one of these (Fig. 9.1(b)) one of the edges is drawn as a curved line while in the other (Fig.9.1(c)) all the edges are drawn as straight line segments. This raises the question whether every planar graph can be drawn in the plane so that each edge is represented by a straight line. Since self-loops or parallel edges in any graph cannot be drawn as straight lines, the answer to the above question is yes for simple planar graphs as stated in the following theorem. The proof of the theorem is beyond the scope of this book.

Theorem1: Every simple planar graph can be drawn in a plane such that every edge is a straight line.

Kuratowski's Two Graphs

A complete graph with five vertices is known as first Kuratowski's graph. The second graph of Kuratowski is $K_{3,3}$. It is a regular connected graph with six vertices and nine edges as shown in Fig. 9.2(b).

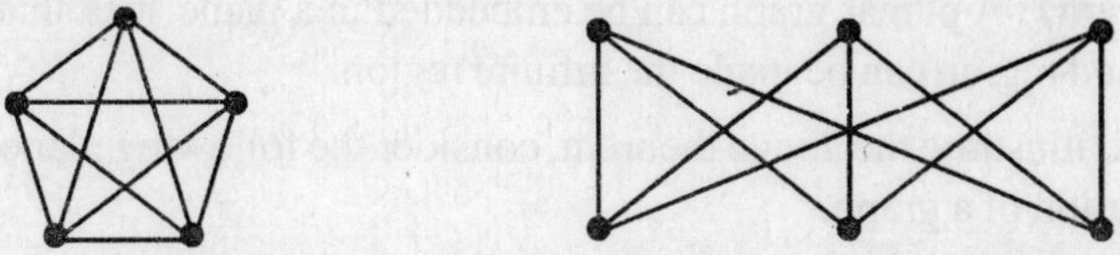

(a) K_5, Kuratowski's first graph

(b) $K_{3,3}$ Kuratowski's second graph

Fig: 9.2

We shall show that both these graphs are non-planar. Kuratowski's second graph is also known as *utility graph*. We may notice that K_5 is the non-planar graph with the smallest number of vertices. In other words, every graph with four vertices is planar. $K_{3,3}$ is the non-planar graph with the smallest number of edges. Thus both are the simplest non-planar graphs.

Regions: A plane representation of a planar graph G divides the plane into **regions**. A region (also called faces or meshes) is characterized by the set of edges forming its boundary. A region of a planar graph is an area of the plane that is bounded by edges and is not further divided into subareas. A region is finite if the area it encloses is finite. Otherwise it is called infinite. For example, the planar graph shown in Fig. 9.3 has six regions (or faces). Regions f_1, f_2, f_3, f_4 and f_5 are finite regions while f_6 is

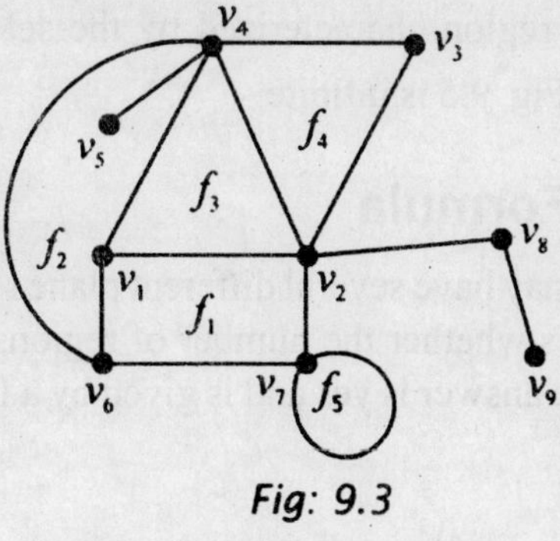

Fig: 9.3

infinite region. Clearly, every planar graph has exactly one infinite region.

Observe that the edges on the boundary of a region contain exactly one circuit and this circuit is said to enclose the region. We now state an important result without proof.

Theorem2: A planar graph can be embedded in a plane such that any specified region can be made the infinite region.

To illustrate the above theorem, consider the following plane representation of a graph.

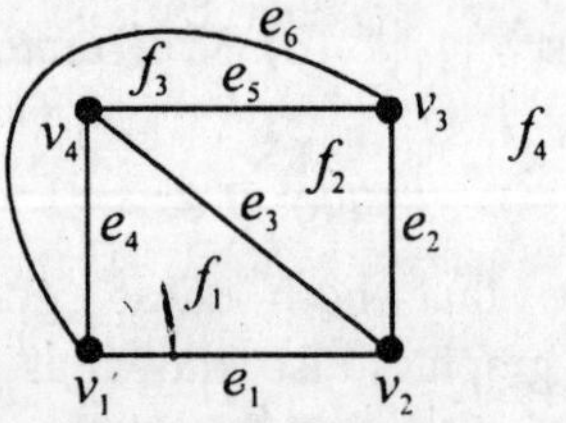

Fig: 9.4

In this plane representation, infinite region f_4 is characterized by the set $\{e_1, e_2, e_6\}$ of edges. Suppose we wish that finite region f_1 characterized by the set $\{e_1, e_3, e_4\}$ be made infinite region. Then we redraw the graph as given below:

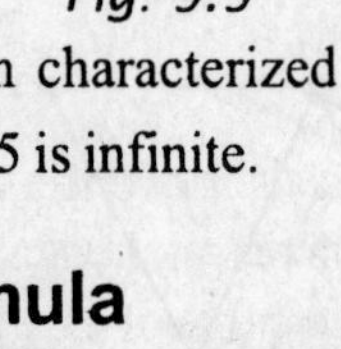

Fig: 9.5

We see that the region characterized by the set $\{e_1, e_3, e_4\}$ of edges in the graph of Fig. 9.5 is infinite.

§ 9.2. Euler's Formula

Since a planar graph may have several different plane representation, a natural question arises whether the number of regions in each representation is same. The answer is yes and is given by a formula, known as *Euler's formula*.

Theorem 3: A connected planar graph with n vertices and e edges has $e-n+2$ regions.

First Proof: Let G be a connected planar graph. We shall prove the result by induction on the number of edges of G. If $e=1$ then $n=1$ or 2 as shown in Fig. 9.6

e=1, n=2 (a) e=1, n=1 (b)

Fig: 9.6

When $e=1$ and $n=2$, we have $e-n+2=1$. Clearly graph in Fig. 9.6(a) has one region. When $e=1$ and $n=1$, then we have $e-n+2=2$. Also graph in figure 9.6(b) has two regions. Thus the result is true for $e=1$.

We assume that the result is true for all graphs with at most $e-1$ edges. Let G be a connected graph with e edges and f regions. If G is a tree then $e=n-1$ and number of faces is 1 (only infinite region). Also by the theorem, number of faces $= e-n+2 = 1$. Hence theorem is true in this case. If G is not a tree then it has some circuits. Let a be an edge in some circuit. Removal of the edge a from the plane representation of G will merge the two regions into one new region. Thus $G-a$ is a connected graph with n vertices, $e-1$ edges and $f-1$ regions (where f is the number of regions in G). By induction hypothesis, we have

$$f-1=(e-1)-n+2$$

$$\Rightarrow \quad f=e-n+2.$$

This completes the proof.

Second proof: It is sufficient to prove the theorem for simple graph because adding a self-loop or a parallel edge simply adds one region to the graph and also increases the value of e by one. We can also disregard (*i.e.* eliminate) all edges that do not form boundaries of any region because removal of any such edge decreases the value of e by one and decreases the value of n by one keeping the quantity $e-n+2$ unchanged.

Since any simple planar graph can have a plane representation such that each edge is drawn as straight line, we draw the given planar graph in such a fashion that each region is a polygon. Let the polygonal graph representing the given graph consist of f regions and let k_p be the number of p-sided regions (for example, the value of k_p in Fig. 9.5 is 4). Since each edge lies on the boundary of exactly two regions,

$$3 \cdot k_3 + 4 \cdot k_4 + 5 \cdot k_5 + \ldots\ldots + r \cdot k_r = 2e \quad \ldots\ldots\ldots\ldots\ldots\ldots(1)$$

where k_r is the number of polygons with maximum edges.

Also, we have

$$k_3 + k_4 + \ldots\ldots + k_r = f \quad \ldots\ldots\ldots\ldots\ldots\ldots\ldots\ldots(2)$$

From geometry, we know that the sum of all interior angles of a p-sided polygon is $(p-2)\,\pi$ and the sum of the exterior angles is $(p+2)\pi$. Moreover, only one region is infinite and remaining $f-1$ regions are finite. Without loss of generality, we may assume that one region out of k_r regions of r-sided polygon is infinite. Thus the sum of all angles subtended at each vertex is

$$(3-2)\pi \cdot k_3 + (4-2)\pi\, k_4 + \ldots\ldots\ldots + (r-2)\pi\,(k_r - 1) + (r+2)\pi = 2n\pi$$

$$\Rightarrow \quad (3k_3 + 4k_4 + \ldots\ldots\ldots + rk_r)\pi - 2\,(k_3 + k_4 + \ldots\ldots\ldots + k_r) + 4\pi = 2n\pi$$

$$\Rightarrow \quad 2e\pi - 2f\pi + 4\pi = 2n\pi \qquad \text{[using (1) and (2)]}$$

$$\Rightarrow \quad f = e - n + 2$$

This completes the proof.

As an example of the above theorem, we consider the following graph in Fig. 9.7. This graph has $n = 8$, $e = 12$ and $f = 6$.

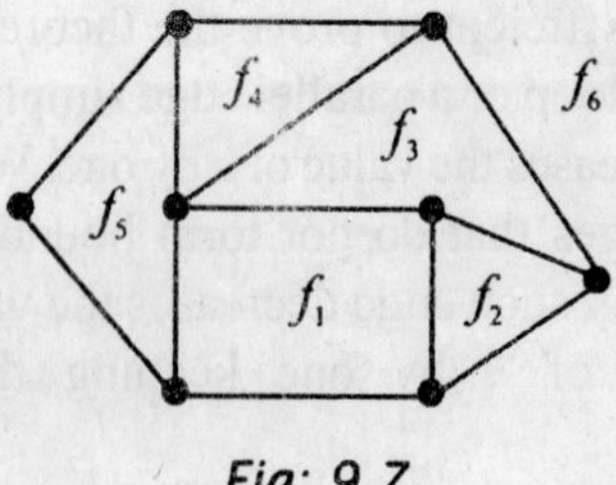

Fig: 9.7

From the Euler's formula, we have

$$f = e - n + 2 = 6$$

Remark: In the graph of Fig. 9.7, we have $k_3 = 3, k_4 = 2, k_5 = 0, k_6 = 1$. Thus $k_3 + k_4 + k_5 + k_6 = 6 =$ number of regions.

Corollary1: If a connected simple planar graph G has $n \geq 3$ vertices, e edges and f regions then

$$3f \leq 2e \quad \ldots\ldots\ldots\ldots\ldots\ldots\ldots (1)$$

and

$$e \leq 3n - 6 \quad \ldots\ldots\ldots\ldots\ldots\ldots\ldots (2)$$

Proof: Since each region in a simple connected graph is bounded by atleast three edges and each edge belongs to exactly two regions, we have

$$3f \leq 2e$$

Now substituting f from Euler's formula, we have

$$3(e - n + 2) \leq 2e$$

$$\Rightarrow \quad e \leq 3n - 6$$

Corollary2: K_5 is non-planar.

Proof: K_5, the complete graph of five vertices has $n = 5, e = 10$.

If K_5 is planar then by corollary 1, we have

$$e = 10 \leq 3n - 6 = 9,$$

a contradiction. Thus K_5 is non-planar

Corollary 3: $K_{3,3}$ is non-planar.

Proof: To prove that $K_{3,3}$ is non-planar, we make use of the fact that no

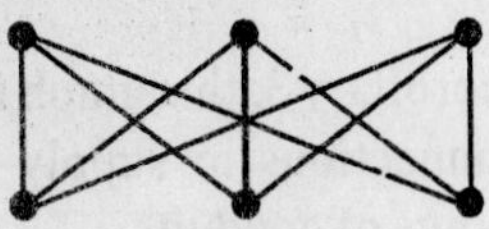

Fig: 9.7 graph $K_{3,3}$

region in $K_{3,3}$ is bounded with fewer than four edges. Hence if this graph were planar, we would have

$$4f \leq 2e$$

and substituting for f from Euler's formula, we get

$$4(e-n+2) \leq 2e$$

$$\Rightarrow \qquad e \leq 2n-4$$

In $K_{3,3,}$ we have $e = 9$ and $n = 6$.

Therefore $9 \leq 2 \times 6 - 4 = 8$, a contradiction.

Hence $K_{3,3}$ is non-planar.

Solved Examples

Example 9.1: (Utility Problem) Suppose there are three houses H_1, H_2 and H_3 and each house is to be connected to each of the three utilities water (W), gas (G) and electricity (E) by means of conduits. Is it possible to make such connections without any crossings of the conduits?

Solution: We can represent this problem by means of a graph as shown in Fig 9.8 below. The conduits are shown as edges while the houses and utility supply centres as vertices.

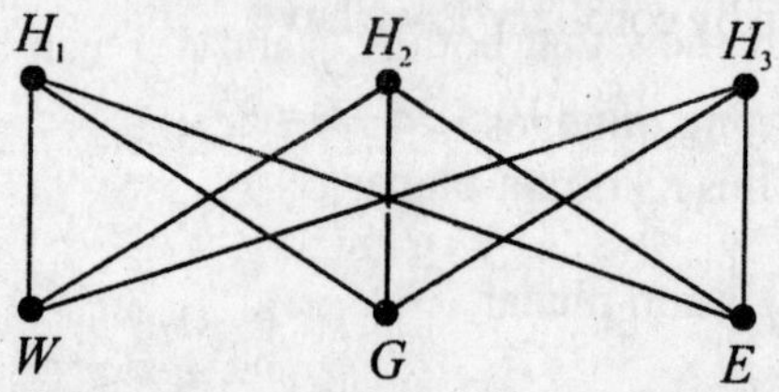

Fig: 9.8

As we have seen in corollary 3, this graph is not planar. Hence it is not possible to make connections to supply three utilities to three houses without any crossings of conduits.

Example 9.2: Show that the graphs in Fig. 9.9 are planar.

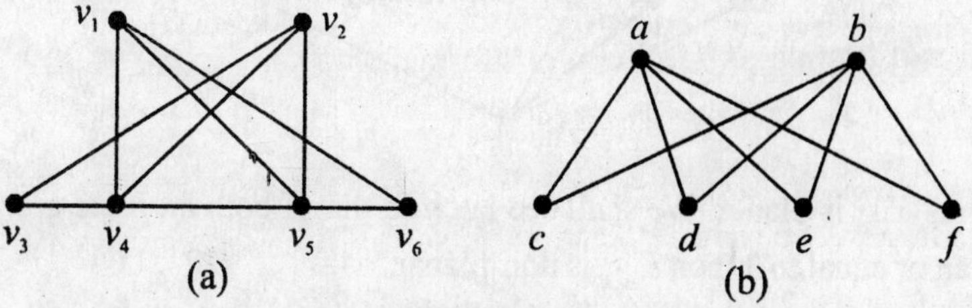

Fig: 9.9

Solution: The graphs in Fig 9.9 are planar because we can draw these graphs in a plane without crossing of edges as shown below:

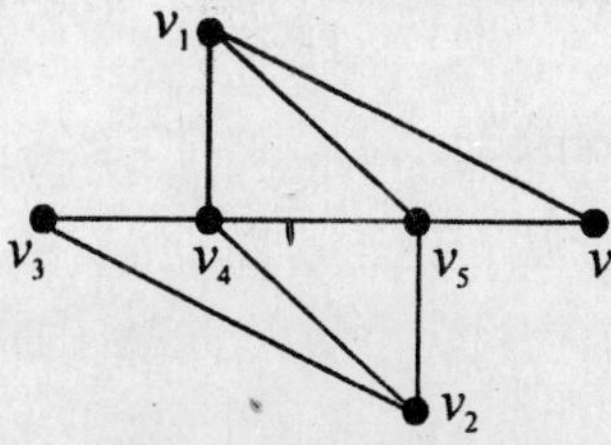

Planar embedding of Fig. 9.9(a)

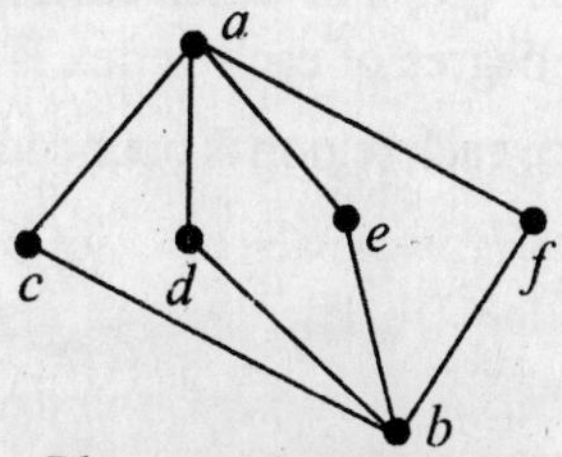

Planar embedding of Fig. 9.9(b)

Example 9.3: Show that a complete bipartite graph $K_{m,n}$ is planar if m or n is less than or equal to 2.

Solution: We shall show that both $K_{1,n}$ and $K_{2,n}$ can be drawn on a plane without crossing of edges.

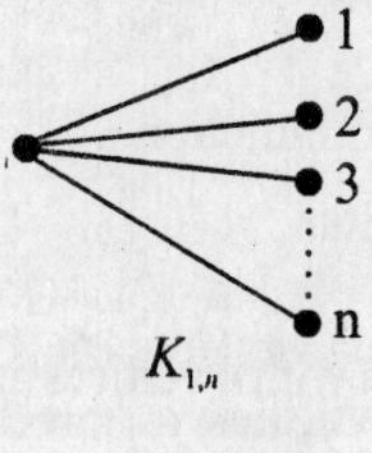

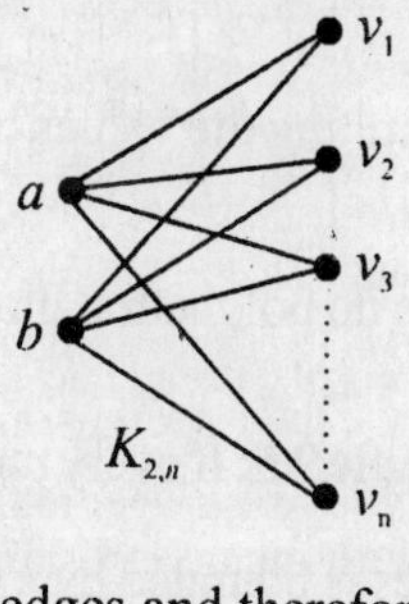

Clearly the graph $K_{1,n}$ is without crossing of edges and therefore it is planar. We redraw $K_{2,n}$ as given below:

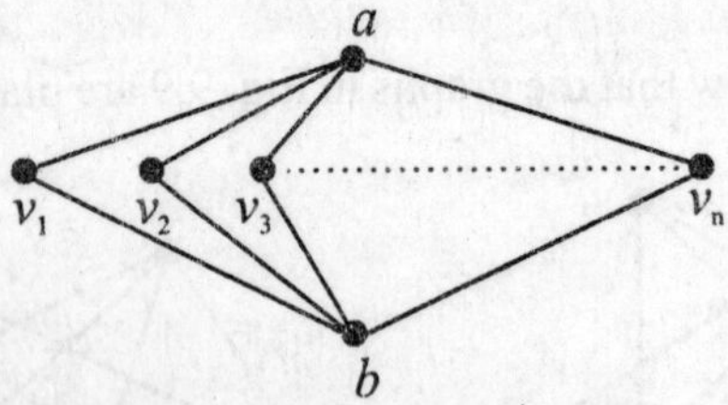

Thus $K_{2,n}$ is planar. We shall see later on that if both m, n are greater than or equal to 3 then $K_{m,n}$ is non-planar.

Example 9.4: Prove that there exists no polyhedral planar graph with exactly 7 edges.

Solution: We know that a polyhedral graph is a regular connected planar graph in which each region is bounded by three or more edges and degree of each vertex is ≥ 3.

Since each region is bounded by 3 or more edges,

$$3f \leq 2e = 2 \times 7 = 14$$

$$\Rightarrow \qquad f \leq \frac{14}{3}$$

Since f must be integer, therefore, $f = 4$.

Again, if the graph has n vertices, then

$$3n \leq 2e = 14$$

$$\Rightarrow \qquad n \leq \frac{14}{3}$$

$\Rightarrow$ n must be less than or equal to 4.

From Euler's formula, we have

$$f + n = e + 2$$

Substituting the values of e, n and f, we get

$$4+4 \leq f+n = 7+2 = 9, \text{ a contradiction.}$$

Hence no polyhedral graph with exactly 7 edges exits.

Example 9.5: If every region of a simple planar graph with n vertices and e edges is bounded by k edges, show that $e = \dfrac{k(n-2)}{(k-2)}$

Solution: Since every region is bounded by k edges, therefore, we have

$$kf = 2e, \text{ where } f \text{ is the number of regions.}$$

Substituting the value of f from this equation into the Euler's formula

$$f = e - n + 2.$$

We have

$$\frac{2e}{k} = e - n + 2,$$

$$\Rightarrow \quad (n-2)k = (k-2)e$$

$$\Rightarrow \quad e = \frac{k(n-2)}{k-2}$$

Note: If in Example 9.5, each region is bounded by k or more edges then $e \leq \frac{k(n-2)}{k-2}$

Example 9.6: In a simple planar graph G of n vertices ($n \geq 3$), there is atleast one vertex of degree ≤ 5.

Solution: Let G be a simple planar graph with n vertices and e edges. If possible, suppose that degree of each vertex of G is greater than 5. Then

$$2e \geq 6n \quad \text{(because sum of degrees of } G \text{ is atleast } 6n\text{)}$$

or

$$e \geq 3n$$

But by corollary 1 of Euler's formula,

$$e \leq 3n - 6$$

These two inequalities contradict one another. Hence some vertex of G has degree ≤ 5.

Example 9.7: For a connected planar graph with $n \geq 3$ vertices, show that

$$5V_1 + 4V_2 + 3V_3 + 2V_4 + V_5 \geq V_7 + 2V_8 + \ldots\ldots + (k-6)V_k + 12$$

where V_r denotes the number of vertices with degree r and k is the largest degree of a vertex in G.

Solution: For a connected planar graph, we have

$$e \leq 3n - 6 \quad \text{[by corollary 1]}$$

$$\Rightarrow \quad 6n - 2e \geq 12 \quad \ldots\ldots\ldots\ldots(1)$$

Also,

$$V_1 + V_2 + V_3 + \ldots\ldots + V_k = n \quad \ldots\ldots\ldots\ldots(2)$$

where V_r denotes the number of vertices with degree r,

and

$$V_1 + 2V_2 + 3V_3 + \ldots\ldots + kV_k = 2e \quad \ldots\ldots\ldots\ldots(3)$$

Substituting n and $2e$ from equations (2) and (3) into equation (1), we get

$$6(V_1 + V_2 + \ldots\ldots + V_k) - (V_1 + 2V_2 + \ldots\ldots + kV_k) \geq 12$$

or, $$5V_1 + 4V_2 + 3V_3 + 2V_4 + V_5 \geq V_7 + 2V_8 + \ldots\ldots + (k-6)V_k + 12$$

Example 9.8: Prove that a simple planar graph of n vertices ($n \geq 4$) has at least four vertices with degree five or less.

Solution: We assume without loss of generality that the graph has no pendant vertices. Moreover creation or elimination of a vertex of degree two does not affect planarity so we assume that the graph has no vertex of degree 2 also, Thus, we have

$V_1 = V_2 = 0$ in Example 9.7.

$$\therefore \; 3(V_3 + V_4 + V_5) \geq 3V_3 + 2V_4 + V_5 \geq 12$$

$$\Rightarrow V_3 + V_4 + V_5 \geq 4$$

Hence the graph has at least four vertices with degree five or less.

Example 9.9: Redraw the graph in Fig. 9.3 such that f_3 becomes the infinite region.

Solution: We can redraw the graph in Fig. 9.3 as given below:

Observe that the region f_3 is characterized by the edges (v_1, v_2), (v_1, v_4) and (v_2, v_4) in Fig. 9.3.

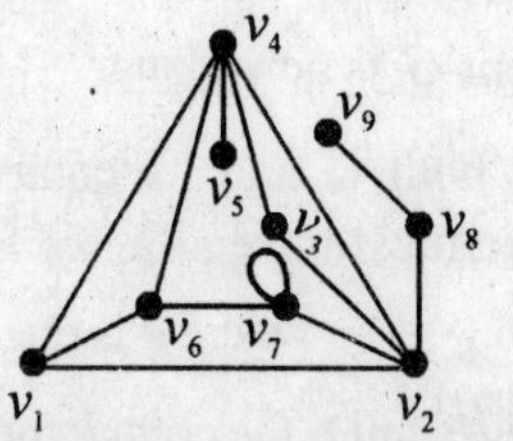

The finite region f_3 has become infinite region in the above figure.

Example 9.10: Show that a simple planar graph with less than 30 edges has a vertex of degree 4 or less.

Solution: Suppose that degree of each vertex is ≥ 5. Then

$$2e \geq 5n \quad \ldots\ldots\ldots\ldots\ldots\ldots\ldots\ldots(1)$$

Further, since graph is simple planar, each region is bounded by atleast 3 edges. Therefore

$$2e \geq 3f \quad \ldots\ldots\ldots\ldots\ldots\ldots\ldots\ldots(2)$$

By Euler's formula, we have

$$e = f + n - 2$$

Substituting n and f from (1) and (2), we get

$$e \leq \frac{2e}{3} + \frac{2e}{5} - 2 = \frac{(16e - 30)}{15}$$

⇒ $e \geq 30$, a contradiction because the given graph has less than 30 edge. Thus it follows that there is at least one vertex of degree ≤ 4.

Example 9.11: Show that in a connected planar simple graph with 6 vertices and 12 edges, each of the regions is bounded by 3 edges.

Solution: In the given graph, we have

$$e = 12, n = 6$$

Therefore, by Euler's formula,

$$f = e - n + 2 = 12 - 6 + 2 = 8$$

Hence the given graph has eight regions. Since the graph is simple, each region is bounded by atleast three edges, we have, $2e \geq 3f$. Since $e = 12$ and $f = 8$, equality holds and hence each region is bounded by 3 edges.

Example 12: Let G be a graph with 11 or more vertices. Show that either G or its complement $\bar{G}$ is non-planar.

Solution: Suppose that both G and its complement $\bar{G}$ are planar. Then for the graph G, we must have

$$e \le 3n-6, \qquad \text{where } e \text{ is the number of edges of } G.$$

Let e' be the number of edges in $\bar{G}$, the complement of G. Since $\bar{G}$ is also planar, we have

$$e' \le 3n - 6$$

Thus $\quad e + e' \le 6n - 12.$

But $\quad e + e' = \dfrac{n(n-1)}{2}$, number of edges in complete graph.

Therefore,

$$\frac{n(n-1)}{2} \le 6n - 12$$

$$\Rightarrow \quad n^2 - 13n + 24 \le 0$$

$\Rightarrow$ n must be less than 11, a contradiction to the given condition that G has 11 or more vertices. Hence both G and its complement together cannot be planar.

Exercise

(1) Verify Euler's formula for

(a) the wheel graph W_8

(b) the complete bipartite graph $k_{2,5}$

(c) the graph of Fig. 9.7

(2) Redraw the graph in Fig. 9.7 with f_3 as the infinite region

(3) The *girth* of a graph G is the length of a shortest circuit in G. If G is a connected planar graph of girth 5(*i.e.*,each region is bounded by 5 or more edges) then show that $e \le 5(n-2)/3$, where n and e are the number of vertices and edges in G.

(4) Use problem 3 above to deduce that the the Petersen graph is non-planer (Hint: each region in Petersen graph is bounded by 5 edges)

(5) Is it possible to obtain planar representation for each of the following graphs? If yes, draw it. If not, explain why not?

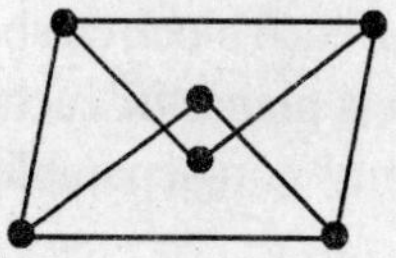

(a)

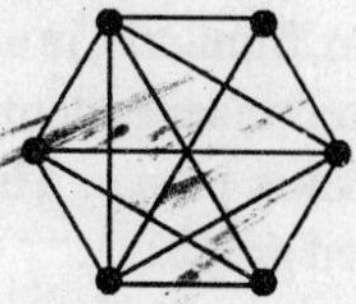

(b)

(6) Prove that no polyhedral graph with exactly 30 edges and 11 regions exists.

(7) For a connected simple planar graph G with n vertices and e edges, show that

(a) $n \geq 2 + \frac{f}{2}$ (b) $f \geq 2 + \frac{n}{2}$

(8) Show that a connected planar graph with 7 vertices and degree of each vertex equal to 4 must have 8 regions of degree 3 and 1 region of degree 4.

(9) Give an example of planar connected graph such that

(a) $e = 3n-6$ (b) $e < 3n-6$

(10) Show that every planar graph of less than 12 vertices has a vertex of degree less than or equal to 4.

(11) Show that a planar connected graph with more than 10 vertices has a complement which is not planar.

Answers

(5) (a) yes,

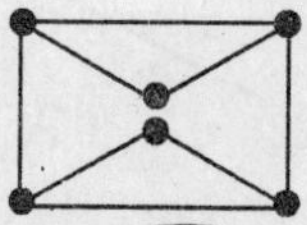

(b) No. Observe that $e \leq 3n-6$ is not satisfied.

(9) (a)

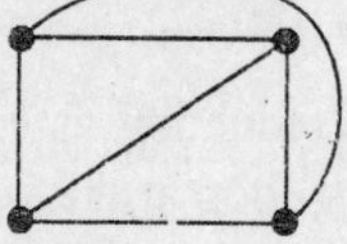

(b)

§ 9.3. Detection of Planarity

It may be noted that every subgraph of a planar graph is planar and that every graph which has a non-planar subgraph is also non-planar. Further, a disconnected graph is planar iff each of its components is planar. Similarly, in a separable graph, planarity of each block can be considered independently. Thus a separable graph is planar iff each of its blocks is planar. Thus we need to consider only non-separable connected graphs.

Let G be a non-separable connected graph. To check its planarity, we simplify (reduce) the graph G to a new graph H without disturbing its planarity using the following steps:

Step1: Since addition or removal of self-loops in a graph does not affect planarity of the graph, we remove all self-loops.

Step2: Since parallel edges do not affect planarity, we remove parallel edges except one between every pair of vertices.

Step3: Removal of a vertex of degree two by merging two edges in series does not affect planarity. Therefore we can merge all edges in series. (Two edges are said to be in series if they have exactly one vertex of degree two in common.)

Repeated use of steps 2 and 3 usually reduces a graph drastically without affecting its planarity. The following example illustrates the series-parallel reduction.

Example 9.13: Reduce (or simplify) the graph in Fig. 9.10 without affecting its planarity

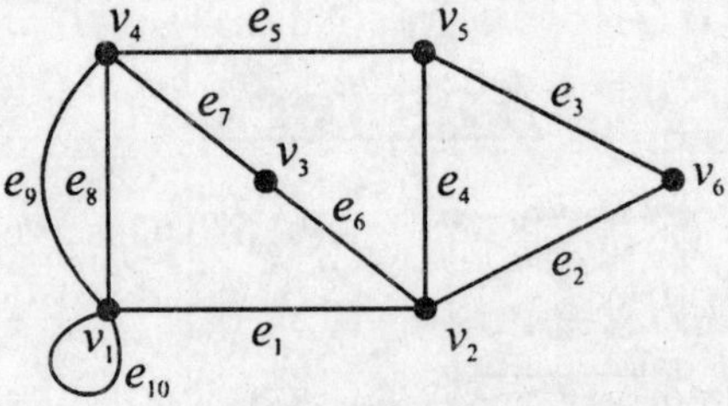

Fig: 9.10

Solution: We apply series-parallel reduction as explained above to this graph. By step1, we remove self-loop at the vertex v_1.

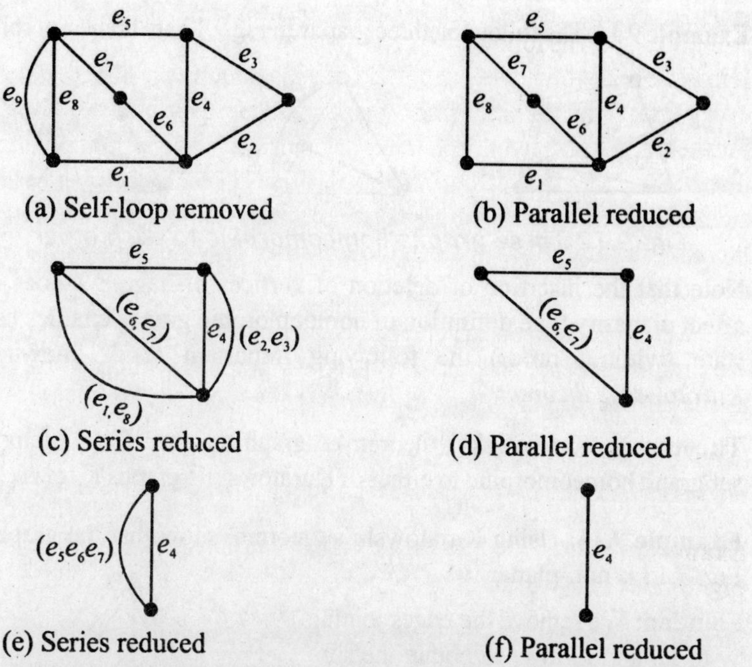

(a) Self-loop removed (b) Parallel reduced

(c) Series reduced (d) Parallel reduced

(e) Series reduced (f) Parallel reduced

Fig: 9.11 Series-Parallel reduction of graph in Fig. 9.10

It can be seen that if a non-separable connected graph G be reduced to a new graph H after applying steps 1 to 3, the graph H will be

(1) a single edge, or

(2) a complete graph of four vertices, or

(3) a non-separable simple graph with $n \geq 5$ and $e \geq 7$.

If the reduced graph H is of categories 1 or 2 then H is planar and hence G is planar. If H is of category 3, then we check whether $e \leq 3n - 6$. If this inequality is not satisfied then H and hence G is non-planar. If this inequality is satisfied, further investigation is needed and for this we first define the following:

Homeomorphic Graphs

Definition: Two graphs G_1 and G_2 are said to be *homeomorphic* if one graph can be obtained from the other by the creation of edges in series (in other words, by insertion of vertices of degree two) or by the merger of edges in series.

Example 9.14: The following three graphs in Fig 9.12 are homeomorphic.

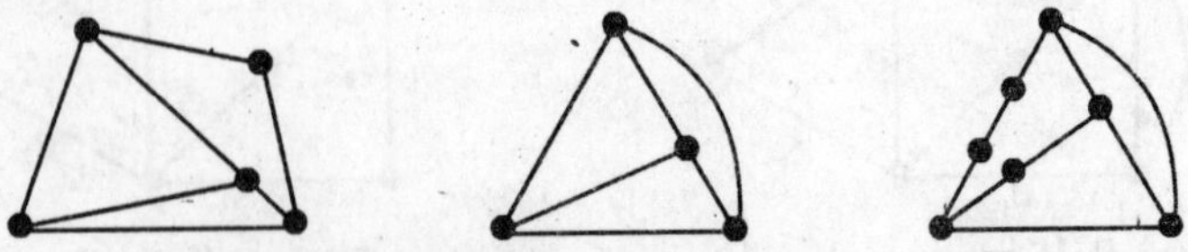

Fig: 9.12 Three graphs homeomorphic to each other.

Note that the insertion or deletion of vertices of degree 2 does not affect planarity. The definition of homeomorphic graphs enables us to state (without proof) the following important result known as *Kuratowski's theorem*

Theorem 1: (Kuratowski's theorem) A graph is planar iff it contains no subgraph homeomorphic to either of Kuratowski's graphs K_5 or $K_{3,3}$.

Example 9.15: Using Kuratowski's theorem, show that the graph in Fig. 9.13 is non-planar.

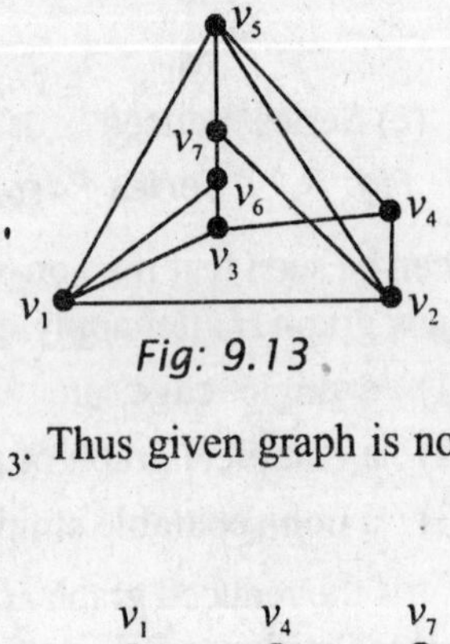

Fig: 9.13

Solution: We remove the edges joining v_1 to v_6, and v_2 to v_5 to obtain the sub-graph in Fig. 9.14(a). This subgraph is homomorphic (merge two series edges at vertex v_6) to the graph in Fig. 9.14(b). The graph in Fig. 9.14(b) can be redrawn as in Fig. 9.14(c) which is $K_{3,3}$. Thus given graph is non-planar by Kuratowski's theorem.

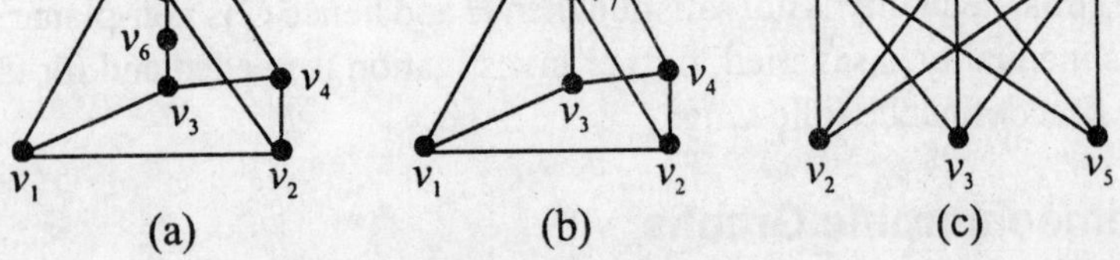

Fig: 9.14

Example 9.16: Show that the Petersen graph is non-planar.

Solution: Petersen graph is shown in Fig. 9.15

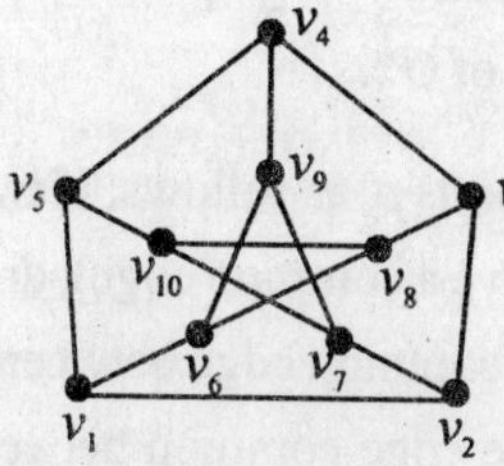

Fig: 9.15

We remove the vertex v_{10} to obtain the subgraph as shown in Fig. 9.16(a)

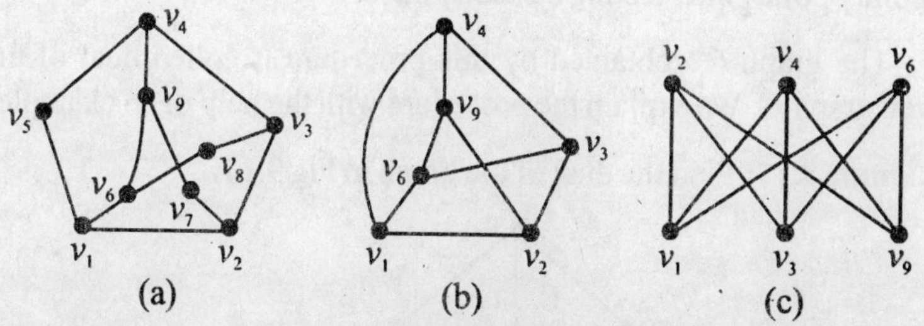

Fig: 9.16

Note that if we remove a vertex then all edges incident on that vertex are also removed. The graph in Fig. 9.16 (a) has three vertices v_5, v_7 and v_8 each of degree 2. This graph is homeomorphic to the graph in Fig 9.16(b). The graph in Fig. 9.16(b) can be redrawn as shown in Fig. 9.16(c) which is $K_{3,3}$. Hence by Kuratowski's theorem, Petersen graph is non-planar.

§ 9.4 Dual Graphs

Kuratowski's theorem gives us necessary and sufficient conditions for a graph to be planar. We now discuss another characterization of planar graphs. For this, we introduce the concept of duality in graphs.

Given a plane representation of a planar graph G, we construct another graph G^* called the (geometric) dual of G as follows:

Step 1: Inside each region F_i of graph G, we take a point p_i. These points p_i are the vertices of G^*.

Step 2: We join these points p_i as follows: If two regions F_i and F_j are adjacent (*i.e* F_i and F_j have a common edge), draw a line joining points p_i and p_j that crosses the common edge between F_i and F_j exactly once. If there is more than one edge common between F_i and F_j, draw one line between points p_i and p_j for each of the common edge.

Step 3: For an edge e lying entirely in one region, say F_k, draw a self-loop at point p_k of F_k intersecting e exactly once.

The graph G^* obtained by this procedure is called dual of the given graph G. We explain the procedure with the help of an example.

Example 9.17: Find the dual of the graph in Fig 9.17.

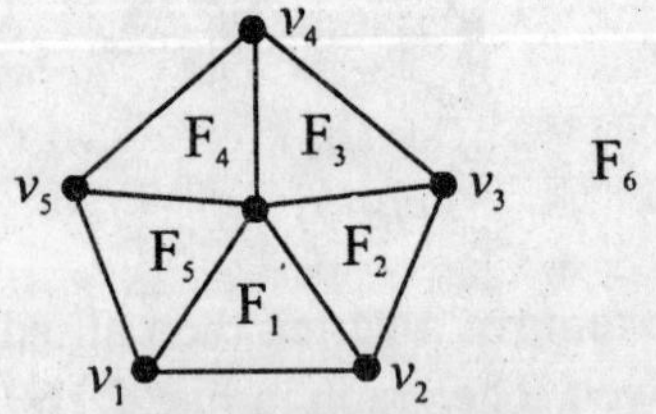

Fig: 9.17

Solution: This graph has six regions – one infinite and five finite.

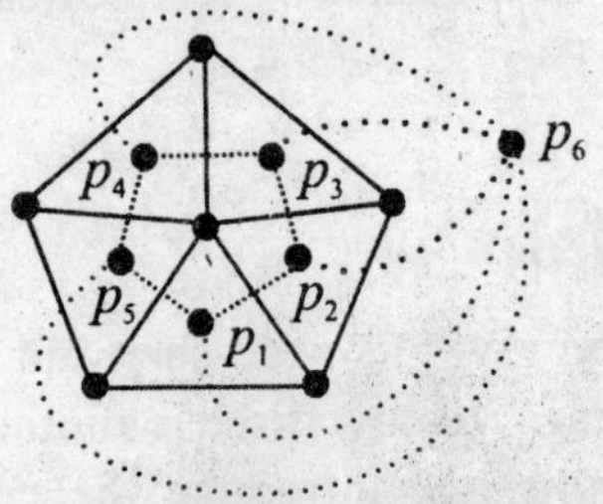

Fig: 9.18

We take a point p_i in each region F_i. The edges of G are shown by solid lines and the edges e^* of dual G^* are shown by dotted lines. We redraw the dual graph below:

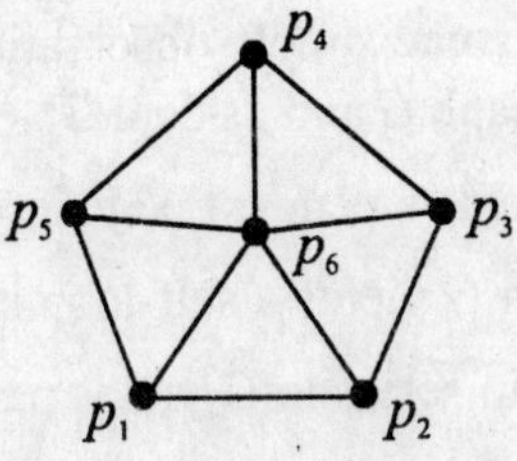

Fig: 9.19

Example 9.18: Find the dual of the complete graph of four vertices.

Solution: The complete graph of four vertices can be drawn on a plane as follows:

We take a vertex in each region and draw dotted lines as per the procedure given above

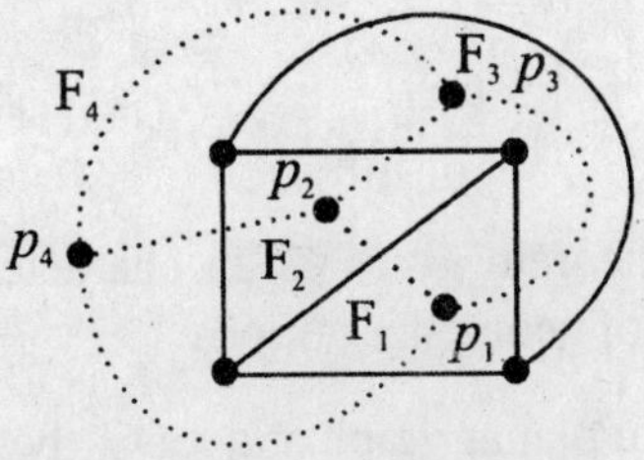

Fig: 9.20

The graph with vertices p_1, p_2, p_3 and p_4 and edges shown by dotted lines is the dual of the given graph. We redraw the dual as follows:

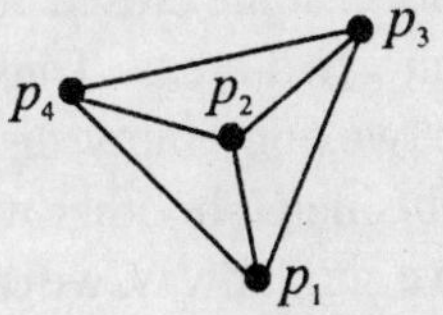

Fig: 9.21

Self-Dual Graphs

If a planar graph is isomorphic to its own dual, it is called *self-dual*. It is clear from the the Examples 9.17 and 9.18 that these graphs are self-dual. Hence the complete graph of four vertices and the wheel graph shown in fig. 9.18 are self-dual.

The following are some simple observations about the relationship between a planar graph G and its dual G^*.

1. Self-loop in G yields a pendant edge in G^*.
2. A pendant edge in G yields a self-loop in G^*
3. Edges which are in series in G yields parallel edges in G^*
4. Parallel edges in G produce edges in series in G^*
5. If a region F_i in G is bounded by k edges then the corresponding vertex p_i of G^* is of degree k and vice versa.
6. If e, n, f, r and μ denote the numbers of edges, vertices, regions, rank and nullity of a connected planar graph G and if e^*, n^*, f^*, r^* and m^* are the corresponding numbers in dual graph G^*, then

$$n^* = f,\ e^* = e,\ f^* = n$$

and

$$r^* = \mu\,,\ \mu^* = r.$$

The following theorem gives us an characterization of duality which is independent of geometric notions.

Theorem1: Let G be a planar graph and let G^* be a (geometric) dual of G. Then a set of edges in G forms a circuit in G if and only if the corresponding set of edges of G^* forms a cut-set in G^*.

Proof: Let C be a circuit in G and let C^* be the corresponding set of edges in G^*. Clearly C will form some closed curve in G dividing the plane representation of G into two areas. Thus the vertices of dual graph G^* are partitioned into two non-empty, disjoint subsets, namely, one inside C and the other outside C. In other words, the set of edges C^* in G^* is a cut-set in G^*. In a similar way, we can show that each cut-set in G^* corresponds to a circuit in G. This proves the theorem.

The next theorem gives another characterization of planar graphs.

Theorem 2: A graph G has a dual if and only if it is planar.

Proof: If the graph is planar the clearly we can draw its dual. Thus every planar graph has a dual graph. Next, we show that if a graph G has a dual then it is planar. Equivalently, we show that if a graph G is non-planar then it has no dual. Let G be a non-planar graph. By Kuratowski's theorem, G contains K_5 or $K_{3,3}$ or a graph homeomorphic to either of these. If G has a dual then $K_{3,3}$ or K_5 should have dual. Thus if we show that neither K_5 nor $K_{3,3}$ has a dual then our result will be proved.

(a) We first show that $K_{3,3}$ has no dual. First we observe that

1. $K_{3,3}$ has no cut-sets of two edges
2. $K_{3,3}$ has circuits of length four or six only
3. $K_{3,3}$ has nine edges

Suppose that $K_{3,3}$ has a dual D. Then by theorem 1, the cut-sets in $K_{3,3}$ corresponds to circuits in D and vice-versa. Since $K_{3,3}$ has no cut-set consisting of two edges, D has no circuit of length two. That is, D has no pair of parallel edges. Since every circuit in $K_{3,3}$ is of length four or six, D has no cut-set with less than four edges. Therefore, the degree of every vertex is at least four. Since D has no pair of parallel edges and degree of each vertex is at least four, therefore D must have at least five vertices, each of degree at least four. Thus D must have at least $\frac{1}{2} \times 5 \times 4 = 10$ edges. This is a contradiction, because $K_{3,3}$ has nine edges and so D should have only nine edges. Thus $K_{3,3}$ has no dual.

(b) We now show that K_5 has no dual. Suppose if possible H is the dual of K_5. Note that K_5 has

1. no cut-set with two edges,
2. cut-sets with four or six edges only,
3. no pair of parallel edges
4. 10 edges

Consequently, dual H must have

1. 10 edges
2. no vertex with degree less than three
3. no pair of parallel edges, and
4. circuits of length four or six only.

Now graph H contains a hexagon (because of a circuit of length six) and no more than three edges can be added to a hexagon without creating a circuit of length three or a pair of parallel edges. Since both of these are not allowed in H and H has 10 edges, therefore H must have atleast seven vertices. The degree of each vertex of H is atleast three, hence H must have at least 11 edges. A contradiction to the fact that K_5 has 10 edges. Hence K_5 has no dual.

§ 9.5 Thickness and Crossings

For a non-planar graph G, it is a natural to know, what is the minimum number of planes required for embedding G?

The *thickness* of a graph G is defined as the minimum number of planar subgraphs whose union is G. The thickness of a planar graph is one. The thickness of each of Kuratowski's graph is two. It can be shown by actually sketching that the thickness of the complete graph of eight vertices is two, whereas that of the nine vertices is three.

The second question which one might ask about a non-planar graph is, what is the minimum number of crossings in a non-planar graph. The **crossing number** of a graph G is the minimum number of crossings necessary in order to draw the graph on a plane. The crossing number of a planar graph is clearly zero and of K_5 or $K_{3,3}$ is one. In general, no formula exists to give the crossing number of an arbitrary graph.

Solved Examples

Example 9.19: Find the dual of the graph given below:

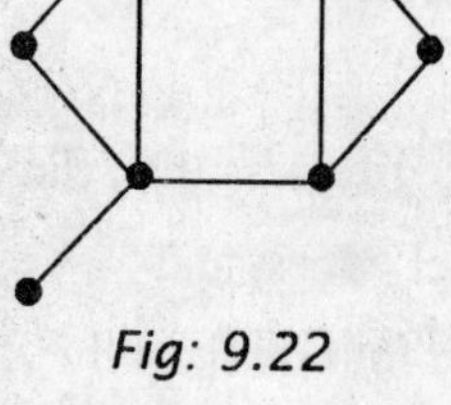

Fig: 9.22

Solution: The dual of the given graph is shown by dotted lines below:

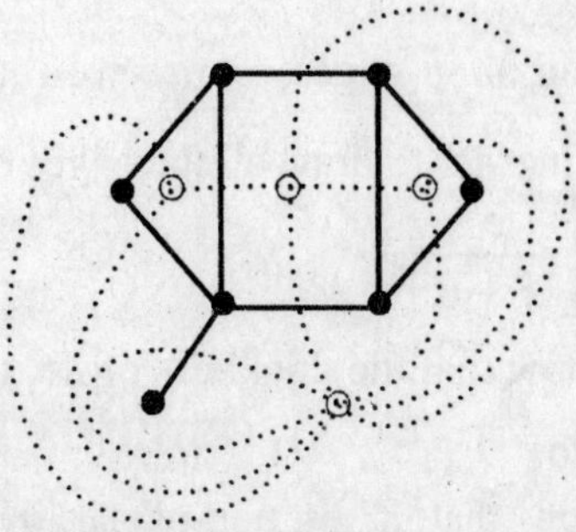

Fig: 9.23 Dual of graph in Fig. 9.22

Example 9.20: Draw the dual of the graph given below:

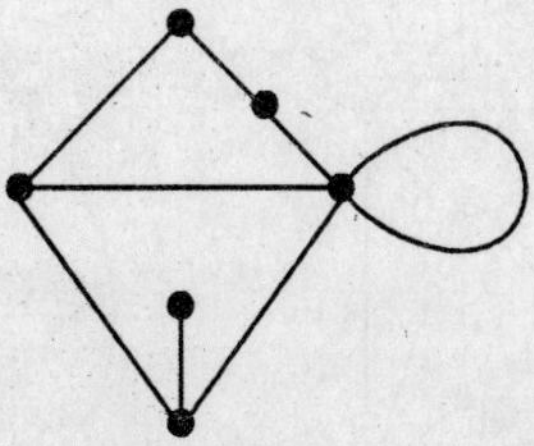

Fig: 9.24

Solution: Dual of the graph in Fig 9.24 is shown below by dotted lines.

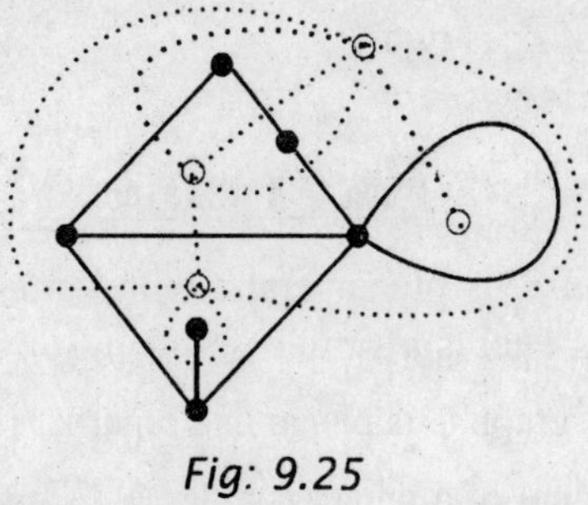

Fig: 9.25

Example 9.21 : Show that for any simple graph G with n vertices and e edges, thickness of $G \geq \left[\frac{e}{3n-6}\right]$

Solution: Let G be a simple graph with n vertices and e edges. By corollary to Euler's formula, we know that if a graph is planar then $e \leq 3n - 6$. That is, we cannot have a simple planar graph with n vertices and more than $3n–6$ edges. Hence we require atleast $\left[\frac{e}{3n-6}\right]$. planar subgraphs in order to draw all the edges of given graph. Hence thickness of $G \geq \left[\frac{e}{3n-6}\right]$..

Example 9.22: Show that the thickness of K_8, the complete graph of eight vertices is two.

Solution: We know that K_8 is non-planar. Hence its thickness is greater than 1. We draw below a pair of two planar subgraphs whose union is K_8.

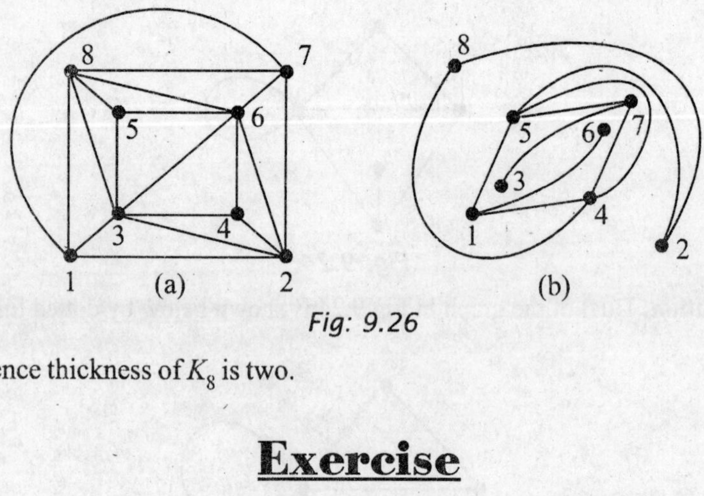

Fig: 9.26

Hence thickness of K_8 is two.

Exercise

(1) Prove that if G is planar graph with vertex connectivity equal to 3 than its dual is a simple graph.

(2) Prove that a graph G is planar and bipartite iff its dual is Eulerian.

(3) Show that dual of a wheel is a wheel. Thus wheel is a self-dual graph. Give another example of a self dual graph.

(4) Find the thickness and crossing number of the Petersen graph.

(5) Verify Euler's formula for the graph of the octahedron.

(6) What is the dual of circuit graph C_n.

Answers

(4) The following drawing of Petersen graph shows that its crossing number is two.

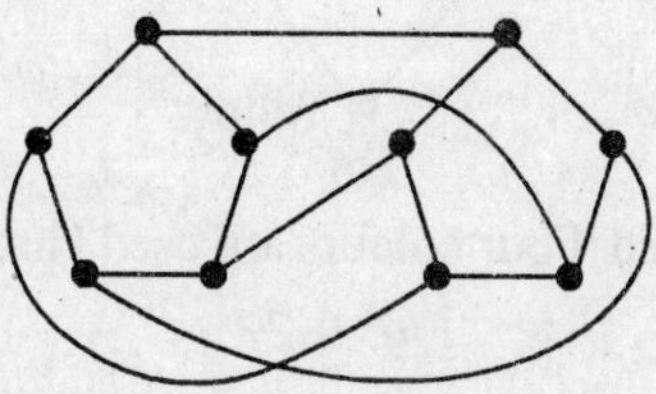

Since Petersen graph is non planar its thickness is at least 2. The following two subgraphs show that Petersen graph can be obtained by taking union of these two planar subgraphs.

Thus thickness of the Petersen graph is 2.

(5) In octahedron, $n = 6$, $e = 12$, $f = 8$. Thus $f = e - n + 2$

(6) Graph with two edges and n parallel edges between them.

§ 9.6 Colouring of Graphs

By colouring of a graph we mean a painting of the vertices of the graph such that adjacent vertices are painted with distinct colours. In other words, a colouring of a graph G is an assignment of colours to its vertices such that no adjacent vertices have the same colour. We say that a graph G is properly coloured (or simply coloured) if the vertices of G are painted such that no two adjacent vertices have the same colour. In general a given graph can be properly coloured in many different ways. Here, we are interested in finding the minimum number of colours with which a given graph can be properly coloured. For example, Fig.9.27 shows three different proper colourings of a graph with different number of colours.

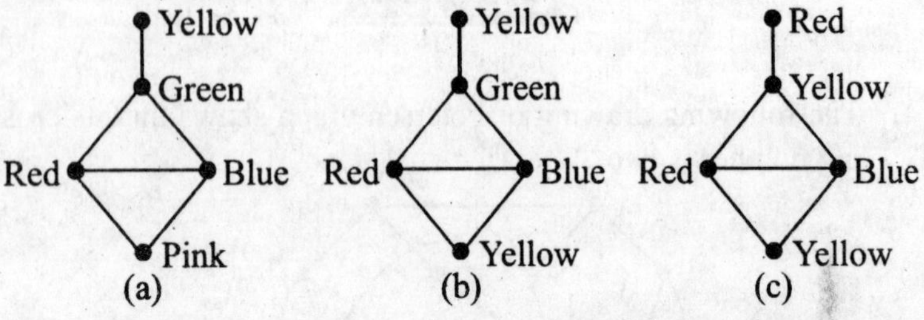

(a) Five colours are used (b) Four colours are used (c) Three colours are used

Fig: 9.27

Chromatic Number

The *chromatic number* of a graph is the minimum number of colours required for proper colouring of the graph. A graph that requires *k* different colours for its proper colouring and no less, is called *k*-chromatic graph. For example, the graph in Fig. 9.27 is 3-chromatic.

If the graph is disconnected then colouring of one component (of the disconnected graph) has no effect on the colouring of other compnents. Therefore we assume that graphs are connected. Further, if a graph has parallel edges then all parallel edges can be replaced by a single edge without affecting adjacency of vertices, therefore in colouring problems, we assume that graphs are simple.

From the definition of chromatic number it is clear that

1. Chromatic number of null graph is 1.
2. Chromatic number of complete graph K_n of n vertices is n.
3. Chromatic number of a graph having one or more edges is atleast two.
4. Chromatic number of a graph having a triangle is at least three. (Because three colours are needed to colour a triangle).
5. If a graph is a circuit with n vertices then
   (i) it is 2-chromatic if n is even
   (ii) it is 3-chromatic if n is odd

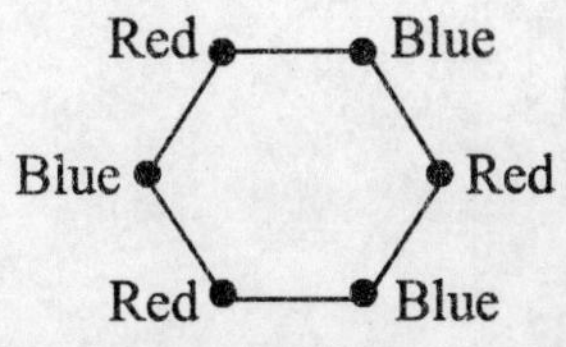

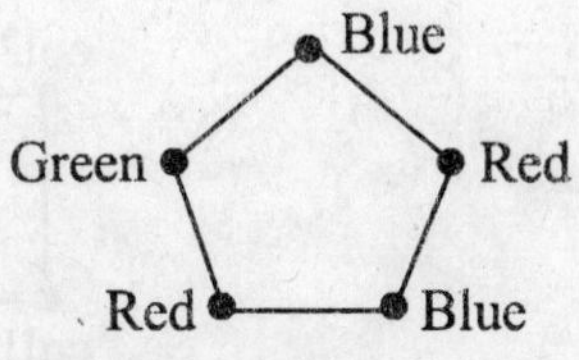

Fig: 9.28

6. If a graph is a circuit and it is given that its chromatic number is even then it has even number of vertices (edges). If a circuit graph is 3-chromatic then number of vertices in the graph is odd.

Theorem1: Every tree with two or more vertices is 2-chromatic but converse is not true.

Proof: Let T be a tree with two or more vertices. Choose any vertex v in the graph T. Paint v with colour 1. Next paint all the vertices adjacent to v with colour 2. Now paint all the vertices adjacent to those vertices which have been coloured with colour 2 using colour 1. Continue this process until every vertex of T has been painted as shown in Fig. 9.29.

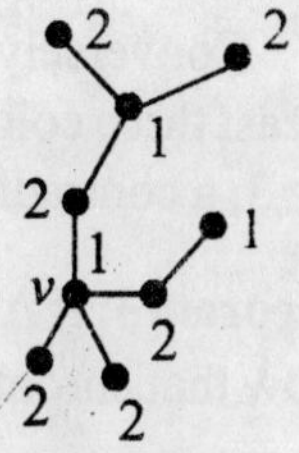

Fig: 9.29

Since there is one and only one path between any two vertices in a tree, no two adjacent vertices will have the same colour.

Thus T has been properly coloured with two colours. Since T has atleast two vertices, it will have atleast one edge. Hence one colour would not be enough. Thus T is 2-chromatic.

The following example shows that not every 2-chromatic graph is tree.

Example 9.23: Give an example of a graph which is 2-chromatic but not a tree

Solution: Consider the graph in Fig. 9.30. Clearly this graph is 2-chromatic: But it is not a tree

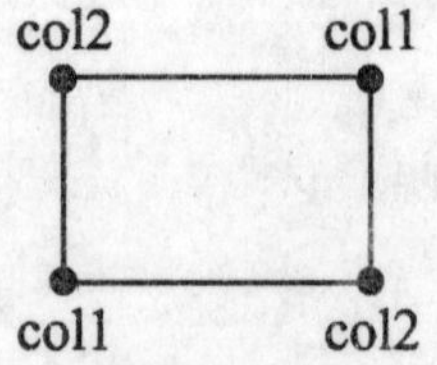

Fig: 9.30

Theorem 2: A graph with edges $e \geq 1$ is 2-chromatic if and only if it has no circuits of odd length.

Proof: Let G be connected graph with circuits of even lengths only. Select any vertex and paint it with colour 1. Next we paint all vertices adjacent to it with colour 2. If a vertex is painted with colour 2 then any vertex adjacent to it will be painted with colour 1. Since every circuit has even length, no adjacent vertices will have the same colour. Hence G is 2-chromatic.

Conversely, if G has a circuit of odd length then we would need atleast three colours just for that circuit. Hence chromatic number of G is ≥ 3, a contradiction. This completes the proof.

Theorem 3: If Δ is the maximum degree of the vertices in a graph G, then show that chromatic number of $G \leq 1 + \Delta$.

Proof: Let G be a graph with n vertices. We shall prove the result by induction on the number of vertices. If G has 1 or 2 vertices then the result is clearly true. We assume that the result is true for all graph having less than n vertices. Remove any vertex v and all edges incident on v. Then $G - v$ is a graph with $n - 1$ vertices and maximum degree of any vertex in $G - v$ is at most Δ. Hence by induction hypothesis, chromatic number of $G - v \leq 1 + \Delta$.

Now we add the vertex v and all edges incident on v to $G - v$. Since degree of v is Δ, atmost Δ colours are needed to colour all vertices adjacent to v. We can assign a different colour to v from $\Delta + 1$ colours. Hence chromatic number of $G \leq 1 + \Delta$.

Theorem4: A graph G with at least one edge is bipartite iff G is 2-chromatic.

Proof: Suppose G in 2-chromatic. Let V_1 be the set of vertices painted with colour 1, and let V_2 be the set of vertices painted with colour 2. Then V_1 and V_2 form a partition of the vertices of G such that every edge in G joins a vertex in V_1 with a vertex in V_2.

Conversely, suppose G be a bipartite graph with at least one edge. Let V_1 and V_2 be the partition of the vertex-set of G such that every edge in G joins a vertex in V_i to a vertex in V_2. If we assign colouur 1 to all vertices in V_1 and colouur 2 to all vertices in V_2 then G has a proper colouring. Since G has at least one edge, two colouurs are necessary. Hence G is 2-chromatic.

Corollary1: Every tree is a bipartite graph.

Corollary2: A graph G is bipartite iff every circuit in G is of even length.

Proof: This follows from above theorem and theorem 2.

§ 9.7. Five Colour Problem

In section 9.6, we have studied proper colouring of vertices of a graph (not necessarily planar). We now consider proper colouring of *regions* in a planar graph. As in the case of colouring of vertices, the regions of a planar graph are said to be properly coloured if no two adjacent regions have the same colour. Two regions are said to be adjacent if they have a common edge between them. Proper colouring of regions is also called *map colouring*.

We know that a graph is planar iff it has a dual. Therefore colouring the regions of a planar graph G is same as colouring the vertices of its dual G^* and vice versa. We first prove that every planar map can be properly coloured with five colours:

Five-Colour Theorem

The vertices of a planar graph can be properly coloured with five colours.

Proof: We shall prove the result by induction on the number of vertices of graph G. If the number of vertices n in a graph is less than or equal to 5 then we can properly colour these vertices with five colours. Hence the result holds. We assume that the result holds for all planar graphs having less than n vertices. Now consider a planar graph with n vertices. Then it must have a vertex v with degree five or less (Ex. 6 of §9.2). Let $G' = G - v$, the graph obtained from G by deleting vertex v. Then by induction hypothesis, G'(of $n - 1$ vertices) requires no more than five colours. Paint the vertices of G' using five colours and now add to it the vertex v along with all edges incident on v. If the degree of v is less than 4, we can assign a proper colour to v and obtain a proper colouring of G.

This leaves only a case when $d(v) = 5$ and all the five colours have been used in colouring the vertices a, b, c, d and e adjacent to v as shown in Fig. 9.31(a).

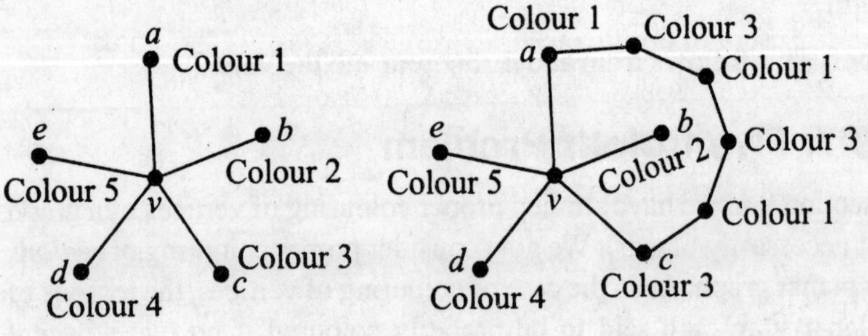

Fig: 9.31

Let H be subgraph of G on the vertices which have been assigned the colour1 and colour 3. Hence a, c are in H. If a and c belong to different components of H, then we can interchange the colours 1 and 3 in the component which contains vertex a without destroying the proper colouring of $G - v$. This interchange will paint vertices a and c with colour 3. Now colour 1 can be assigned to vertex v and thus G is 5-chromatic. On the other hand, if a and c are in the same component of H. Then there is a path P from a to c whose vertices are painted alternately with colours 1 and 3 as shown in Fig. 9.31(b). The path P together with the edges (a,v) and (c,v) form a circuit C which encloses b (or d but not both) as shown in Fig. 9.31(b). Consider now the subgraph K of G on the vertices painted with colours 2 or 4. Since C encloses v_2 or v_4 but not both, vertices b and d belong to different components of K. Therefore we can interchange the colours 2 and 4 in the component containing b without destroying the proper colouring of $G-v$. This interchange will paint vertices b and d with colour 4. Now colour 2 can be assigned to vertex v. Hence the theorem.

Four Colour Theorem

Five colour theorem states that every planar graph is 5-colourable (*i.e.*, can be properly coloured using 5 colours). It is natural to ask whether this result can be strengthened and this leads to four colour conjecture that "every planar graph can be properly coloured with four colours".

This conjecture was one of the most famous unsolved problems in graph theory and was finally proved by *K*. Appel and *W*. Haken in 1976. We now state this result, known as the Four Colour Theorem.

Four Colour Theorem: Every planar graph can be properly coloured using 4 colours only.

The proof of this theorem is beyond the scope of this book.

§ 9.8. Chromatic Partitioning and Independent Sets

A proper colouring of a graph induces a partitioning of the vertices into different subsets. For example, the colouring in the graph of Fig. 9.32 gives the partitioning $\{v_1\}$, $\{v_2, v_3\}$, $\{v_4, v_5\}$.

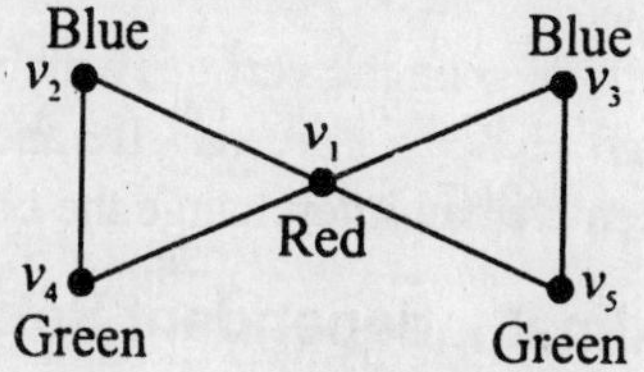

Fig: 9.32

This leads to the following definition.

Definition: A set of vertices in a graph G is said to be an *independent set* of vertices or simply an **independent set** if no two vertices in the set are adjacent. An independent set is also called a *stable* set.

A single vertex in any graph is always an independent set. The set of vertices $\{v_2, v_5\}$ in graph of Fig. 9.32 is an independent set.

Maximal Independent Set

A *maximal independent* set is an independent set to which no other vertex can be added without destroying its independence property. For example, the set $\{a, e\}$ in Fig. 9.33 is a maximal independent set. The set $\{c, d, f\}$ is also a maximal independent set. The set $\{a, f\}$ is another

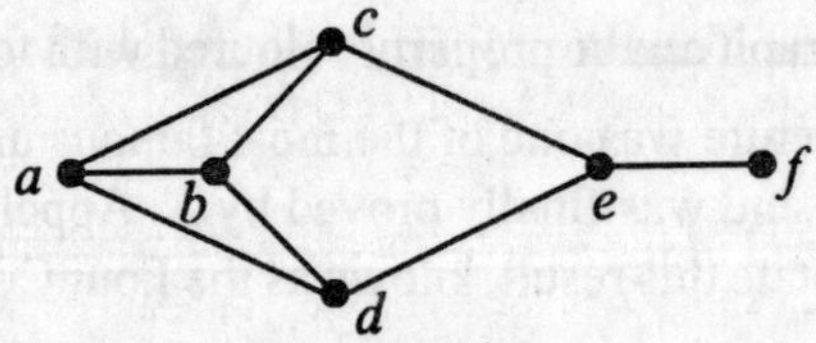

Fig: 9.33

maximal independent set. In general, a graph has many different maximal independent set and they are of different sizes. The number of vertices in the largest independent set of a graph is called the *independence number*. We shall denote this number by $\beta(G)$.

Let G be a graph with n vertices and let G be k-chromatic. By definition of independence number, the largest number of vertices in any proper colouring of G having the same colour cannot exceed the independence number, therefore we have

$$k \cdot \beta(G) \geq n$$

or

$$\beta(G) \geq n/k$$

Finding all Maximal Independent Sets in a Graph

We discuss below a method of obtaining all maximal independent sets in a graph G using Boolean algebra. In this method we shall treat all the vertices of the graph as Boolean variables. We define the two binary operations '+' and '·' and a unary operation ''' on the set of vertices of the graph as follows:

$a+b$=vertex a or b or both is/are included in a subset of vertex set of G

$a \cdot b$ = both vertices a and b are included in a subset of vertex set of G

a' = vertex a is not included in a subset of vertex set of G.

In order to find a maximum independent set in a graph G, we must find a largest subset of vertices that does not include the two end vertices of any edge in G. If x and y are the end vertices of an edge then we express this edge by a product xy. We take sum of all such products in G to get a boolean expression.

$$\varphi = \Sigma xy \text{ for all edges } (x, y) \text{ in } G$$

Taking complement and simplifying, we can write

$$\varphi' = f_1 + f_2 + \ldots + f_m.$$

Now, a subset of the vertex set of G is a maximal independent set if and only if $\varphi = 0$.

But $\varphi = 0$ if and only if $\varphi' = 1$

if and only if $f_i = 1$ for some i

if and only if each vertex appearing in f_i is excluded from the subset of vertex set of G.

$\Rightarrow$ each f_i will yield a maximal independent set and all maximal independent set will be produced by this method. We explain the above procedure with the help of an example

Example 9.24: Find all maximal independent sets of the graph in Fig. 9.34

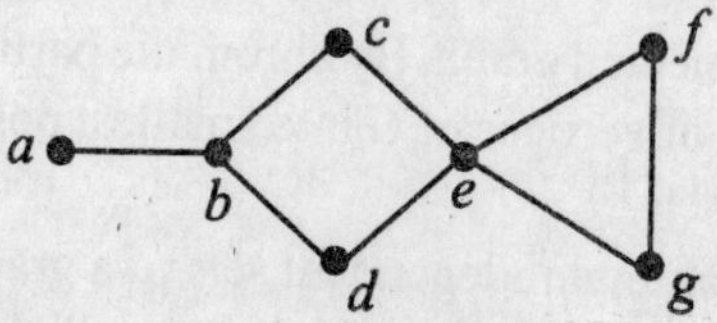

Fig: 9.34

Solution: This graph has 8 edges. We make the function

$$f = ab + bc + bd + ce + de + ef + eg + gf$$

by expressing each edge (x, y) as product xy.

By De-Morgan's rule, we have

$$\varphi' = (a'+b')\,(b'+c')\,(b'+d')\,(c'+e')\,(d'+e')(e'+f')\,(e'+g')\,(g'+f')$$

$\Rightarrow\quad \varphi' = (b'+a'c'd')(e'+c'd'f'g')(g'+f')$ using $(a+b)(a+c) = a+bc$

$\Rightarrow\quad \varphi' = (b'e'+b'c'd'f'g'+a'c'd'e'+a'c'd'f'g')(g'+f')$ using $aa = a$

$\Rightarrow\quad \varphi' = b'e'g+b'e'f'+b'c'd'f'g'+a'c'd'e'g'+a'c'd'e'f'+a'c'd'f'g'$

using $a + a = a$ and $aa = a$

Now if we exclude from the vertex set of G the vertices appearing in any of the six terms, we get a maximal independent set. The six maximal independent sets are

$$\{a, c, d, f\}, \{a, c, d, g\}\ \{a, e\}, \{b, f\}, \{b, g\}, \{b, e\}.$$

Finding Independence Number

After obtaining all maximal independent sets of a graph, we find the size of a maximal independent set having maximum number of elements to get the independence number. For example, there are two maximal independent set having maximum number of elements in the above example each of size 4. Thus the independence number of the graph in Fig. 9.34 is four.

Finding Chromatic Number: After obtaining all maximal independent sets of a graph G, we find the minimum number of these sets whose union is the vertex set of G. This minimum number is the chromatic number of G. For example, the graph in Fig. 9.34 is 3-chromatic because atleast three sets out of six maximal independent sets are necessary to include all the vertices of G.

Chromatic Partitioning

Let G be a simple connected graph. By chromatic partitioning of G, we mean the partitioning of vertices of G into smallest possible number of disjoint independent sets.

After obtaining all maximal independent sets of a graph, we select the smallest number of maximal independent sets which include all vertices of the graph. For example, the following are some chromatic partitions of the graph in Fig. 9.34.

$$\{\{a, c, d, f\}, \{b, g\}, \{e\}\}$$

$$\{\{a, c, d, g\}, \{b, f\}, \{e\}\}$$

$$\{\{a, c, d, f\}, \{b, e\}, \{g\}\}$$

Uniquely Colourable Graphs

A graph that has only one chromatic partition of its vertices is called a uniquely colourable graph. The graph in Fig. 9.34 is not uniquely colourable because it has more than 1 chromatic partitions. The graph in the following example is uniquely colourable.

Example 9.25: Show that the graph in Fig. 9.35 is uniquely colourable.

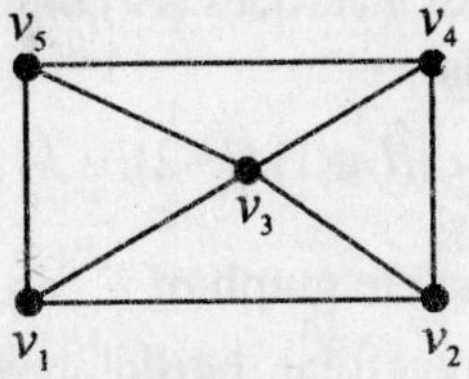

Fig: 9.35

Solution: This graph has 8 edges. We first find all maximal independent sets of this graph. As in Example 9.24,

$$\varphi = v_1 v_2 + v_1 v_3 + v_1 v_5 + v_2 v_3 + v_2 v_4 + v_3 v_4 + v_3 v_5 + v_4 v_5$$

$$\Rightarrow \varphi' = (v_1' + v_2')(v_1' + v_3')(v_1' + v_5')(v_2' + v_3')(v_2' + v_4')(v_3' + v_4')$$
$$(v_3' + v_5')(v_4' + v_5')$$
$$= (v_1' + v_2'v_3'v_5')(v_2' + v_3'v_4')(v_3' + v_4'v_5')(v_4' + v_5')$$
$$= v_1'v_2' + v_1'v_3'v_4' + v_2'v_3'v_5' + v_2'v_3'v_4'v_5')(v_3'v_4' + v_3'v_5' + v_4'v_5')$$
$$= v_1'v_2'v_3'v_4' + v_1'v_2'v_3'v_5' + v_1'v_2'v_4'v_5' + v_1'v_3'v_4' + v_1'v_3v'_4v_5' +$$
$$v_1'v_3'v_4'v_5' + v_2'v_3'v_4'v_5' + v_2'v_3'v_5' + v_2'v_3'v_4'v_5'$$
$$= v_1'v_3'v_4' + v_2'v_3'v_5' + v_1'v_2'v_4'v_5' \qquad \text{using } a + ab = a$$

Thus three maximal independent sets are

$$\{v_2, v_5\}, \{v_1, v_4\} \text{ and } \{v_3\}.$$

Thus $\{\{v_2, v_5\}, \{v_1, v_4,\} \{v_3,\}\}$ is the only chromatic partition. Since there is only one chromatic partition of G, it is uniquely colourable. Moreover this graph is 3-chromatic.

§ 9.9 Chromatic Polynomial

In this section we discuss the question of counting the number of distinct ways of proper colouring of a graph using λ-colours. We can express this number by means of a polynomial. This polynomial is called chromatic polynomial.

Definition: Chromatic polynomial $P_n(\lambda)$ of a graph with n vertices is a polynomial in λ which gives the number of ways of properly colouring the graph using λ or fewer colours.

Theorem 1: A graph G of n vertices is a complete graph if and only if its chromatic polynomial is

$$P_n(\lambda) = \lambda(\lambda-1)(\lambda-2)\dots\dots(\lambda-n+1).$$

Proof: Let G be a complete graph of n vertices $v_1, v_2, \dots, v_n$. Given λ colours, the first vertex v_1 can be coloured with any one of the λ colours, vertex v_2 with any one of remaining $\lambda-1$ colours and so on. Thus

$$P_n(\lambda) = \lambda(\lambda-1)(\lambda-2)\dots\dots(\lambda-n+1).$$

Conversely, suppose $P_n(\lambda) = \lambda(\lambda-1)(\lambda-2)\dots\dots(\lambda-n+1)$ is a chromatic polynomial of a graph G with n vertices: By definition of chromatic polynomial, there are λ different ways of colouring any selected vertex. A second vertex can be coloured property in $\lambda-1$ ways, the third in $\lambda-2$ ways, ..., and the nth in $\lambda-n+1$ if every vertex is adjacent to every other vertex. Thus the graph is complete.

Theorem 2: A graph G with n vertices is a tree iff its chromatic polynomial is

$$p_n(\lambda) = \lambda(\lambda-1)^{n-1}$$

Proof: We shall prove the result by induction. Suppose the graph G is tree with n vertices. If G has 1 or 2 vertices then $p_n(\lambda) = \lambda$ or $\lambda(\lambda-1)$. Hence result is true for $n = 1, 2$. We assume that result is true for all trees having less than n vertices. Now consider a tree T with n vertices. Then T has atleast two pendant vertices. Let v be a pendant vertex of T. Consider $T^* = T - v$. Then T^* is a tree with $n-1$ vertices. Hence by induction hypothesis, chromatic polynomial of T^* is

$$P_{n-1}(\lambda) = \lambda(\lambda-1)^{n-2}.$$

Let u be the vertex adjacent to v in T. Now add v to T^*. Suppose u is coloured with any one of the λ colours. Then v can be coloured with any one of the remaining $\lambda-1$ colours.

Thus, the number of ways of colouring T is

$$P_n(\lambda) = \lambda(\lambda-1)^{n-2} \cdot (\lambda-1) = \lambda(\lambda-1)^{n-1}.$$

Hence the result follows by induction.

Conversely, suppose G is a graph of n vertices whose chromatic polynomial is

$$P_n(\lambda) = \lambda(\lambda-1)^{n-1}$$

We shall prove by induction on n that G is tree. If $n=1$ then $P_1(\lambda) = \lambda$. Hence the vertex is isolated and thus is a tree. If $n=2$, then $P_2(\lambda) = \lambda(\lambda-1)$. The corresponding graph is clearly an edge joining the two vertices which is a tree. Hence for $n=1, 2$, result is true. We assume that the result is true for $n-1$. That is, we assume that if T is a graph with $n-1$ vertices and chromatic polynomial $P_{n-1}(\lambda) = \lambda(\lambda-1)^{n-2}$ then T is tree. Let G be a graph with n vertices whose chromatic polynomial is

$$P_n(\lambda) = \lambda(\lambda-1)^{n-1}$$

From the chromatic polynomial of G, it is clear that every vertex of G except one vertex can be coloured in $\lambda-1$ ways. Remove one such vertex v_0. Then $G-v_0$ is a graph with $n-1$ vertices whose chromatic polynomial is

$$P_{n-1}(\lambda) = \lambda(\lambda-1)^{n-2}$$

By induction hypothesis, $G-v_0$ is a tree. Now add v_0 along with edges incident on v_0 to the graph $G-v_0$. Clearly v_0 can not be isolated vertex. Hence v_0 is connected to some vertex of $G-v_0$. We show that v_0 cannot be connected to more than one vertex in $G-v_0$. Suppose if possible v_0 is joined to two vertices in $G-v_0$. Since $G-v_0$ is a tree with chromatic polynomial

$$P_{n-1}(\lambda) = 1(\lambda-1)^{n-2}$$

and v_0 is joined to two vertices in $G-v_0$, therefore v_0 can be coloured in $(\lambda-2)$ ways only. But then chromatic polynomial of G would be

$$P_n(\lambda) = \lambda\,(\lambda-1)^{n-2}\,(\lambda-2)$$

which is not true. Hence v_0 is joined to only one vertex in $G-v_0$. Thus G is a tree.

Theorem 3: Let u and v be to non-adjacent vertices in a graph G. Let G' be a graph obtained by adding an edge e between u and v. Let G'' be a simple graph obtained from G by fusing the vertices u and v together and replacing parallel edges with single edge. Then

$$P_n(\lambda) \text{ of } G = P_n(\lambda) \text{ of } G' + P_n(\lambda) \text{ of } G''.$$

Proof: The number of ways of properly colouring of G can be divided into two cases, one such that vertices u and v are of the same colour and the other such that u and v are of different colours. Since the number of ways of properly colouring G such that u and v have different colours = number of ways of properly colouring have different colours = number of ways of properly colouring G', and number of ways of properly colouring G such that u and v have the same colour = number of ways of properly colouring G''. Thus we have

$$P_n(\lambda) \text{ of } G = P_n(\lambda) \text{ of } G' + P_n(\lambda) \text{ of } G''.$$

Corollary: If $e = (u, v)$ is an edge of a simple graph G then

$$P_n(\lambda) \text{ of } G = P_n(\lambda) \text{ of } G - e - P_n(\lambda) \text{ of } G''$$

where $G - e$ is the graph obtained from G by removing an edge e and G'' is the graph as defined in above theorem.

Theorem 3 is very useful in obtaining the chromatic polynomial of a graph. If we repeatedly apply the formula given in Theorem 3 on a graph G then chromatic polynomial of G is obtained in terms of the chromatic polynomials of complete graphs. The procedure is illustrated in the following example

Example 9.26: Find the chromatic polynomial of the graph in Fig. 9.36.

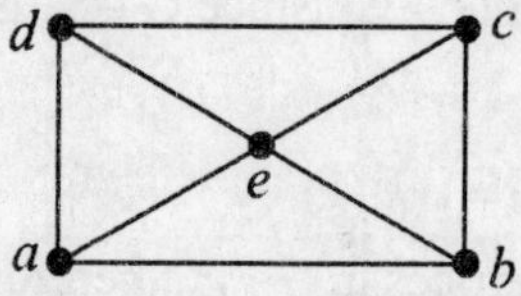

Fig: 9.36

Solution: We shall use Theorem 3 to obtain the chromatic polynomial of this graph. Since this graph has non-adjacent vertices, choose any pair of of non-adjacent vertices in the graph. Suppose we choose a and c. We now draw an edge between a and c to obtain the graph G'. We will also fuse vertices a and c to obtain G'' as shown in the following figure:

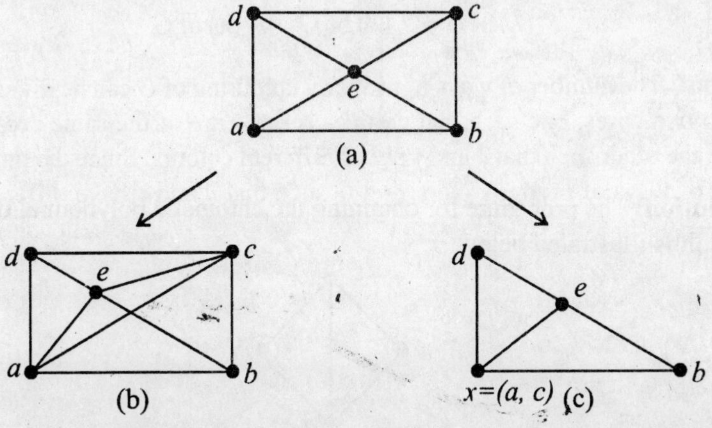

Again taking b and d as non-adjacent vertices in Fig.(b) and (c) and applying the same procedure we get

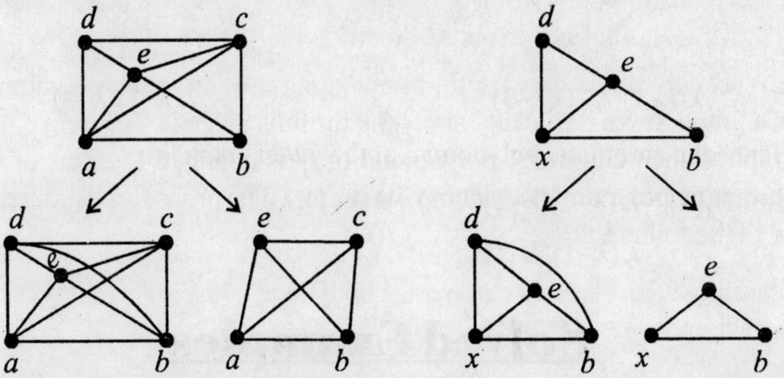

All the four graphs are now complete graphs, therefore chromatic polynomial of the given graph is the sum of chromatic polynomials of these four complete graphs. Chromatic polynomial of a complete graph is given by Theorem 1.

Hence chromatic polynomial of the given graph is

$$p_5(\lambda) = \lambda(\lambda-1)(\lambda-2)(\lambda-3)(\lambda-4)+2\lambda(\lambda-1)(\lambda-2)(\lambda-3)+\lambda(\lambda-1)(1-2)$$

$$= \lambda(\lambda-1)(\lambda-2)(\lambda^2-5\lambda+7).$$

Example 9.27: Find the chromatic polynomial of the graph given below:

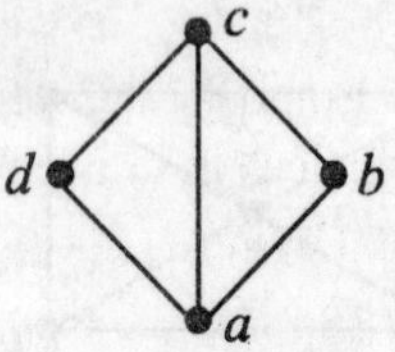

Fig: 9.37

Solution: The procedure for obtaining the chromatic polynomial of a graph is illustrated below:

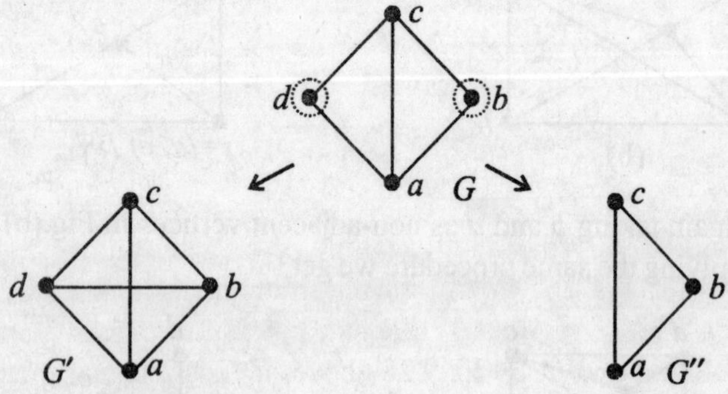

$\lambda(\lambda-1)(\lambda-2)(\lambda-3)$ $\lambda(\lambda-1)(\lambda-2)$

Hence the chromatic polynomial of the given graph is

$$P_4(\lambda) = \lambda(\lambda-1)(\lambda-2)(\lambda-3) + \lambda(\lambda-1)(\lambda-2)$$

$$= \lambda(\lambda-1)(\lambda-2)^2$$

Solved Examples

Example 9.28: Show that the regions of a planar graph can be 2-coloured if each vertex is of even degree.

Solution: Let G be a graph in which each vertex is of even degree.

Therefore G is an Euler graph and hence all cut-sets of G have even number of edges (see Ex.8.10 of §6 chapter 8). Thus all circuits in dual G^* of G are of even length and hence vertices of G^* can be 2-coloured. Thus the regions of G can be two-coloured.

Example 9.29: Show that a simple graph with 7 vertices each of degree 4 is non-planar.

Solution: Let G be a simple graph with 7 vertices each of degree 4. Then sum of degrees of all vertices in G is equal to $7 \times 4 = 28$. Thus, number of edges e in G is given by

$$2e = 28$$

$$\Rightarrow \qquad e = 14$$

If the graph is planar then by Euler's formula

$$f = e - n + 2 = 14 - 7 + 2 = 9$$

Suppose that there are x regions of degree 3 and $(9-x)$ regions of degree 4. Then

$$3x + 4(9-x) = 28$$

$$\Rightarrow \qquad x = 8$$

Hence there would be 8 regions of degree 3 and 1 region of degree 4 (observe that G cannot have any region of degree ≥ 5). Any region of degree 3 requires 3 colours to colour its vertices. Hence G is not 2-colourable. But by Ex.9.28 above, if G is a planar graph with each vertex of even degree then it is 2-colourable. Contradiction proves that G is not planar.

Note: By degree of a region we mean number of edges forming the boundary of the region.

Example 9.29: Let G be a simple graph with n vertices which is regular of degree r. Prove that chromatic number of $G \geq \dfrac{n}{n-r}$

Solution: Let C_i be the number of vertices in any proper colouring of G which receives the colour i, where $1 \leq i \leq k$, the chromatic number of G. Then

$$C_i \leq n - d \qquad \text{for each } 1 \leq i \leq k.$$

Now $n = C_1 + C_2 + ... + C_k$.

$$\le (n-d) + (n-d) + ...(n-d) \quad k \text{ times}$$

$$= k(n-d)$$

Thus $k \ge \dfrac{n}{n-d}$

Example 9.30: Find the chromatic polynomial of the complete bipartite graph $K_{2,5}$.

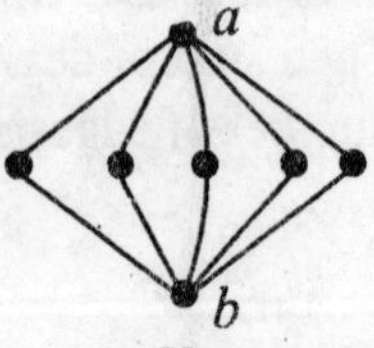

Solution: If vertices a and b have the same colour then total number of ways of proper colouring $K_{2,5}$ is $k(k-1)^5$. If a and b have different colour then there are $k(k-1)(k-2)^5$ different ways of colouring. Thus

$$P_n(\lambda) = k(k-1)^5 + k(k-1)(k-2)^5.$$

Exercise

1. Write down the chromatic polynomial of the complete graph of 6 vertices
2. Write down the chromatic polynomial of a tree having four vertices
3. Using maximal independent sets or otherwise find the chromatic number of each graph given below:

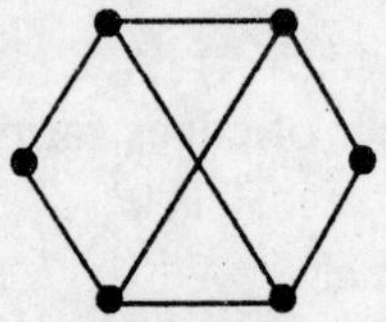

(a)

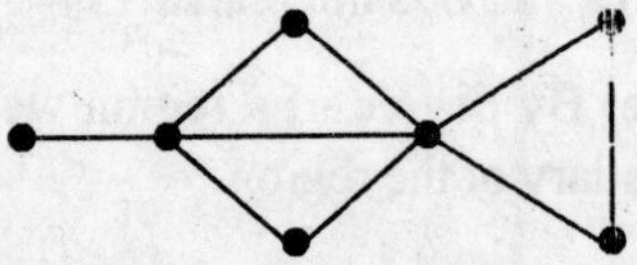

(b)

4. Find the chromatic polynomial of the graph

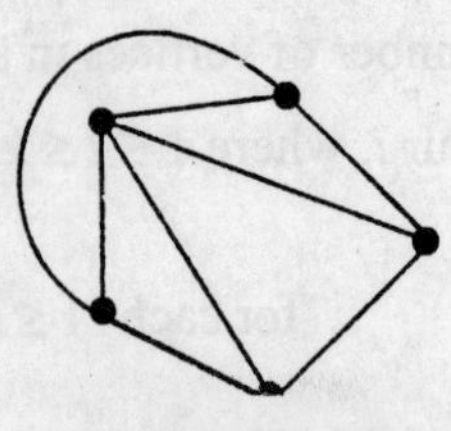

5. Find three graphs with chromatic polynomial

$$P_n(\lambda) = k^5 - 4k^4 + 6k^3 - 4k^2 + k$$

(Hint: $P_n(\lambda) = k(k-1)^4$. Hence the graph is tree with 5 vertices)

6. Prove that the chromatic polynomial of $K_{2,n}$ is

$$P_n(\lambda) = k(k-1)^n + k(k-1)(k-2)^n.$$

Answers

1. $P_6(\lambda) = \lambda(\lambda-1)(\lambda-2)(\lambda-3)(\lambda-4)(\lambda-5)$
2. $P_4(\lambda) = \lambda(\lambda-1)^3$
3. (a) 2 (b) 3
4. $P_5(\lambda) = \lambda(\lambda-1)(\lambda-2)(\lambda^2-5\lambda+7)$
5. 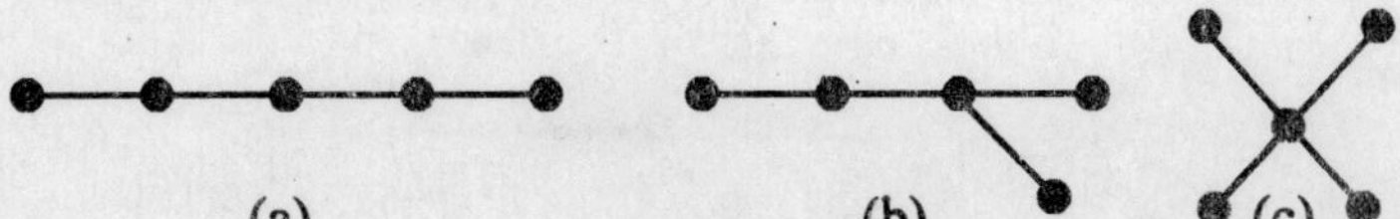

(a) (b) (c)

CHAPTER TEN

Matrices Associated with Graphs

A diagrammatic representation of a graph is very convenient but it has its own limitations. To analyze (or manipulate) large graphs using computer we need to store graphs in computer. In this chapter we will discuss the most common ways of representing graphs using matrices. Matrix representation of a graph has several advantages. It is easy to store and manipulate matrices and hence the graphs represented by them in a computer. Matrix algebra will also provide us an opportunity to calculate paths, circuits and other properties of a graph.

Essentially, there are two ways of representing and storing a graph in the memory of a computer. One is by incidence matrix and other is using adjacency matrix.

§ 10.1. Incidence Matrix

Let G be a graph with n vertices $v_1, v_2, \ldots, \ldots, v_n$ and m edges $e_1, e_2, \ldots, e_m$ and no self-loops. We define an $n \times m$ matrix $A = [a_{ij}]$ whose n rows correspond to the n vertices and m columns correspond to the m edges as follows:

$a_{ij} = 1$, if j^{th} edge e_j is incident on i^{th} vertex v_i and

$= 0$, otherwise

The matrix $A = [a_{ij}]$ is called *incidence matrix*. Sometimes incidence matrix A of a graph G is also written as $A(G)$. Any element of the incidence matrix is either 0 or 1. That is why it is also called (0,1)-matrix or a bit matrix.

For a given graph $G = (V, E)$, an incidence matrix depends upon the ordering of the vertices of G and that of edges of G. For different ordering of vertices and edges we get different incidence matrix of the same graph G. However, any one of the incidence matrix of a graph G can be obtained from another incidence matrix of the same graph by interchanging some of the rows and columns of the matrix.

Example 10.1: Write the incidence matrix of the graph in Fig.10.1

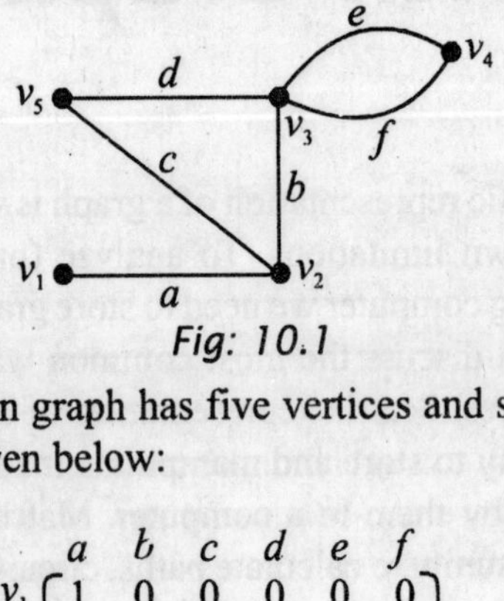

Fig: 10.1

Solution: The given graph has five vertices and six edges. The incidence matrix in given below:

$$\begin{array}{c} \\ v_1 \\ v_2 \\ v_3 \\ v_4 \\ v_5 \end{array} \begin{array}{c} \begin{array}{cccccc} a & b & c & d & e & f \end{array} \\ \begin{bmatrix} 1 & 0 & 0 & 0 & 0 & 0 \\ 1 & 1 & 1 & 0 & 0 & 0 \\ 0 & 1 & 0 & 1 & 1 & 1 \\ 0 & 0 & 0 & 0 & 1 & 1 \\ 0 & 0 & 1 & 1 & 0 & 0 \end{bmatrix} \end{array}$$

Properties of Incidence Matrix

1. Each column of incidence matrix A of a graph without self loops has exactly two 1's because each edge is incident on exactly two vertices.

2. Given a incidence matrix, we can draw the corresponding graph easily.
3. The sum of each row gives the degree of the corresponding vertex.
4. If a graph has parallel edges then the corresponding columns in the incidence matrix are identical.
5. A row with all zero's represents an isolated vertex.
6. If a graph G is disconnected and H_1 and H_2 are its components then the incidence matrix $A(G)$ of G can be written as.

$$A(G) = \left[\begin{array}{c:c} A(H_1) & 0 \\ \hdashline 0 & A(H_2) \end{array}\right]$$

7. In changing any two rows or columns in an incidence matrix simply corresponds to relabelling the vertices and edges of the same graph. Thus two graphs G_1 and G_2 are isomorphic if the incidence matrix of one graph can be obtained from the incidence matrix of the other by interchanging some of the rows and the columns.
8. Since each column of incidence matrix A contains exactly two 1's. Therefore we can obtain any row of A from the remaining $n-1$ rows. Thus any $n-1$ rows of A contain all information about A. In other words, the rows of A are linearly dependent.

§ 10.2. Adjacency Matrix

Adjacency matrix is another way of representing a graph. Adjacency matrix of a graph with n vertices and no parallel edges (self-loops are allowed) is an $n \times n$ matrix $X = [x_{ij}]$ where

$$\begin{aligned} x_{ij} &= 1, \text{ if there is an edge between } i^{th} \text{ and } j^{th} \text{ vertices and} \\ &= 0, \text{ if there is no edge between them.} \end{aligned}$$

Notice that while adjacency matrix as defined above allow us to represent loops, it does not allow us to represent parallel edges.

Example 10.2: Write the adjacency matrix of the following graph.

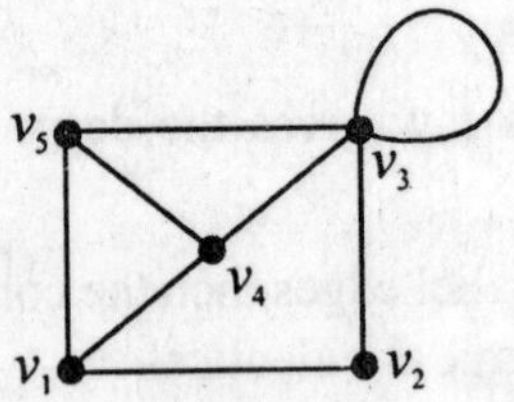

Fig: 10.2

Solution: The adjacency matrix is given below:

$$X = \begin{array}{c} \\ v_1 \\ v_2 \\ v_3 \\ v_4 \\ v_5 \end{array} \begin{array}{c} \begin{array}{ccccc} v_1 & v_2 & v_3 & v_4 & v_5 \end{array} \\ \begin{bmatrix} 0 & 1 & 0 & 1 & 1 \\ 1 & 0 & 1 & 0 & 0 \\ 0 & 1 & 1 & 1 & 1 \\ 1 & 0 & 1 & 0 & 1 \\ 1 & 0 & 1 & 1 & 0 \end{bmatrix} \end{array}$$

Properties of Adjacency Matrix

1. Adjacency matrix $X(G)$ of a graph G is symmetric.
2. If graph has no self-loops then the diagonal entries of $X(G)$ are zero and vice versa.
3. If the graph has no self-loops then sum of ith row (or sum of ith column) is the degree of v_i.
4. Two graphs G_1 and G_2 are isomorphic iff the adjacency matrix of one can be obtained from the adjacency matrix of the other by interchanging some of the rows and the corresponding columns.
5. Given any square, symmetric (0,1)-matrix X of order n, we can draw a graph G with n vertices and no parallel edges such that X is the adjacency matrix of G.
6. If G has two components H_1 and H_2 then we can write adjacency matrix $X(G)$ as

$$X(G) = \left[\begin{array}{c:c} X(H_1) & 0 \\ \hdashline 0 & X(H_2) \end{array}\right]$$

Definition: Let $G = (V, E)$ be a graph. An **edge sequence** of length n from a vertex x to a vertex y is a set of edges S such that

$$S = \{(x, v_1), (v_1, v_2), \ldots, (v_{n-1}, y)\}$$

This definition allows repetition of edges. Thus an edge sequence from a vertex x to a vertex y is a sequence of edges in which each edge (except the first and last) has one vertex in common with the edge preceding it and one vertex in common with the edge following it. A walk is an edge sequence but converse need not be true. Observe that an edge may appear more than once in an edge sequence.

Theorem 1: If X is an adjacency matrix of a simple graph G then the ij^{th} entry in X^r is the number of edge sequences of length r between vertices v_i and v_j.

Proof: We will use induction to prove the result. If $r = 1$ then $X^1 = X$ and ij^{th} entry in X is 1 if there is an edge between vertices v_i and v_j which is an edge sequence of length 1, and 0 otherwise. Thus the theorem is true for $r = 1$. We assume that the result is true for $r-1$. Now

$$X^r = X^{r-1}.X.$$

Let us consider the ij^{th} entry in X^r.

$$ij^{\text{th}} \text{ entry of } X^r = \sum_{k=1}^{n} s_{ik}.x_{kj}$$

where s_{ik} are the elements of ith row of X^{r-1} and x_{kj} are the elements of jth column of X. By induction s_{ij} gives the number of edge sequences of length $r-1$ between v_i and v_j. Now x_{kj} is either 0 or 1. If x_{kj} is 0 then there is no edge between v_k and v_j and therefore $s_{ik}\, x_{kj} = 0$ edge sequences of length r between v_i and v_j. If $x_{kj} = 1$ then there is an edge between v_j and v_k. Therefore, $s_{ik}\, x_{kj} = s_{ij}$ edge sequences of length r between v_i and v_j. Summing over all k, we can count all edge sequences of length r between v_i and v_j. Thus ij^{th} entry in X^r gives the number of edge sequences of length r between v_i and v_j. This completes the proof.

Example 10.3: Write adjacency matrix of the following graph and verify above theorem for X^2 and X^3 between vertices v_1 and v_5.

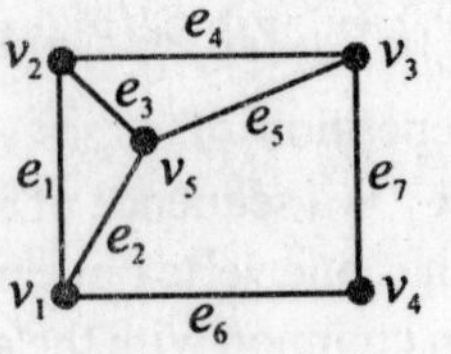

Fig: 10.3

Solution: The adjacency matrix of the graph is given below:

$$X = \begin{array}{c} \\ v_1 \\ v_2 \\ v_3 \\ v_4 \\ v_5 \end{array} \begin{array}{c} \begin{array}{ccccc} v_1 & v_2 & v_3 & v_4 & v_5 \end{array} \\ \begin{bmatrix} 0 & 1 & 0 & 1 & 1 \\ 1 & 0 & 1 & 0 & 1 \\ 0 & 1 & 0 & 1 & 1 \\ 1 & 0 & 1 & 0 & 0 \\ 1 & 1 & 1 & 0 & 0 \end{bmatrix} \end{array}$$

Multiplying, we get

$$X^2 = \begin{bmatrix} 3 & 1 & 3 & 0 & 1 \\ 1 & 3 & 1 & 2 & 2 \\ 3 & 1 & 3 & 0 & 1 \\ 0 & 2 & 0 & 2 & 2 \\ 1 & 2 & 1 & 2 & 3 \end{bmatrix} \quad \text{and} \quad X^3 = \begin{bmatrix} 2 & 7 & 2 & 6 & 7 \\ 7 & 4 & 7 & 2 & 5 \\ 2 & 7 & 2 & 6 & 7 \\ 6 & 2 & 6 & 0 & 2 \\ 7 & 5 & 7 & 2 & 4 \end{bmatrix}$$

If the graph has no self-loops then value of each diagonal entry in X^2 equals the degree of the corresponding vertex. The value of the non-diagonal entry in X^2 is given as follows:

$(i, j)^{th}$ entry in X^2 $(i \neq j)$ = number of vertices which are adjacent to both i^{th} and j^{th} vertices

= number of different paths of length 2 between i^{th} and j^{th} vertices.

For example, consider $(1,5)^{th}$ element of X^2. For the matrix for X^2, we have

$$(1, 5)^{th} \text{ entry in } X^2 = 1.$$

This single edge sequence of two edges between v_1 and v_5 is

$\{(v_1, v_2), (v_2, v_5)\}$.

We may also write above the edge sequence as $\{e_1, e_3\}$.

Let us now consider the $(i, j)^{th}$ entry in X^3.

$(i, j)^{th}$ entry in X^3 = product of ith row in X^2 and j^{th} column of X.

$= \sum_{k=1}^{n} (i,k)^{th}$ entry of X^2. $(k,j)^{th}$ entry of X

$= \sum_{k=1}^{n}$ number of all different edge sequences of three edge from i^{th} to j^{th} vertex via k^{th} vertex.

= number of different edge sequences of three edges between i^{th} and j^{th} vertices.

For example, consider $(1, 5)^{th}$ entry of X^3.

$(1,5)^{th}$ entry of X^3 = product of 1st row of X^2. 5th column of X

$= 7$

These seven different edge sequences of length 3 between v_1 and v_5 are (from graph)

$\{e_2, e_2, e_2\}$, $\{e_1, e_1, e_2\}$, $\{e_1, e_4, e_5\}$, $\{e_6, e_7, e_5\}$, $\{e_6, e_6, e_2\}$, $\{e_2, e_3, e_3\}$, $\{e_2, e_5, e_5\}$.

The following corollary of the theorem 1 discussed above is very useful in determining the connectedness of a graph.

Corollary: If X is the adjacency matrix of graph G with n vertices and

$$Y = X + X^2 + \ldots + X^{n-1}$$

then G is disconnected if and only if there exists at least one entry in matrix Y that is zero.

Remark: We can extend the idea of matrix representation to graphs having parallel edges. If G is a graph (self-loops and parallel edges are also allowed). Then we define adjacency matrix for such graphs as follows:

Define:

$x_{ij} = k$, if there are k edges between vertices v_i and v_j

$= 0$, if there is no edge between vertices v_i and v_j.

Example 10.4: Draw the graph corresponding to the following adjacency matrix.

$$X = \begin{bmatrix} 0 & 2 & 0 & 1 \\ 2 & 1 & 1 & 1 \\ 0 & 1 & 0 & 1 \\ 1 & 1 & 1 & 0 \end{bmatrix}$$

Solution: A graph with the given adjacency matrix contains four vertices. The graph is given below:

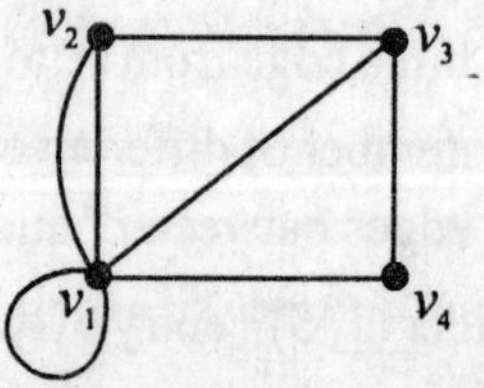

Fig: 10.4

§ 10.3. Path Matrix

Another (0, 1)-matrix which is useful in communication and transportation networks is the path matrix. A path matrix is defined for a given pair of vertices in a graph, say, (u, v) and is written as $P(u, v)$. Path matrix $P(u, v) = [p_{ij}]$ is defined as follows:

$p_{ij} = 1$, if j^{th} edge lies in i^{th} path and

$= 0$, otherwise.

Here rows in $P(u, v)$ correspond to different paths between vertices u and v, and the columns correspond to the edges in G.

Example 10.5: Find a path matrix between v_2 and v_4 in Fig. 10.5.

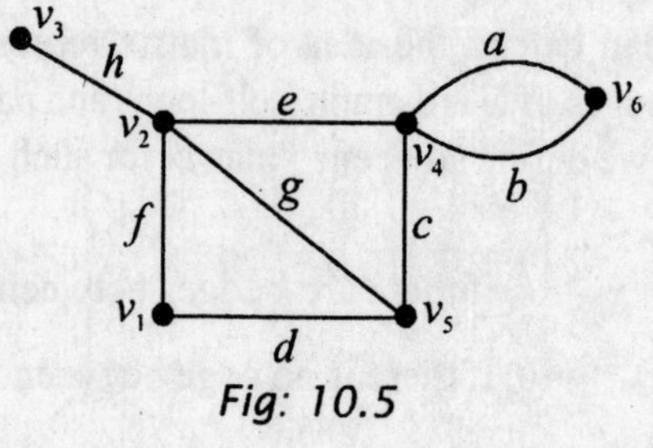

Fig: 10.5

Solution: There are three different paths $\{h, e\}$, $\{h, g, c\}$ and $\{h, f, d, c\}$ between v_3 and v_4. Let us call these paths as 1, 2 and 3 respectively. Since there are eight edges in the given graph, we get a 3×8 path matrix $P(v_3, v_4)$:

$$P(v_3, v_4) = \begin{matrix} \\ 1 \\ 2 \\ 3 \end{matrix}\begin{matrix} \begin{matrix} a & b & c & d & e & f & g & h \end{matrix} \\ \begin{bmatrix} 0 & 0 & 0 & 0 & 1 & 0 & 0 & 1 \\ 0 & 0 & 1 & 0 & 0 & 0 & 1 & 1 \\ 0 & 0 & 1 & 1 & 0 & 1 & 0 & 1 \end{bmatrix} \end{matrix}$$

Properties of a Path Matrix *P*(*u*,*v*)

1. A column of all zeros corresponds to an edge that does not lie in any path between *u* and *v*.
2. A column of all one's corresponds to an edge that lies in every path between *u* and *v*.
3. There can not be row with all zeros.

§ 10.4. Circuit Matrix

Let *G* be a graph with *e* edges and *q* different circuits. Then a circuit matrix $B = [b_{ij}]$ of *G* is a $q \times e$ matrix defined as follows:

$$b_{ij} = 1, \text{ if } i^{\text{th}} \text{ circuit includes } j^{\text{th}} \text{ edge, and}$$
$$= 0, \text{ otherwise.}$$

In order to find circuit matrix of a graph, we first arrange the circuits in an order.

Example 10.6: Find circuit matrix of the graph in Fig. 10.5.

Solution: The graph in Fig. 10.5 has four different circuits

$\{a, b\}$, $\{c, e, g)$, $\{d, f, g\}$ and $\{c, d, f, e\}$. The circuit matrix is

$$B(G) = \begin{matrix} \\ 1 \\ 2 \\ 3 \\ 4 \end{matrix}\begin{matrix} \begin{matrix} a & b & c & d & e & f & g & h \end{matrix} \\ \begin{bmatrix} 1 & 1 & 0 & 0 & 0 & 0 & 0 & 0 \\ 0 & 0 & 1 & 0 & 1 & 0 & 1 & 0 \\ 0 & 0 & 0 & 1 & 0 & 1 & 1 & 0 \\ 0 & 0 & 1 & 1 & 1 & 1 & 0 & 0 \end{bmatrix} \end{matrix}$$

where rows 1, 2, 3 and 4 corresponds to circuits $\{a, b\}$, $\{c, e, g\}$, $\{d, f, g\}$ and $\{c, d, f, e\}$, respectively.

Properties of a Circuit Matrix

1. If an edge does not belong to any circuit then the corresponding column of circuit matrix contains all zeros.
2. A circuit matrix is capable of representing a self-loop.
3. The number of one's in a row is equal to the number of edges in the corresponding circuit.
4. If graph G is separable (or disconnected) and consists of two blocks (or components) H_1 and H_2 then the circuit matrix $B(G)$ of G can be written as:

$$B(G) = \left[\begin{array}{c:c} B(H_1) & 0 \\ \hdashline 0 & B(H_2) \end{array}\right]$$

   where $B(H_1)$ and $B(H_2)$ are the circuit matrices of H_1 & H_2. This follows from the fact that circuits in H_1 and H_2 are edge-disjoint.
5. Let A and B be the incidence matrix and the circuit matrix respectively of a self-loop free graph whose columns are arranged using the same order of edges. Then

$$A \cdot B^T = B \cdot A^T = 0 \pmod 2$$

where A^T and B^T are the transposes of matrix A and B respectively and sum of the corresponding product of entries in A and B^T is taken over addition modulo 2.

Example 10.7: Show that $A \cdot B^T = 0$ for the graph in Fig. 10.5, where A and B are incidence matrix and circuit matrix of the graph.

Solution: From the graph and arranging the edges in A and B in the same order, we have

$$A = \begin{array}{c} \\ v_1 \\ v_2 \\ v_3 \\ v_4 \\ v_5 \\ v_6 \end{array} \begin{array}{c} a\ b\ c\ d\ e\ f\ g\ h \\ \begin{bmatrix} 0 & 0 & 0 & 1 & 0 & 1 & 0 & 0 \\ 0 & 0 & 0 & 0 & 1 & 1 & 1 & 1 \\ 0 & 0 & 0 & 0 & 0 & 0 & 0 & 1 \\ 1 & 1 & 1 & 0 & 1 & 0 & 0 & 0 \\ 0 & 0 & 1 & 1 & 0 & 0 & 1 & 0 \\ 1 & 1 & 0 & 0 & 0 & 0 & 0 & 0 \end{bmatrix} \end{array} \text{ and } B = \begin{array}{c} \\ 1 \\ 2 \\ 3 \\ 4 \end{array} \begin{array}{c} a\ b\ c\ d\ e\ f\ g\ h \\ \begin{bmatrix} 1 & 1 & 0 & 0 & 0 & 0 & 0 & 0 \\ 0 & 0 & 1 & 0 & 1 & 0 & 1 & 0 \\ 0 & 0 & 0 & 1 & 0 & 1 & 1 & 0 \\ 0 & 0 & 1 & 1 & 1 & 1 & 0 & 0 \end{bmatrix} \end{array}$$

(For matrix B, see Ex. 10.6)

Now

$$A.B^{T} = \begin{bmatrix} 0 & 0 & 0 & 1 & 0 & 1 & 0 & 0 \\ 0 & 0 & 0 & 0 & 1 & 1 & 1 & 1 \\ 0 & 0 & 0 & 0 & 0 & 0 & 0 & 1 \\ 1 & 1 & 1 & 0 & 1 & 0 & 0 & 0 \\ 0 & 0 & 1 & 1 & 0 & 0 & 1 & 0 \\ 1 & 1 & 0 & 0 & 0 & 0 & 0 & 0 \end{bmatrix} \begin{bmatrix} 1 & 0 & 0 & 0 \\ 1 & 0 & 0 & 0 \\ 0 & 1 & 0 & 1 \\ 0 & 0 & 1 & 1 \\ 0 & 1 & 0 & 1 \\ 0 & 0 & 1 & 1 \\ 0 & 1 & 1 & 0 \\ 0 & 0 & 0 & 0 \end{bmatrix}$$

$$= \begin{bmatrix} 0 & 0 & 2 & 2 \\ 0 & 2 & 2 & 2 \\ 0 & 0 & 0 & 0 \\ 2 & 2 & 0 & 2 \\ 0 & 2 & 2 & 2 \\ 2 & 0 & 0 & 0 \end{bmatrix} = \begin{bmatrix} 0 & 0 & 0 & 0 \\ 0 & 0 & 0 & 0 \\ 0 & 0 & 0 & 0 \\ 0 & 0 & 0 & 0 \\ 0 & 0 & 0 & 0 \\ 0 & 0 & 0 & 0 \end{bmatrix} \pmod{2}$$

§ 10.5. Cut-Set Matrix

Similar to a circuit matrix, we define another matrix called **cut-set matrix** $C = [c_{ij}]$, where the rows correspond to the cut-sets and the columns to edges of the graph as follows:

$C_{ij} = 1$, if i^{th} cut-set contains j^{th} edge, and

$= 0$, otherwise.

Example 10.8: Find the cut-set matrix for the graph in Fig. 10.5.

Solution: Consider all cut-sets of the graph as shown in the figure 10.6 and arrange them in some order. Now cut-set matrix is given below(after fig 10.6)

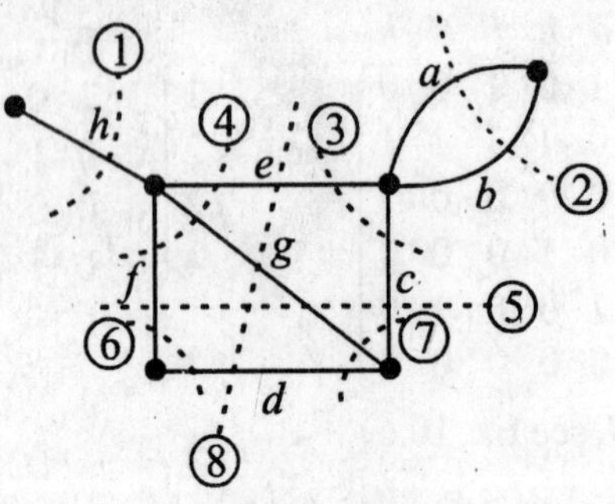

Fig: 10.6

$$C = \begin{array}{c} \\ 1 \\ 2 \\ 3 \\ 4 \\ 5 \\ 6 \\ 7 \\ 8 \end{array}\begin{array}{c} \begin{array}{cccccccc} a & b & c & d & e & f & g & h \end{array} \\ \begin{bmatrix} 0 & 0 & 0 & 0 & 0 & 0 & 0 & 1 \\ 1 & 1 & 0 & 0 & 0 & 0 & 0 & 0 \\ 0 & 0 & 1 & 0 & 1 & 0 & 0 & 0 \\ 0 & 0 & 0 & 0 & 1 & 1 & 1 & 0 \\ 0 & 0 & 1 & 0 & 0 & 1 & 1 & 0 \\ 0 & 0 & 0 & 1 & 0 & 1 & 0 & 0 \\ 0 & 0 & 1 & 1 & 0 & 0 & 1 & 0 \\ 0 & 0 & 0 & 1 & 1 & 0 & 1 & 0 \end{bmatrix} \end{array}$$

Properties of Cut-set Matrix

1. Interchange of rows or columns in a cut-set matrix corresponds to relabeling of the cut-sets and edges respectively.
2. If a column in cut-set matrix has all zeros then the corresponding edge is a self-loop in the graph and vice-versa.
3. Two edges in a graph are parallel edges iff corresponding columns in cut-set matrix are identical.
4. We know that number of edges common to a cut-set matrix C and circuit matrix B are arranged in same order then

$$B \cdot C^T = C \cdot B^T = 0 \pmod 2$$

5. For a connected graph G, rank of cut-set matrix = rank of incidence matrix = rank of the graph G.

CHAPTER ELEVEN

Directed Graphs

In the last four chapters we discussed several basic results in the theory of undirected graphs. No direction was assigned to the edges in the graph. Undirected graphs are not adequate for representing several situations. For example in the graph of a traffic network where an edge may represent a one-way street, we have to assign directions to the edges to indicate the permissible direction of traffic flow. As another example, a computer program can be represented by a graph in which an edge represents the flow of control from one set of instructions to another. In such a representation of a program, directions have to be assigned to the edges to indicate the directions of the flow of control.

Most of the concepts and terminology of undirected graphs are also applicable to directed graphs. In this chapter, we study several basic results in the theory of directed graphs.

§ 11.1. Basic Definitions and Concepts

Directed graph

A directed graph $G = (V, E)$ consists of two sets: a finite set V of elements called vertices and a finite set E of elements called edges.

Moreover, each edge is associated with (or identified by) an *ordered* pair of vertices.

If an edge $e_k = (v_i, v_j)$ then this edge is represented in the directed graph (in short digraph) by a line segment between v_i and v_j with an arrow directed from v_i to v_j. For example, Fig. 11.1 is a directed graph (digraph) with four vertices and seven edges.

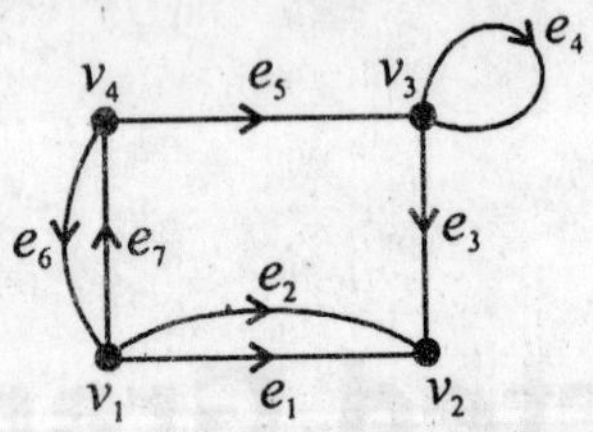

Fig:11.1

Suppose $e_k = (v_i, v_j)$ is a directed edge in a digraph G. Then v_i and v_j are called ***end vertices*** of the edge e_k. The vertex v_i is called ***initial vertex*** and the vertex v_j is called ***terminal vertex***.

Parallel Edges in a Digraph

All edges having the same pair of initial and terminal vertices are called parallel edges. For example, edges e_1 and e_2 are parallel in Fig. 11.1 while edges e_6 and e_7 are not parallel. Thus parallel edges in digraph are those edges which agree in direction of their arrows in addition to being parallel in the sense of undirected graphs.

Self-loop in a Digraph

An edge for which the initial and terminal vertices are the same forms a self-loop. For example, edge e_4 in Fig. 11.1 forms a self-loop.

Let $e_k = (v_i, v_j)$ be an edge in a digraph. Then the edge e_k is said to be **incident out** of its initial vertex v_i and **incident into** its terminal vertex v_j. For example, in Fig. 11.1, edge e_5 is incident out of v_4 and incident into v_3.

A vertex is called **isolated vertex** if no edge is incident out or incident into it.

Two vertices are adjacent if they are the end vertices of some edges. Similarly, if two edges have a common end vertex then they are said to be adjacent edges.

In-degree and Out-degree of a Vertex

The number of edges incident into a vertex v_i is called *in-degree* of the vertex v_i and is denoted by $d^-(v_i)$. Similarly, the number of edges incident out of a vertex v_i is called the *out-degree* of v_i and is denoted by $d^+(v_i)$. Thus

Out-degree of $v_i = d^+(v_i)$ = number of vertices in $\{w \mid (v_i, w) \in E\}$

In-degree of $v_i = d^-(v_i)$ = number of vertices in $\{w \mid (w, v_i) \in E\}$

For example, in the digraph of Fig. 11.1,

$$d^-(v_1) = 1, d^+(v_1) = 3$$

$$d^-(v_3) = 2, d^+(v_3) = 2$$

It may be seen that

$$d(v) = d^+(v) + d^-(v)$$

for any vertex v in a digraph. Here $d(v)$ denote the degree of the vertex v in the graph obtained by disregarding the directions of the edges in the digraph.

Thus isolated vertex in a digraph is that vertex whose in-degree and out-degree both are zero. A vertex v in a digraph is called ***pendant*** vertex if

$$d^+(v) + d^-(v) = 1.$$

Observe that every edge in a digraph contributes 1 to the sum of the in-degrees as well as to the sum of the out-degrees. Hence we have the following theorem.

Theorem 1: In a directed graph with e edges, sum of the in-degrees = sum of the out-degrees = e. In other words,

$$\sum_{i=1}^{n} d^-(v_i) = \sum_{i=1}^{n} d^+(v_i) = e.$$

Directed Walk, Directed Paths and Directed Circuits

Let G be a directed graph. The graph obtained from a directed graph after ignoring the orientations of edges in the digraph is called **undirected** graph corresponding to G. Keeping orientations of edges in the digraph, we define the following:

Directed Walk

A directed walk from a vertex v_i to v_j is an alternating sequence of vertices and edges, beginning with v_i and ending with v_j, such that each edge is oriented (directed) from the vertex preceding it to the vertex following it. Moreover no edge can appear more than once in a directed walk. However, a vertex may appear more than once.

Semi-walk: A semi-walk in a directed graph is a walk in the corresponding undirected graph but is not a directed walk.

Walk in digraph: By walk in a digraph we mean a directed walk or a semi-walk.

Directed path: An open directed walk in which no edge appears more than once is called a directed path. Thus a directed path is a directed walk in which all vertices and edges are distinct.

Semi-directed path: A semi-directed path in a directed graph is a path in the corresponding undirected graph but is not a directed path.

Directed Circuit

A closed directed walk in which no vertex (except the end vertices) appears more than once is called a directed circuit.

Semi-directed Circuit

A semi-directed circuit in a directed graph is a circuit in the corresponding undirected graph but is not a directed circuit.

For example, in Fig. 11.1, $v_1\, e_7\, v_4\, e_6\, v_1$ is a directed circuit but $v_1\, e_1\, v_2\, e_3\, v_3\, e_5\, v_4\, e_6\, v_1$ is a semi-directed circuit (not directed circuit).

Connected Directed Graphs

We know that an undirected graph is connected if there exists a path between every pair of vertices.

In digraph, there are two different types of paths (directed path and semi-directed path). That is why, we have two different types of connectedness in digraphs.

Strongly Connected: A digraph G is said to be *strongly connected* if there exists at least one directed path from every vertex to every other vertex.

Weakly Connected: A digraph G is said to be *weakly connected* if the corresponding undirected graph is connected but G is not strongly connected.

The statement that a digraph G is connected simply means that the corresponding undirected graph is connected. Thus G may not be strongly connected. A digraph which is not connected is called disconnected.

Accessibility

We say a vertex b is accessible (or reachable) from a vertex a if there is a directed path from a to b. Thus we can redefine strongly connectedness in digraph as follows.

A digraph G is strongly connected iff every vertex in G is accessible from every other vertex.

Example 11.1: The digraph in Fig. 11.1 is weakly connected but not strongly connected. The digraph is weakly connected because the corresponding undirected graph is connected. The digraph is not strongly connected because there exist no directed path from vertex v_2 to v_3.

Example 11.2: Consider the following digraphs.

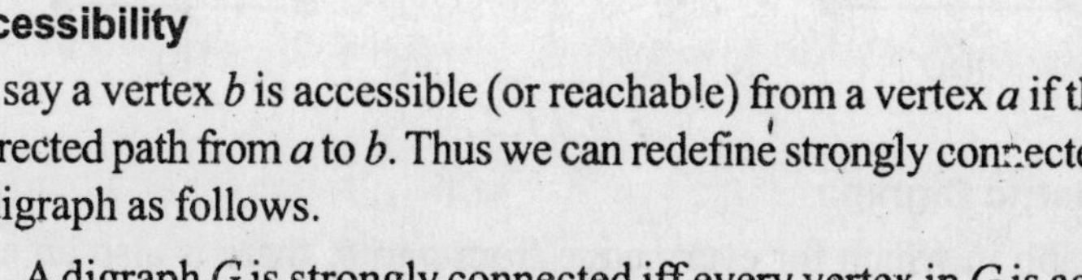

Fig:11.2

Show that digraph in Fig (a) is strongly connected while in (b) is not strongly connected.

Solution: It is easy to observe here that there exists a directed path from every vertex to every other vertex in Fig. 11.2(a).

The digraph in Fig. 11.2(b) is not strongly connected because there exists no directed path from vertex v_1 to vertex v_4.

Some Types of Digraphs

Simple Digraph

A digraph that has no self-loop or parallel edges is called a simple digraph.

The following digraphs are simple digraphs.

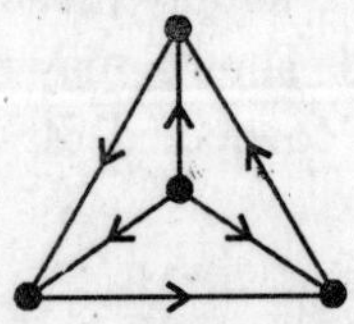

(a)

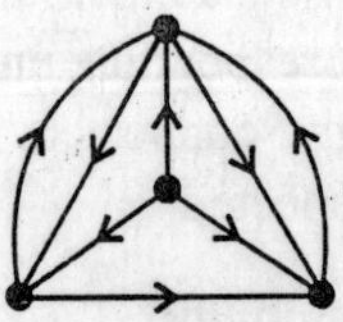

(b)

Fig:11.3

Symmetric Digraph

A digraph in which for every edge from a to b, there is also an edge from b to a is called symmetric digraph.

The digraphs in Fig. 11.3 are not symmetric. The following digraph in Fig. 11.4 is symmetric

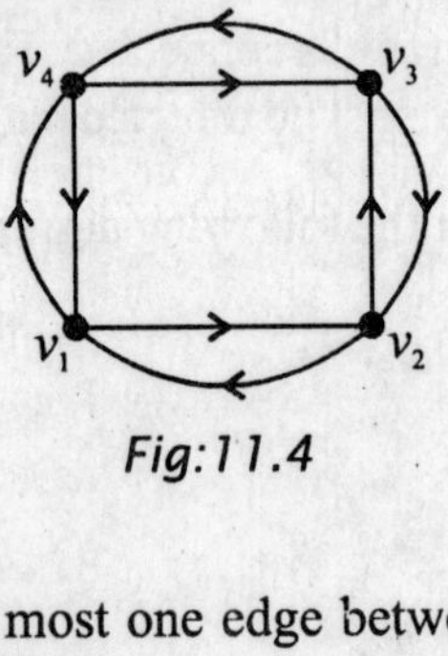

Fig:11.4

Asymmetric Digraph

A digraph which has at most one edge between a pair of vertices is called asymmetric digraph.

The digraphs in Fig. 11.2 are asymmetric where as the digraphs in Fig. 11.3(b) or 11.4 are not asymmetric.

Complete Digraphs

There are two types of complete digraphs namely a complete symmetric digraph and a complete asymmetric digraph. A *complete symmetric* digraph is a simple digraph in which there is exactly one edge directed from every vertex to every other vertex. A *complete asymmetric* digraph is an asymmetric graph in which there is exactly one edge between every pair of vertices.

A complete asymmetric digraph of n vertices has $\frac{n(n-1)}{2}$ edges.

A complete symmetric digraph of n vertices has $n(n-1)$ edges.

Tournament: A complete asymmetric digraph is called a *tournament*.

Digraphs and Binary Relations

In chapter 2, we have studied binary relations in details. We have also represented a binary relation with the help of a graph (Ex. 2.27). Recall that a binary relation on a set $X = \{x_1, x_2, ..., x_n\}$ is a collection of ordered pairs (x_i, x_j) of elements of X. The ordered pair (x_i, x_j) means that x_i is related to x_j. As already seen, a digraph is a most convenient way of representing a binary relation R on a set X. Each $x_i \in X$ is represented by a vertex x_i and every pair (x_i, x_j) in R is represented by a directed edge from x_i to x_j.

Example 11.3: Draw a digraph that represents the relation "is greater than" on a set $X = \{1, 3, 5, 7, 8\}$.

Solution: A digraph representing the given relation is shown below. Note that the relation is given by

$\{(8, 1), (8, 7), (8, 5), (8, 3), (8, 1), (7, 5), (7, 3), (7, 1) (5, 3), (5, 1) (3, 1)\}$

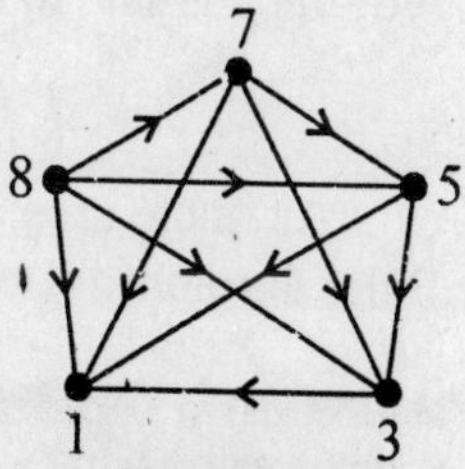

Fig:11.5

The directed graph representing a reflexive relation is called a reflexive digraph. In a similar way, symmetric and transitive digraphs are defined. If the digraph is reflexive then there will be a self-loop at each vertex. The digraph of a symmetric relation is a symmetric digraph. In a similar way, a digraph representing an equivalence relation is called equivalence digraph.

Isomorphic Digraphs

Two digraphs are said to be *isomorphic* if their corresponding undirected graphs are isomorphic and the directions of the corresponding edges are also same.

For example, the two digraphs in Fig. 11.2 are not isomorphic (although corresponding undirected graphs are isomorphic).

§ 11.2. Euler Digraphs

A closed directed walk is a directed walk that starts and ends at the same vertex.

Directed Euler Line

A directed Euler line in a digraph G is a closed directed walk that traverses every edge of G exactly once.

Euler Digraph

A digraph is called *Euler digraph* or directed Euler graph if it contains a directed Euler line.

For example, the digraph in Fig. 11.4 is Euler digraph while the digraphs in Fig. 11.2 are not Euler.

Balanced Digraph

A digraph is said to be *balanced* if for every vertex v_i, we have

$$d^+(v_i) = d^-(v_i) \quad \text{for each } v_i.$$

As in the case of undirected graph we have a simple criterion for Euler digraphs.

Theorem 1: A connected digraph G is Euler digraph iff $d^-(v) = d^+(v)$ for every vertex v in G.

Proof: Proof is exactly the same as in the case of undirected graph.

Theorem 2: Every Euler digraph is strongly connected but converse need not be true.

Proof: Since C is an Euler digraph, it possesses directed Euler line which traverses every vertex. Hence there is a directed path from every vertex to every other vertex. Thus the digraph is strongly connected.

The following example shows that the converse of Theorem 2 is not true in general.

Example 11.4: Draw a digraph which is strongly connected but not Euler.

Solution: The digraph in Fig. 11.2(a) is strongly connected but not Euler digraph.

§ 11.3. Rooted Trees and Binary Trees

A tree for undirected graphs was defined as a connected graph without any circuit. We now define trees in digraphs.

Directed Tree

A directed graph is said to be a *directed tree* if the corresponding undirected graph (graph in which directions of the edges are ignored) is a tree. Thus directed tree is a connected digraph that has no circuits-neither a directed circuit nor a semi-circuit.

Rooted tree

A directed tree is called a *rooted tree* if there is exactly one vertex whose indegree is 0 and the indegrees of all other vertices are 1.

The vertex with in-degree 0 is called the root of the rooted tree.

In a rooted tree, a vertex whose out-degree is 0 is called a **leaf** and a vertex whose out-degree is non-zero is called a **branch node** or an **internal node.**

Let a be a branch node in a rooted tree T. A vertex c is said to be a **descendant** of a if there is a directed path from a to c. Also, a is said to be an **ancester** of c. A vertex b is said to be a **son** (or **child**) of a if there is an edge from a to b.

The following digraph is a rooted tree. In this rooted tree a is a branch node while x is a leaf.

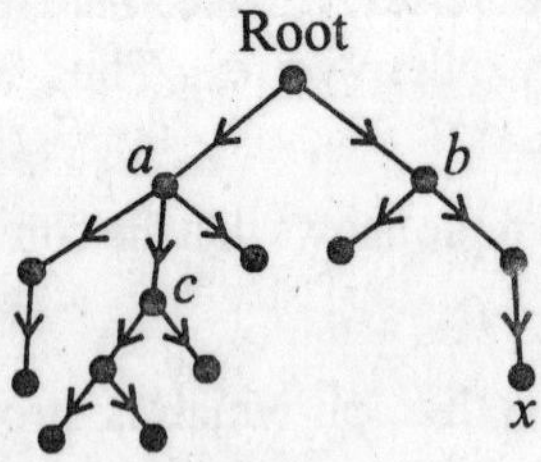

Fig:11.6

Further x is a descendant of b. Observe that the root is a branch node.

Subtree

Let a be a branch node in a rooted tree T. The subtree with a as root is a subgraph T' of T such that vertex set of T' is the set of vertices in T containing a and all of its descendants and the edge set of T' is the set of edges in all directed paths starting from a.

Ordered Tree

A rooted tree T is called on *ordered tree* if the edges incident from each branch nodes are ordered (labeled).

An ordered tree is usually drawn so that edges are ordered from left to right.

Binary Trees

A special class of rooted trees, called binary rooted trees is of particular interest, since they are extensively used in the study of computer science.

Definition: A rooted tree in which out-degree of the root is 2 and each of the remaining internal vertices (branch nodes) has out-degree at most 2 is called a *binary tree*. In other words, a rooted tree in which every branch node has at most two sons (children) is called a binary tree.

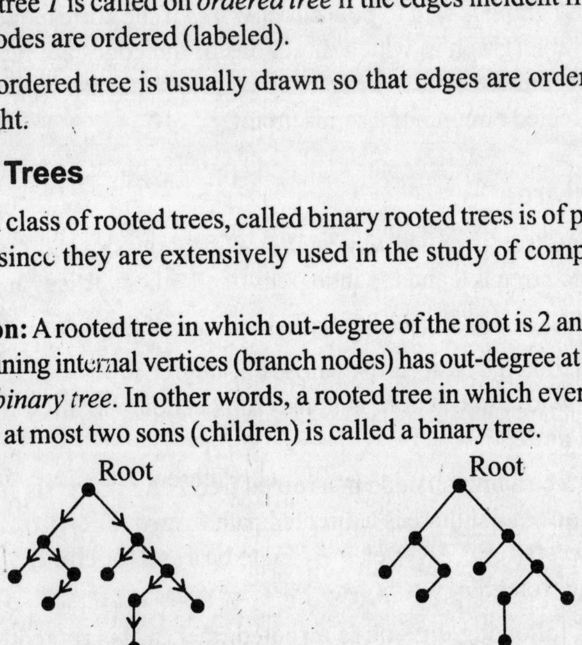

(a) A binary tree (b)

Fig:11.7

While sketching a rooted tree, if we consistently follow the convention of placing the sons of a branch node below it then directions of the edges may be omitted because they will be understood to be pointing downward. That is why, rooted trees are also sometimes considered in undirected graphs. Fig. 11.7(b) is an example of undirected rooted tree.

Full Binary Tree

If in a rooted tree, the outdegree of every node is exactly two then it is called full or complete binary tree. Full binary trees are also sometimes called regular binary trees. Thus in a complete binary tree, every branch node has exactly 2 sons.

Fig. 11.7 is a binary tree but not full binary tree. Tree in Fig. 11.8 is a full (or complete) binary tree.

Two properties of a complete binary trees follow directly from the definition:

1. The number of vertices n in a complete binary tree is always odd. This is because there is exactly one vertex is of even degree and the remaining $n-1$ vertices are of odd degrees and we know that the number of vertices of odd degrees is even. Thus $n-1$ is even and hence n is odd.

2. Number of pendant vertices in any complete binary tree with n vertices is $\frac{n+1}{2}$.

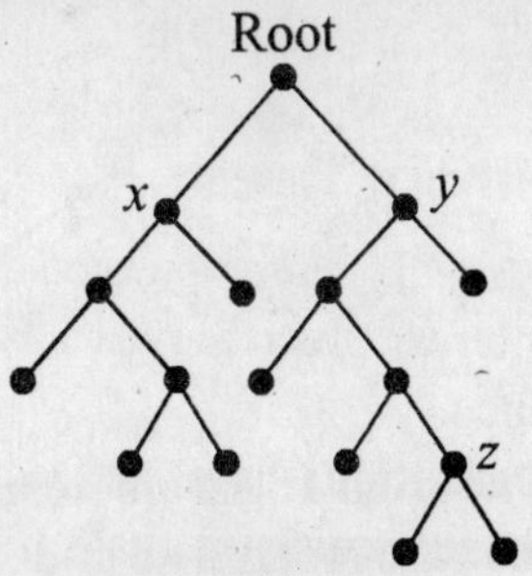

Complete binary tree

Fig: 11.8

Let p be the number of pendant vertices in a complete binary tree T. Then $n-p-1$ is the number of vertices of degree three. Therefore, the number of edges in T equals

$$\frac{1}{2}\left[p+3(n-p-1)+2\right]=n-1$$

$$\Rightarrow \; p=\frac{n+1}{2}.$$

Level of a Vertex

In a rooted tree T vertex v is said to be at **level** (or depth) l if v is at a distance of l from the root. Since there exists only one path between any two vertices, therefore the level of a vertex v in a rooted tree can also be defined as the length of the path from the root to the vertex v.

Clearly the root is at level 0. In Fig. 11.7, level of $x=1$, level of $z=4$.

Height of a Binary Tree

The maximum level, l_{max}, of any vertex in a binary tree is called the *height* of the binary tree. The height of binary tree in Fig. 11.7 is 5.

If the *height* of a binary tree is k then the binary tree is called k-level binary tree. In a rooted tree, all the vertices at any particular level are shown on a horizontal line.

Arborescences

Definition: A digraph G is to be an **arborescence** if G contains no circuit (neither directed nor semi-directed) and G has exactly one vertex v of zero in-degree.

The vertex v with zero in-degree is called root of the arborescence. An arborescence is shown below:

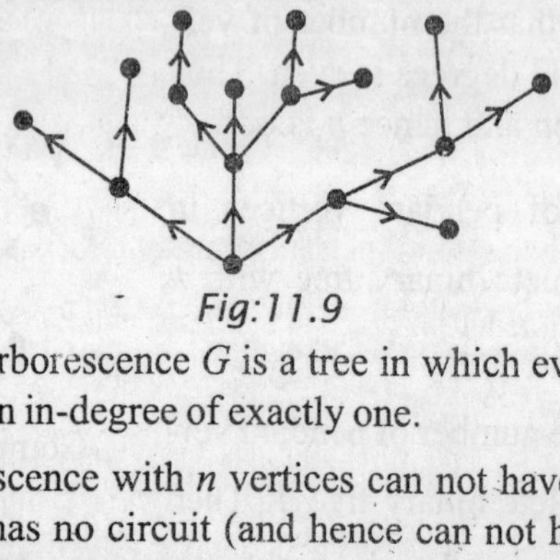

Fig:11.9

Theorem 1: An arborescence G is a tree in which every vertex other than the root has an in-degree of exactly one.

Proof: An arborescence with n vertices can not have more than $n-1$ edges because it has no circuit (and hence can not have more edges than a tree). Consider the sum of in-degrees of all the vertices in G. Since G has at most $n-1$ edges, this sum is $\leq n-1$. That is,

$$d^-(v_1) + d^-(v_2) + ... + d^-(v_n) \leq n-1.$$

Now because there is exactly one vertex in G whose in-degree is zero, others $n-1$ terms must be positive integers. Hence, they must all be equal to 1. Now, since there are exactly $n-1$ vertices of in-degree 1 and one vertex of in-degree zero, G has exactly $n-1$ edges. Thus G must be connected ($\because$ it is circuitless). Hence G is a tree.

Theorem 2: In an arborescence there is a directed path from the root R to every other vertex.

Proof: Consider a path P which starts from the root R and traverses as far as possible in an arborescence. P can end only at a pendant vertex because otherwise we get a vertex whose in-degree is two or more. But then it would be a contradiction. Hence P ends at a pendant vertex. Now since an arborescence is a connected and every directed path from R ends at a pendant vertex, every vertex lies on some directed path from the root R.

Theorem 3: A circuit-less digraph G is an arborescence if there is a vertex v in G such that every other vertex is accessible from v and v is not accessible from any other vertex.

Proof: G is circuit-less and every vertex in G is accessible from v, therefore G is connected and hence G is a tree. Further we are given that v is not accessible from any other vertex, therefore $d^{-}(v) = 0$. Since every vertex is accessible from v, therefore in-degree of every vertex other than v is at least one. Since G is a tree, it has $n-1$ edges, where n is the number of vertices in G. Therefore sum of in-degrees of all the vertices in G. Now there are $n-1$ vertices whose in-degree is atleast one. From this, we conclude that in-degree of each of $n-1$ vertices (other than v) is one. By Theorem 1, G is an arborescence.

Algebraic Expressions and Polish Notations

Consider the arithmetic expression

$$a+b-c.d/(gx+f)$$

If we want to evaluate this expression with the help of a calculating machine then it must perform the arithmetic operations in a certain order. Thus to evaluate the expression, the machine will have to scan the expression back and forth to find the sequence of operations to be performed.

To avoid the scanning back and forth, the computer first converts such expressions into the Polish notation (invented by a Polish mathematician). This notation has the advantage that the operations appear exactly in the same order as they are performed. Therefore, no back and forth scanning is required during the computation and thereby saving the time.

Binary Tree Representation of an Expression

We shall show how an arithmetic expression can be represented the expression $a * b$, where $*$ is any arithmetic operation such as $+, -, \times, \div$ or $\uparrow$ ($a \uparrow b = a^b$) and a and b are two operands by a binary tree given below:

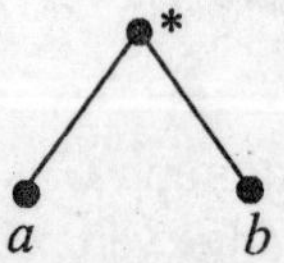

Fig:11.10 Binary tree for expression a∗b

In this tree each pendant vertex represents on operand (variable or constant) and operator appears at the internal vertex. Observe that this binary tree must be ordered since $b–a$ and $a–b$ yield the same tree but not the same ordered tree.

The expression $(a*b)$ o c where $*$ and o are two arithmetic operations and a, b, c are operands can be represented as a binary tree shown in Fig. 11.11, where the operators (namely, $*$ and o) appear in the internal vercices and operands appear in the pendant vertices. Continuing in this way a given arithmetic expression can be represented by means of a binary tree in which each operand (constant or variable) appearing in the expression is represented by a pendant vertex and each internal vertex represents an operator having the two subtrees as its operands.

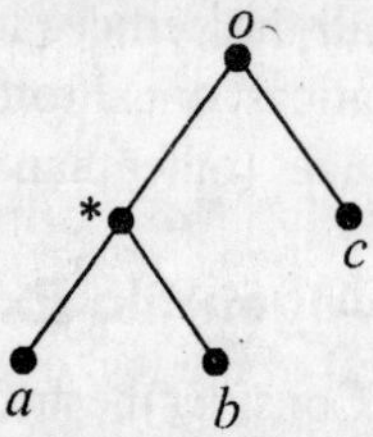

Binary tree for expression (a∗b)oc
Fig:11.11

Example 11.5: Construct the tree of the expression

$$ab + \left(c + \frac{d}{e}\right)f$$

Solution: The tree corresponding to the given expression is shown in Fig. 11.12

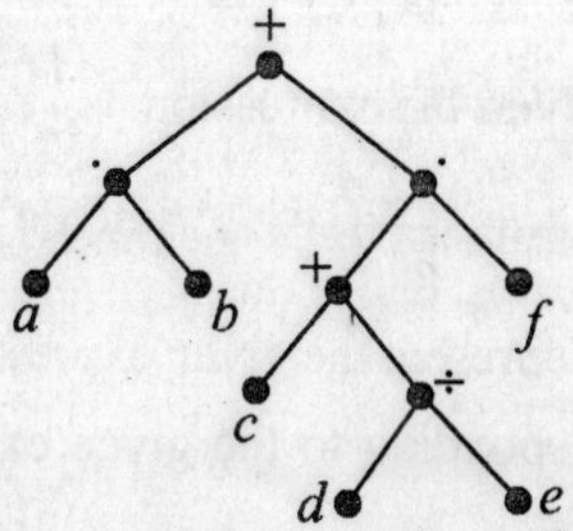

Binary tree for $ab+\left(c+\frac{d}{e}\right)f$

Fig:11.12

Example 11.6: Construct the tree for the expression

$$(a+b-c.d)\div(g^3-f)$$

Solution: The tree corresponding to the given expression is given below:

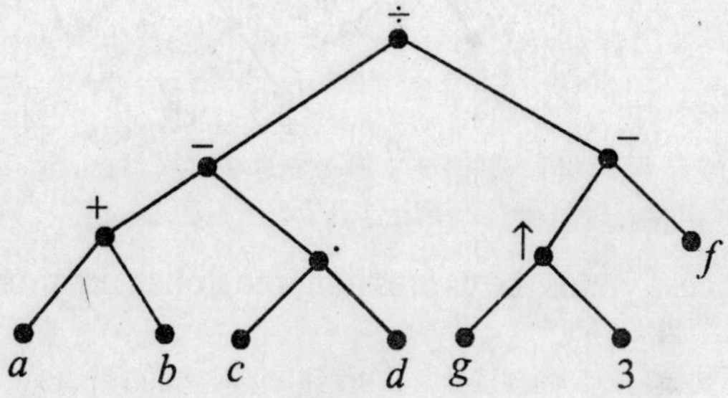

Fig:11.13

Here ↑ denotes exponentiation.

Polish Notations

The basic idea in Polish notation is that an operator appears just to the left of the two operands rather than in between the two operands. For examples

$a+b$ is written as $+ab$ in Polish notation.

and $x\div y$ is written as $\div xy$ in Polish notation.

To obtain the expression in Polish notation (in prefix form) we first represent the given expression by means of a binary tree. Then we traverse the rooted tree starting from the root from left to right and top to bottom as shown by the dotted lines in Ex. 11.7.

Example 11.7: Convert the expression

$$a+b-\frac{c.d}{g^3-f} \quad \text{into Polish notation.}$$

Solution: We first represent the given expression by a rooted tree.

The tree corresponding to the given expression $a + b - c \,.\, d \div (g^3 - f)$ is

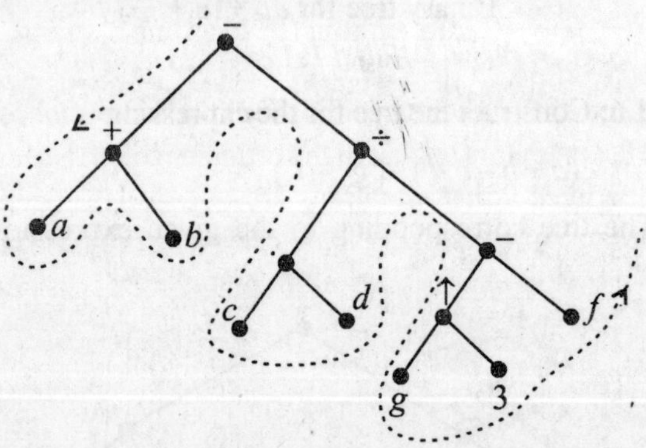

Fig:11.14

The process yields the expression into Polish notation as

$$-+ab\div\cdot cd-\uparrow g3f$$

Remark: Note the difference in expression's and corresponding trees in Ex. 11.6 and 11.7.

Evaluation of an Expression in Polish Notations

We evaluate an expression in Polish notation as described below:

We start from the extreme right and move to the left (of expression in Polish notation). Whenever an operator is found we perform that operation between the two operands immediately to the right of it. After an operation is performed, we regard the resultant as one operand for the next operation. We continue till we reach to the extreme left.

If we use this procedure then the expression

$$-+ab\div\cdot cd-\uparrow g3f$$

is $\quad = -+ab\div\cdot cd-g^3f$

$= -+ab\div .cd\ (g^3-f)$ Regarding g^3 as one operand

$$= -+ab \div (c.d)(g^3 - f)$$

$$= -+ab\left(\frac{cd}{g^3 - f}\right)$$ Regarding $(c \cdot d)$ as one operand and $(g^3 - f)$ as another operand.

$$= -(a+b)\left(\frac{cd}{g^3 - f}\right)$$

$$= a + b - \frac{cd}{g^3 - f}$$ Regarding $(a+b)$ as one operand

The advantage of Polish notation is that in this form there are no parentheses and the operators appear in the order in which they are to be acted upon. Sometimes Polish notations are also called parenthesis-free notations because it contains no parentheses. It may be observed that Polish notation of an expression is not unique.

Note: To obtain the expression in **reverse Polish notation**, we traverse the rooted tree starting from the root from right to left and top to bottom. In reverse Polish notation (or post fix from) we obtain the expression in Ex. 11.7. as

$$ab + cd \cdot g3 \uparrow f - \div - .$$

In this case $a + b$ is written as $ab +$.

§ 11.4. Acyclic Digraphs

Definition: A digraph G is said to be *acyclic* if G has no directed circuit (although it may have semi circuits)

For example, the digraph in Fig. 11.3(a) is acyclic. Every directed tree is also an acyclic digraph but the converse is not true. The digraph in Fig. 11.3(a) is acyclic but it is not a tree.

Theorem 1: Every acyclic digraph G has atleast one vertex with zero in-degree and atleast one vertex with zero out-degree.

Proof: Let P: $v_1, e_1, v_2, e_2, ..., v_p$ be a maximal directed path in G. We claim that the indegree of v_1 and the out-degree of v_p are both equal to zero. Since G is acyclic, v_1 and v_p are distinct.

If the in-degree of v_1 is not zero then there is a vertex w such that the edge $e = (w, v_1)$ is in G. The following two cases arise.

Case 1: Suppose $w \neq v_i$ for any i, $1 \leq i \leq p$. Then there is a directed path P': $w, e, v_1, e_1, \ldots, v_p$ which contains P. This contradicts that P is a maximal directed path.

Case 2: Suppose $w = v_i$ for some i, then there is a directed circuit C: $v_1, e_1, v_2, e_2, \ldots, v_i, e, v_1$. This is again a contradiction.

Thus the in-degree of v_1 is zero. A similar argument shows that the out-degree of v_p is zero.

§ 11.5. Matrices in Digraphs

The matrices associated with a digraph are quite similar to those for an undirected graph. We shall define incidence matrix and adjacency matrix for a digraph.

Incidence Matrix

Let G be a digraph with n vertices, e edges and having no self-loops. The incidence matrix $A = [a_{ij}]$ of G has n rows, one for each vertex and e columns, one for each edge. The element a_{ij} of A is defined as follows:

$$\begin{aligned} a_{ij} &= 1, \text{ if the } j^{\text{th}} \text{ edge is incident out of } i^{\text{th}} \text{ vertex} \\ &= -1, \text{ if the } j^{\text{th}} \text{ edge is incident into } i^{\text{th}} \text{ vertex} \\ &= 0, \text{ if } j^{\text{th}} \text{ edge is not incident on } i^{\text{th}} \text{ vertex} \end{aligned}$$

A digraph and its incidence matrix are in Fig. 11.15.

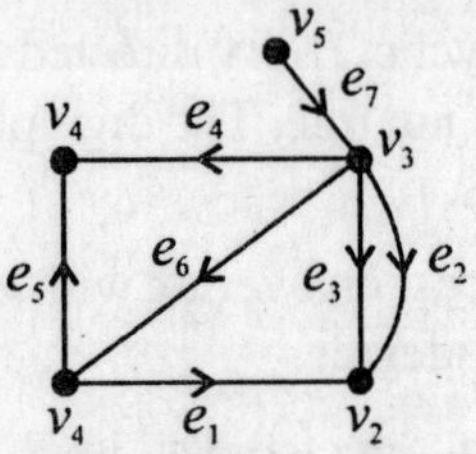

(a) digraph G

$$\begin{array}{c} \\ v_1 \\ v_2 \\ v_3 \\ v_4 \\ v_5 \end{array} \begin{array}{c} \begin{array}{ccccccc} e_1 & e_2 & e_3 & e_4 & e_5 & e_6 & e_7 \end{array} \\ \begin{bmatrix} 1 & 0 & 0 & 0 & 1 & -1 & 0 \\ -1 & -1 & -1 & 0 & 0 & 0 & 0 \\ 0 & 1 & 1 & 1 & 0 & 1 & -1 \\ 0 & 0 & 0 & -1 & -1 & 0 & 0 \\ 0 & 0 & 0 & 0 & 0 & 0 & 1 \end{bmatrix} \end{array}$$

(b) Incidence matrix

Fig: 11.15

It is clear from the definition of incidence matrix that each column of A contains exactly two non-zero entries, 1 and −1. Thus sum of each column is zero. Other properties are similar to those for undirected graph.

Adjacency Matrix of a Digraph:

Let G be a digraph with n vertices $v_1, v_2, \ldots, v_n$.

The adjacency matrix $X = [x_{ij}]$ of G is an $n \times n$ matrix with x_{ij} defined as follows:

$$x_{ij} = \begin{cases} k, & \text{if there are } k \text{ edges directed from } i^{th} \text{ vertex to } j^{th} \text{ vertex.} \\ 0, & \text{otherwise.} \end{cases}$$

If the digraph has no parallel edges, then $k = 1$.

A digraph and its adjacency matrix are shown in Fig. 11.16.

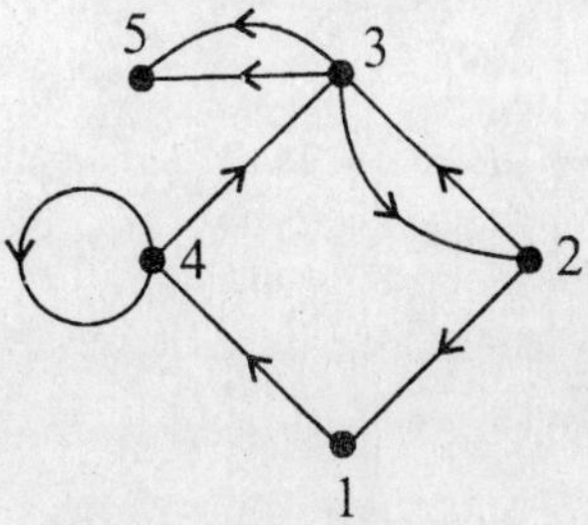

$$X = \begin{array}{c} \\ 1 \\ 2 \\ 3 \\ 4 \\ 5 \end{array} \begin{array}{c} \begin{matrix} 1 & 2 & 3 & 4 & 5 \end{matrix} \\ \begin{bmatrix} 0 & 0 & 0 & 1 & 0 \\ 1 & 0 & 1 & 0 & 0 \\ 0 & 1 & 0 & 0 & 2 \\ 0 & 0 & 1 & 1 & 0 \\ 0 & 0 & 0 & 0 & 0 \end{bmatrix} \end{array}$$

Fig:11.16

Properties of Adjacency Matrix

1. If G is a digraph without parallel edges then X is symmetric if and only if G is a symmetric graph.
2. The sum of each row equals the out-degree of the corresponding vertex.
3. The sum of each column equals the in-degree of the corresponding vertex.
4. The sum of entries in X equals the number of edges in G.
5. If G is a digraph with parallel edges then X is not (0, 1) – matrix
6. Every $n \times n$ matrix X with non-negative integer entries uniquely defines a directed graph with n vertices.
7. The $(i, j)^{th}$ entry in X^r equals the number of different, directed edge sequences of r edges from vertex v_i to v_j.

We now state a result which is very important for determining whether a digraph is strongly connected. The result follows from property 7 given above.

Theorem 1: Let X be the adjacency matrix of a digraph with n vertices and let

$$Y = X + X^2 + X^3 + ... + X^n$$

Then G is strongly connected if and only if Y has no zero entries.

Example 11.8: Determine whether the following digraph G is strongly connected.

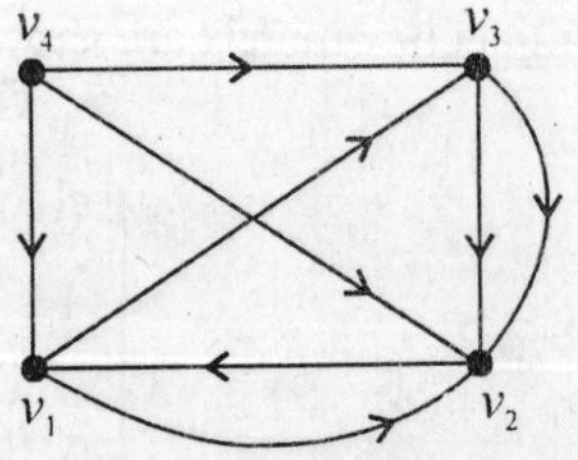

Fig: 11.17

Solution: We use Theorem 1 to determine the strong connectivity of G. Here, number of vertices is 4. First we write adjacency matrix of G.

$$X = \begin{matrix} & 1 & 2 & 3 & 4 \\ 1 \\ 2 \\ 3 \\ 4 \end{matrix} \begin{bmatrix} 0 & 1 & 1 & 0 \\ 1 & 0 & 1 & 0 \\ 0 & 1 & 0 & 0 \\ 1 & 1 & 1 & 0 \end{bmatrix}$$

Now multiplying X with itself repeatedly, we get

$$X^2 = \begin{bmatrix} 1 & 1 & 1 & 0 \\ 0 & 2 & 1 & 0 \\ 1 & 0 & 1 & 0 \\ 1 & 2 & 2 & 0 \end{bmatrix}, \quad X^3 = \begin{bmatrix} 1 & 2 & 2 & 0 \\ 2 & 1 & 2 & 0 \\ 0 & 2 & 1 & 0 \\ 2 & 3 & 3 & 0 \end{bmatrix}, \text{ and}$$

$$X^4 = \begin{bmatrix} 2 & 3 & 3 & 0 \\ 1 & 4 & 3 & 0 \\ 2 & 1 & 2 & 0 \\ 3 & 5 & 5 & 0 \end{bmatrix}$$

Now $$Y = X + X^2 + X^3 + X^4 = \begin{bmatrix} 4 & 7 & 7 & 0 \\ 4 & 7 & 7 & 0 \\ 3 & 4 & 4 & 0 \\ 7 & 11 & 11 & 0 \end{bmatrix}$$

Since Y has zero entries, given digraph is not strongly connected.

Exercise

(1) Show that the maximum number of vertices in a binary tree of height k is $2^{k+1} - 1$.

[Hint: There can be only one vertex at level 0, at most two vertices at level 1, at most four vertices at level 2 and so on. Therefore maximum number of vertices possible in a binary tree of height k is

$$n \leq 2^0 + 2^1 + ... + 2^k = 2^{k+1} - 1]$$

(2) Let G be a binary tree with n vertices. Show that minimum possible height of G is $[\log_2 (n+1) - 1]$ and maximum possible height is $\frac{n-1}{2}$. For $n = 11$, draw two binary trees exhibiting both the cases.

(3) How many leaves (pendant vertices) does a full binary tree with k level have?

(4) Let G be a digraph. A vertex v in G is said to be ***source*** if in-degree $d^-(v) = 0$. A vertex v is called a ***sink*** if out-degree $d^+(v) = 0$.

Consider the graph in the following figure (see the next page)

(a) Express the graph by its adjacency matrix

(b) Does it have any sources or sinks?

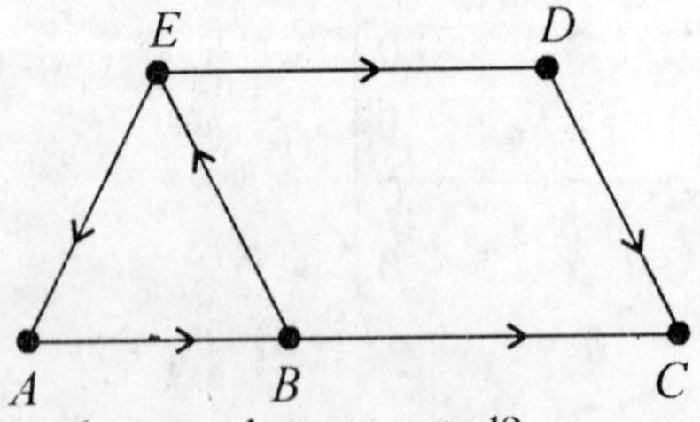

(c) Is this graph strongly connected?

(5) A digraph G is said to be *unilaterally connected* if given any two vertices u and v in G, either there exists a directed path from u to v or there exists a directed path from v to u.

(a) Show that every strongly connected graph is unilaterally connected.

(b) Show that graph in question 4 is unilaterally connected.

(c) Give an example of a graph which is weakly connected but not unilaterally connected.

(6) Show that the following digraph has an Eulerian circuit.

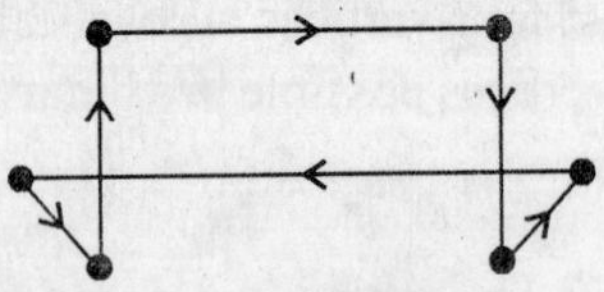

(7) Represent the relation "is greater than" on the set {3, 4, 7, 5, 8} by a digraph.

(8) Determine whether the relation R whose digraph is given below is reflexive?, symmetric?

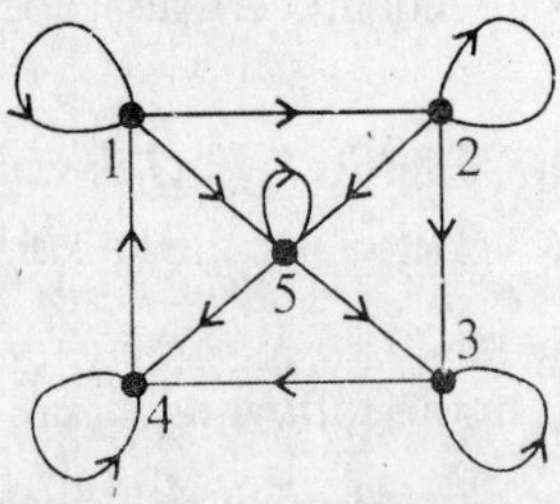

(9) Draw the digraph corresponding to the adjacency matrix:

$$X = \begin{bmatrix} 1 & 3 & 0 & 0 \\ 0 & 0 & 1 & 1 \\ 0 & 1 & 0 & 2 \\ 1 & 0 & 1 & 0 \end{bmatrix}$$

(10) Draw a tree corresponding to the expression

(a) $(a - b) \div c$

(b) $a - b \div c$

(11) Draw the tree which corresponds to the expression

$$E = (2x + y)\,(5a - b)^3.$$

Hence convert E into Polish notation.

(12) Draw the ordered rooted tree T using an ($\uparrow$) for exponentiation of the expression

$$E = \frac{(3x - 5z)^4}{a(2b + c^2)}$$

Use tree T to write E in Polish notation.

Answers

(3) 2^k

(4) (a)

$$\begin{array}{c|ccccc} & A & B & C & D & E \\ \hline A & 0 & 1 & 0 & 0 & 0 \\ B & 0 & 0 & 1 & 0 & 1 \\ C & 0 & 0 & 0 & 0 & 0 \\ D & 0 & 0 & 1 & 0 & 0 \\ E & 1 & 0 & 0 & 1 & 0 \end{array}$$

(b) No source, C is a sink. (C) No

(5) (c)

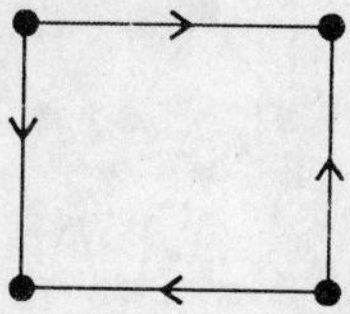

(7)

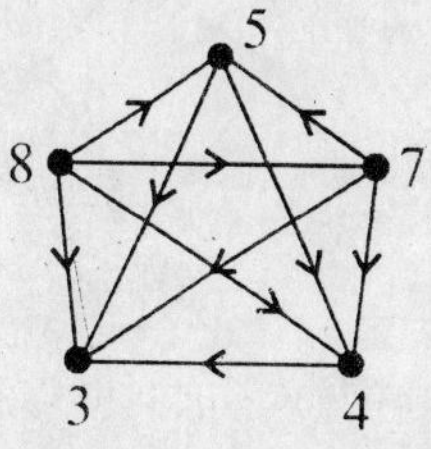

(8) Reflexive but not symmetric

(9)

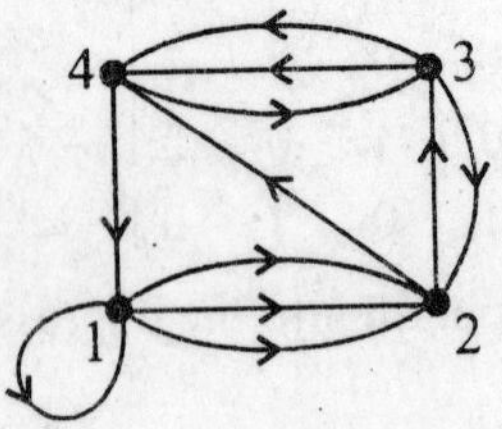

(10) (a)

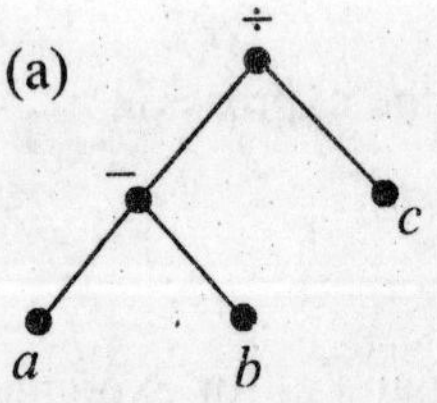

(b)

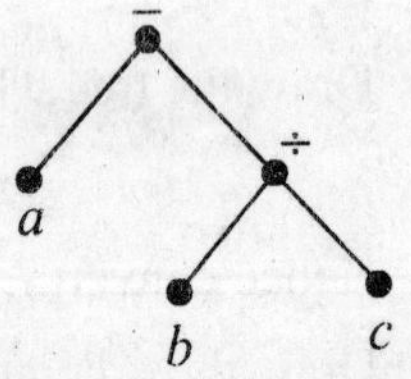

(11)

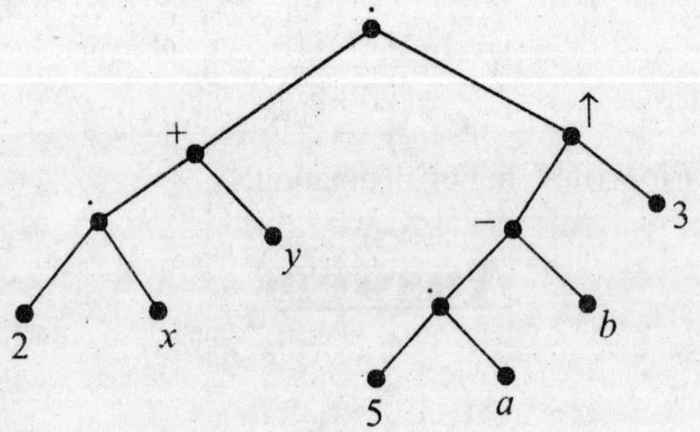

Polish notation is $\cdot + \cdot\, 2xy \uparrow - \cdot\, 5\, ab3$

(12)

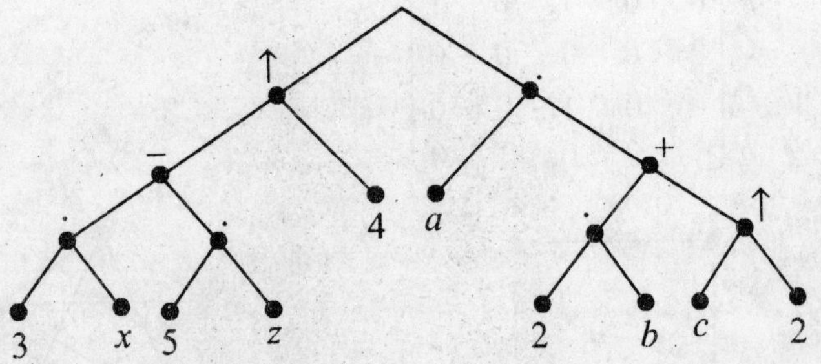

$E = \div \uparrow - \cdot\, 3x \cdot 5z4 \cdot a + \cdot\, 2b \uparrow c2$

CHAPTER TWELVE

General Counting Methods

Combinatorics is a branch of mathematics that deals with counting. In this chapter, we shall study some topics which are essentially the methods for counting the ways in which various combination of objects can arise.

12.1 Sum and Product Rules

We begin by stating two basic principles for counting the number of ways events can happen.

Sum Rule : If one job can be done in m ways and another job can be done in n ways and if there is no way common to both jobs then the total number of ways in which either of the two jobs can be done is equal to $m+n$.

In set-theoretic notations, sum rule states that if A and B are two finite sets such that $A \cap B = \phi$ then

$$|A \cup B| = |A| + |B|,$$

where $|X|$ denotes the number of elements in X.

Product Rule : If one job can be done in m ways and following this another job can be done in n ways then the total number of ways in which both the jobs can be done in the stated order is $m \times n$.

In set-theoretic notations, product rule says that

$$|A \times B| = |A| \times |B|,$$

where A and B are finite sets.

Example 12.1 : Suppose an institute offers seven different courses in the morning shift and six different courses in the evening shift.

(a) How many ways are there for students who want admission in one course only ?

(b) How many ways are there for students who want admission in one course in the morning shift and one in the evening shift ?

Solution :

(a) By sum rule, students will have $7 + 6 = 13$ choices if they want admission in only one course.

(b) By product rule there will be $7 \times 6 = 42$ choices for students who want to take admission in one course in the morning shift and one in the evening shift.

Example 12.2 : There are five different Hindi books, six different English books and eight different Sanskrit books. How many ways are there to pick two books not both in the same language ?

Solution : There are three cases :

(*i*) 1 Hindi and 1 English book

(*ii*) 1 Hindi and 1 Sanskrit book

(*iii*) 1 English and 1 Sanskrit book

Since there are five Hindi books and six English books, we can select one Hindi and one English book in 5×6 ways. Similarly, one Hindi and one Sanskrit book can be chosen in $5 \times 8 = 40$ ways and one English and one Sanskrit book can be chosen in $6 \times 8 = 48$ ways. Since the above three types of selections are disjoint, therefore by sum rule, there are $30 + 40 + 48 = 118$ ways in all.

Example 12.3 : If A and B are finite sets then

1. $|A - B| = |A| - |A \cap B|$

2. $|A - B| = |A| - |B|$ if $B \subset A$

Solution :

1. We know from set-theory that

$$(A - B) \cup (A \cap B) = (A \cap B') \cup (A \cup B)$$
$$= A \cap (B \cup B')$$
$$= A$$

and
$$(A - B) \cap (A \cap B) = (A \cap B') \cap (A \cap B)$$
$$= A \cap B \cap B' = \phi$$

Therefore, by sum rule,

$$|A| = |(A - B) \cup (A \cap B) = |A - B| + |A \cap B|$$
$$\Rightarrow \quad |A - B| = |A| - |A \cap B|$$

2. It follows from (1) because $B \subset A$ implies $A \cap B = B$. Substituting $A \cap B = B$ in (1), we get $|A - B| = |A| - |B|$.

12.2 Permutations

Definition : A permutation is an arrangement of a finite set of objects in a particular order.

For example, there are six different permutations of the set $\{a, b, c\}$. They are *abc*, *acb*, *bac*, *bca*, *cab* and *cba*.

Any arrangement of n distinct objects taken r at a time $(r \leqslant n)$ is called r-permutation. The number of permutations of n distinct objects taken r at a time is given by

$$^nP_r = n(n-1)(n-2)\ldots(n-r+1)$$
$$= \frac{n!}{(n-r)!}$$

Permutations with Repetitions

If out of n objects in a set, p objects are exactly alike of one kind, q objects exactly alike of second kind and r objects exactly alike of third kind and the remaining objects are all different then the number of permutations of n objects taken all at a time is

$$= \frac{n!}{p!\,q!\,r!}$$

Example 12.4 : How many ways are there to arrange the nine letters in the word 'ALLAHABAD'.

Solution : Since the word ALLAHABAD contains $4A$'s and $2L$'s, therefore there are

$$= \frac{9!}{4!\,2!} = 7560 \text{ ways.}$$

12.3 Combinations

Definition : Any selection which can be made by taking some or all of the objects at a time out of a given number of objects is called a *combination.*

For example, a team of two players selected out of five players is a combination of 5 players taken 2 at a time.

In the case of a combination, the order of the objects does not matter. The objects in a combination when arranged give rise to a permutation. The number of combinations of n distinct objects taken r $(r \leqslant n)$ at a time is given by

$$^nC_r = \frac{n!}{r!\,(n-r)!}$$

It is clear from the definitions or ^nP_r and ^nC_r that

$$^nP_r = {}^nC_r \times r\,!$$

Example 12.5 : Find the number of diagonals which can be drawn by joining the angular points of a heptagon.

Solution : A heptagon has seven angular points (vertices) and seven sides. The join of two angular points is either a side or a diagonal. The number of lines joining the angular points

$$^7C_2 = \frac{7!}{2!\,5!} = 21$$

Since this number includes the seven sides also, therefore, number of diagonals = 21 – 7 = 14.

Example 12.6 : In how many ways can a committee of 5 persons be formed from 6 men and 4 women so as to include at least 2 women ?

Solution : There are the following three cases. The committee may consist of

(*i*) 3 men and 2 women,

(*ii*) 2 men and 3 women,

(*iii*) 1 man and 4 women.

We can select 3 men and 2 women in

$$^6C_3 \times {}^4C_2 = 20 \times 6 = 120 \text{ ways}$$

We can select 2 men and 3 women in

$$^6C_2 \times {}^4C_3 = 15 \times 4 = 60 \text{ ways}$$

We can select 1 man and 4 women in

$$^6C_1 \times {}^4C_4 = 6 \times 1 = 6 \text{ ways.}$$

Since all the three cases are disjoint, hence by sum rule, the required number of ways

$$= 120 + 60 + 6 = 186.$$

12.4 The Pigeonhole Principle

It states that if there are more pigeons (objects) than the pigeonholes (boxes), then some pigeonhole (box) must contain two or more pigeons (objects).

In set-theoretic terms, pigeonhole principle is equivalent to the following :

Let A and B be finite sets. If $|A| > |B|$ then any function $f: A \to B$ cannot be one-to-one. That is, there exists at least 2 elements x and y in A such that $f(x) = f(y)$.

Although the pigeohole principle itself is trivial but when applied cleverly, it can yield nontrivial results. The pigeonhole principle is also called the **Dirichlet drawer principle**. To apply the pigeonhole principle, we must identify pigeons (objects) and pigeonholes (categories of the desired characteristic) and be able to count the number of pigeons and the number of pigeonholes.

Example 12.7 : Show that among 13 people, there are at least two people who were born in the same month.

Solution : If we take 13 people as pigeons and the 12 months (January, February, March, ..., December) as the pigeonholes then by the pigeonhole principle there will be at least two people who were born in the same month.

Example 12.8 : Show that in a set of five distinct integers, there must exist two integers with the same remainder when divided by 4.

Solution : We know that when any integer is divided by 4 then the possible remainders are 0, 1, 2 and 3. Assuming 5 given integers as pigeons and 4 remainders as pigeonholes, we get, by the pigeonhole principle, that there are at least two integers with same remainder (when divided by 4.)

Example 12.9 : Suppose 14 students in a class appear at an examination. Prove that there exist at least two among them whose seat numbers differ by a multiple of 13.

Solution : Let A be the set of seat numbers of 14 students. We know that when we divide any integer by 13 then the remainder can only have the values 0, 1, 2, ..., 12. So, we define a function

$$f : A \rightarrow \{0, 1, 2, ..., 12\} \text{ by}$$

$f(x)$ = remainder when x is divided by 13.

Since $|A| = 14$ and $|\{0, 1, 2, ..., 12\}| = 13$, therefore by the pigeonhole principle, function f cannot be one-to-one. Thus, there exists two distinct seat numbers x and y such that $f(x) = f(y)$. Now x and y can be written as (by Euclidean algorithm)

$$x = 13a + f(x) \quad \text{and} \quad y = 13b + f(y)$$

where a and b are integers.

$$\Rightarrow \quad x - y = 13(a - b) \qquad \because f(x) = f(y)$$

Since $a - b$ is an integer, $x - y$ is a multiple of 13.

Hence, there exist at least two students whose seat numbers differ by a multiple of 13.

Example 12.10 : If any 51 integers are chosen from the set {1, 2, 3, ..., 100} then show that among the chosen integers there exist two integers such that one is multiple of the other.

Solution : We know that every positive integer x can be written as $= 2^p . r$, where r is odd and $p \geqslant 0$ (for example, $12 = 2^2.3$ and $13 = 2^0.13$). We shall call this odd integer r the odd part of integer x.

If x is in {1, 2, ..., 100} then its odd part r can only be in 1, 3, 5, ..., 99}.

Let S be any set of 51 integers chosen from the set {1, 2, 3, ..., 100}. Define a function

$$f : S \rightarrow \{1, 3, 5, 7, ..., 99\} \text{ by}$$
$$f(x) = \text{odd part of } x.$$

Since $|S| = 51$ and $|\{1, 3, 5, 7, ..., 99)\}| = 50$, it follows from the pigeonhole principle that f cannot be one-to-one. Thus, there exist two integers x and y in S such that their odd parts $f(x)$ and $f(y)$ are same. That is, $f(x) = f(y)$. Therefore, we have

$x = 2^p.r$ and $y = 2^q.r$ for some integers p and q.

If $p \geqslant q$ then $x = 2^{p-q}. y$ and therefore x is a multiple of y. If $p < q$ then y is a multiple of x. Thus, either x is multiple of y or y is multiple of x.

Example 12.11 : Find the minimum number of elements to be chosen from the set $S = \{1, 2, 3, ..., 9\}$ such that two of them should add up to 10.

Solution : We first construct all possible different set $\{x, y\}$ of two elements from the given set S such that $x + y = 10$. They are {1, 9}, {2, 8}, {3, 7}, {4, 6}. Now consider the following partition of S :

$$\{1, 9\} \{2, 8\}, \{3, 7\}, (4, 6\} \text{ and } \{5\}$$

Observe that each element of S belongs to one and only one of the above five sets. If we choose any six elements of the set S then it follows by the pigeonhole principle that two of them must belong to the same set which add up to 10.

The Pigeonhole Principle can be generalized in many way. We give below two generalizations. Another generalization is given in problem set.

***Theorem* 1 :** If $r \geqslant 2$ is an integer and if $n(r-1)+1$ objects are placed in n boxes, then there exists a box with at least r objects.

Proof : Suppose if possible, every box contains less than r objects. Then there will be atmost

$$n.(r-1) \text{ objects.}$$

But this is a contradiction to our assumption that there are $n(r-1)+1$ objects. Hence, there exists a box with atleast r objects.

***Theorem* 2 :** (Generalized pigeonhole principle). If there are n pigeons and m pigeonholes $(n > m)$ then some pigeonholes must contain at least $\left[\frac{(n-1)}{m}\right]+1$ pigeons, where $[x]$ denotes the largest integer not greater than x.

Proof : If possible, suppose that none of the pigeonholes contains more than $\left[\frac{(n-1)}{m}\right]$ pigeons. Then there will be atmost

$$m.\left[\frac{(n-1)}{m}\right] \leqslant m.\frac{(n-1)}{m} = n-1 \text{ pigeons}$$

This is a contradiction to the assumption that there are n pigeons.

Corollary : If pigeons are more than k times the pigeonhole then some pigeonhole must contain at least $k+1$ pigeons.

Proof : Put $n > km$ in the above theorem to get the result.

Example 12.12 : Show that if seven colours are used to paint 50 cars, atleast eight cars will have the same colour.

Solution : Here 50 cars (pigeons) are to assign 7 colours (pigeonholes). Hence, by the generalized pigeonhole principle, at least $\left[\frac{(50-1)}{7}\right]+1$ $= 8$ cars will have the same colour.

Example 12.13 : How many letters one must choose from the set of 15A's, 20B's and 25C's so that 12 identical letters will always be include in the selection.

Solution : Here there are three types of letters (pigeonholes) each having more than 12 letters. We have to find minimum number of letters (pigeons) n such that

$$\left[\frac{(n-1)}{3}\right]+1=12.$$

$$\Rightarrow \qquad \left[\frac{(n-1)}{3}\right]=11$$

$$\Rightarrow \qquad n=11\times 3+1=34$$

Thus, 34 is the minimum number of letters to be selected from the set having 15 A's, 20 B's and 25 C's so that 12 identical letters will always be included in the selection.

Example 12.14 : Show that at any party with six people, there either exists a set of three mutual friends or a set of three mutual strangers.

Solution : Let x be any person at the party. Let S_1 be the set of persons who are friends of x and S_2 be the set of persons who are strangers to x. Then the remaining 5 persons (pigeons) can be either in S_1, or in S_2. By the generalized pigeonhole principle either S_1 or S_2 contains at least $\left[\frac{5-1}{2}\right]+1=3$ persons. We first consider the case when S_1 contains 3 persons a, b and c. If any two of a, b, c are friends then these two together with x form a set of three mutual friends. On the other hand if no two of a, b, c are friends of each other then $\{a, b, c\}$ is a set of mutual strangers. In the second case suppose S_2 contains 3 persons a, b and c. If two of a, b, c are strangers to each other then these two together with x form a set of three mutual strangers. If no two of a, b, c are strangers to each other then they form a set of three mutual friends.

Example 12.15 : Show that any set of seven distinct integers contains two integers x and y such that either $x + y$ or $x - y$ is divisible by 10.

Solution : Let $A = \{a_1, a_2, ..., a_7\}$ be any set of seven distinct integers. Let r_i be the remainder when any element a_i of A is divided by 10. Consider the following partition of A :

$$S_1 = \{a_i : a_i \in A \text{ and remainder } r_i = 0\}$$

$$S_2 = \{a_i : a_i \in A \text{ and remainder } r_i = 5\}$$

$$S_3 = \{a_i : a_i \in A \text{ and remainder } r_i = 1 \text{ or } 9\}$$

$$S_4 = \{a_i : a_i \in A \text{ and remainder } r_i = 2 \text{ or } 8\}$$

$$S_5 = \{a_i : a_i \in A \text{ and remainder } r_i = 3 \text{ or } 7\}$$

$$S_6 = \{a_i : a_i \in A \text{ and remainder } r_i = 4 \text{ or } 6\}$$

We know that when we divide any integer by 10, then the remainder can only have the values 0, 1, 2, ..., 9. Therefore, if we divide any integer by 10 then its remainder must belong to one of the sets S_i, $i = 1, 2, ..., 6$.

Taking seven distinct integers a_i, $i = 1, 2, ..., 7$ as pigeons and six sets S_i, $i = 1, 2, ... 6$ as pigeonholes we conclude from the pigeonholes principle that some S_i must contain at least two integers a_i and a_j from A. If both a_i and a_j are in S_1 or S_2 then $a_i + a_j$ and $a_i - a_j$ both are divisible by 10. If a_i and a_j are in one of the other four subsets then

$$a_i = 10x + r_i \qquad \text{and} \qquad a_j = 10y + r_j,$$

where r_i and r_j are in one of the sets S_3, S_4, S_5, S_6.

Now

$$a_i - a_j = 10(x - y) + (r_i - r_j) \quad \text{and} \quad a_i + a_j = 10(x + y) + (r_i + r_j).$$

If r_i and r_j are equal then $a_i - a_j$ is divisible by 10. If $r_i \neq r_j$ then $r_i + r_j = 10$ and hence $a_i + a_j$ is divisible by 10.

Thus either $a_i - a_j$ or $a_i + a_j$ in divisible by 10.

Example 12.16 : In a tournament in which each player plays against every other player, and each player wins at least once, show that there are at least two players having the same number of wins.

Solution : Let the number of players in a tournament be n. Since each player wins at least once, the number of wins for a player is at least 1 and at most $n - 1$. Let us define a function f from the set of players to the set $\{1, 2, ..., n - 1\}$ by $f(x)$ = number of wins of the player x.

Since there are n players, it follows from the pigeonhole principle that atleast two players have the same number of wins.

Example 12.17 : Show that any sequence of $n^2 + 1$ distinct integers contains an increasing subsequence of length $n + 1$ or a decreasing sequence of length $n + 1$.

Solution : Let $a_1, a_2, a_3, ..., a_{n^2+1}$ be a sequence of $n^2 + 1$ distinct integers. Corresponding to each a_k, let us consider an ordered pair (x_k, y_k), where x_k is the length of a longest increasing subsequence starting at a_k and y_k be the length of a longest decreasing subsequence starting at a_k. Suppose if possible there is no increasing or decreasing subsequence of length $n + 1$ in the sequence a_i of $n^2 + 1$ integers. Then the set $\{(x_k, y_k)\}$ has atmost n^2 distinct ordered pairs. It follows by the pigeonhole principle that there must exist two elements a_i and a_j in the sequence which correspond to the same ordered pair. In other words, $(x_i, y_i) = (x_j, x_j)$. But this is not possible because if $a_i < a_j$ then we must have $x_i > x_j$ and if $a_i > a_j$ then we must have $y_i > y_j$. Contradiction proves that there is either an increasing subsequence or a decreasing subsequence of length $n + 1$.

Example 12.18 : Given any five points in the interior of an equilateral triangle of side 1, show that there exists two points with in a distance of at most 1/2.

Solution : Divide the given equilateral triangle into four equal triangles as shown in the figure.

Now the given five points will be placed in these four triangles. By the pigeonhole principle atleast two of them must belong to one of the four small triangles. The distance between these two points cannot exceed the side of the triangle which is $\frac{1}{2}$. Thus, there exists two points within a distance of atmost $\frac{1}{2}$.

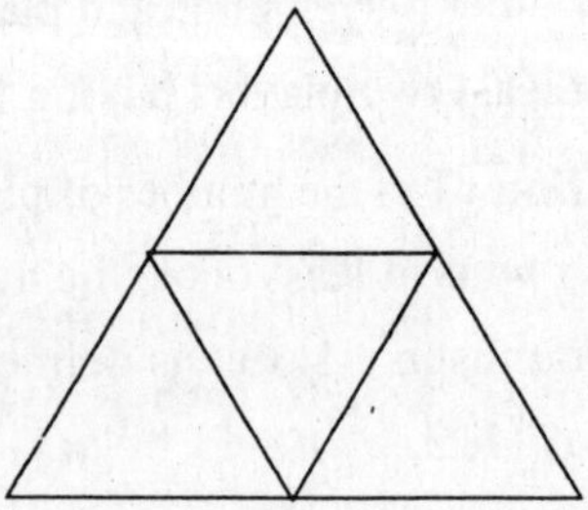

Exercise

1. Prove that if $q_1 \geqslant 1$, $q_2 \geqslant 1$, ..., $q_n \geqslant 1$ are integers and if $q_1 + q_2 +$ ... $q_n - n + 1$ objects are put into n boxes, then either the first box contains atleast q_1 objects or the second box contains atleast q_2 objects, ..., or the nth box contains atleast q_n objects.

2. Show that at a party of n people ($n > 1$), there are 2 people who have the same number of friends.

   (**Hint :** Note that friendship is a symmetric relation. A person can have 0, 1, 2, ... upto $n - 1$ friends. If any person at the party has 0 friend then no person can have $n - 1$ friends and if any person has $n - 1$ friends then no person can have 0 friend. Thus, number of friends of a person are either $\{0, 1, 2, ..., n - 2\}$ or $\{1, 2, ..., n - 1\}$. In either case we have n people and $n - 1$ choices for number of friends.)

3. Show that any subset of $n + 1$ different integers between 1 and $2n$ ($n \geqslant 2$) always contains a pair of integers with no common divisor.

4. Given 10A's, 20B's, 8C's, 15D's and 25E's, how may letters must be chosen to guarantee that there are 12 identical letters.

   [Ans. 52]

5. A professor tells three jokes in his class each year. How many jokes does the professor require in order to never repeat the exact same triple of jokes over a period of 12 years.

[Ans. 6]

6. Find the minimum number of elements to be chosen from $\{1, 2, 3, ..., 8\}$ such that two of them will add up to 9.

[Ans. 5]

7. A student wants to prepare for his final examination by solving some unsolved question papers in 77 days. He decides to solve at least one paper a day but not more then 132 papers altogether. Show that there is a period of consecutive days with in which he solves exactly 21 papers.

8. How many times must we throw two dice in order to be sure that we obtain the same total score atleast 7 times ? [Ans. 67]

9. How many cards must be drawn from a pack of 52 cards to be sure that you have seven cards of one suit. [Ans. 25]

10. Given any five points inside a square of side 1, show that there exists two points within a distance of at most $1/\sqrt{2}$.

11. Given any ten points inside an equilateral triangle of side 1, show that there are two points within a distance of at most 1/3. [**Hint** : Divide the given triangle into 9 equal triangles]

12. If there are 104 different pairs of people who know each other at a party of 30 people, show that some person has at most 6 acquaintances.

13. If p is a prime number then show that there exists integers a and b such that p divides $a^2 + b^2 + 1$.

12.5 The Inclusion–Exclusion Principle

If $A \cap B \neq \phi$, then sum rule

$$|A \cup B| = |A| + |B|$$

does not hold. For example if

$$A = \{a, b, c\}, B = \{c, d, e, f\}$$

then $|A| = 3$ and $|B| = 4$ but $|A \cup B| = 6$ not 7. The general formula which is true for any finite sets A and B is

$$|A \cup B| = |A| + |B| - |A \cap B| \qquad \text{.......(1)}$$

Here to find $|A \cup B|$ we included (added) $|A|$ and $|B|$ and we excluded (subtracted) $|A \cap B|$. We shall see that formula (1) is a special case of the inclusion-exclusion principle. Before stating the general inclusion-exclusion principle for n sets we give it for three sets.

***Theorem* 1 :** For any finite sets A, B, C we have

$$|A \cup B \cup C| = |A| + |B| + |C| - |A \cap B| - |A \cap C| - |B \cap C| + |A \cap B \cap C|$$

Proof : $$|A \cup B \cup C| = |(A \cup B) \cup C|$$
$$= |A \cup B| + |C| - |(A \cup B) \cap C|$$
$$= |A| + |B| - |A \cap B| + |C| - |(A \cap C) \cup (B \cap C)|$$
$$= |A| + |B| + |C| - |A \cap B| - \{|A \cap C| + |B \cap C| - |A \cap B \cap C|\}$$
$$= |A| + |B| + |C| - |A \cap B| - |A \cap C| - |B \cap C| + |A \cap B \cap C|$$

We now state and prove the inclusion-exclusion principle.

***Theorem* 2 :** (The inclusion-exclusion principle) Let $A_1, A_2, \ldots, A_n$ be n finite sets. Then.

$$|A_1 \cup A_2 \cup A_3 \ldots \cup A_n| = \sum_{i=1}^{n} |A_i| - \sum_{1 \leqslant i < j \leqslant n} |A_i \cap A_j| + \sum_{1 \leqslant i < j < k \leqslant n} |A_i \cap A_j \cap A_k| \ldots + (-1)^{n-1} |A_1 \cap A_2 \cap A_3 \ldots \cap A_n|.$$

Proof : We shall prove the formula by showing that each element of $A_1 \cup A_2 \cup \ldots \cup A_n$ is counted exactly once on the right hand side of the equation. Let x be any element in $A_1 \cup A_2 \cup \ldots \cup A_n$. Suppose x belongs in exactly m sets, $1 \leqslant m \leqslant n$. Now, in $\sum_{i=1}^{n} |A_i|$ it is counted m times. In

$\sum_{1 \leqslant i < j \leqslant n} |A_i \cap A_j|$, x is counted ${}^mC_2$ times and so on. Thus, the element x is counted exactly

$$m - {}^mC_2 + {}^mC_3 \ldots + (-1)^{m-1} \cdot {}^mC_m$$
$$= {}^mC_0 - [{}^mC_0 - {}^mC_1 + {}^mC_2 + \ldots\ldots (-1)^m \, {}^mC_m]$$
$$= 1 - 0 = 1$$

because ${}^mC_0 - {}^mC_1 + {}^mC_2 - {}^mC_3 + \ldots\ldots (-1)^m \, {}^mC_m = 0.$

Thus, each element of $A_1 \cup A_2 \ldots \cup A_n$ is counted exactly once on the right-hand side of the equation. This completes the proof.

Note : The above principle can also be proved by induction on n.

Alternative form of the inclusion-exclusion principle

Let S be a finite set and $A_1, A_2, \ldots, A_n$ be subsets of S. Let A_i' denote the complement of A_i. Then we have the following theorem :

***Theorem* 3 :** If $A_1, A_2, \ldots, A_n$ are subsets of a finite set S then

$$|A_1' \cap A_2' \cap A_3' \ldots \cap A_n'| = |S| - \sum_{i=1}^{n} |A_i| + \sum_{1 \leqslant i < j \leqslant n} |A_i \cap A_j| + \ldots + (-1)^n |A_1 \cap A_2 \cap \ldots \cap A_n|$$

Proof : By De Morgan's laws, we have

$$\bigcap_{i=1}^{n} A_i' = S - \bigcup_{i=1}^{n} A_i$$

$$\Rightarrow \quad \left| \bigcap_{i=1}^{n} A_i' \right| = |S| - \left| \bigcup_{i=1}^{n} A_i \right|$$

$$\because \quad [A \subset X \Rightarrow |X - A| = |X| - |A|]$$

If we substitute the value of $\left| \bigcup_{i=1}^{n} A_i \right|$ from Theorem 2, we get the result.

Example 12.19 : Find the number of integers between 1 and 250 that are divisible by any of the integers 2, 3, and 7.

Solution : Let A_1 denote the set of all integers between 1 and 250 that are divisible by 2, A_2 denote the set of integers that are divisible by 3 and A_3 denote the set of integers that are divisible by 7. Then

$$|A_1| = \left[\frac{250}{2}\right] = 125, \quad |A_2| = \left[\frac{250}{3}\right] = 83 \text{ and } |A_3| = \left[\frac{250}{7}\right] = 35,$$

where $[x]$ denotes the largest integer not greater than x.

Now $|A_1 \cap A_2|$ = no. of integers between 1 and 250 that are divisible by both 2 and 3.

= no. of integers that are divisible by 6.

$$= \left[\frac{250}{6}\right] = 41$$

Similarly, $|A_1 \cap A_3| = \left|\frac{250}{2 \times 7}\right| = 17$ and $|A_2 \cap A_3| = \left|\frac{250}{3 \times 7}\right| = 11$

$$|A_1 \cap A_2 \cap A_3| = \left|\frac{250}{2 \times 3 \times 7}\right| = 5$$

$$\therefore \quad |A_1 \cup A_2 \cup A_3| = 125 + 83 + 35 - 41 - 17 - 11 + 5 = 179$$

Example 12.20 : How many integer solutions are there to the equation

$$x_1 + x_2 + x_3 + x_4 = 13, \quad 0 \leqslant x_i \leqslant 5$$

Solution : Let S be the set of all integer solutions of the given equation with each $x_i \geqslant 0$ and let A_i be the set of integer solutions with $x_i > 5$ or equivalently $x_i \geqslant 6$. Then number of solutions with

$0 \leqslant x_i \leqslant 5$ will be $|A_1' \cap A_2' \cap A_3' \cap A_4'|$

Now $|S| = {}^{13+4-1}C_{13} = {}^{16}C_{13} = 560$

$$|A_i| = {}^{(13-6)+4-1}C_{13-6} = {}^{10}C_7 = 120$$

$$|A_i \cap A_j| = {}^{(13-6-6)+4-1}C_{13-6-6} = {}^{4}C_1 = 4$$

Since each $x_i \geqslant 6$, therefore sum of three x_i's would exceed 13.

Therefore,

$$|A_i \cap A_j \cap A_k| = 0.$$

Similarly, $|A_1 \cap A_2 \cap A_3 \cap A_4|$ is also zero.

Now from Theorem 3, we have

$$|A_1' \cap A_2' \cap A_3' \cap A_4'| = |S| - \sum_{i=1}^{4}|A_i| + \sum_{1 \leqslant i < j \leqslant 4} |A_i \cap A_j|$$

$$= 560 - 4 \times 120 + {}^4C_2 \times 4 = 104$$

Note : This question can also be done with the help of generating functions.

Onto Functions : We know that the total number of functions from a set A having n elements to a set B having m elements is m^n. If $m \geqslant n$ then number of one-one functions is $m\,!/(m-n)!$. With the help of the inclusion-exclusion principle we obtain the number of onto functions in the following theorem.

***Theorem* 4 :** Let A and B be two sets having n and m elements respectively. Then the number of onto functions from A to B is

$$m^n - {}^mC_1\,(m-1)^n + {}^mC_2\,(m-2)^n - \ldots\ldots + (-1)^{m-1}\,{}^mC_{m-1}$$

Proof : Let S be the set of all functions from A to B Then $|S| = m^n$. Let $B = \{b_1, b_2, \ldots, b_m\}$. For each $i = 1, 2, \ldots\ldots m$, let S_i be the set of all functions from A to $B - \{b_i\}$. That is, S_i contains all functions $f_{b_i} : A \to B - \{b_i\}$. Since $|B - \{b_i\}| = m - 1$,

$$|S_i| = (m-1)^n.$$

For $i \neq j$, $S_i \cap S_j$ is the set of functions which takes values in the set $B - \{b_i, b_j\}$. Again $|B - \{b_i, b_j\}| = m - 2$, therefore,

$$|S_i \cap S_j| = (m-2)^n.$$

In general, for any k-tuple $\{i_1, i_2, \ldots, i_k\}$ with $1 \leqslant i_1 < i_2 < i_3 \ldots < i_k \leqslant m$, we have

$$|S_{i_1} \cap S_{i_2} \cap \ldots \cap S_{i_k}| = (m-k)^n.$$

We observe that a function from A to B is onto iff it does not belong to any S_i' s, $i = 1, 2, ..., m$. Applying the principle of inclusion-exclusion (Theorem 3), we get

$$|S_1' \cap S_2' \cap ... \cap S'_m| = |S| - \sum_{i=1}^{m} |S_i| + {}^mC_2 |S_i \cap S_j|$$

$$- {}^mC_3 |S_i \cap S_j \cap S_k| + ... + (-1)^m |S_1 \cap S_2 ... \cap S_m|$$

$$= m^n - {}^mC_1 (m-1)^n + {}^mC_2(m-2)^n .... + (-1)^{m-1} {}^mC_{m-1} . 1^m + 0$$

$$= m^n - {}^mC_1 (m-1)^n + {}^mC_2 (m-2)^n - ... + (-1)^{m-1} {}^mC_{m-1}.$$

Example 12.21 : In how many ways six different jobs can be assigned to four different persons if each person is assigned at least one job.

Solution : Let A be the set of six jobs and B be the set of five persons. Now we have to assign jobs to persons such that each person gets at lest one job. This is equivalent to finding onto functions from the set A to the set B. Here $|A| = 6$ and $|B| = 5$. Therefore, there are

$$5^6 - {}^5C_1 (5-1)^6 + {}^5C_2 (5-2)^6 - {}^5C_3(5-3)^6 + {}^5C_4(5-4)^6 = 1800$$

ways to assign the jobs to the persons in the desired manner.

Derangements

Definition : A derangement of a sequence $a_1, a_2, ..., a_n$ of n objects is an arrangement in which no object a_i is left in its original position (that is, no object a_i is still in the ith position).

For example if 1, 2, 3, 4, 5 is a sequence then 2, 1, 4, 5, 3 is a derangement of 1, 2, 3, 4, 5. Another derangement of 1, 2, 3, 4, 5 is 3, 1 2, 5, 4.

***Theorem* 5 :** The number of derangements of n objects is equal to

$$D_n = n! \left[1 - \frac{1}{1!} + \frac{1}{2!} - \frac{1}{3!} + ... + (-1)^n \frac{1}{n!}\right]$$

Proof : Let A be the set of n objects and S be the set of all permutations of the n objects. So $|S| = n!$

Let A_i be the set of all those permutations in which object i remains at its own place. Then

$|A_i| = (n-1)!$ because keeping object i fixed, we can change the places of remaining $n-1$ objects among themselves and this can be done in $(n-1)!$ ways. Similarly, $|A_i \cap A_j| = (n-2)!$ because it is the number of permutations of n objects in which objects i and j remain at their own places. In general,

$$|A_{i_1} \cap A_{i_2} \cap \ldots \cap S_{i_n}| = (n-m)!.$$

Since there are ${}^nC_m$ ways to select m elements from n elements, we have

$$\sum |A_i| = {}^nC_1 (n-1)!$$

$$\sum |A_i \cap A_j| = {}^nC_2 .(n-2)! \qquad \text{and in general}$$

$$\sum |A_{i_1} \cap A_{i_2} \cap \ldots \cap A_{i_m}| = {}^nC_m . (n-m)!$$

The number of derangements is the number of permutations in which no object remains at their own place. Thus,

$$D_n = |A_1' \cap A_2' \cap A_3' \ldots \cap A_n'| = |S| - \sum_{i=1}^{n} |A_i| + \sum |Ai \cap \text{Aj}|$$

$$- \ldots + (-1)^n |A_1 \cap A_2 \cap A_3 \ldots \cap A_n|$$

$$= n! - {}^nC_1(n-1)! + {}^nC_2(n-2)! + \ldots\ldots + (-1)^n\, {}^nC_n\, 0!$$

$$= n! - {}^nC_1(n-1)! + {}^nC_2(n-2)! + \ldots\ldots + (-1)^n$$

$$= n!\left[1 - \frac{1}{1!} + \frac{1}{2!} + \ldots + \frac{(-1)^n}{n!}\right]$$

Example 12.22 : Suppose 6 people randomly arrive at a dark room to retrieve their hats. Find the number of ways such that no person picks his own hat.

Solution : Here $n = 6$. Therefore, the required number is

$$= 6!\left[1 - \frac{1}{1!} + \frac{1}{2!} - \frac{1}{3!} + \frac{1}{4!} - \frac{1}{5!} + \frac{1}{6!}\right] = 265.$$

Example 12.23 : Find the number of ways of putting 5 letters in five addressed envelopes such that no letter is placed in the right envelope.

Solution : Here $n = 5$. Hence, the number of derangements of 5 objects is

$$D_5 = 5!\left[1 - \frac{1}{1!} + \frac{1}{2!} - \frac{1}{3!} + \frac{1}{4!} - \frac{1}{5!}\right]$$

$$= 60 - 20 + 5 - 1 = 44$$

Exercise

1. In how many ways can the four walls of a room be painted with three colours so that no two adjacent walls have the same colour ?
(**Hint :** $3^4 - {}^4C_1.3^3 + 4C_2.3^2 - {}^4C_3.3 + 3$)

2. Suppose 8 persons are given greeting cards. These cards are collected, shuffled and again distributed to the persons, one to each. Find the number of ways in which no person gets the same card again.
(**Hint :** Find the derangements of 8 objects)

3. In which polygon the number of sides and diagonals are equal.

4. Find how many functions are there from X to Y in the following cases
(a) $X = \{1, 2, 3\}$, $Y = \{a, b, c\}$
(b) $X = \{1, 2, 3, 4\}$, $Y = \{a, b, c\}$
(*i*) How many of these functions are one-to-one ?
(*ii*) How many of these functions are onto ?
(*iii*) How many of these functions are one-to-one and onto ?

5. How many solutions are there to the equation
$x_1 + x_2 + x_3 = 13, 0 \leqslant x_1 \leqslant 4, 0 \leqslant x_2 \leqslant 5$ and $0 \leqslant x_3 \leqslant 6$.

6. How many numbers from 1 to 1000 are not divisible by 2, 3 and 5 ?

7. Suppose 6 people randomly arrive at a dark room to retrieve their hats. Find the number of ways such that exactly two persons pick up their own hats.
(Hint : ${}^6C_2 \times D_4$)

8. In a town of 10,000 families, it was found that 40% families buy newspaper A, 20% families buy newspaper B and 10% families buy newspaper C. Further 5% families buy both A and B, 3% buy B and C and 4% buy A and C. If 2% buy all the three newspapers, find the number of families which buy
(*i*) A only
(*ii*) none of A, B and C
(Hint : $|A \cap B' \cap C'| = |A \cap (B \cup C)'|$
$= |A - (B \cup C) = |A| - |A \cap (B \cup C)|$
$= |A| - |(A \cap B) \cup (A \cap C)|$
$= |A| - |A \cap B| - |A \cap C|$
$+ |A \cap B \cap C|$

9. Find the number of integers n, $1 \leqslant n \leqslant 1000$, not divisible by 4, 5 or 6.

10. Find the number of primes $\leqslant 100$.

Answers

1. 18

2. $8!\left[\frac{1}{2!} - \frac{1}{3!} + \frac{1}{4!} - \frac{1}{5!} + \frac{1}{6!} - \frac{1}{7!} + \frac{1}{8!}\right]$

3. Pentagon

4. **(a)** $3^3 = 27$

(*i*) one-to-one functions = 6

(*ii*) onto functions = 6

(*iii*) both one-to-one and onto = 6.

(b) $3^4 = 81$

(*i*) one-to-one functions = 0

(*ii*) onto function = 36

(*iii*) both one-to-one and onto = 0.

5. 6

6. 266

7. 135

8. (*i*) 3300 (*ii*) 4000

9. 535

10. 25.

CHAPTER THIRTEEN

Formal Languages, Grammars and Finite State Machines

Formal languages were developed in the 1950's by a linguist *N.* Chomsky in an attempt to describe natural languages such as English, French etc. with an objective to make translation using computers from one language to another easier. Although attempts in this direction have only been partially successful but these descriptions have found a number of applications in computer science such as compiler construction for high-level programming languages.

In earlier chapters we discussed switching circuits in which output at any time depended only on the input. These circuits have no memory. Such circuits are called *combinatorial* circuits. Here we shall discuss circuits in which the output depends not only on the input but also on the state of the system at the time of input. The state of the system is determined by previous processing. In this sense, these circuits have memory. Such circuits are called *sequential* circuits. Finite state machines will be studies as an abstract model for sequential circuits.

13.1 Languages and Grammars

In the context of languages we often use the term alphabet. By an *alphabet* we mean a finite set of elements (or symbols). An element of an alphabet is called a *letter*, a *character*, or a *symbol.*

String

A string over an alphabet A is an ordered set of symbols from the alphabet. A string is also called a *sequence*, a *word* or a *sentence* depending upon its nature.

A string consisting of n symbols is called a string of length n. The length of a string α is denoted by $|\alpha|$. A string containing no element is called an *empty* string or a *null* string and is denoted by λ. The set of all strings over a set A is denoted by A^*.

Language

A language over an alphabet A is a subset of the set A^*, the set of all strings over A.

Example 13.1 : Let $A = \{a, b, c\}$ be an alphabet. Then

$L = \{a, b, ab, aa, ac, abc, cab\}$ is a language over A. Similarly, $L' = (a^i cb^i : i \geqslant 1\}$ is also a language over A.

Grammar

Definition : A *phase-structure grammar* or simply a *grammar* G is a 4-tuple (T, N, S, P), where

1. T is a non-empty finite set whose elements are called *terminals.*
2. N is a non-empty finite set whose elements are called non-terminals and $T \cap N = \phi$.
3. S is a special element of N called the *starting symbol.*
4. P is a finite set whose elements are of the form $\alpha \to \beta$, where α and β are strings over $T \cup N$ and α has at least one symbol from N. Elements of P are called *productions* or product rules or rewriting rules.

Note : The set $T \cup N$ consisting of terminal and non-terminal symbols is called the *vocabulary* of the language.

Let us first explain the above definition of grammar.

1. The terminals in T are symbols used to make up sentences in the language.
2. The elements in N(*i.e.*, non-terminals) are intermediate symbols used to describe the structure of the sentences.
3. The productions are grammatical rules that specify how sentences in the language are formed. A production $\alpha \to \beta$ specifies that string α can be transformed into string β.

Definition 1 : Let $G = (T, N, S, P)$ be a grammar and let w_1 and w_2 be two strings over $T \cup N$. We say w_1 *directly derives* w_2 (denoted as $w_1 \Rightarrow w_2$) if there are strings u and v over $T \cup N$ and a production $\alpha \to \beta$ such that $w_1 = u\,\alpha\,v$ and $w_2 = u\,\beta\,v$.

Note that in the above definition, u and v may be empty strings. In particular if $\alpha \to \beta$ is a production then $\alpha \Rightarrow \beta$.

Definition 2 : Let $G = (T, N, S, P)$ be a grammar. We say string w_1 produces string w_2 (written as $w_1 \overset{*}{\Rightarrow} w_2$) if there are finite number of strings $u_1, u_2, \ldots u_n$ such that

$$w_1 \Rightarrow u_1,\ u_1 \Rightarrow u_2,\ u_2 \Rightarrow u_3,\ \ldots\ldots,\ u_n \Rightarrow w_2.$$

Example 13.2 : Let $T = \{0, 1\}$ $N = \{S\}$ and $P = \{S \to 0S1, S \to \lambda\}$. Then $G = (T, N, S, P)$ is a grammar, here λ denotes the null string.

Language of a Grammar

The language generated by a grammar G (denoted by $L(G)$ is the set $\{w \in T^* : S \overset{*}{\Rightarrow} w\}$.

The elements of $L(G)$ are usually called sentences.

Example 13.3 : If $G = (\{0, 1\}, \{S\}, S, \{S \to 0S1, S \to \lambda\})$ is a grammar, find $L(G)$.

Solution : Since $S \to \lambda$ is a production therefore $S \Rightarrow \lambda$. So $\lambda \in L(G)$. Also, for $n \geqslant 1$

$$S \Rightarrow 0S1 \Rightarrow 00S11 = 0^2S1^2 \Rightarrow 00^2S1^2\,1 = 0^3S1^3 \Rightarrow \ldots\ldots$$
$$\ldots \ldots \Rightarrow 0^n S 1^n \Rightarrow 0^n\, 1^n.$$

(Note that in the last derivation we applied $S \to \lambda$)

Therefore,

$$0^n 1^n \in L(G) \text{ for } n \geqslant 0.$$

we claim $L(G) = \{0^n 1^n : n \geqslant 0\}$. Let w be any string in $L(G)$. The derivation of w starts with S. If $S \to \lambda$ is applied, we get empty string λ which is of the form $0^n 1^n$ for $n = 0$. If the first production to be applied is $S \to 0S1$ then the string in T^* is obtained only by applying $S \to \lambda$. Thus, the derivation of w is of the form $S \overset{*}{\Rightarrow} 0^n\, S1^n \Rightarrow 0^n 1^n$. Thus,

$$L(G) \subseteq \{0^n 1^n : n \geqslant 0\}.$$

Therefore

$$L(G) = \{0^n 1^n : n \geqslant 0\}.$$

Example 13.4 : Find the language $L(G)$ over $\{a, b\}$ generated by the grammar $G = (\{a, b\}, \{S, C\}, S, P)$ where P consists of

$$S \to aCa,\ C \to aCa \text{ and } C \to b.$$

Solution : Since $S \to aCa$, we have $S \Rightarrow aCa$.

Since $C \to b$, we have $aCa \Rightarrow aba$.

Thus, $S \Rightarrow aCa \Rightarrow aba$. So $aba \in L(G)$

Again $S \Rightarrow aCa$ $\qquad (\because S \to aCa)$

$aCa \Rightarrow aaCaa = a^2Ca^2$ $\qquad (\because C \to aCa)$

Applying $C \to aCa$ $(n-1)$ times, we get

$$S \overset{*}{\Rightarrow} a^n C a^n \Rightarrow a^n b\, a^n \qquad (\because C \to b)$$

Hence, $a^n b\, a^n \in L(G)$, $n \geqslant 1$.

Therefore, $\{a^n\, b\, a^n : n \geqslant 1\} \subseteq L(G)$

Let $w \in L(G)$. Then $w \in T^*$ such that $S \overset{*}{\Rightarrow} w$. Since $S \to aCa$ is the only production having left hand S, therefore $S \to aCa$ is the first production we have to apply to derive any string in $L(G)$. If we apply $C \to b$, we get $S \overset{*}{\Rightarrow} aba$. Hence, $aba \in L(G)$. If we apply $C \to aCa$ repeatedly, we get $S \overset{*}{\Rightarrow} a^n C a^n$. Now applying $C \to b$, we get

$S \overset{*}{\Rightarrow} a^n\, b\, a^n$. Thus, any element $w \in L(G)$ is of the form $a^n\, b\, a^n$ where $n \geqslant 1$. So

$$L(G) = \{a^n\, b\, a^n : n \geqslant 1\}$$

Example 13.5 : Construct a grammar for the language

$$L = \{a^i\, b^{2i} : i \geqslant 1\}.$$

Solution : Let $T = \{a, b\}$, $N = \{S\}$ with S as the starting symbol. Let the set P of productions be

$$S \to aSbb \quad \text{and} \quad S \to abb.$$

It is easy to see the idea behind the construction. Any string in L in generated as follows :

(*i*) $abb \in L$

(*ii*) if $x \in L$ then $axbb \in L$.

So, we have the productions $S \to abb$ and $S \to aSbb$.

Thus, if $G = \{T, N, S, P\}$ then $L = L(G)$.

Example 13.6 : Let $G = (\{a, b\}, \{S, A, B\}, S, P)$, where P consists of $S \to aABa$, $A \to baABb$, $B \to Aab$, $aA \to baa$, $bBb \to abab$, be a grammar. Test whether $w = babaaaba$ is in $L(G)$.

Solution : Since $L(G) = \{w \in T^* : S \overset{*}{\Rightarrow} w\}$, therefore we have to start with a production whose left hand side is S. At only stage we apply a suitable production which in likely to derive w.

$$\begin{aligned} S &\Rightarrow aABa && (\because\ S \to aABa) \\ &\Rightarrow baaBa && (\because\ aA \to baa) \\ &\Rightarrow baaAaba && (\because\ B \to Aab) \\ &\Rightarrow babaaaba && (\because\ aA \to baa) \end{aligned}$$

Hence, $S \overset{*}{\Rightarrow} babaaaba$

Thus, $w \in L(G)$

13.2 Types of Grammars and Languages

In the following discussion, we shall use capital letters A and B to denote arbitrary nonterminals, lower case letters a and b to denote arbitrary terminals and α and β to denote arbitrary strings over $T \cup N$, where T and N are disjoint sets of terminals and nonterminals. Depending upon the form of productions, grammars are classified into four types.

Definition :

1. A grammar is called a *type*-3 *grammar* or *regular* grammar if all productions in the grammar are of the forms $A \to a$ and $A \to aB$.

   In other words, in any production the left hand side string is a single nonterminal and right hand string is either a terminal or a terminal followed by a nonterminal.

2. A grammar is called a *type-2grammar* if every production in the grammar is of the form $A \to \alpha$.

   In other words , in any production the left hand side string is always a single nonterminal and right hand side string is any string on $T \cup N$.

   A type-2 grammar is also called a *context free* grammar. It may be seen that a type-3 grammar is also a type-2 grammar.

3. A grammar is called a *type*-1 *grammar* if for every production $\alpha \to \beta$ in the grammar, we have $|\alpha| \leqslant |\beta|$

   A type-1 grammar is also called *context-sensitive* grammar. Trivially, a type-3 or a type-2 grammar is a type-1 grammar.

4. A grammar with no restriction is called a *type*-0 *grammar*.

   Since languages are generated by a grammar, therefore corresponding to above four types of grammar, there are four types of languages also.

Definitions : A language is said to be *a type-k*, where $k = 0, 1, 2, 3$, and 4 language if it can be specified by a type-k grammar but cannot be specified by a type-$(k + 1)$ grammar.

Example 13.7 : Language $L = \{a^i b^i : i \geqslant 1\}$ is a type-2 language because it can be specified by the type-2 grammar with productions of the from $A \to aAb$ and $A \to ab$. It may be seen that L cannot be specified by a type-3 grammar.

Backus Nour Form

This is an alternative way to denote productions of a grammar. In *BNF* (Backus Nour Form) every Non-terminal symbol is enclosed in a bracket of the form $< >$ and the symbol $\rightarrow$ is replaced by $::=$. Thus the production $A \rightarrow aB$ is written as $<A>::=a<B>$. Moreover different productions of the form $S::=\beta_1, S::=\beta_2, \ldots, S::=\beta_n$ having the same left hand side and different right hand sides are written as

$$S::=\alpha_1 \mid \alpha_2 \mid \alpha_3 \mid \ldots \mid \alpha_n.$$

Thus, | be used to separate different productions. *BNF* has been used in many programming languages such as PASCAL, ALGOL etc.

Example 13.8 : Write the following productions in *BNF*.

$$S \rightarrow aA, S \rightarrow B, S \rightarrow aBb.$$

Solution : The *BNF* of the given productions is

$$<S>::=a<A>, <S>::=<B> \quad \text{and} \quad <S>::=a<B>b$$

which can be combined as

$$<S>::=a<A> \mid <B> \mid a<B>b.$$

Example 13.9 : Let G be a grammar with vocabulary $V = \{S, 0, 1\}$, set of terminals $T = \{0, 1\}$, the starting symbol S and the productions given by $S \rightarrow 1 \mid <S> \mid 0$. Find $L(G)$.

Solution : Since we are given vocabulary and terminals, we can easily find nonterminals. Thus, the grammar G is given by $(\{0, 1\}, \{S\}, S, P)$ where P consists of productions

$$S \rightarrow 11S \quad \text{and} \quad S \rightarrow 0.$$

Since $S \rightarrow 0$ therefore $S \Rightarrow 0$ and so $S \overset{*}{\Rightarrow} 0$. Thus, $0 \in L(G)$. Again we can apply production $S \rightarrow 11S$ repeatedly to obtain $S \overset{*}{\Rightarrow} 1^{2n} S$ where $n \geqslant 1$. Now applying production $S \rightarrow 0$, we get

$$S \overset{*}{\Rightarrow} 1^{2n} 0.$$

Hence, $\quad 1^{2n} 0 \in L(G), n \geqslant 1.$

Thus, $L(G) = \{1^{2n} 0 : n \geqslant 0\}$ because $0 \in L(G)$.

Alternately, we can white $L(G) = \{0, 110, 1111110, \ldots\}$. Hence, $L(G)$ consists of all strings in which even number of 1's are followed by a 0.

Example 13.10 : Find a grammar that generates the language $L = \{0^n 1^n : n \geqslant 0\}$. Also state the type of this grammar.

Solution : Let $T = \{0, 1\}$, $N = \{S\}$ where S is the starting symbol. Since empty string λ belongs to L, we must take production $S \to \lambda$. Moreover we see that if $x \in L$ then $0x1 \in L$. Therefore, we take production $S \to 0S1$ in our grammar. Now it may be seen easily that

$$G = (\{0, 1\}, \{S\}, S, \{S \to \lambda, S \to 0S1\})$$

generates the language L. Clearly it is a type-2 grammar.

13.3 Regular Sets and Regular Languages

In this section we first define regular expressions over the set of alphabet A. Each regular expression describes a particular set. We take here A as any set of alphabets but A will be taken as the set of input alphabets when we will discuss finite-state machines.

Regular Expressions

Regular expressions over A are defined as :

1. λ and ϕ are regular expressions, where λ and ϕ denote empty string and empty expression respectively.
2. Any element (symbol) of A is a regular expression.
3. If R_1 and R_2 are regular expressions then $R_1 + R_2$ is also regular expression, where $R_1 + R_2$ denotes the union of regular expressions R_1 and R_2.
4. If R_1 and R_2 are regular expressions then $R_1R_2 = \{\alpha\beta : \alpha \in R_1$ and $\beta \in R_2\}$ is also regular expression.
5. If R is a regular expression then R^* and (R) are also regular expressions.

All regular expression are obtained recursively by applying 1 to 5 above. Thus we see that regular expressions are strings which are defined over the symbols of A and the five symbols : (), $*$, $+$, λ, ϕ.

Regular Sets

Let R be a regular expression over A. A subset of A^* is called a regular set corresponding to regular expression R if it is constructed according to following rules :

1. The expression ϕ corresponds to the empty set ϕ.
2. The expression λ corresponds to the set $\{\lambda\}$, where λ is the empty string over A.
3. If $x \in A$ then regular expression x corresponds to the set $\{x\}$.
4. If the regular expressions R_1 and R_2 correspond to the sets M_1 and M_2 of A^* then
   (a) the regular expression R_1R_2 corresponds to the set $M_1M_2 = \{xy : x \in M_1 \text{ and } y \in M_2\}$.
   (b) the regular expression $R_1 + R_2$ corresponds to the set $M_1 \cup M_2$.
5. It the regular expression R corresponds to a subset M of A^* then the regular expression R^* corresponds to the set M^*.

Example 13.11 : Let $A = \{0, 1\}$. Show that the following expressions are regular expressions over A.

(*i*) $0^* (0 + 1)$ (*ii*) $00^* (0 + 1)^*1$ (*iii*) $(01)^* (01 + 1^*)$

Also find regular sets corresponding to these regular expressions.

Solution :

(*i*) Since $0, 1 \in A$, therefore 0 and 1 are regular expressions by condition (2) of regular expression.

Now $0 + 1$ is a regular expression by (3) and 0^* and $(0 + 1)^*$ are regular expressions by (5). Finally we see that $0^*(0 + 1)^*$ is regular by (4).

(*ii*) We know that 0, 1 and $0^*(0 + 1)^*$ are regular expressions. Therefore, $00^*(0 + 1)^*1$ must be regular in view of condition (4).

(*iii*) Since 0 and 1 are regular therefore 01 is regular and hence $(01)^*$ is regular. Again since 1^* and 01 is regular, therefore $(01 + 1^*)$ is regular. Now $(01)^*$ and $(01 + 1^*)$ are regular gives $(01)^* (01 + 1^*)$ is regular.

The regular set corresponding to (*i*) $0^*(0 + 1)^*$ consists of all sequences of 0's and 1's. Thus, the regular set is A^*.

The regular set corresponding to (*ii*) 00*(0 + 1)*1 must contain all sequences of 0's and 1's which begin with at least one 0 and end with at least one 1.

The regular set corresponding to regular expression (iii) (01)* (01 + 1*) is the set of all sequences of the form $\alpha\beta$ where $\alpha = (01)^n$ for $n \geqslant 0$ and $\beta = 1^m$ for $m \geqslant 0$. Thus, this set contains the strings of the form 1111, 01, 010101, 010101011111 and 011.

Regular language

The regular language $L(R)$ over A corresponding to regular expression R over A is defined by the following rules :

1. $L(\lambda) = \{\lambda\}$ and $L(()) = \phi$, the empty set.
2. $L(a) = \{a\}$ for each $a \in A$.
3. $L(R^*) = (L(R))^*$ for all regular expression R
4. $L(R_1 + R_2) = L(R_1) \cup L(R_2)$
5. $L(R_1R_2) = L(R_1).L(R_2)$.

Definition : A language L over A is said to be *regular language* over A if there exists a regular expression R over A such that : $L = L(R)$.

Note : If there is no chance of confusion, we shall omit parenthesis whenever possible.

Example 13.12 : Let $R_1 = aa^*$, $R_2 = a + b^*$ and $R_3 = (a + b)^*$ be regular expressions over $\{a, b\}$. Find corresponding languages $L(R_i)$, $i = 1, 2, 3$.

Solution : Let $R_1 = aa^*$. Then R_1 contains all strings a^n where $n \geqslant 1$. Thus, $L(R_1)$ consists of all positive powers of a.

Let $R_2 = a + b^*$. Then $L(R^2)$ consists of a and all strings b^n where $n \geqslant 0$. That is,

$$L(R_2) = \{a, \lambda, b, b^2, \ldots\}$$

Let $R_3 = (a + b)^*$. We know that $L(a + b) = L(a) \cup L(b)$

$$= \{a\} \cup \{b\} = \{a, b\} = A$$

Hence, $L(R_3) = L((a + b)^*) = (L(a + b))^* = A^*$.

Thus $L(R_3)$ contains all strings over A.

Example 13.13 : Describe the following languages by regular expressions R over $A = \{a, b\}$

(a) $L_1 = \{a^m b^n : m, n > 0\}$

(b) $L_2 = \{b^m a b^n : m, n > 0\}$

Solution :

(a) L_1 consists of all strings beginning with one or more a's followed by one or more b's. Hence, we can take $R = aa^*bb^*$ or $R = a^*abb^*$.

(b) L_2 consists of all strings which begin with one or more b's followed by a single a and then followed by one or more b's. Hence, we can take $R = bb^* abb^*$ or b^*bab^*b.

13.4 Finite State Machines

Finite state machine is a model that captures some important characteristics of a computer.

Definition : A *finite state machine M* is a 6-tuple (S, I, O, f, g, s_0), where

1. S is a finite set $(s_0, s_1, s_2, \ldots\}$ whose elements are called *states* of the machine.
2. s_0 is a special element of S called the *initial state* of the machine.
3. $I = \{i_1, i_2, \ldots\}$ is a finite set of input alphabet.
4. $O = (o_1, o_2, \ldots\}$ is a finite set of output alphabet.
5. F is function from $S \times I$ to S called the *transition* function.
6. g is function from S to O called the *output* function.

It should be noted that at any instant of time, a finite state machine is in some state. On receiving an input symbol, the machine moves to a next state according to the transition function. Moreover, at each state the machine produces an output according to the output function. Initially, the machine is in its initial state s_0.

Example 13.14 : Let $S = \{s_0, s_1, s_2\}$, $I = \{0, 1\}$, $O = \{0, 1\}$ and the two functions f and g are given by a table called *state table* shown below :

State	Input 0	Input 1	Output
⇒ s_0	s_1	s_1	0
s_1	s_2	s_0	1
s_2	s_2	s_0	1

Fig. 13.1

Then $M = (S, I, O, f, g, s_0)$ is a finite state machine.

In the above table, the double arrow pointing at s_0 indicates that s_0 is the initial state of the machine. It is clear from the table that the machine M begins in state s_0 which has an output of 0. If the first input is 0, the next state of the machine is s_1 and the output is 1. Now if next input symbol is 1 then the machine remains in state s_1 with output 1. It can be seen that if we input the sequence of symbols 011101 (starting from left) then we get output 0111110. Note that the initial 0 (from left) of the output string is not the result of any input. It merely reflects the starting state.

We can describe a finite state machine with the help of directed graph also. In this directed graph, each vertex corresponds to a state of a machine. The output associated with a state is written below the state name. The transition function (also called next state function) is described by directed edges of the graph labelled with the corresponding input letters. The directed graph describing a finite state machine is called a *state graph.* The state graph for the finite state machine of Ex. 13.14 is given below :

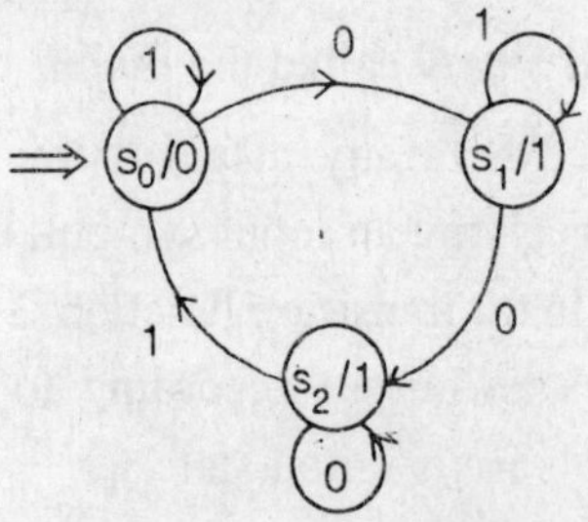

Fig. 13.2 : State graph of machine in Ex. 13.14.

Remark : It may be noted that a finite state machine produces an output of the starting state even before the arrival of the first input. Consequently, there is always one more symbol in the output sequence than in the corresponding input sequence.

The following example shows that a finite state machine can be used to model a physical system.

Example 13.15 : Design a finite state machine that receives the set {0, 1, 2} as input and produces an output such that the output is equal to the modulo 3 sum of the digits in the input sequence. (By modulo 3 sum of some integers we mean the remainder when the sum of the integers is divided by 3).

Solution : We shall use states of the machine to represent the following cases.

1. The output should be 0 when the modulo 3 sum of all input digits is 0.
2. The output should be 1 when the modulo 3 sum of all input digits is 1.
3. The output should be 2 when the modulo 3 sum of all input digits is 2.

We define state s_0 (initial state) of the machine corresponding to situation 1, state s_1 corresponding to situation 2 and state s_2 corresponding to situation 3.

After considering all cases, we have the state table and state graph of the machine as given below :

State	Input 0	Input 1	Input 2	Output
⇒ s_0	s_0	s_1	s_2	0
s_1	s_1	s_2	s_0	1
s_2	s_2	s_0	s_1	2

State table

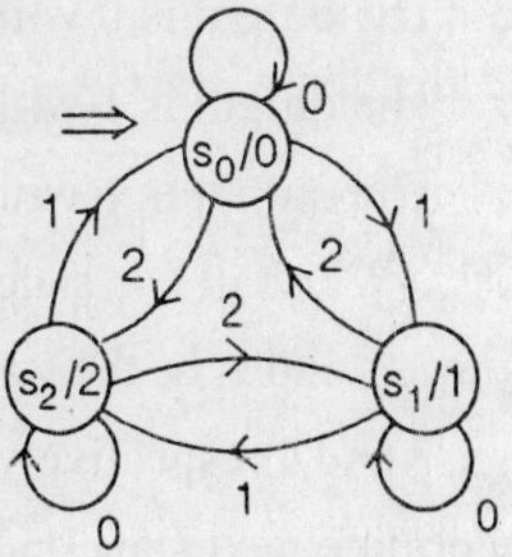

State graph

Fig. 13.3

Example 13.16 : Design a finite state machine *M* which can add two binary numbers :

Solution : Since we can write as many zeros as required to the left of a given number without changing its value (for example, 10 can be written as 0010), we can assume that binary numbers to be added have the same number of digits. From binary addition, we know that.

$$0 + 0 = 0$$
$$0 + 1 = 1$$
$$1 + 0 = 0$$
$$1 + 1 = 10$$

Hence, if the inputs are 110101 and 011110 then the machine should give the output as follows :

$$\begin{array}{r} 110101 \\ +\ 011110 \\ \hline 1010011 \end{array}$$

If we assume that the digits of the two numbers to be added come in one by one with least significant digits coming in first then the input numbers are 00, 01, 10 and 11 where the two digits in each pair are the corresponding digits in the numbers being added. Thus, the set of input alphabet is {00, 01, 10, 11} and the output alphabet is {0, 1}.

From the above sum, we see that there are four cases :

1. The output is 0 with no carry.
2. The output is 0 with a carry to next column.
3. The output is 1 with no carry.
4. The output is 1 with a carry to next column.

We shall use states of the machine to represent these cases. Let s_0, s_1, s_2, and s_3 be the states of the machine corresponding to situations 1, 2, 3 and 4 respectively. We take state s_0 as the starting state. Now we determine next state on the basis of present state and the input.

After considering all cases, we have the state table and state graph of the machine as follows :

State	Input 00	01	10	11	Output
⇒ s_0	s_0	s_2	s_2	s_1	0
s_1	s_2	s_1	s_1	s_3	0
s_2	s_0	s_2	s_2	s_1	1
s_3	s_2	s_1	s_1	s_3	1

Fig. 13.4 (a) : State table

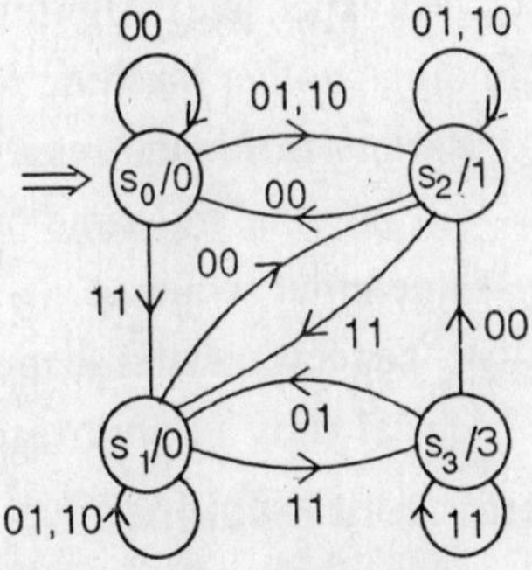

Fig. 13.4(*b*) : State graph

Example 13.17 : In the machine constructed for adding two binary numbers in Ex. 13.16, what is the next state if the present state is s_2 and input is 11 ?

Solution : Consider the machine in Ex. 13.16. Suppose present state is s_2 and input is 11. The output of the present state s_2 is 1 and there is no carry. If the input is 11 then we have to add 0 + 1 + 1 which is equal to 10. Thus, the next state is s_1 whose output is 0 and there is a carry.

Example 13.18 : Compute the sum of 110110 and 010101 by using the machine of Ex. 13.16.

Solution : The operation of the machine in adding the given numbers (with the lower-order digits coming in first) is as follows :

The input is 01, 10, 11, 00, 11 and 10. In the begining the machine is in state s_0 and input is 01. The output is 0 and machine will go to state s_2. Now the present state is s_2 and input is 10. The output of the present state is 1 and machine is in state s_0 and input is 01. The output is 0 and machine will remain in state s_2. Continuing in this manner, the final output is 10010110. Here the initial 0 does not reflect the action of any input. Ignoring this zero, we obtain the sum as 1001011.

13.5 Machine Minimization

If we wish to construct a physical machine corresponding to a finite state machine the number of internal states of the machine is a factor in the cost of construction (a machine with fewer states is likely to be less expensive). Minimization of a given machine M is a process of finding another machine M' with the following properties :

1. Starting from their respective start (initial) states both M and M' will produce the same output sequence when they are given the same input sequence.
2. M' has fewer states than M, if possible (if this is not possible then M is already a minimal machine and cannot be further reduced).

Equivalent Machines

Two finite state machines M and M' are said to be *equivalent* if starting from their respective start states both M and M' will produce the same output sequence when they are given the same input sequence. In other words, equivalent machines have identical behaviours (with respect to output) even though their internal structures (such as states) might be different.

Example 13.19 : Show that the two machines described by their state tables in Fig. 13.5 **(a)** and **(b)** are equivalent.

Solution : The reader can verify that the machines described in Fig 13.5(a) and (b) will produce the same output when they are given the same input. This verification would require an exhaustive examination of all possible input sequences of arbitrary length.

State	Input 1	Input 2	Output
⇒ s_0	s_1	s_2	0
s_1	s_5	s_3	0
s_2	s_6	s_4	0
s_3	s_7	s_1	0
s_4	s_1	s_5	1
s_5	s_3	s_7	0
s_6	s_4	s_1	0
s_7	s_1	s_2	1

Fig. : 13.5 (a)

State	Input 1	Input 2	Output
⇒ s_0	s_1	s_2	0
s_1	s_2	s_3	0
s_2	s_3	s_4	0
s_3	s_4	s_1	0
s_4	s_1	s_2	1

Fig. 13.5 (b)

From the above example, we see that it is practically impossible to determine whether two finite state machines are equivalent using the definition given above. Hence, we need to develop an algorithm to identity equivalent machines. For this purpose, we first introduce the notion of *equivalent states.*

Definition : (Equivalent states). Two states s_i and s_j of a finite state machine $M = (S, I, O, f, g, s_0)$ are said to be equivalent if for any $x \in I^*$, $f(s_i, x) = f(s_j, x)$, where I is the set of input alphabet and f is the transition function of M.

Thus, equivalent states of a machine produce identical output sequences for any input sequence. Clearly two equivalent states can be combined into one state without changing the terminal behaviour of the machine. Thus, if s_i and s_j are equivalent states, we can modify the transition function by eliminating state s_j and replacing all transitions into state s_j with state s_i.

For the purpose of determining whether two states s_i and s_j are equivalent, we introduce the following :

Definition : (0-equivalent states). Two states s_i and s_j of a machine are said to be 0-*equivalent* if the states s_i and s_j have the same output.

For example, the states s_0, s_1, s_2, and s_3 of the machine in Fig. 13.5(b) are 0-equivalent to each other.

Definition : (k-equivalent states). Two states s_i and s_j of a finite state machine $M = (S, I, O, f, g, s_0)$ are said to be k-*equivalent* if they have the same output and for every input letter, their successors (next states) are $(k-1)$-equivalent.

Thus, we see that two states s_i and s_j are 1-equivalent if they have the same output and for every $x \in I$, their next states $f(s_i, x)$ and $f(s_j, x)$ are 0-equivalent.

For example, in the finite state machine in Fig. 13.5(a), states s_2 and s_5 are 0-equivalent. States s_4 and s_7 are 1-equivalent because they have the same output namely 1 and for input letters 1 and 2 their next states are s_1, s_1 and s_5, s_2 respectively whose output are same.

It may be seen from the definition that if two states s_i and s_j of a machine are k-equivalent then for any $x \in I^*$ such that

$$|x| \leqslant k,\ f(s_i, x) = f(s_j, x).$$

Hence, we can say that two states are equivalent if they are k-equivalent for all k.

***Theorem* 1 :** Let $M = (S, I, O, f, g, s_0)$ be a finite state machine. Then the relation 'k-equivalence on the set S of all states of M' is an equivalence relation.

Proof : From the definition of k-equivalent states, it is trivially true that every state s_i is k-equivalent to itself. Hence the relation of being k-equivalent states on the set S of states is reflexive. Again, from the definition, we see that if a state s_i is k-equivalent to a state s_j, then s_j is also k-equivalent to s_i. Hence, the relation is symmetric.

Finally we suppose that s_i is k-equivalent to s_j and s_j is k-equivalent to s_h. Then s_i and s_h have the same output and for any input letter, their next states are $(k-1)$ equivalent. Hence, the relation is an equivalence relation.

We know that every equivalence relation on any set X partitions the set X into disjoint equivalence classes. Hence, the relation of being two states k-equivalent induces a partition on the set S of all states of a machine M. We shall denote this partition of S by π_k. In other words, π_k is the collection of all disjoint equivalence classes of the relation 'k-equivalence'.

Example 13.20 : Find π_0, π_1 and π_2 (That is, partitions corresponding to 0-equivalence, 1-equivalence and 2-equivalence relations) for the finite state machine of Fig. 13.6

	State	Input 0	Input 1	Output
⇒	s_0	s_1	s_5	0
	s_1	s_0	s_5	0
	s_2	s_6	s_0	0
	s_3	s_7	s_1	0
	s_4	s_0	s_6	0
	s_5	s_7	s_2	1
	s_6	s_0	s_3	1
	s_7	s_0	s_2	1

Fig. 13.6

Solution : Here set S of states is $\{s_0, s_1, s_2, s_3, s_4, s_5, s_6, s_7\}$. We denote the equivalence class of a state s_i under the relation 'k-equivalence' by $[s_i]_k$. Then

$$[s_0]_0 = \{s_i : s_0 \text{ is 0-equivalent } s_i\}$$
$$= \{s_i : s_0 \text{ and } s_i \text{ have the same output}\}$$
$$= \{s_0, s_1, s_2, s_3, s_4\}$$

Since $[s_0]_0 \neq S$, we take an element of S not in $[s_0]_0$.

$$[s_5]_0 = \{s_i : s_5 \text{ is 0-equivalent to } s_5\}$$
$$= \{s_5, s_6, s_7\}$$

Thus, $\pi_0 = \{\{s_0, s_1, s_2, s_3, s_4\}, \{s_5, s_6, s_7\}\}$

Similarly, $\pi_1 = \{\{s_0, s_1, s_4\}, \{s_2, s_3\}, \{s_5\}, \{s_6, s_7\}\}$

and $\pi_2 = \{\{s_0, s_1\} \{s_2, s_3\}, \{s_4\}, \{s_5\}, \{s_6, s_7\}\}$

The following theorem will provide us a method for obtaining the partitions $\pi_0, \pi_1, \pi_2, \ldots \pi_k$ successively :

***Theorem* 2 :** Two states s_i and s_j will be in the same equivalence class in π_k if and only if s_i and s_j are in the same equivalence class in π_{k-1} and for any input letter x, their next states $f(s_i, x)$ and $f(s_j, x)$ are in the same equivalence class in π_{k-1}.

Proof : Let s_i and s_j be two k-equivalent states. From the definition of k-equivalent states, we see that s_i and s_j will have the same output and for any input letter, their next states are $(k-1)$ equivalent. Since s_i and s_j are k-equivalent, they are also $(k-1)$-equivalent. Thus we conclude that two states s_i and s_j are k-equivalent if and only if they are $(k-1)$ equivalent and for any input letter x, their next states are also $(k-1)$-equivalent. This completes the theorem.

Example 13.21 : Find the π_k for all $k \geqslant 0$ for the finite state machine in Fig. 13.6.

Solution : We first find π_0 exactly in the same way as in Ex. 13.20.

Thus, $\pi_0 = \{\{s_0, s_1, s_2, s_3, s_4\}, \{s_5, s_6, s_7\}\}$

We shall now obtain π_1 whose equivalence classes are 1-equivalent. From the above theorem, two states are 1-equivalent if and only if they are 0-equivalent and their next states are 0-equivalent for all $a \in \{0, 1\}$. This step is performed by splitting equivalence classes of π_0 whenever their next-states do not all fall with in an equivalence class of π_0. We observe that s_0 and s_1 are in the same equivalence class in π_0 and for input 0 their next states are s_1 and s_0 which are in the same equivalence class in π_0 and for input 1, their next states are both s_5, which is in the same equivalence class of π_0. Hence, we conclude that s_0, s_1 will be in the same equivalence class in π_1. Continuing in this way, we obtain

$$\pi_1 = \{\{s_0, s_1, s_4\}, \{s_2, s_3\} \{s_5\}, \{s_6, s_7\}\}$$

Similarly, $\pi_2 = \{\{s_0, s_1\}, \{s_2, s_3\}, \{s_4\}, \{s_5\}, \{s_6, s_7\}\}$

and $\pi_3 = \{\{s_0, s_1\}, \{s_2, s_3\}, \{s_4\}, \{s_5\}, \{s_6, s_7\}\}$.

Since $\pi_2 = \pi_3$ therefore $\pi_k = \pi_2$ for all $k \geqslant 2$.

Thus, we have calculated all π_k for $k \geqslant 0$.

Note : The procedure of finding π_k from π_{k-1} terminates when $\pi_k = \pi_{k-1}$. Moreover, when $\pi_k = \pi_{k-1}$, all the states that are k-equivalent are equivalent.

For a given finite machine, we use the above procedure to determine the equivalent states. We then replace all equivalent states by a single state to obtain an equivalent machine that might have fewer states.

Example 13.22 : Minimize finite state machine M where M is given by the state table 13.7.

	State	Input 0	Input 1	Output
⇒	s_0	s_3	s_1	1
	s_1	s_4	s_1	0
	s_2	s_3	s_0	1
	s_3	s_2	s_3	0
	s_4	s_1	s_0	1

Fig. 13.7

Solution : Since output of states s_0, s_2, s_4 is same as 0 and output of states s_1 and s_3 is same as 1, therefore

$$\pi_0 = \{\{s_0, s_2, s_4\}, \{s_1, s_3\}\}$$

From π_0, we calculate π_1 in the same way as in Ex. 13.21.

$$\pi_1 = \{\{s_0\}, \{s_2, s_4\}, \{s_1, s_3\}\}$$

Similarly, $\pi_2 = \{\{s_0\}, \{s_2, s_4\}, \{s_1, s_3\}\}$

Since $\pi_1 = \pi_2$, we have $\pi_k = \pi_1 \ \forall\ k \geqslant 1$. Thus, we can eliminate s_3 and s_4 to obtain the reduced machine as in Fig. 13.8.

	State	Input 0	Input 1	Output
⇒	s_0	s_1	s_1	1
	s_1	s_2	s_1	0
	s_2	s_1	s_0	1

Fig. 13.8

13.6 Finite State Machines as Language Recognizers

In this section we shall see that finite state machines can be used as a devices to recognize (or accept) sentences in a language. Such machines are called acceptor (or recognizer). A finite state machine can be built to recognize a sentence by producing an output of 1 when the sentence received as input is a sentence in a certain language; otherwise the output is 0.

Consider a finite state machine M with output alphabet $O = \{0, 1\}$. To avoid writing down output for each state, we define the following :

Accepting states : A state of a finite state machine M (with output alphabet $O = \{0, 1\}$) is said to be accepting state if its output is 1.

Rejecting states : A state of a finite state machine M (with output alphabet $O = \{0, 1\}$) is said to be rejecting state if its output is 0.

Thus, the states of the machine M are divided into only two classes-accepting and rejecting states. An input sequence is said to be accepted (rejected) by a finite state machine if it takes the machine from initial state to an accepting state (a rejecting state).

Accepting states of a finite state machines when it is used as an acceptor will be denoted by double circles (concentric circles) in the state graph of the machine.

Example 13.23 : Show that the finite state machine in Fig. 13.9 accepts all binary sequences that end with the digits 011.

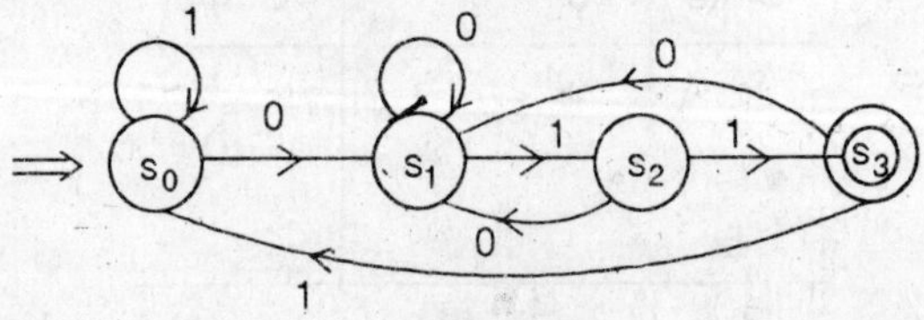

Fig. 13.9

Solution : In this finite state machine state s_3 is the only accepting state (denoted by double circles). Let $\alpha = x_1\, x_2 \ldots x_i 011$ be any sequence of binary numbers which end with the digits 011. The finite state machine is initially in state s_0. On receiving α as input, machine will move to other states according to state graph of Fig. 13.9. We divide α in to two segments as shown below :

$$\alpha = \overline{x_1\, x_2 \ldots x_i} \;\; \overline{001}$$

After receiving the first segment $x_1\, x_2 \ldots x_i$ of α, the finite state machine will be in one of the states s_0, s_1, s_2 and s_3. Suppose, for example, the machine is in state s_1 after receiving the first segment $x_1\, x_2 \ldots x_i$ of α. The next input symbol is 0. The next state for s_1 with input 0 is s_1 itself. The next input symbol of α is 1. From state graph, the next state for s_1 is s_2 which goes to s_3 on receiving last input 1. Since s_3 is the accepting state, therefore α is accepted by our machine.

Example 13.24 : Design a finite state machine having an output of 1 exactly when the input sequence ends with the digits 101.

Solution : We want to design a finite state machine having an output 1 when input sequence α ends with the digits 101. In particular the machine will produce output 1 when input sequence is 101. Let s_0 be the starting state of our machine. Then the machine whose state graph is given below will accept 101.

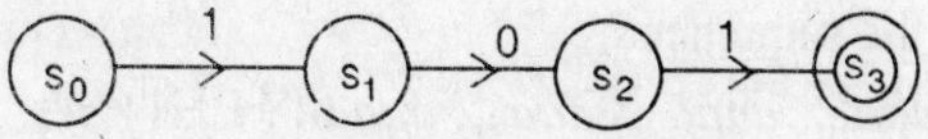

It is clear from the figure that we (machine) should be in state s_2 whenever we find symbols 10 in the given sequence (regardless of what came before). In particular a sequence 1010 should put us in state s_2. Hence, the next state of s_3 with input 0 is s_2. Similarly, we should be in state s_1 whenever we come across symbol 1 in the sequence and such that next two symbols 01 take us to state s_3. In particular, 1011 should take us in state s_1. Hence, the next state of s_3 with input 1 is s_1. Similarly, we can determine the next states of s_2 and s_1 for input symbols 0 and 1. Thus, we arrive at the following state graph.

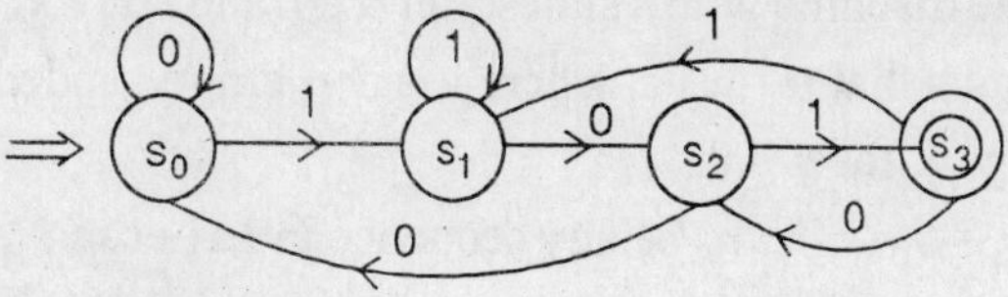

Finite State Language

A language is said to be a *finite state language* if there is a finite state machine that accepts all the sentences in the language and no other. In other words, a language L is finite state language if we can find a finite state machine M such that M accepts a sentence α if and only if $\alpha \in L$.

A finite state language is also known as *regular* language.

Clearly every finite state machine defines a finite state language. But a given language may or may not be finite state language. The following example shows that there is a language which is not finite state language.

Example 13.25 : The language $L = \{a^m : m = n^2, n \geqslant 1\}$ is not a finite state language.

Solution : Suppose if possible that L is a finite state language. Then there exists a finite state machine which accepts exactly all the sentences in L. Suppose this machine has N states. Corresponding to this N, we choose an integer n such that

$$(n+1)^2 - n^2 > N.$$

Now consider the sequence

$$aaaa \ldots aaaa \ldots aaa \ldots aaa \text{ of } (n+1)^2 \text{ } a\text{'s}.$$

Since $(n+1)^2 - n^2$ is greater than N, there are more than N a's between the n^2th a and the $(n+1)^2$th a in the above sequence. Therefore, during the application (execution) of this sequence, the machine will visit a certain state s_m more than once. Therefore, removal of the a's between these two visits to state s_m will yield a sentence that will also be accepted by the finite state machine. But this sentence is not in the language L because it contains more than n^2 but less than $(n+1)^2$ a's. Hence, L is not a finite state language.

***Theorem* 1 :** (Pumping lemma) Let L be a finite state language accepted by a finite state machine with k states. Let $\alpha \in L$ and $|\alpha| \geqslant k$. Then there exists x, y, z such that $\alpha = x\,y\,z$, where y is non-empty, and $x\,y^q\,z$ is also in the language L for $q \geqslant 0$.

Proof : Let $\alpha = a_1 a_2 a_3 \ldots a_k$ be any sequence in L. Let $s_0, s_1, s_2, \ldots\ldots, s_k$ denote the states the machine visits while receiving the sequence α, where s_0 is the initial state and s_k is an accepting state. Since our machine has only k states, therefore among the $k+1$ states $s_0, s_1, s_2, \ldots, s_k$, at least two states must be same. We assume that states, s_i and s_j are same. That is,

$$s_i = s_j \text{ where } 0 \leqslant i < j \leqslant k.$$

We can divide the sequence α into three segments as shown below :

$$\alpha = \overline{a_1 a_2 \ldots a_3 \ldots a_i} \quad \overline{a_{i+1} \ldots a_j} \quad \overline{a_{j+1} \ldots a_k}$$

Let $x = a_1 a_2 \ldots a_i$, $y = a_{i+1}\, c_{i+2}, \ldots a_j$ and $z = a_{j+1} \ldots a_k$.

Since $i < j$, y is nonempty and $\alpha = xyz$. Machine starts from the initial state s_0. On applying the string x, it reaches the state s_i. On applying the string y, it reaches the state s_j which is same as s_i. Thus, on applying y after x the machine comes back to state s_i. So after application of y^q for each $q \geqslant 0$, the machine is in the state $s_i (= s_j)$. On applying z at this state, machine reaches s_k, an accepting state. Hence $xy^qz \in L$ for each $q \geqslant 0$.

Example 13.26 : Show that $L = \{0^i : 1^i : i \geqslant 1\}$ is not regular.

Solution : Suppose, if possible, L is a regular language. Then there exists a finite state machine that accepts the sentences in L.

Let this machine has N states. Let $w = 0^N 1^N$. Then $w \in L$ and $|w| = 2N > N$. By pumping lemma we can write $w = xyz$ with y nonempty. We want to find q such that $xy^qz \notin L$ for getting contradiction. The string y can be in one of the following three forms :

Case (i) : $y = 0^k$ for some $k \geqslant 1$

Case (ii) : $y = 1^k$ for some $k \geqslant 1$

Case (iii) : $y = 0^k1^j$ for some $k, j \geqslant 1$.

In case (i), we take $q = 0$. Then by the pumping lemma, $xy^0z = xz = 0^{N-k}1^N$ must belong to L. But $k \geqslant 1$ gives $N - k \neq N$ and so $xz \notin L$, a contradiction.

In case (ii), we take $q = 0$. Hence, by the pumping lemma, xy^0z must belong to L. But $xy^0z = xz = 0^N1^{N-k} \notin L$ because $N \neq N - k$.

In case (iii), we take $q = 2$. Hence, by the pumping lemma, xy^2z must belong to L. But $xy^2z = 0^{N-k}(0^k 1^j)^2 1^{N-j}$

$$= 0^N 1^j 0^k 1^N \notin L$$

because xy^2z is not of the form $0^i 1^i$.

Thus, in all the cases we arrive at a contradiction. Hence, L is not regular.

Kleen's Theorem

Recall that a language is said to be type-3 language if it is generated by a type-3 grammar. A grammar is called type-3 grammar if all productions in the grammar are of the forms

$$A \to a \quad \text{and} \quad A \to aB,$$

where capital letters denote nonterminals (states) and lower case letters denote terminals (input letters). Type-3 languages are also called regular.

We give below Kleen's Theorem without proof :

***Theorem* 2 :** Class of finite state languages is same as the class of type-3 languages.

In other words, Kleen's theorem states that every finite state language is a regular language and every regular language is finite state language. Because of this result, finite state languages are also called regular languages.

Exercise

1. Find all equivalent states and obtain an equivalent finite state machine with smallest number of states for the finite state machine given below :

State	Input 0	Input 1	Output
⇒ A	F	B	0
B	D	C	0
C	G	B	0
D	E	A	1
E	D	A	0
F	A	G	1
G	C	H	1
H	A	H	1

Fig. 13.9

2. Show that the language $L = \{a^i b^i : i \geqslant 1\}$ is not regular.

3. Minimize the machine whose state table is given below :

	State	Input 0	Input 1	Output
⇒	s_0	s_3	s_6	1
	s_1	s_4	s_2	0
	s_2	s_4	s_1	0
	s_3	s_2	s_0	1
	s_4	s_5	s_0	1
	s_5	s_3	s_5	0
	s_6	s_4	s_2	1

4. Let $G = (\{A_0, A_1\}, \{a, b\}, P, A_0)$ where P consists of $A_0 \to aA_1$, $A_1 \to bA_1, A_1 \to a, A_1 \to bA_0$. Construct a state graph of a machine accepting $L(G)$.

5. Show that $L = \{a^{2n} : n \geqslant 1\}$ is a regular language.

6. Let $A = \{a, b\}$. Construct a machine M which will accept precisely those words from A which end in two b's.

7. Construct a machine M which will accept all those words over $\{a, b\}$ in which number of b's is divisible by 3.

Answers

1. A and C; B and E; and F, G and H are equivalent.

State	Input 0	Input 1	Output
A	F	B	0
B	D	A	0
D	B	A	1
F	A	F	1

3. $A = \{s_0\}, B = \{s_1, s_2, s_5\}, C = \{s_3, s_4\}$, and $D = \{s_6\}$

	Input		
State	0	1	Output
A	*C*	*D*	1
B	*C*	*B*	0
C	*B*	*A*	1
D	*C*	*B*	1

4.

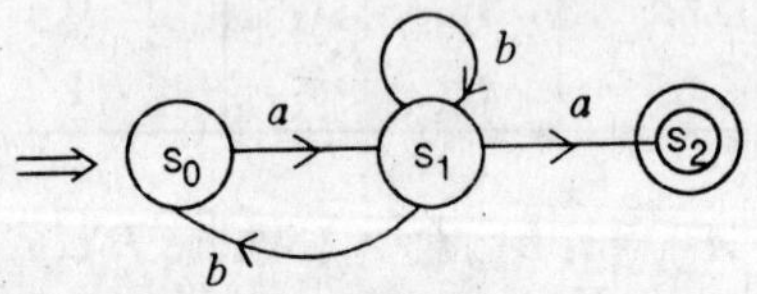

6.

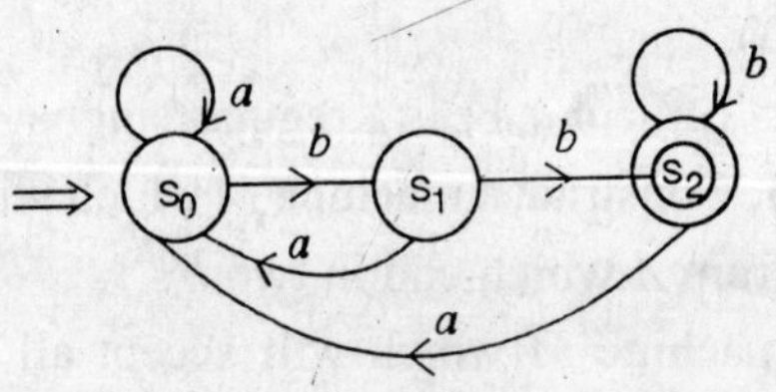

7.

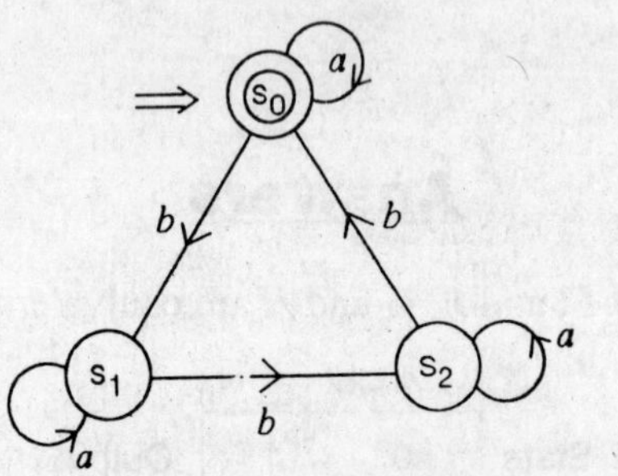

CHAPTER FOURTEEN

Semigroups and Monoids

In section 2.7 we have defined and studied binary operations. Recall that a binary operation on a set X is a mapping from $X \times X$ to X. A binary operation is usually denoted by means of a symbol such as $+, \times, \cdot, *,$ o, $\oplus$. If we denote a binary operation on a set X by $*$ then the result of $*$ on the elements x, y of X is expressed by $x * y$ (instead of $*(x, y)$).

14.1 Algebraic Structure

We first define n-ary operations for $n = 1, 2, ...,$ on a set X.

Definition : Let X be a non-empty set. A mapping $f: X^n \rightarrow X$ is called an n-ary operation for $n = 1, 2, ...$ on the set X.

For $n = 1$, such an operation is called a **unary** operation. Thus, any function from X to itself is an unary operation on X. For example, $f: N \rightarrow N$ defined by $f(x) = x + 2$ for all $x \in N$ is a unary operation on the set N of natural numbers.

For $n = 2$, n-ary operation is called binary operation. Thus, binary operation is a function from $X \times X$ to X. Similarly, a function from $X \times X \times X$ into X is called 3-ary (or ternary) operation.

We now define algebraic structure (or system).

Definition : A non-empty set X together with one or more n-ary operations ($n = 1, 2, ...$) is called an *algebraic structure* (or *system*).

An algebraic structure is denoted by $(X, +, \times, \cdot, ...)$, where X is non-empty set and $+, \times, \cdot, ...$ are operations on X. In this chapter, we shall study some of those algebraic structures which involve binary operations only. Some simple algebraic structures are $(N, +)$, $(N, \cdot)$ and $(N, +, \cdot,)$. Note that $(N, -)$, is not an algebraic structure because subtraction of two natural numbers is not necessarily a natural number and so $-$ is not a binary operation on N.

We now define groupoid which is the simplest of all algebraic structures.

Groupoid (Definition) : If $*$ is a binary operation on a nonempty set S then the pair $(S, *)$ is called a *groupoid*.

Example 1 : $(N, +)$, where N is the set of natural numbers and $+$ is the operation of addition on N, is a groupoid. Note that $(N, -)$ is not a groupoid.

14.2 Semigroups

Before defining semigroups, we recall the following definitions.

Definition : A binary operation $*$ on a nonempty set S is called.

(a) **commutative** if $a * b = b * a$ for all $a, b \in S$.

(b) **associative** if $(a * b) * c = a * (b * c)$ for all $a, b, c \in S$.

Definition : Let $(S, *)$ be an algebraic system where $*$ is a binary operation on S. An element e in S is said to be an *identity* element if for all $a \in S$,

$$a * e = e * a = a$$

Example 2 : In the algebraic system $(I, \times)$, where I is the set of integers and $\times$ is ordinary multiplication of integers, 1 is the identity element.

Example 3 : In the algebraic system $(N, +)$, there is no identity element.

Invertible elements

Definition : Let $(S, *)$ be an algebraic system with identity e. An element a of S is said to be **invertible** if there exists an element b in S such that

$$a * b = b * a = e$$

and b is called inverse of a. It is clear from the definition that if b is inverse of a then a is inverse of b. Also, identity element e is always invertible because

$$e * e = e * e = e.$$

The algebraic structures which satisfy the property of associativity are called semigroups. They are useful in the study of sequential machines, formal languages and automata theory.

Definition (*Semigroups*) : Let S be a non-empty set and $*$ be a binary operation on S. Then $(S, *)$ is said to be *semigroup* if the operation $*$ is associative.

In other words, $(S, *)$ is a semigroup if

$x * y \in S$ for all x, y in S.

and $(x * y) * z = x * (y * z)$ for all $x, y, z \in S$.

Example 4 : $(N, +)$, $(N, \cdot)$ and $(R, +)$ are all semigroups.

Example 5 : Let R be the set of real numbers and let operation $*$ on R be defined as follows :

$$a * b = | a - b | \text{ for all } a, b \in R$$

Then $*$ is a commutative binary operation on R but $(R, *)$ is not a semigroup.

Solution : Since $| a - b |$ is always a positive real number, $a * b \in R$ whenever $a, b \in R$. Thus, $*$ is binary operation on R.

Since $| a - b | = | b - a |$ for all $a, b \in R$

We have $a * b = b * a$ for all $a, b \in R$

Thus, $*$ is a commutative operation.

$*$ is not associative because

$$1 * (2 * 3) = 1 * | 2 - 3 | = 1 * 1 = | 1 - 1 | = 0$$

but $(1 * 2) * 3 = | 1 - 2 | * 3 = 1 * 3 = | 1 - 3 | = 2$

Hence, $1 * (2 * 3) \neq (1 * 2) * 3$.

Example 6 : The power set $P(S)$, where S is a non-empty set, together with the operation $\cup$ of union of two sets is a semigroup.

Example 7 : Let A be a non-empty set of symbols. We recall from section 13.1 that A^* is the set of all strings over an alphabet A. That is, A^* consists of all words that can be formed from the alphabet A including empty string λ. Let $\alpha, \beta \in A^*$. An operation $\cdot$ on A^* is defined as follows :

$\alpha.\beta$ = the string obtained by writing the string α on the left of the string β.

For example, if $\alpha = a_1a_2 \ldots\ldots a_n$ and $\beta = b_1b_2 \ldots\ldots b_k$ then

$\alpha.\beta = a_1a_2 \ldots\ldots a_n\, b_1b_2 \ldots\ldots b_k$. It is easy to verify that $\cdot$ is a binary operation on A^*. The operation $\cdot$ on A^* is called **concatenation** or simply catenation. It is clear that the concatenation is associative. Thus, $(A^*, \cdot)$ is a semigroup. The semigroup $(A^*, \cdot)$ is known as **free semigroup** generated by the alphabet A.

Subsemigroups

We say that a set X is **closed under the operation** $*$ if $*$ is binary operation on X. Thus, if $*$ is a binary operation on a set X then we say that X is closed under the operation $*$.

Definition : Let $(S, *)$ be a semigroup and let T be a subset of S. If the set T is closed under the operation $*$ (that is, $a * b \in T$ whenever a and b are in T) defined on S then $(T, *)$ is called a *sub semigroup* of $(S, *)$.

Observe that the associative property holds in any subset of a semigroup. Therefore, a sub semigroup $(T, *)$ of a semigroup $(S, *)$ in itself a semigroup.

Example 8 : If $(S, *)$ is a semigroup then $(S, *)$ itself is a subsemigroup of $(S, *)$. Such a subsemigroup is called trivial subsemigroup.

Example 9 : Let E denote the set of even positive integers. Then $(E, \cdot)$, where $\cdot$ is ordinary multiplication of integers, is a subsemigroup of semigroup $(N, \cdot)$.

Powers of an element in a semigroup

In a semigroup $(S, *)$, the powers of an element $a \in S$ are defined as

$a^1 = a$, $a^2 = a * a$ and $a^{n+1} = a^n * a$ for all n in N.

Moreover, if $(S, *)$ contains identity element e of $*$, we also define

$$a^0 = e, \text{ the identity element}$$

Since $*$ is associative, it is easy to see that

$a^m * a^n = a^{m+n}$ for all non-negative integers m and n.

Example 10 : Let $(S, *)$ be a semigroup and $a \in S$. If

$$T = \{a^n : n \in N\}$$

Then $(T, *)$ is a subsemigroup of $(S, *)$. $(T, *)$ is often called *cyclic* semigroup.

Example 11 : Let $S = \{1, 2, 3, 4, 5, 6\}$. Then it can be shown by constructing composition table that S in a semigroup with respect to the operation 'multiplication modulo 7'. If $T = \{2, 4, 6\}$ then T is not a subsemigroup of S because T is not closed with respect to multiplication modulo 7.

Direct product of semigroups

From any two given semigroups, we can generate another semigroup called the direct product of the semigroup. Recall that if A and B are two non-empty sets then

$$A \times B = \{(a, b) : a \in A \text{ and } b \in B\}.$$

Definition : Let $(S, *)$ and (T, o) be two semigroups. The direct product of $(S, *)$ and (T, o) is the system $(S \times T, \oplus)$, where the operation $\oplus$ on $S \times T$ is defined as follows :

$$(a, b) \oplus (c, d) = (a * c, b \text{ o } d) \text{ for all } (a, b), (c, d) \in S \times T.$$

***Theorem* 1 :** The direct product of any two semigroups is a semigroup.

Proof : Let $(S, *)$ and (T, o) be two semigroups. We have to show that $(S \times T, \oplus)$, where $\oplus$ is defined as $(a, b) \oplus (c, d) = (a * c, b \text{ o } d)$ for all $(a, b), (c, d) \in S \times T$ is a semigroup.

Since $a * c \in S$ for all $a, c \in S$ and $b \text{ o } d \in T$ for all $b, d \in T$, therefore $(a, b) \oplus (c, d) \in S \times T$. Hence, $\oplus$ is a binary operation on $S \times T$. Next we show that $\oplus$ is associative on $S \times T$. Let (a, b), (c, d) and (e, f) be any three elements in $S \times T$. Then

$$((a, b) \oplus (c, d)) \oplus (e, f) = (a * c, b \circ d) \oplus (e, f)$$
$$= ((a * c) * e, (b \circ d) \circ f)$$
$$= (a * (c * e), b \circ (d \circ f))$$

because both $*$ and o are associative

$$= (a, b) \oplus ((c * e, d \circ f)$$
$$= (a, b) \oplus ((c, d) \oplus (e, f))$$

Thus, $\oplus$ is associative operation. Hence, $(S \times T, \oplus)$ is a semigroup.

***Theorem* 2** : If $(S, *)$ and (T, o) are commutative semigroups then their product is also commutative semigroup.

Proof : We have already shown that if $(S, *)$ and (T, o) are semigroups then their product is semigroup. We now show that product $S \times T$ is commutative. Let (a, b), (c, d) be any two elements in $S \times T$. Then

$$(a, b) \oplus (c, d) = (a * c, b \circ d)$$
$$= (c * a, d \circ b)$$

because both $*$ and o are commutative.

$$= (c, d) + (a, b)$$

Thus, $\oplus$ is a commutative operation on $S \times T$. Hence, $(S \times T, \oplus)$ is commutative semigroup.

14.3 Homomorphism of Semigroups

Definition : Let $(S, *)$ and (T, o) be any two semigroups. A mapping $f: S \to T$ is called **semigroup homomorphism** (or just homomorphism) if

$$f(a * b) = f(a) \circ f(b) \qquad \text{for all } a, b \in S$$

A homomorphism of a semigroup $(S, *)$ into itself is called endomorphism. Thus, endomorphism is a mapping $f: S \to S$ such that

$$f(a * b) = f(a) * f(b) \text{ for all } a, b \in S.$$

Definition : Let $(S, *)$ and (T, o) be two semigroups A semigroup homomorphism $f: S \to T$ is called :

(a) **semigroup epimorphism** if f is onto also.

(b) **semigroup monomorphism** if f is one-to-one also.

(c) **semigroup isomorphism** if f is bijective (i.*e*., one-to-one and onto) also.

Definition (*Automorphism*) : A semigroup isomorphism from a semigroup $(S, *)$ onto itself is called *semigroup automorphism*. In other words, a semigroup automorphism is a mapping from a semigroup $(S, *)$ onto itself which is homomorphism and bijective.

Isomorphic semigroups

Definition : Two semigroups $(S, *)$ and (T, o) are said to be isomorphic if there exists a semigroup isomorphism from S to T (or from T to S).

If two semigroups $(S, *)$ and (T, o) are isomorphic then they are structurally identical in the sense that they can differ only in the nature of their elements and the operations involved.

Example 12 : Show that there exists a semigroup homomorphism from the semigroup $(N, +)$ of natural numbers under addition to the semigroup $(\{0, 1, 2, 3\}, +_4)$, where $+_4$ denotes the operation of addition modulo 4 on the set $\{0, 1, 2, 3\}$.

Solution : Let us define a mapping $f: N \to \{0, 1, 2, 3\}$ as follows :

$f(a) = a(\text{mod } 4)$ for all $a \in N$.

$= $ the remainder r, $0 \leq r < 4$, when a is divided by 4.

For any $a, b \in N$, let $f(a) = i$ and $f(b) = j$. Then

$$\begin{aligned} f(a+b) &= (a+b) \quad (\text{mod } 4) \\ &= (i+j) \quad (\text{mod } 4) \\ &= i +_4 j \\ &= f(a) +_4 f(b) \end{aligned}$$

Thus, f is a homomorphism.

Theorem **1 :** Let $f: S \to T$ be an onto mapping from a semigroup $(S, *)$ to an algebraic structure (T, o), where o is a binary operation on T. If f is semigroup homomorphism then (T, o) is a semigroup.

Proof : In order to prove that (T, o) is a semigroup, we must show that o is an associative operation on T.

Let x, y, z be any three elements in T. Since f onto mapping, there exist a, b, c in S such that $x = f(a)$, $y = f(b)$ and $z = f(c)$.

Now,

$$
\begin{aligned}
(x \text{ o } y) \text{ o } z &= (f(a) \text{ o } f(b)) \text{ o } f(c) \\
&= f(a * b) \text{ o } f(c) && \because f \text{ is homomorphism} \\
&= f((a * b) * c) && \because f \text{ is homomorphism} \\
&= f(a * (b * c)) && \because * \text{ is associative} \\
&= f(a) \text{ o } f(b * c) && \because f \text{ is homomorphism} \\
&= f(a) \text{ o } (f(b) \text{ o } f(c)) && \because f \text{ is homomorphism} \\
&= x \text{ o } (y \text{ o } z)
\end{aligned}
$$

Hence, o is associative and so (T, o) is a semigroup.

Idempotent elements

Definition : An element a in a semigroup $(S, *)$ is called an idempotent if $a^2 = a$.

Example 13 : Let N be the set of natural numbers and $*$ be a binary operation defined on N as follows :

$$a * b = \max \{a, b\} \quad \text{for all } a, b \in N.$$

Then $(N, *)$ is a semigroup in which every element is an idempotent element.

***Theorem* 2 :** Let f be a semigroup homomorphism from semigroup $(S, *)$ to semigroup (T, o). If $a \in S$ is an idempotent element then $f(a)$ is idempotent in (T, o).

Proof : Let $(S, *)$ and (T, o) be two semigroups and let $f : S \to T$ be a semigroup homomorphism. Let a be any idempotent element in S. Then $a^2 = a$.

$$
\begin{aligned}
(f(a))^2 &= f(a) \text{ o } f(a) \\
&= f(a * a) && \because f \text{ is homomorphism} \\
&= f(a^2) \\
&= f(a) && \because a^2 = a
\end{aligned}
$$

Thus, $f(a)$ is an idempotent element. Therefore, the property of idempotency is preserved under the semigroup homomorphism.

***Theorem* 3 :** Prove that every finite semigroup has an idempotent element (That is, an element a such that $a^2 = a$).

Proof : Let x be any element in a finite semigroup S. Consider the positive integral powers of x,

$$x, x^2, x^3, x^4, \ldots\ldots .$$

All these powers of x cannot be distinct ($\because$ S is finite). Therefore, there exist positive integers r and s with $r > s$ such that $x^r = x^s$.

Let $k = r - s$. Then $k \geqslant 1$ and we can write $x^r = x^s$ as

$$x^{s+k} = x^s \qquad \ldots\ldots(1)$$

Now $\quad x^{2s+k} = x^s \cdot x^{s+k}$

$= x^s \cdot x^s \quad$ using (1)

$= x^{2s}$

Similarly, $\quad x^{3s+k} = x^{3s}$

and we can write by principle of induction

$$x^{ms+k} = x^{ms} \text{ for any } m \in N \qquad \ldots\ldots(2)$$

Again $\quad x^{ms+2k} = x^{ms+k} \cdot x^k$

$= x^{ms} \cdot x^k \quad$ using (2)

$= x^{ms+k}$

$= x^{ms} \quad$ using (2) $\qquad \ldots\ldots(3)$

and

$x^{ms+3k} = x^{ms+2k} \cdot x^k$

$= x^{ms} \cdot x^k \quad$ using (3)

$= x^{ms+k}$

$= x^{ms} \quad$ using (2)

Using principle of induction, we can write

$$x^{ms+nk} = x^{ms} \text{ for any } n \in N \qquad \ldots\ldots(4)$$

If we take $m = k$ and $n = s$ in (4), we obtain

$x^{ks+sk} = x^{ks}$

or $\quad x^{2ks} = x^{ks} \qquad \ldots\ldots(5)$

Let $a = x^{ks}$. Then $a \in S$ because ks is a positive integer. From (5), we get

$$a^2 = a.$$

***Theorem* 4 :** If f is a homomorphism from a commutative semigroup $(S, *)$ onto a semigroup (T, o) then (T, o) is also commutative.

Proof : Let x and y be any two elements of T. Since f is onto therefore there exist elements a and b in S such that $x = f(a)$ and $y = f(b)$.

Now,

$$\begin{aligned} x \circ y &= f(a) \circ f(b) \\ &= f(a * b) \quad \because f \text{ is homomorphism} \\ &= f(b * a) \quad \because (S, *) \text{ is commutative} \\ &= f(b) \circ f(a) \\ &= y \circ x \end{aligned}$$

Hence, (T, o) is commutative.

***Theorem* 5 :** Let f be a homomorphism from a semigroup $(S, *)$ to semigroup (T, o). If H is a subsemigroup of $(S, *)$ then

$$f(H) = \{t \in T : t = f(h) \text{ for some } h \in H\}$$

(called the image of H under f) is a sub-semigroup of (T, o).

Proof : Let x and y be any two elements of $f(H)$. By definition of $f(H)$, there must exist a and b in H such that

$$x = f(a) \quad \text{and} \quad y = f(b)$$

Now,

$$\begin{aligned} x \circ y &= f(a) \circ f(b) \\ &= f(a * b) \\ &= f(c) \end{aligned}$$

where $c = a * b$ belongs to H. Hence, $f(c) \in f(H)$. Thus, $f(H)$ is closed under the operation o of T. Hence, $f(H)$ is a subsemigroup of (T, o).

Exercise

1. Show that the set N of natural numbers is a semigroup under the operation $*$ defined by $x * y = \max\{x, y\}$.
2. Let $X = \{a, b, p, q\}$ and the operation $*$ be defined by the following table :

$*$	a	b	p	q
a	q	p	b	a
b	b	b	b	b
p	p	p	p	p
q	a	b	p	q

Show that $(X, *)$ is a semigroup which is not commutative.

(**Hint :** Observe that $a * a = q$ and $a * b = p$)

3. Let $S = \{1, 2, 3, 6, 12\}$ and the operation $*$ be defined as follows :

$$a * b = \text{greatest common divisor of } a \text{ and } b.$$

Show that $(S, *)$ in a semigroup. Find idempotent elements in $(S, *)$.

4. Let Z_n denote the set of integers $\{0, 1, 2, ..., n-1\}$. Let $\times_n$ be the binary operation such that

$$a \times_n b = \text{the remainder of } ab \text{ divided by } n.$$

(a) Construct the composition table for $\times_n$ when $n = 5$.

(b) Show that $(Z_n, \times_n)$ is a semigroup for any n.

5. Let Q be the set of rational numbers and $*$ be the operation on Q defined by

$$a * b = a + b - ab.$$

Show that $(Q, *)$ is a commutative semigroup.

6. Let $S = \{a, b\}$. If $(S, *)$ is a semigroup in which $a * a = b$ then show that

(a) $a * b = b * a$

(b) $b * b = b$

7. Let $(S, *)$ be a commutative semigroup. If a and b are idempotent elements then $a * b$ is also idempotent.

8. Prove that every finite semigroup has an idempotent element.

9. Show that the non-empty intersection of two sub-semigroups of a semigroup is again a sub-semigroup.

10. Let $(S, *)$, (T, o) and $(V, \oplus)$ be semigroups and $f : S \to T$ and $g : T \to V$ be semigroup homomorphisms. Prove that $g \circ f : S \to V$ is a semigroup homomorphsim from $(S, *)$ to $(V, \oplus)$.

11. Prove that the set of all semigroup endomorphism of a semigroup is a semigroup under the operation of composition of functions.

12. Let A = $\{a, b, c\}$ and consider the semigroup $(A^*, \cdot)$ where $\cdot$ is the concatenation. If $\alpha = abac$, $\beta = cba$ and $\gamma = babc$ then compute $(\alpha . \beta) \cdot \gamma$, $\gamma \cdot (\alpha \cdot \alpha)$ and $(\gamma \cdot \beta) \cdot \alpha$.

13. Let $(S, *)$ and (T, o) be semigroups. Prove that $S \times T$ and $T \times S$ are isomorphic semigroups.

14. Show that the union of two sub semigroups of a semigroup need not be a subsemigroup.

15. Let $f: S \to T$ be an isomorphism of semigroup $(S, *)$ and (T, o). Show that if x is a zero of S then $f(x)$ is a zero of (T, o).
16. Let $(S, *)$ be a semigroup. Show that the set S^S of all functions from S to itself is a semigroup under the operation o of composition of functions.

14.4 Monoids

Let $*$ be a binary operation on a non-empty set S. An element $e \in S$ is said to be an identity element (or a neutral element) for the operation $*$ if

$$a * e = e * a = a \quad \forall\, a \in S$$

Observe that a semigroup $(S, *)$ may or may not have an identity element with respect to the operation $*$. For example, $(E, +)$, where E is the set of positive even integers and $+$ is the operation of addition on E is a semigroup which does not contain identity element. Semigroups which contain identity elements are called monoids.

Definition : A semigroup $(M, *)$ with an identity element with respect to the operation $*$ is called monoid.

In other words, an algebraic system $(M, *)$ is called a monoid if the following three conditions are satisfied.

1. M is closed with respect to $*$. That is, if $x, y \in M$ then $x * y \in M$.
2. $*$ is an associative operation. That is, for any x, y, $z \in M$, $x * (y * z) = (x * y) * z$
3. Existence of identity element. That is, there exists an element $e \in M$ such that
$$e * x = x * e = x \quad \text{for any } x \in M.$$

Remark : If each element of a monoid is invertible then the monoid is called a *group*.

Example 1 : Consider the power set $P(S)$ of any non-empty set S together with the operation $\cup$ of union of two sets. then $(P(S), \cup)$ is a monoid with empty set ϕ as identity element.

Example 2 : Let $M = \{0, 1, 2, 3, \ldots, n-1\}$. Let $+_n$ and $\times_n$ denote the operation of addition modulo n and multiplication modulo n respectively. Then $(M, +_n)$ and $(M, \times_n)$ both are monoids. 0 and 1 are the identity elements with respect to the operations $+_n$ and $\times_n$ respectively.

Example 3 : Let $E = \{2, 4, 6, ...\}$ and let + be the operation of addition on E. Then $(E, +)$ is a semigroup but it is not a monoid because E does not contain the identity element for the operation +.

Commutative Monoid

A monoid $(M, *)$ is said to be *commutative* if the operation $*$ is commutative on M.

Cyclic Monoid : A monoid $(M, *)$ is said to be *cyclic* monoid if there exists an element $a \in M$ such that every element of M can be written as some non-negative powers of a. That is, there exists an element $a \in M$ such that every element of M can be written as a^n for some $n \in N$. In such a case, the cyclic monoid $(M, *)$ is said to be generated by the element a and the element a is called **generator** of the cyclic monoid.

Example 4 : The set $M = \{0, 1, 2, 3, 4, 5\}$ with the operation $+_6$ is a cyclic monoid. This monoid is generated by 1. It may be noted that 5 is also generator of M. It can be seen that

$$1^1 = 1,\ 1^2 = 1 +_6 1 = 2,\ 1^3 = 1^2 +_6 1 = 3,\ 1^4 = 1^3 +_6 1 = 4,$$
$$1^5 = 1^4 +_6 1 = 5 \text{ and } 1^6 = 0$$

Thus, $M = \{1, 1^2, 1^3, 1^4, 1^5, 1^6 = 0\}$.

Example 5 : $(N, +)$ is an infinite cyclic monoid generated by $1 \in N$.

Example 6 : Let A^* be the set of all strings over an alphabet A including empty string λ. Then A^* is a monoid with respect to the operation of concatenation (see Ex. 7 of section 2). The empty string λ plays the role of identity element. If A contains more than one element then A^* with concatenation operation is not commutative monoid.

Monoid Consisting of Partitions of a Set

We shall now describe a monoid consisting of partitions of a set. This monoid is useful in the study of sequential machines.

Let S be any non-empty set and let $\Pi(S)$ denote the set of all partitions of S. Recall that a collection $\{A_1, A_2, ...\}$ of non-empty subsets of S is called a *partition* of S if

1. $A_1 \cup A_2 \cup .... = S$
2. If $A_i \neq A_j$ then $A_i \cap A_j = \phi$.

For example, if $S = \{0, 1\}$ then $\{\{0\}, \{1\}\}$ and $\{\{0, 1\}\}$ are only two partitions of S. Thus,

$$\Pi(S) = \{\{\{0\}, \{1\}\}, \{\{0, 1\}\}\}.$$

Similarly, if $T = \{0, 1, 2\}$ then

$$\Pi(T) = \{\{\{0\}, \{1\}, \{2\}\}, \{\{0, 1\}, \{2\}\}, \{\{0, 2\}, \{1\}\}, \{\{1, 2\}, \{0\}\}, \{\{0, 1, 2\}\}\}.$$

Example 7 : Let S be any non-empty set and let $\Pi(S)$ denote the set of all partitions of S. Let $P, Q \in \Pi(S)$. Define a binary operation $*$ on $\Pi(S)$ as follows :

$P \cdot Q$ = the set of intersections of every element of P with every element of Q leaving out empty sets.

It is easy to verify that $*$ is an associative operation. Also, the partition $\{S\}$ is the identity of the operation $*$. Hence, $(\Pi(S), *)$ is a monoid. The operation $*$ is called the product of partitions.

We explain below the above example with the help of a particular case. Let $S = \{a, b, c, d\}$. If

$P = \{\{a, b\}, \{c, d\}\}$ and $Q = \{\{a, c\}, \{b, d\}\}$ then

$P * Q = \{\{a\}, \{b\}, \{c\}, \{d\}\}$ which is also a partition of S. Since intersection $\cap$ is an associative operation on sets, $*$ is also associative. Moreover the partition $\{\{a, b, c, d\}\}$ plays the role of identity element.

Some Properties of Monoids

***Theorem* 1 :** Every cyclic monoid is a commutative monoid.

Proof : Let $(M, *)$ be a cyclic monoid generated by an element a of M. Let x, y be any two elements of M. Then

$$x = a^m \text{ and } y = a^n \text{ for some } m, n \in N.$$

Now

$$x * y = a^m * a^n$$
$$= a^{m+n}$$

$$= a^{n+m}$$
$$= a^n * a^m$$
$$= a * x$$

Thus, $(M, *)$ is commutative.

***Theorem* 2 :** The identity element in any monoid is unique.

Proof : Let $(M, *)$ be any monoid. If possible, suppose e and e' are two identity elements in $(M, *)$. Since e is an identity element, therefore

$$e * e' = e' * e = e' \qquad ....(1)$$

Again, since e' is an identity element, therefore,

$$e' * e = e * e' = e \qquad ....(2)$$

From (1) and (2), we have

$$e = e'$$

Thus, the identity element in a monoid is unique.

Submonoids

Definition : Let $(M, *)$ be a monoid with identity element e and let T be a non-empty subset of M. If T is closed under the operation $*$ and identity $e \in T$ then $(T, *)$ is called a submonoid of $(M, *)$.

Since associative property holds in a monoid $(M, *)$, therefore it holds in any subset of monoid M. Thus, a submonoid of a monoid is also a monoid.

Example 8 : If $(M, *)$ is a monoid with identity e and $T = \{e\}$ then $(T, *)$ is a submonoid. Moreover $(M, *)$ is also a submonoid of $(M, *)$.

Example 9 : If $(M, *)$ is a monoid, $a \in M$ and

$$T = \{a^i : i \in N \quad \text{or} \quad i = 0\}$$

Then $(T, *)$ is a submonoid of $(M, *)$.

Example 10 : If T is the set of all even positive integers then $(T, \cdot)$ is a sub-semigroup of monoid $(N, \cdot)$, where $\cdot$ denotes the ordinary multiplication on N. Here, $(T, \cdot)$ is not a submonoid of $(N, \cdot)$ since the identity of N, namely 1, does not belong to T.

***Theorem* 1 :** Prove that the intersection of two submonoids of a monoid $(M, *)$ is again a submonoid of $(M, *)$.

Proof : Let $(T_1, *)$ and $(T_2, *)$ be two submonoids of a monoid $(M, *)$. Let e be the identity element in M. Then by definition of submonoid, $e \in T_1$ and $e \in T_2$. Thus, $T_1 \cap T_2 \neq \phi$.

1. $T_1 \cap T_2$ is closed with respect to $*$.

   Let $x, y \in T_1 \cap T_2$

$$\Rightarrow x, y \in T_1 \quad \text{and} \quad x, y \in T_2$$

$$\Rightarrow x * y \in T_1 \quad \text{and} \quad x * y \in T_2$$

$$(\because T_1 \text{ and } T_2 \text{ are submonoids})$$

$$\Rightarrow x * y \in T_1 \cap T_2$$

   Thus, $T_1 \cap T_2$ is closed with respect to $*$.

2. We now show that $e \in T_1 \cap T_2$.

   Since T_1 and T_2 are submonoids therefore both T_1 and T_2 contain identity element.

   Now,

$$e \in T_1 \quad \text{and} \quad e \in T_2 \quad \Rightarrow e \in T_1 \cap T_2$$

   Hence, $T_1 \cap T_2$ contains identity element e of monoid $(M, *)$. Thus $(T_1 \cap T_2, *)$ is a submonoid.

***Theorem* 2 :** If $(M, *)$ is a commutative monoid then the set of all idempotent elements of M forms a submonoid.

Proof : Let S be the set of all idempotent elements of M. That is,

$$S = \{x \in M : x^2 = x\}$$

Since the identity element $e \in M$ is idempotent, we have, $e \in S$. We now show that S is closed with respect to $*$. Let a, b be any two elements of S. Then

$$a^2 = a \quad \text{and} \quad b^2 = b$$

$$\begin{aligned} \text{Now} \quad (a * b)^2 &= (a * b) * (a * b) \\ &= a * (b * a) * b && \because * \text{ is associative} \\ &= a * (a * b) * b && \because * \text{ is commutative} \\ &= (a * a) * (b * b) && \because * \text{ is associative} \\ &= a^2 * b^2 \\ &= a * b && \because a^2 = a \text{ and } b^2 = b \end{aligned}$$

Thus, $a * b$ is idempotent element of M. Hence, $a * b \in S$ and so $(S, *)$ is a submonoid.

14.5 Homomorphism of Monoids

Definition : Let $(M, *)$ and (T, o) be two monoids with identity elements e and e' respectively. A mapping $f: M \to T$ is said to be a **monoid homomorphism** or simply a homomorphism if for any two elements $a, b \in M$,

$$f(a * b) = f(a) \text{ o } f(b)$$

and $$f(e) = e'$$

A homomorphism of a monoid into itself is called a monoid endomorphism.

Definition : Let $(M, *)$ and (T, o) be two monoids with identities e and e' respectively. A homomorphism $f: M \to T$ is called :

(a) **monoid epimorphism** if f is onto also.

(b) **monoid monomorphism** if f is one-to-one also.

(c) **monoid isomorphism** if f is bijective also.

Isomorphic monoids

Definition : Two monoids $(M, *)$ and (T, o) are said to be isomorphic monoids if there exists a monoid isomorphism from M to T (or from T to M).

In other words, monoids $(M, *)$ and (T, o) are said to be isomorphic if there exists a mapping $f: M \to T$ which in bijective and homomorphsim.

Automorphisms

Definition : An isomorphism of a monoid onto itself is called a monoid automorphism. Thus, a mapping

$$f: (M, *) \to (M, *)$$

is a monoid isomorphism of M if

1. f is bijective
2. $f(a * b) = f(a) * f(b) \ \forall \ a, b \in M$
3. $f(e) = e$

Example 11 : Show that monoids $(I, +)$ and $(E, +)$ are isomorphic, where I is the set of integers and E is the set of even integers.

Solution : Let us define a mapping

$$f : I \to E$$

by setting $f(x) = 2x \quad \forall \quad x \in I.$

We shall show that f is an isomorphism.

***f* is one-to-one**

Let $x, y \in I$

Suppose $\quad f(x) = f(y)$

$\Rightarrow \quad 2x = 2y$

$\Rightarrow \quad x = y$

Thus, $f(x) = f(y) \Rightarrow x = y$. Hence, f is one-to-one.

***f* is onto**

Let y be any element in E. Then y is an even integer. Therefore, $y/2$ is an integer. Let $x = y/2$ then $x \in I$ such that

$$f(x) = 2x = 2 \times y/2 = y$$

Thus, f is onto.

***f* is homomorphism**

Let $x, y \in$ I. Then

$$f(x + y) = 2(x + y)$$
$$= 2x + 2y$$
$$= f(x) + f(y)$$

Hence, f is a homomorphism. Moreover, $f(0) = 0$. Thus, monoids $(I, +)$ and $(E, +)$ are isomorphic.

***Theorem* 1 :** Let $(M, *)$ and (T, o) be two monoids with identities e and e' respectively. If f in an onto mapping from M onto T such that

$$f(a * b) = f(a) \circ f(b) \qquad \forall \; a, b \in M$$

then $\quad f(e) = e'$.

Proof : Let y be any element of T. Since f is onto, there exists an element $x \in M$ such that

$$f(x) = y$$

Now $\quad y = f(x) = f(x * e) \qquad \because \qquad e$ is the identity of $(M, *)$

$$= f(x) \circ f(e)$$
$$= y \circ f(e)$$

Similarly,

$$y = f(x) = f(e * x)$$
$$= f(e) \circ f(x)$$
$$= f(e) \circ y$$

Thus,

$$f(e) \circ y = y \circ f(e) = y$$

$\Rightarrow f(e)$ is the identity for T. Since identity element in a monoid is unique, we have, $e' = f(e)$.

Remark : In view of above theorem, a onto mapping f from a monoid $(M, *)$ onto monoid $(T, \circ)$ is called homomorphism if for any two elements $a, b \in M$

$$f(a * b) = f(a) \circ f(b).$$

This is because $f(e) = e'$ is a consequence of f being onto and satisfying $f(a * b) = f(a) \circ f(b)$ for all $a, b \in M$.

***Theorem* 2 :** Let $(M, *)$ be a monoid with identity element e and $(T, \circ)$ be any algebraic structure (o is binary operation on T). If a mapping $f : M \to T$ is onto and satisfies

$$f(a * b) = f(a) \circ f(b) \quad \forall\ a, b \in M$$

then $(T, \circ)$ is a monoid with $f(e)$ as its identity element.

Proof : We first show that o is associative. Let x, y and z be any three elements in T. Since f is onto, there exist a, b and c in M such that

$$x = f(a),\ y = f(b) \text{ and } z = f(c).$$

Now,

$$x \circ (y \circ z) = f(a) \circ (f(b) \circ f(c))$$
$$= f(a) \circ f(b * c) \quad \because f(b * c) = f(b) \circ f(c)$$
$$= f(a * (b * c))$$
$$= f((a * b) * c) \quad \because \quad * \text{ is associative on } M$$
$$= f(a * b) \circ f(c)$$
$$= (f(a) \circ f(b)) \circ f(c)$$
$$= (x \circ y) \circ z$$

Thus, o is an associative operation on T.

Next, we show that $f(e)$ is the identity element for T. Let x be any element of T. Since f is onto, there exists an element $a \in M$ such that $f(a) = x$.

Now, $x = f(a)$

$= f(a * e)$

$= f(a) \text{ o } f(e)$

$= x \text{ o } f(e)$

Similarly, $x = f(a) = f(e * a) = f(e) \text{ o } f(a) = f(e) \text{ o } x$

Thus,

$$x = x \text{ o } f(e) = f(e) \text{ o } x$$

This implies that $f(e)$ is the identity element for (T, o). Hence, (T, o) is a monoid with $f(e)$ as its identity element.

***Theorem* 3 :** If f is a homomorphism from a commutative monoid $(M, *)$ onto a monoid (T, o), then (T, o) is also commutative.

Proof : The proof is similar to the corresponding result in semigroups. Students are advised to reproduce it.

Direct Product of Monoids

***Theorem* 4 :** Let $(S, *)$ and (T, o) be two monoids with identities e and e' respectively. Prove that $(S \times T, \oplus)$ is a monoid with identity (e, e'), where $\oplus$ is defined by

$$(a, b) \oplus (c, d) = (a * c, b \text{ o } d) \text{ for all } (a, b), (c, d) \in S \times T.$$

Proof : Let $(S, *)$ and (T, o) be two given monoids with identities e and e' respectively. We have to show that $(S \times T, \oplus)$ is a monoid with identity (e, e'). Let (a, b), (c, d) be any two elements in $S \times T$. Since $a * c \in S$ and $b \text{ o } d \in T$, therefore $(a, b) \oplus (c, d) = (a * c, b \text{ o } d) \in S \times T$. Hence, $\oplus$ is a binary operation on $S \times T$.

$\oplus$ is associative on $S \times T$

Let (a, b), (c, d) and (p, q) be any three elements in $S \times T$. Then

$((a, b)) \oplus (c, d)) \oplus (p, q) = (a * c, b \text{ o } d) \oplus (p, q)$

$= ((a * c) * p, (b \text{ o } d) \text{ o } q)$

$= (a * (c * p), b \text{ o } (d \text{ o } q))$

(because both $*$ and o are associative operations)

$= (a, b) \oplus (c * p, d \text{ o } q)$

$= (a, b) \oplus ((c, d) \oplus (p, q))$

Thus, $\oplus$ is associative operation on $S \times T$.

Existence of identity : We show that (e, e') is the identity element in $(S \times T, \oplus)$. Let (a, b) be any element in $S \times T$.

$$(a, b) \oplus (e, e') = (a * e, b \circ e')$$
$$= (a, b) \quad \because \ e \text{ and } e' \text{ are identities in } (S, *) \text{ and } (T, \circ) \text{ respectively.}$$

Similarly,

$$(e, e') \oplus (a, b) = (a, b)$$

Thus, $(a, b) \oplus (e, e') = (e, e') \oplus (a, b) = (a, b) \quad \forall (a, b) \in S \times T$

Hence, (e, e') is the identity element of $(S \times T, \oplus)$ and so $(S \times T, \oplus)$ is a monoid.

14.6 Congruence Relation and Quotient Semigroups

Recall that (section 2.12) a relation R on a non-empty set S is called an equivalence relation if it is reflexive, symmetric and transitive. Every equivalence relation R induces a partition of the set S into equivalence classes. The equivalence class containing the element a in S is denoted by $[a]$ and the set of equivalence classes is the quotient set S/R.

We now discuss equivalence relations on a semigroup.

Congruence relation

Definition : An equivalence relation R on the semigroup $(S, *)$ is called a **congruence relation** if

$$a\,R\,a' \quad \text{and} \quad b\,R\,b' \Rightarrow (a * b)\,R\,(a' * b')$$

Example 12 : Show that the equivalence relation R defined by

$$a\,R\,b \text{ if and only if } a \equiv b \pmod 2$$

on the semigroup $(I, +)$ of integers under addition is a congruence relation on $(I, +)$.

Solution : Suppose that $a, b, a', b' \in I$ such that $a\,R\,a'$ and $b\,R\,b'$.

$$a\,R\,a' \Rightarrow a \equiv a' \pmod 2$$
$$\Rightarrow a - a' \text{ in divisible by } 2$$

Similarly,

$b\,R\,b' \Rightarrow b - b'$ in divisible by 2

Therefore, we can white

$$a - a' = 2m \quad \text{and} \quad b - b' = 2n$$

where m and n are integers

Now $\quad (a - a') + (b - b') = 2(m + n)$

or $\quad (a + b) - (a' + b') = 2(m + n)$

$\Rightarrow \quad (a + b) \equiv (a' + b') \pmod 2$

$\Rightarrow \quad (a + b)\,R\,(a' + b')$

Hence, the relation R is a congruence relation.

Example 13 : Let $A = \{0, 1\}$ and A^* be the set of all strings over alphabet A. Consider the free semigroup $(A^*, \cdot)$, where $\cdot$ is concatenation operation. Show that the relation R defined on A^* by

$\alpha\,R\,\beta$ if α and β have the same number of 1's,

is a congruence relation.

Solution : We first show that R is an equivalence relation.

1. ***R* is reflexive :** Since α and α have the same number of 1's, therefore $\alpha\,R\,R$ for any $\alpha \in A^*$.
2. ***R* is symmetric :** Suppose $\alpha\,R\,\beta$ for $\alpha, \beta \in A^*$. Then α and β have the same number of 1's. Therefore β and α also have the some number of 1's. Thus $\beta\,R\,\alpha$.
3. ***R* is transitive :** Suppose that $\alpha\,R\,\beta$ and $\beta\,R\,\gamma$.

   $\alpha\,R\,\beta \Rightarrow \alpha$ and β have the same number of 1's.

   $\beta\,R\,\gamma \Rightarrow \beta$ and γ have the same number of 1's.

   Thus, $\alpha\,R\,\beta$ and $\beta\,R\,\gamma \Rightarrow \alpha$ and γ have the same number of 1's.

   $$\Rightarrow \alpha\,R\,\gamma$$

   Thus, R is an equivalence relation.

   We next show that R is congruence relation. Suppose that $\alpha\,R\,\alpha'$ and $\beta\,R\,\beta'$.

   $\alpha\,R\,\alpha' \Rightarrow \alpha$ and α' have the same number of 1's.

   $\beta\,R\,\beta' \Rightarrow \beta$ and β' have the same number of 1's.

   From the definition of concatenation operation, it is clear that the number of 1's in $\alpha.\beta$ = the number of 1's in α + the number of

1's in β.

Since $\alpha R \alpha'$ and $\beta R \beta'$, we have,

the number of 1's in $\alpha.\beta$ = the number of 1's in $\alpha'.\beta'$

$$\Rightarrow \quad (\alpha . \beta) R (\alpha' . \beta')$$

$$\Rightarrow \quad R \text{ is a congruence relation.}$$

If R is an equivalence relation on a set S then the quotient set $S/R = \{[a] : a \in S\}$, where $[a] = \{x \in S : a R x\}$ is a equivalence class containing a in S.

***Theorem* 1 :** Let R be a congruence relation on a semigroup $(S, *)$. Then quotient set S/R is a semigroup with respect to the operation $\oplus$ defined by.

$$[a] \oplus [b] = [a * b] \quad \text{for all } a, b \in S,$$

where $[a]$ denotes the equivalence class of element a in S corresponding to the relation R.

Proof : Let $(S, *)$ be a semigroup and let R be a congruence relation on S. For any $a \in S$, the equivalence class $[a]$ is a set containing all those elements of S which are related to a under the relation R. For any $a, b \in S$, we define an operation $\oplus$ on S/R as follows :

$$[a] \oplus [b] = [a * b]$$

We first show that the operation $\oplus$ is well defined on S/R. Suppose that $[a] = [a']$ and $[b] = [b']$. In order to show that $\oplus$ is well defined on S/R, we must show that $[a] \oplus [b] = [a'] \oplus [b']$. That is, $[a * b] = [a' * b']$. Since $[a] = [a']$ and $[b] = [b']$ therefore $a R a'$ and $b R b'$. Now since R is a congruence relation, therefore

$$a R a' \quad \text{and} \quad b R b' \Rightarrow a * b R a' * b'$$

$$\Rightarrow \quad [a * b] = [a' * b']$$

$$\Rightarrow \quad \oplus \text{ is well defined on } S/R$$

Thus, $\oplus$ is a binary operation on S/R

Next we show that $\oplus$ is an associative operation. For any $a, b, c \in S$, we have

$$[a] \oplus ([b] \oplus [c]) = [a] \oplus [b * c]$$

$$= [a * (b * c)]$$
$$= [(a * b) * c] \quad \because * \text{ is associative on } S$$
$$= [a * b] \oplus [c]$$
$$= ([a] \oplus [b]) \oplus [c]$$

Thus, $\oplus$ is associative and hence S/R is a semigroup.

Remark : The semigroup $(S/R, \oplus)$ is called **quotient semigroup** or **factor semigroup** of semigroup $(S, *)$ by the congruence relation R. The operation $\oplus$ an S/R is sometimes called quotient binary relation.

Corollary : Let R be a congruence relation on the monoid $(M, *)$. Then $(M/R, \oplus)$ is a monoid, where $\oplus$ on M/R is defined by

$$[a] \oplus [b] = [a * b]$$

Proof : Let $(M, *)$ be a monoid with identity element e. Then by above theorem, $(M/R, \oplus)$ is a semigroup. We show that $(M/R, \oplus)$ has identity element. We shall show $[e]$ is the identity element in $(M/R, \oplus)$. Let $[a]$ be any element in M/R where $a \in M$. Then

$$[a] \oplus [e] = [a * e]$$
$$= [a] = [e * a]$$
$$= [e] \oplus [a]$$

Thus, $[e]$ is the identity in M/R. Hence, $(M/R, \oplus)$ is a monoid.

***Theorem* 2 :** Let $(S, *)$ be a semigroup and let R be a congruence relation on $(S, *)$. Then show that there exists a homomorphism from $(S, *)$ onto quotient semigroup $(S/R, \oplus)$.

Proof : We define a function $f : S \to S/R$ as follows :

$f(a) = [a]$, the equivalence class of $a \in S$ under R.

We shall show that f is onto and homomorphism.

***f* is onto :** Let $[a]$ be any element in S/R. Then $a \in S$ and $f(a) = [a]$. Thus, f is onto.

***f* is homomorphism :** Let a, b be any two elements of S. Then

$$f(a * b) = [a * b]$$
$$= [a] \oplus [b] \quad \text{By definition of } \oplus$$
$$= f(a) \oplus f(b)$$

so f is a homomorphism.

Remark : The mapping $f : S \to S/R$ defined by $f(a) = [a]$ is called *natural homomorphism.*

***Theorem* 3 :** Let $f : S \to T$ be a semigroup homomorphism from the semigroup $(S, *)$ to the semigroup (T, o). Then prove that the relation R on $(S, *)$ defined by

$$a\,R\,b \text{ if and only if } f(a) = f(b) \text{ for all } a, b \in S$$

is a congruence relation on $(S, *)$ corresponding to f.

Proof : We first show that R is an equivalence relation.

1. **R is reflexive :** Since $f(a) = f(a)$, we have, $a\,R\,a$ for all $a \in S$.
2. **R is symmetric :** Suppose that $a\,R\,b$. Then $f(a) = f(b)$ which can be written as $f(b) = f(a)$. Hence, $b\,R\,a$.
3. **R is transitive :** Suppose that $a\,R\,b$ and $b\,R\,c$.

$$a\,R\,b \Rightarrow f(a) = f(b)$$
$$b\,R\,c \Rightarrow f(b) = f(c)$$

Thus, $a\,R\,b$ and $b\,R\,c \Rightarrow f(a) = f(c)$

$$\Rightarrow a\,R\,c$$

Thus, R is an equivalence relation. We now show that R is a congruence relation. Suppose that

$$a\,R\,a' \quad \text{and} \quad b\,R\,b'$$

Then $f(a) = f(a')$ and $f(b) = f(b')$

Now $f(a * b) = f(a) \text{ o } f(b)$ $\quad \because f$ is homomorphism

$$= f(a') \text{ o } f(b')$$
$$= f(a' * b')$$
$$\Rightarrow \quad (a * b)\,R\,(a' * b')$$

Thus, we see that

$$a\,R\,a' \text{ and } b\,R\,b' \Rightarrow (a * b)\,R\,(a' * b')$$

Hence, R is a congruence relation on $(S, *)$.

***Theorem* 4 :** (Fundamental theorem homomorphism, of semigroups)

Let $(S, *)$ and (T, o) be semigroups. If $f : S \to T$ is a semigroup homomorphism, then semigroup (T, o) is isomorphic to some quotient semigroup of $(S, *)$.

Proof : Let $f: S \to T$ be a semigroup homomorphism of the semigroup $(S, *)$ onto the semigroup (T, o). Let R be the congruence relation on $(S, *)$ corresponding to the homomorphism f. Then R is defined by

$a \, R \, b$ iff $f(a) = f(b)$ for all $a, b \in S$. We shall show that (T, o) is isomorphic to quotient semigroup $(S/R, \oplus)$ of $(S, *)$.

If $a \in S$ then $[a] \in S/R$ and $f(a) \in T$. Let us define a mapping $\psi : S/R \to T$ by setting

$$\psi([a]) = f(a) \text{ for all } a \in S.$$

We shall show that ψ is a semigroup isomorphism. We first show ψ is well defined. Suppose that $a, a' \in S$ and $[a] = [a']$.

$$\begin{aligned}[a] = [a'] \quad &\Rightarrow a \, R \, a' \\ &\Rightarrow f(a) = f(a') \\ &\Rightarrow \psi([a]) = \psi([a'])\end{aligned}$$

Thus, ψ is well defined

ψ is one-to-one : Suppose that $\psi([a]) = \psi([b])$. Then,

$$\begin{aligned}f(a) &= f(b) \\ &\Rightarrow a \, R \, b \\ &\Rightarrow [a] = [b]\end{aligned}$$

Hence, ψ is one-to-one.

ψ is onto : Let b be any element of T. Since f is onto, there exists an element a in S such that $b = f(a)$. Now

$$\psi([a]) = f(a) = b$$

Hence, ψ is onto.

Finally we show that ψ is semigroup homomorphism.

Let $[a]$, $[b]$ be any two elements in S/R. Then

$$\begin{aligned}\psi([a] \oplus [b]) &= \psi([a * b]) \\ &= f(a * b) \\ &= f(a) \, o \, f(b) \qquad \because f \text{ is homomorphism} \\ &= \psi([a]) \, o \, \psi([b])\end{aligned}$$

Hence, ψ is an isomorphism of S/R onto T. Thus, $(S/R, \oplus)$ and (T, o) are isomorphic.

Example 14 : Let A be a non-empty set of symbols and let A^* be the free semigroup generated by A under the operation of concatenation. Show that the function

$$f: A^* \to N \text{ defined by}$$
$$f(\alpha) = l(\alpha), \text{ the length of } \alpha$$

where $l(\alpha)$ denotes the number of symbols in α (each symbol is counted as many times as it appears in α) is a homomorphism and if R is the relation induced by f then show that $(A^*/R, \oplus)$ is isomorphic to $(N, \oplus)$.

Solution : Let α, β be any two elements (words) in A^*. Then

$$\begin{aligned} f(\alpha \,.\, \beta) &= l(\alpha \,.\, \beta) \\ &= l(\alpha) + l(\beta) \\ &= f(\alpha) + f(\beta) \end{aligned}$$

Thus, f is a homomorphism.

By theorem 3, f induces a congruence relation on A^* defined by $\alpha\, R\, \beta$ if and only if $l(\alpha) = l(\beta)$

Since f is anto ($\because$ if n is any positive integer in N then $\alpha = a^n \in A^*$ for any $a \in A$ such that $f(\alpha) = n$), therefore by Theorem 4, $(A^*/R, \oplus)$ is isomorphic to $(N, +)$.

***Example* 15 :** Let $A = \{0, 1\}$ and let A^* be the free semigroup generated by A. Let $(N, +)$ be the semigroup of natural numbers. Find a semigroup homomorphism $f: A^* \to N$ which induces congruence relation R on A^* defined by $\alpha\, R\, \beta$ iff α and β have the same number of 1's.

Solution : We have already seen (Ex. 13) that the given relation R on A^* is a congruence relation. Define a mapping $f: A^* \to N$ as follows :

$$f(\alpha) = \text{the number of 1's in } \alpha \text{ for all } \alpha \in A^*.$$

Clearly, for any $\alpha, \beta \in A^*$.

$$\begin{aligned} f(\alpha \,.\, \beta) &= \text{the number of 1's in } \alpha \,.\, \beta \\ &= \text{the number of 1's in } \alpha + \text{the number of 1's in } \beta \\ &= f(\alpha) + f(\beta) \end{aligned}$$

Thus, f is a homomorphism from $(A^*, \cdot)$ to $(N, +)$.

Now by theorem 3, for any $\alpha, \beta \in A^*$

$$\alpha R \beta \text{ if and only if } f(\alpha) = f(\beta).$$

Thus, the congruence relation R is induced by the homomorphism f.

Exercise

1. Consider the set N of natural numbers under the operation $*$ defined by
$$x * y = \max\{x, y\}$$
Is $(N, *)$ a monoid ?
2. Show that the set of all invertible elements of a monoid forms a monoid under the same operation as that of the monoid.
3. Let I be the set of integers and let $\cdot$ denote the ordinary multiplication of integers. Show that $(I, \cdot)$ is a monoid. Moreover, show that $(\{0\}, \cdot)$ is a subsemigroup but not a submonoid.
4. Show that the monoids $(\{0, 1, 2, 3\}, +_4)$ and $(\{1, 3, 7, 9\}, X_{10})$ are isomorphic.
5. Consider the semigroup $(I, +)$ of integers under ordinary addition. Determine whether the relation R on $(I, +)$ defined by $a R b$ if an only if $a + b$ is even is a congruence relation.
6. Show that the intersection of two congruence relations on a semigroup is a congruence relation.
7. Let $(M, *)$ be a monoid. Show that there exists a subset T of M^M (the set of all functions from M to M) such that $(M, *)$ is isomorphic to the monoid (T, o), where o denotes the operation of composition of two functions.
(**Hint** : For any $a \in M$, let us define $f : M \to M^M$ by setting $f(a) = f_a$, where $f_a : M \to M$ is defined by $f_a(b) = a * b$ for any $b \in M$. Show that f is homomorphism and take $T = f(M) \subset M^M$).